DESIGN, SAFETY AND OPERATION OF DAMS

Cover photography:
Naret arch dam in Switzerland, height 80 m, year of commissioning 1969 (photo Giosanna Crivelli)

Anton J. Schleiss, Henri Pougatsch

DESIGN, SAFETY AND OPERATION OF DAMS

EPFL PRESS

EPFL PRESS is an imprint owned by the Presses polytechniques et universitaires romandes, a Swiss academic publishing company whose main purpose is to publish the teaching and research works of the Ecole polytechnique fédérale de Lausanne (EPFL).

PPUR, EPFL – Rolex Learning Center, CM Station 10, CH-1015 Lausanne, info@epflpress.org, tel. : +41 21 693 21 30, fax : +41 21 693 40 27.

www.epflpress.org

© 2022, First edition in English, EPFL Press
ISBN 978-2-88915-485-2
Printed in Czech Republic

All rights reserved, including those of translation into other languages.
No part of this book may be reproduced in any form – by photoprint, microfilm, or any other means – nor transmitted or translated into a machine language without written permission from the publisher.

Preface

Dams—often called 'useful pyramids'—comprise some of the greatest works ever made by humankind. For thousands of years, people have used them both as a means to harness the power of water and also to protect themselves against it. It is therefore no exaggeration to state that dams play and will always play an essential role, allowing humans to guarantee their basic needs in water, food, and energy, while respecting the environment and ensuring the sustainable management of resources. From time immemorial, the con-struction of dams has contributed to human development and economic prosperity.

Dams present a compelling picture of civil engineering. This is particularly the case in Switzerland, which has so many fine examples of dam structures. Despite their impressive size, they in fact only represent one aspect of civil engineering, which also includes other types of works such as bridges, tunnels, roads, and buildings. A dam engineer truly feels as if he or she is part of something exceptional when they are called on to participate in the construction of such a work. As well as the disciplines specific to civil engineering, such as structural engineering, hydraulics, and soil and rock mechanics, an engineer today also joins a multi-disciplinary team comprising geologists, hydrologists, mechanical and electrical engineers, surveyors, environmentalists, economists, and even legal experts.

With some 220 large dams operating in our country, Switzerland has a long history of planning and constructing dams. Several engineering firms operate in this field, with clients around the world, and Swiss engineers have thus acquired an enviable global reputation. Since the 1960s, more than 180 large dams have been built across the world with the participation of Swiss experts. Switzerland can today be considered a center for excellence in this sector, with engineering teams having access to an unrivalled wealth of experience not only in the design and construction of dams, but also in their operation and maintenance.

This book charts the practical experience gained from dams in Switzerland and beyond our borders. It also aims to contribute to the upkeep of this long tradition and highlight the scientific advances in the construction of these large water infrastructures. This book was first developed as a French version from printed handouts that were provided to masters' students in the module on dams given by the civil engineering section at the École polytechnique fédérale de Lausanne (EPFL). The subject matter of this book, translated and updated from the second French edition of 2020, goes into much more depth and surpasses the classroom context. It provides useful and complementary elements for the development of semester projects and practical work for various diplomas. The book is also intended as a reference manual for practicing engineers and any person involved or interested in the world of dams. It will allow them to further develop their skills and knowledge in the design of useful works integrated into the natural environment throughout the world.

The authors would like to sincerely thank Laurent Mouvet and Christoph Oehy for the important contribution they made to the writing of the very first French version of the course notes handed out in class, as well as Rudolf W. Müller and Alfred Kobelt for their documentary research for the book. We are grateful to Jo Nicoud-Garden for the translation from the second French version into this English version. Finally, we would also like to thank the EPFL School of Architecture, Civil and Environmental Engineering, the Swiss Committee on Dams, Alpiq SA, AXPO Power AG, AFRY Switzerland Ltd, Gruner-Stucky Ltd, IM Maggia Engineering Ltd and BG Consulting Engineers for their financial support of the English translation.

Lausanne, September 2022 *Anton J. Schleiss, Henri Pougatsch*

The English translation of the book was financially supported by:

School of Architecture, Civil and Environmental Engineering

Contents

	Preface	V

I. GENERAL POINTS

1. Introduction 3

1.1	Definitions	5
1.2	Unique Structures	5
1.3	The Role of Dams	5
1.4	The Functions of Dams	6
1.5	Particularity of Dam Construction	8
1.6	Assessment Criteria for a Water Storage Scheme	9

2. The Development and Future of Dams 11

2.1	Dam Construction Over the Centuries	13
	2.1.1 Introduction	13
	2.1.2 Dams in the past	13
	2.1.3 The construction of dams in the twentieth century	15
	2.1.4 Dams today	18
	2.1.5 Overview of the development of dam technology	20
2.2	The Development of Dams in Switzerland	22
	2.2.1 Introduction	22
	2.2.2 Main stages in the construction of dams	23
	2.2.3 Strengthening and rehabilitation works	29
	2.2.4 Resolution of problems related to the reservoir	33
	2.2.5 The development of concrete pouring and placing	34
	2.2.6 The development of monitoring systems	36
	2.2.7 Legislative basis	36
	2.2.8 Measures to guarantee public safety	38
2.3	Looking to the Future	39
	2.3.1 Introduction	39
	2.3.2 Future activities in Switzerland	39
	2.3.3 Dams and sustainable development in the twenty-first century	42

3. Different Types of Dams 47

3.1	Introduction	49
3.2	Concrete and Masonry Dams	49
	3.2.1 Introduction	49
	3.2.2 Gravity dams	52
	3.2.3 Buttress dams	53
	3.2.4 Arch dams	56

		3.2.5 Dams built with roller compacted concrete (RCC)	59
		3.2.6 Post-tensioned concrete dams	61
3.3	Embankment Dams		61
3.4	Advantages and Specificities of Different Types of Dams		64

4. Impacts on the Environment ... 67

4.1	Introduction	69
4.2	Examples of Impacts	69
	4.2.1 Introduction	69
	4.2.2 Impacts related to the atmosphere	70
	4.2.3 Impacts related to the creation of a barrier to water flow	71
	4.2.4 Impacts related to the creation of a reservoir	72
	4.2.5 Impacts related to the modification of a downstream flow regime	73
	4.2.6 Human and sociological aspects	74
	4.2.7 Impacts outside of the reservoir	75
4.3	Description of Impacts	75
4.4	Technical Measures	76
4.5	Stance of International Organizations	76
4.6	Education	78

II. DAM SAFETY

5. Dam Safety Concept ... 81

5.1	Introduction	83
5.2	Structural Safety	83
	5.2.1 Introduction	83
	5.2.2 Service criteria agreement and basis of design	84
	5.2.3 Structural provisions	85
	5.2.4 Operating dams	87
5.3	Surveillance and Maintenance	88
	5.3.1 Introduction	88
	5.3.2 Main activities	88
	5.3.3 Organizational structure	89
5.4	Emergency Planning	91
	5.4.1 General information	91
	5.4.2 Potential hazards	91
	5.4.3 Management of proven hazards	99
	5.4.4 Planning evacuation measures	100
	5.4.5 Alarm systems	102
	5.4.6 General organization	103

6. Hazards and Damage Affecting Dams and Their Foundations ... 105

6.1	Description of Weak Points	107
6.2	Description of Causes	109

7. Types of Failure ... 111
7.1 General Information ... 113
7.2 Types of Failure in Concrete Dams ... 114
7.3 Types of Failure in Embankment Dams ... 116

8. Protection Measures ... 119

9. Risk Analysis ... 123
9.1 Introduction ... 125
9.2 Fields of Application for Dams ... 125
9.3 Definition of Risk ... 126
9.4 Theoretical Basis for Risk Analysis ... 126
    9.4.1 General information ... 126
    9.4.2 Risk analysis outline ... 129
    9.4.3 Defining a framework for analysis ... 130
    9.4.4 Risk analysis ... 130
    9.4.5 Risk estimation ... 136
    9.4.6 Risk evaluation ... 137
    9.4.7 Risk management ... 141

III. BASIS FOR THE DESIGN AND CONSTRUCTION OF A PROJECT

10. Identifying Conditions Related to the Site ... 145
10.1 Topographical Conditions ... 147
    10.1.1 Topographical survey ... 147
    10.1.2 Topographical criteria ... 147
10.2 Geological and Geotechnical Surveys ... 152
    10.2.1 The importance of geology ... 152
    10.2.2 Description of field investigations ... 153
    10.2.3 Description of types of geomechanical and geotechnical investigations and tests ... 157
    10.2.4 Types of possible dams depending on the nature of the foundations ... 161
    10.2.5 Interventions by a geologist during the works ... 163
10.3 Searching for Material ... 164
    10.3.1 Aggregates for concrete dams ... 164
    10.3.2 Materials for embankment dams ... 166
10.4 Seismicity ... 168
    10.4.1 Seismic elements ... 168
    10.4.2 Seismicity in Switzerland ... 174
    10.4.3 Earthquakes and dams ... 180
10.5 Climatic Conditions ... 181
10.6 Hydrological Study ... 181

		10.6.1	Flood control	181
		10.6.2	Definition of a flood	182
		10.6.3	Floods considered	183
		10.6.4	Methods for flood estimation	188
		10.6.5	Historical methods	188
		10.6.6	Empirical formulas	188
		10.6.7	Probabilistic methods	189
		10.6.8	Hypotheses for synthetic hydrographs	191
		10.6.9	Procedure for calculating the flood safety level based on the design flood	192
		10.6.10	Deterministic methods (PMP-PMF)	192

11. Actions and Forces ... 197

11.1	Types of Loading	199
11.2	Combining Loads	200
11.3	Permanent Loads	201
	11.3.1 Self weight	201
	11.3.2 Earth pressure (downstream backfill)	201
	11.3.3 Anchoring forces	202
11.4	Variable Loads	202
	11.4.1 Water pressure	202
	11.4.2 Sediment pressure	203
	11.4.3 Uplift	205
	11.4.4 Concrete temperature	213
	11.4.5 Flow pressure from percolation water	216
	11.4.6 Pore pressure	216
	11.4.7 Snow	216
	11.4.8 Ice load	217
	11.4.9 Bearing loads	217
11.5	Exceptional Loads	218
	11.5.1 Floods	218
	11.5.2 Earthquakes	218
	11.5.3 Avalanches	228
	11.5.4 Debris flows	228
	11.5.5 Other individual actions	229

12. Administrative Procedures and Requirements ... 231

12.1	Granting of a Concession and Construction Permit	233
	12.1.1 Request for concession	233
	12.1.2 Construction permit	234
12.2	Environmental Impact Assessment (EIA)	235
	12.2.1 Process for completing an EIA	235
	12.2.2 Environmental aspects	236
	12.2.3 Reviewing the EIA	238

IV. CONCRETE DAMS

13. Gravity dams 241
- 13.1 General Shape 243
- 13.2 Loads and Key Verifications 245
- 13.3 Safety with Regard to Overturning 246
- 13.4 Safety with Regard to Sliding 252
  - 13.4.1 Definitions and principles of calculation 252
  - 13.4.2 Slip surfaces and their strength 253
  - 13.4.3 Improvement in sliding safety 256
  - 13.4.4 Safety with regard to floating 260
- 13.5 Safety with Regard to Failure 260
  - 13.5.1 Loads and stresses in the dam 261
  - 13.5.2 Principal stresses 262
  - 13.5.3 Load-bearing capacity 264
- 13.6 Loading During an Earthquake 265
  - 13.6.1 True behavior of a dam during an earthquake 265
  - 13.6.2 Pseudo-static analysis 267
  - 13.6.3 Safety during an earthquake 268
  - 13.6.4 Dynamic analysis 269
  - 13.6.5 Brief description of methods suited to dynamic analysis 271
- 13.7 Earthquake Verification of Gravity and Masonry Dams 273
  - 13.7.1 Methods of calculation 273
  - 13.7.2 Process for the verification of category II water-retaining facility 274
- 13.8 The Effect of Temperature 286
  - 13.8.1 Solar radiation 286
  - 13.8.2 Concrete heating during hardening 290
- 13.9 3D Stability Analyses 294
- 13.10 Specific Construction Aspects 296
  - 13.10.1 Shape and dimension of blocks 296
  - 13.10.2 Precautions for major earthquakes 297
  - 13.10.3 Concrete quality, cement content 298
  - 13.10.4 Construction joints and waterproofing system 298
  - 13.10.5 Freeboard 300
- 13.11 Gravity dam heightening 301
  - 13.11.1 Reasons and prerequisites 301
  - 13.11.2 Heightening methods 301
  - 13.11.3 Use of post-tensioning 305
- 13.12 Strengthening Gravity Dams 313

14. Hollow Gravity Dams and Buttress Dams 315
- 14.1 From the Traditional Gravity Dam to the Buttress Dam 317
  - 14.1.1 Traditional gravity dam vs. hollow gravity dam 317
  - 14.1.2 Buttress dam 319

	14.1.3	Buttress dam with closed downstream facing	321
	14.1.4	Comparing different profiles through figures	322
	14.1.5	Other types of buttress dams	324
14.2	Stresses in Buttress Dams		325
	14.2.1	Stresses on the faces	325
	14.2.2	Vertical stresses	326
	14.2.3	Stresses inside the buttress	326
	14.2.4	Shape of the upstream head	327
	14.2.5	Optimizing the diamond head	328
14.3	Overturning and Sliding Safety		330
	14.3.1	Methods of calculation	330
	14.3.2	Hypothesis for considering uplift	331
	14.3.3	Buckling of the buttress web	331
14.4	Earthquake Behavior		331
	14.4.1	Transversal loading	332
	14.4.2	Longitudinal loading	333
14.5	The Effects of Temperature		334
14.6	Specific Issues		335
	14.6.1	Buttress foundations	335
	14.6.2	Construction stages	335
	14.6.3	Sealing system	336

15. Arch Dams 339

15.1	General Shape and Advantages		341
15.2	Main Types of Arch Dams		341
	15.2.1	Single curvature arch dams	341
	15.2.2	Double curvature arch dams	342
15.3	Choosing the Initial Shape		344
	15.3.1	Dam height	344
	15.3.2	Shape of horizontal sections	345
	15.3.3	Shape of the vertical sections	349
	15.3.4	Thickness of the cantilever at the crown	352
	15.3.5	Slenderness coefficient	355
	15.3.6	Stability of the rock abutments	356
15.4	Methods of Calculation		357
	15.4.1	General information	357
	15.4.2	Tube formula, membrane	357
	15.4.3	Trial-load method	358
	15.4.4	Finite element method (FEM)	384
15.5	The Effects of Temperature		388
	15.5.1	Consequences of meteorological conditions	388
	15.5.2	Hypotheses relative to the internal distribution of temperature	388
	15.5.3	The effects of two modes of temperature load	390
15.6	Loads and Stresses		392

		15.6.1	Localizing compressed zones and zones under tension	394
		15.6.2	Effects of load	397
		15.6.3	Effect of shear force	397
	15.7	Structural Details		400
		15.7.1	Configuration of the toe of the dam	400
		15.7.2	Galleries and shafts	402
		15.7.3	Artificial cooling of hard concrete (post cooling)	405
		15.7.4	Treatment and grouting of joints	405
	15.8	Heightening Arch Dams		407
		15.8.1	Reasons for heightening	407
		15.8.2	Heightening the Mauvoisin dam (VS)	407
		15.8.3	Heightening of the Luzzone dam (TI)	409
		15.8.4	Heightening of the Vieux-Emosson dam (VS)	410

16. Multiple Arch Dams ... 413

16.1	General Introduction		415
16.2	Principal Characteristics and Sizes		416
	16.2.1	The arches	416
	16.2.2	The buttresses	418
	16.2.3	The foundations	419
16.3	General Stability Assessment		419
	16.3.1	Sliding stability	419
	16.3.2	Overturning stability	421
16.4	Loads and Stresses		421
	16.4.1	Arches	422
	16.4.2	Buttresses	424

17. Roller Compacted Concrete Dams (RCC) ... 425

17.1	General Description		427
	17.1.1	A brief history	427
	17.1.2	Main characteristics and construction process	427
17.2	Design and Structural Layout		430
	17.2.1	Site conditions	430
	17.2.2	Choice of profile for an RCC gravity dam	430
	17.2.3	Integrating appurtenant structures	431
	17.2.4	Facing and impermeability of upstream face	432
	17.2.5	Finishing and facing of the downstream face	436
	17.2.6	Treatment of horizontal work surfaces	437
	17.2.7	Vertical transverse contraction joints	438
	17.2.8	Galleries and shafts	439
	17.2.9	Drainage and seepage control	440
	17.2.10	Grouting	440

17.3	Essential Aspects in Design Analysis		440
	17.3.1	Analysis of general stability	440
	17.3.2	Loads and stresses	441
	17.3.3	Cracking	442
	17.3.4	The effect of RCC's internal temperature	443
17.4	Constructing RCC Dams		443
	17.4.1	Composition of RCC	443
	17.4.2	Typical properties of RCC	447
	17.4.3	Tests	448
	17.4.4	Placing RCC	450
17.5	RCC Arch Dams		454
	17.5.1	Brief historical overview	454
	17.5.2	Key design elements	454
	17.5.3	Structural aspects	454
	17.5.4	Monitoring system	457
17.6	Hard Embankment Dams		457
	17.6.1	A brief history	457
	17.6.2	The new concept of a symmetrical profile with hardfill	459
	17.6.3	Characteristics of a hardfill dam	459
	17.6.4	Controlling seepage through the hardfill dam	460
17.7	Rock-filled Concrete Dams		460
17.8	Other Uses for RCC		461
	17.8.1	Protection against dam overtopping	461
	17.8.2	Dam strengthening	461
	17.8.3	Protection against erosion	461
	17.8.4	Adapting foundations	461
	17.8.5	Cofferdams	462
	17.8.6	Heightening concrete dams	462

V. CONCRETE

18. Concrete Technology 465

18.1	Concrete Components		467
	18.1.1	Aggregate	467
	18.1.2	Binders	468
	18.1.3	Mixing water	468
	18.1.4	Admixtures	468
	18.1.5	Occluded air content (void content)	468
18.2	Formulating Concrete		469
	18.2.1	Particle size	469
	18.2.2	Cement content	471
	18.2.3	Moisture content and the W/C ratio	471
	18.2.4	Admixture content	472
	18.2.5	Occluded air content	472

18.3	Properties of Fresh and Hardened Concrete		472
	18.3.1	Fresh concrete	472
	18.3.2	Hardened concrete	473
18.4	Manufacture and Placement		478
18.5	Concrete Testing		479
	18.5.1	Tests during the design phase	480
	18.5.2	Control testing during construction	481
	18.5.3	Control testing during operation	481
18.6	Using Test Results on Hardened Concrete		482
	18.6.1	Considerations relative to concrete strength	482
	18.6.2	Factor of safety	484
	18.6.3	Determining cement content	489
	18.6.4	Worksite laboratory	490
18.7	Research on Concrete		491
	18.7.1	Behavior of unreinforced concrete under dynamic forces	491
	18.7.2	Size of samples and scaling effect	496

19. Concrete Behavior and Observed Phenomena 497

19.1	Development of Deformation Over Time		499
19.2	Cracking		499
	19.2.1	How cracks form	499
	19.2.2	Development of cracks	500
	19.2.3	Observations made concerning concrete dams	501
19.3	Creep and Shrinkage		502
	19.3.1	Description and characteristics of creep and shrinkage	502
	19.3.2	Implications for concrete dams	502
	19.3.3	Laboratory testing	503
19.4.	Swelling		504
	19.4.1	Description and characteristics of swelling	504
	19.4.2	Alkali-aggregate reaction (AAR)	504
	19.4.3	Identification and analysis of swelling	505
	19.4.4	Measuring displacement and visual observations	506
	19.4.5	Laboratory analyses and tests	506
	19.4.6	Implications for concrete dams	507
	19.4.7	Effects on different types of dams	508
	19.4.8	Preventive measures	509
	19.4.9	In cases of rehabilitation	510
19.5	The Effects of Frost		510

VI. EMBANKMENT DAMS

20. Overview 515

201.1 Background 517

20.2	Criteria for Choosing a Site	518
20.3	Types of Embankment Dams	519
20.4	Foundation	521
20.5	Behavior of Embankment Dams	523
20.6	Appurtenant Structures	524

21. Homogeneous Earthfill Dams 529

21.1	Overall Layout	531
21.2	Main Characteristics of the Materials	531
21.3	Factors Concerning the Design and Operation of the Dam	531

22. Zoned Embankment Dams 535

22.1	Overall Layout	537
22.2	Embankment Dam with Central Core	537
22.3	Embankment Dam with Inclined Core	539
22.4	Description and Characteristics of the Materials	539
	22.4.1 Dam shell	539
	22.4.2 Clay core	540
	22.4.3 Filter	541

23. Embankment Dam with Asphalt Concrete Core 543

23.1	Overall Layout	545
23.2	Description of the Asphalt Core	546
	23.2.1 Main characteristics	546
	23.2.2 Components of asphalt concrete	547
	23.2.3 Testing and monitoring	547
	23.2.4 Placement	547

24. Embankment Dam with an Upstream Facing 549

24.1	Overall Layout	551
24.2	Embankment Dam with a Concrete Upstream Facing	552
	24.2.1 Main characteristics	552
	24.2.2 Constructing a concrete facing	554
24.3	Constructing an Asphalt Concrete Facing	560
	24.3.1 General information	560
	24.3.2 Technical characteristics	560
24.4	Other Types of Upstream Facing	562
	24.4.1 Geomembrane	562
	24.4.2 Lining based on stabilized ground	563
24.5	Comparison Between an Upstream Facing and a Central Core	563

25.	**Construction and Behavior of Embankment Dams**	565
25.1	Placement and Compaction of Rockfill	567
	25.1.1 Cohesive materials	567
	25.1.2 Noncohesive materials	570
25.2	Trial Areas	572
25.3	Inspections and Measurements During Construction	572
25.4	Behavior During and After Construction	572
	25.4.1 Vertical deformation	572
	25.4.2 Conditions with a full reservoir	576
	25.4.3 Reservoir operation	578
26.	**Stability Analysis**	579
26.1	Principles of Analysis	581
26.2	Methods of Analysis	581
26.3	Load Cases	581
26.4	Factor of Safety	583
26.5	Slope Angles	583
26.6	Safety in the Case of an Earthquake Based on a Pseudo-static Analysis	584
26.7	Verification of Embankment Dams with Regard to Earthquakes	584
	26.7.1 Verification basis and requirements	584
	26.7.2 Flowchart for the calculation process	586
	26.7.3 Summary of geological and geotechnical conditions of the foundation as well as typical values of materials comprising the body of the embankment dam and the foundation soil	587
	26.7.4 Assessment of the potential increase in pore pressure due to an earthquake	587
	26.7.5 Simplified analysis of seismic stability	593
	26.7.6 Simplified calculation for sliding displacement	595
	26.7.7 Analysis of the increase in pore pressure due to an earthquake	597
	26.7.8 Simplified analysis of seismic stability or calculation of sliding displacement by considering the increase in pore pressure due to an earthquake	598
	26.7.9 Stability analysis after an earthquake while considering an increase in pore pressure due to the earthquake	598
26.8	The Processes of Internal Erosion and Their Consequences	598
	26.8.1 Characteristics of the phenomena	598
	26.8.2 Description of the process of internal erosion	600
	26.8.3 Repair and available methods	604
	26.8.4 Surveillance and monitoring	604
27.	**Structural Details**	607
27.1	Determining Freeboard	609
	27.1.1 Definition of freeboard	609
	27.1.2 Effects of wind and waves	609
	27.1.3 Required freeboard	612

27.2	Crest		613
27.3	Berms		614
27.4	Grouting and Monitoring Galleries		614
27.5	Contact between Sealing Elements and the Subsurface		615
27.6	Slope Protection		616
	27.6.1	Protection of the upstream slope	616
	27.6.2	Protection of the downstream slope	617
	27.6.3	Construction near to fill	618

VII. TREATMENT OF FOUNDATIONS

28.	**Excavations**		621
29.	**Grouting of Rock Foundations**		625
29.1	Objectives		627
29.2	Geological Knowledge		627
29.3	Construction Method for Rock Grouting		628
	29.3.1	Drilling methods	628
	29.3.2	Grouting method	629
	29.3.3	Types of grouting material	630
	29.3.4	Grouting pressure	631
29.4	The GIN Method		632
	29.4.1	Principle	632
	29.4.2	Monitoring grouting works	634
	29.4.3	Summary of essential aspects of the GIN method	635
29.5	Structural Arrangement of the Grout Cut-off		636
	29.5.1	Overall layout	636
	29.5.2	Number of grouting lines	638
	29.5.3	Position of the grout cut-off	638
	29.5.4	Spacing between boreholes	638
	29.5.5	Orientation of holes	639
	29.5.6	Depth of grout cut-off	639
	29.5.7	Extension past the dam	640
	29.5.8	Test zones	641
	29.5.9	Depiction of grouting results	641
30.	**Other Methods for Treating Foundations**		643
30.1	Drainage Systems for Concrete Dams		645
	30.1.1	Drainage holes	645
	30.1.2	Drainage galleries	645
30.2	Jet Grouting		646
30.3	Alluvial Grouting		647
30.4	Vertical Diaphragm		647

30.5	Abutment Consolidation	650

VIII. OPERATION

31.	**Necessary Operating Documents**	657
31.1	Surveillance Guidelines	659
31.2	Dam File	659
31.3	Dam Monograph	660
32.	**Surveillance and Maintenance**	661
32.1	Aim and Organization	663
32.2	Visual Inspections	664
32.3	Testing of Water-Release Structures	665
32.4	General Overview of the Monitoring System	666
	32.4.1 Purpose of a monitoring system	666
	32.4.2 Parameters to be tracked	666
	32.4.3 Some key principles	667
	32.4.4 Choice and characteristics of measuring devices	668
	32.4.5 Automation and transmission of measurements	671
	32.4.6 Description of instruments	672
	32.4.7 Frequency of measurements	695
32.5	Analysis and Interpretation of Measurements	695
	32.5.1 Checking raw data	696
	32.5.2 Processing of measurement data	699
	32.5.3 Comments regarding the interpretation of data from geodesic measurements	703
32.6	Reporting System	706
33.	**Emergency Planning and Public Safety**	709
33.1	The Importance of the Emergency Plan	711
33.2	Strategy	712
33.3	Possible Actions in the Event of Abnormal Dam Behavior	712
33.4	Preparation	713
	33.4.1 Emergency regulations	713
	33.4.2 Extent of the submerged zone	713
33.5	Alarm Systems and Sirens Available in Switzerland	714
	33.5.1 Facilities in submerged zones	714
	33.5.2 Analysis of hazards affecting key points in the flood alarm system	715
33.6	Communication and Information	716
33.7	Emergency Organization	716
33.8	Emergency Action Plan	716

IX. RESERVOIR SEDIMENTATION

34. Issues Created by Reservoir Sedimentation and Their Mitigation 719
   34.1 Overview 721
   34.2 Surface Erosion in Alpine Catchment Areas 722
   34.3 Turbidity Current as the Principal Cause of Sediment Transportation in Reservoirs ... 723
   34.4 Measures Against Sedimentation 725
        34.4.1 Measures in the catchment area 726
        34.4.2 Measures in the reservoir 727
        34.4.3 Measures at the dam 728
   34.5 Controlling Turbidity Currents 728
   34.6 Example taken from the Grimsel Reservoir 731
   34.7 Venting Turbidity Currents 733
   34.8 Evacuating Suspended Fine Sediment through the Water Intake Using Jets 733
   34.9 Effect of Cycles of Pumped Storage Operation on Sedimentation in Reservoirs 735
   34.10 Summary 736

   Bibliography 737

   Notations 753

   Index 761

   Index of Dams 768

I. GENERAL POINTS

By virtue of their complexity, dams can be regarded as exceptional works. Indeed, the particularity of these imposing civil engineering structures is that they can be assigned different functions. They have, however, two major roles: they store a supply of water in order to meet the basic and economic needs of the population (drinking water, irrigation, energy supply, navigation) and thus provide a solution in times of water shortage, and they play a protective role against the destructive force of water (flood control, sediment retention, avalanche protection). Chapter 1 outlines general aspects of dams and their use.

Chapter 2 deals with the development and future of dam construction in Switzerland and the world. To give a brief historical overview, the first large dams were constructed by ancient civilizations, in particular along the Nile Valley, in Mesopotamia, China, and South Asia.

Over the course of time, the techniques for dam construction naturally developed, and significant progress was made with regard to their implementation and safety. The number and height of dam constructions has continued to grow.

In Switzerland, small dams designed for fisheries and mill operations were first built in around the fifteenth century. Toward the eighteenth and nineteenth centuries, the industrial revolution and a period of economic growth furthered the development of dams. And finally, throughout the twentieth century, particularly between 1950 and 1970, the construction of dams with an integrated hydroelectric component contributed to a strong increase in the production of electricity.

The material generally used for the construction of dams includes natural loose soil or rockfill (embankment dams), conventional concrete (gravity dams, arch dams, buttress dams), and roller compacted concrete (mainly gravity dams). Several options for construction therefore exist depending on the chosen material, which must be available in sufficient quantity near to the site. Chapter 3 provides a description of different types of dams.

Chapter 4 addresses the impact of dams on the environment, as their construction entails the formation of a barrier to the flow of water, the creation of a reservoir, and the modification of upstream water flow, not to mention human and socioeconomic aspects. However, there are several measures for countering these effects, and today it is common practice to carry out an environmental impact assessment.

1. Introduction

The Solis arch dam in Switzerland, height 61 m, year commissioned 1986
(Courtesy Mateo-Matthias Kunfermann)

1.1 Definitions

Dams are, by definition, hydraulic structures that block a section of valley across its entire width, thus creating a geologically watertight artificial basin. Generally speaking, and in most cases, the dam height extends above the uppermost level of water reached during periods of severe flooding.

Fundamentally, dams have two characteristic effects:

- The water reservoir, created by the presence of the dam, can usually contain most of the direct or diverted inflow of water, as well as ice, snow, or material transported by water
- The dam raises the upstream water level

Under Swiss regulations (WRFO, 2012),[1] a water-retaining facility includes a dam (retaining structure) and a reservoir, as well as the ancillary infrastructure. Besides the dam itself, safety requirements for water-retaining facilities also take into account its foundation and the banks of the reservoir.

1.2 Unique Structures

Dams are unique civil engineering structures in many ways:

- They are complex structures that must be considered as a system. Their study and construction take into account much data and many parameters. Each dam has to be considered as a prototype. No single, defined procedure for determining the best solution exists. The approach must therefore be pragmatic, systematic, recursive, and be able to evolve. It requires the formulation of many hypotheses, which are perfected and verified over the course of the project.
- The behavior of a dam during its life cycle is complex. It depends on several phenomena and factors that are more or less well defined: the change in material characteristics (aging), foundation stability (often only partially known), meteorological and thermal conditions (variable), the chemical effects of water, seismic effects (unpredictable), hydrological risks, and the manner in which the reservoir is operated. This complexity is managed by the implementation of models appropriate to the structure and its foundations, as well as to the various environmental factors that affect it.
- Finally, the requirements in terms of safety are extremely stringent. They are present in all project phases: planning, design, construction, and operation. The period during which the dam is in operation is undoubtedly the most delicate in terms of the safety of local populations. For this reason, almost all countries in the world have imposed formal regulations for the supervision of dams through continuous monitoring and behavior analysis.

1.3 The Role of Dams

By constructing dams, humans have exerted a major influence on the natural flow of watercourses. Four main reasons may justify this intervention:

- *The creation of a reservoir*
  Depending on the reservoir's active storage capacity, water inflow, and the way the stored water is used, we can distinguish daily, weekly, seasonal, or interseasonal storage.
- *Regulating flow*
  In most regions in the world, precipitation mainly occurs over short time periods. Inflow from precipitation is often irregular from one year to another, yet water needs are spread more evenly across

---

[1] Water Retaining Facilities Ordinance (WRFO, 2012).

the year. This results in a succession of periods of water shortage and excess that can only be managed by the creation of a reservoir. Furthermore, regulating flow can prevent flooding downstream.

- *The raising of the water level of a river*
  Constructing a dam across a watercourse results in the raising of water level upstream. This effect is, of course, used for hydropower generation, but it can also be used to manage the diversion of river water toward an inlet, and then to a headrace channel for irrigation or for the supply of drinking water.
- *The creation of a lake*
  The creation of an artificial lake means that a body of water becomes available for leisure activities, tourism, fisheries, navigation, and fire protection, among other uses.

Table 1.1 indicates the variability of precipitation in subtropical and arid countries, where rain occurs mainly over very short periods. The construction of reservoirs in these countries is the only way to develop irrigation and therefore agriculture.

Table 1.1 Temporal distribution of precipitation in the world in subtropical and arid countries

	Precipitation in mm		
	Yearly average	Driest month	Wettest month
Marrakesh, Morocco	253	5	40
Beirut, Lebanon	893	0	190
Zanzibar, Tanzania	1486	35	335
Kozhikode, India	3085	10	830
Cherrapunji, India	10824	10	2560

In alpine regions, the flow of water in rivers and streams is higher in summer months. This is due to the distribution of precipitation throughout the year and especially snow and glacier melt. However, electricity consumption is higher in winter. To offset this difference, large alpine dams accumulate water in high altitude reservoirs over summer. The energy potential of this water is used in winter to produce electricity. This way of using reservoirs typically corresponds to seasonal accumulation or storage.

1.4 The Functions of Dams

Water is a vital element, but it also contains great power for destruction. This is why reservoirs can be divided into two main categories, depending on the desired objective:

- Water storage in view of ulterior use
- Protection against flood water and debris

Table 1.2 lists the various main possible functions of reservoirs. When two or several uses can be combined, these are referred to as multipurpose schemes.

Table 1.2 Main functions of a reservoir

Water as a vital element	*Water as a destructive element*
Water storage	Protection structures
Hydropower	Reservoir for flood control (inundations, erosion)
Water and industrial supply, firewater retention	Flood control dike
Irrigation	Sediment trap
Fishing, fish farming (economic necessity in some countries)	Check dam
	Avalanche protection structure
Artificial snow production	Regulating dam
Low flow support (minimum guaranteed flow)	Ice control structure (Canada, Scandinavian countries)
River navigation (minimum guaranteed draft)	

Other uses can be added to those in Table 1.2, such as biotopes and reservoirs remaining from former dam structures, as well as leisure activities (fishing, swimming, diving) and tourism (walking, boating). Finally, it should be noted that dams can also fulfil ancillary functions, particularly through the construction of a road connection, for the passage of industrial pipelines (gas, hydrocarbons, water, telecommunications, electricity) or as a support for other structures (masts a.s.o.) (SwissCoD, 2000b).

It should be noted that artificial reservoirs locally modify the water balance in a catchment area. The flow is therefore affected and controlled by objectives for water use and protection. Firstly, reservoirs protect communities not only from excess water, but also from water shortages during droughts (Figure 1.3, right). Excess water causes flood events and therefore inundation and soil erosion. Water shortages harm agricultural activities and can lead to a lack of drinking water, as well as having a damaging effect on the microclimate. Secondly, water stored in a reservoir can be used for hydropower, irrigation, and transport (navigation) (Figure 1.3, left). Ship transport becomes possible with the regulation of water flow upstream of the reservoir. Most reservoirs in the world are used for irrigation and energy production at the same time. The presence of a reservoir or artificial lake is naturally suited for tourism and leisure activities, and after some arrangements, fish farming can also benefit.

There are currently nearly 60,000 dams (height > 15 m) in the world today whose reservoirs contain a total volume of 7,500 km^3, of which approximately 4,000 km^3 are used for regulating water inflows (useful storage). As a comparison, the total volume stored in watercourses around the world is in the order of 1,000 to 2,000 km^3. As a result, artificial reservoirs play an important role in the supply of water for basic human needs in energy, food, water, and transport.

1.5 Particularity of Dam Construction

Through their size and impact on the environment, dams are exceptional infrastructures, whose construction methods sometimes differ greatly from other civil engineering structures.

First of all, preliminary studies are particularly sizeable and costly. They cover a vast range of fields including hydrology, geology, hydrogeology, hydraulics, materials science, topography, geography, biology, chemistry, rural economics, energy economics, water economics, sociology, public law, development

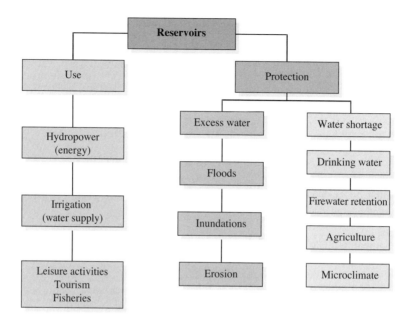

Figure 1.3 Functions of reservoirs

politics, finance, and many more. These studies are often characterized by their duration, frequently requiring more than ten years.

Alongside these studies, in-depth reconnaissance work must also be carried out: topographical mapping, geological surveys (geophysics, boreholes, exploration tunnels), geotechnical tests (wells, on-site tests, laboratory tests).

Environmental impact assessment begins very early on in order to evaluate the various solutions and implement impact mitigation or compensation measures.

The means required for construction are a challenge in terms of logistics and construction site organization. From the beginning, a number of factors relative to each site must be taken into account: access, supply, work-site installation, equipment, skill level of laborers, duration of construction, and so on.

Furthermore, the immense volume of material required has a considerable impact on the duration, resources required, and technical parameters of the project. Unlike other types of structures, it is not a question of finding satisfactory material based on the given criteria or specifications for a project, but of designing a structure that can be built with the types of material available. Each dam is thus unique since it is built out of material that is different in each case.

A dam construction site lasts for several years. The various works must be executed in a precise order, often determined by hydrological and meteorological conditions. These conditions require the work to be split into phases depending on available resources. This split can have a direct impact on the project.

And finally, no matter the type of dam, safety requirements across the whole lifetime of the structure are primordial. Dams are continually assessed and carefully monitored. The results of this examination and observations from monitoring are constantly analyzed as part of standard control procedures.

1.6 Assessment Criteria for a Water Storage Scheme

The assessment criteria for a storage scheme comprising a reservoir created by a dam are generally the following (Figure 1.4):

- *Technical*: can the purely technical aims be achieved with this scheme?
- *Economic*: if the scheme is completed, do the economic benefits outweigh the option in which the scheme is not built?
- *Financial*: are there sufficient financial means (during the construction of the scheme and during its operation)?
- *Political*: is the project supported at a political level and by the affected communities?
- *Societal*: will potential users be able to benefit from the project?
- *Environmental*: can the impact of the project on the environment and surrounding area be defended?

The feasibility of the storage scheme is based on the results of these assessment criteria.

The contraindications to the construction of a storage scheme can be disregarded after examining whether the project is:

- Technically feasible
- Economically justifiable
- Socially acceptable
- Environmentally defensible

Finally, the following points must be taken into account:

- The physical condition of the site must be such that the construction of the works, with the technical means available, is possible and the planned objectives can be reached
- The construction, operation, and maintenance costs must be in line with the gain from energy generation, food production and other benefits from the water stored in the reservoir
- Any harm to the natural environment or negative impact on communities must be reconciled against the gain

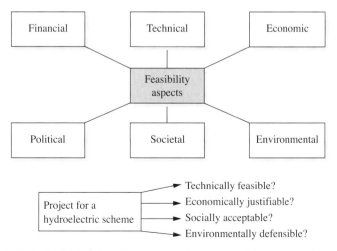

Figure 1.4 Assessment criteria for a water storage scheme comprising a reservoir created by a dam

2. The Development and Future of Dams

Montsalvens arch dam in Switzerland, height 52 m, year commissioned 1921 (Courtesy D. Quinche)

2.1 Dam Construction Over the Centuries

2.1.1 Introduction

Throughout the ages, humans have sought access to water, not only for basic and economic needs, but also to protect against floods and erosion. The aim of early dams was therefore to provide water for irrigation and to supply townships with drinking water. Water came from directly from seasonal flooding or was diverted. Much later, as industry developed, some factories (paper mills, tanneries, chemical industries, etc.) needed water in order to operate. Toward the late nineteenth century, hydraulic energy was used to meet the need for hydroelectric power. The increase in large dams went hand in hand with the growth of cities and the rise in industry.

Alongside this, works were also undertaken to reconfigure waterways, supply water canals, and compensate for periods of low water flow.

2.1.2 Dams in the past[1]

The first large dams were constructed by early civilizations, in particular along the Nile Valley, and in Mesopotamia, China, and South Asia. They were undoubtedly the traces of ancient peoples that were the easiest for archaeologists to find. The earliest known remains come from the Sadd-el Kafara dam, built in Egypt between 2950 and 2750 BC. This construction, 14 meters high and 113 meters along the crest, was designed with a central core comprising loose earthfill material (silty sand and gravel) and two rockfill shoulders. The water stored in the 0.5 million m^3 reservoir during the wet season was used for irrigation during the dry season (Figure 2.1).

For many centuries, humans had relied on empiricism, but they displayed great mastery in the construction of water-retaining structures. One such dam, the Mala'a dam in the Nile Valley, operated for thousands of years. The upstream face of the dam was in masonry and was strengthened by abutments downstream. Among some of the most remarkable ancient structures is the Marib dam in Yemen, which was built in the ancient kingdom of Saba from the eighth century BC and was only finally abandoned 1,300 years later in 572 AD following an exceptional flood (Figure 2.2). In 1986, Yemen constructed a new 40–meter-high embankment dam that now stores some 398 million m^3 of water.

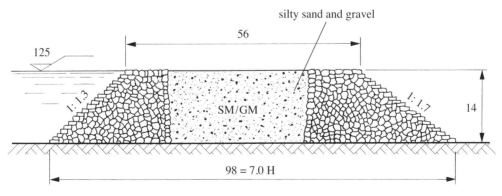

Figure 2.1 The Sadd-el Kafara dam (from Schnitter, 1994)

---

[1] Based on the book *A History of Dams: The Useful Pyramids* (Schnitter, 1994).

Figure 2.2 Marib dam (Yemen): historical intake structure (right) and new dam with bottom outlet in operation (left)

It has been observed that by using loose material, ancient constructors had sought to ensure the impermeability of their structures and to protect against superficial erosion by lining the faces. The Jawa dam in Jordan comprises an impermeable section formed by two walls and a two-meter-thick earthen core. Stability was provided by a supporting structure made of rockfill. In Sri Lanka, where the oldest structure dates from 380 BC, single-fill embankment dams were constructed with gentle upstream and downstream slopes and facing protected the upstream side against waves. The Romans constructed simple masonry walls with the same thickness and foundations built on rock (Figure 2.3). In cases of insufficient stability, the walls were reinforced with abutments or backfill downstream. The Romans were also the creators of the first arch and multiple-arch dams. To build these, they transported material with handbarrows and carts. They were the first to use hoisting devices and develop concrete (made of sand, gravel, burnt silt, water, and volcanic ash).

Masonry dams were built in other regions, particularly in Iran, including the Kebar and Kurit dams which were cylindrical and whose thickness diminished from the base to the crest. In Japan, several dams with single-material fill were built from the sixth century onward.

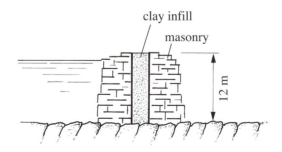

Figure 2.3 Cross-section of a Roman dam

Finally, some ancient structures had a spillway, either by using a naturally occurring depression in the rock or carved out of a block of rock, for example, allowing water to flow down the sides.

It was in Europe, during the industrial revolution of the Middle Ages in the eleventh and twelfth centuries, that technology using hydraulic force developed. The waterwheel, which had been in use from the end of the second century BC in China and the Mediterranean basin, began to be installed in diversion channels in conjunction with river weirs (Figure 2.4). In Central Europe, some embankment dams designed for fisheries were built from 1298 to 1590. In the mining region of Harz, Germany, many embankment dams were constructed to harness hydraulic power from the sixteenth century. This was also where the first embankment dams with a central core began to appear from 1715 (Figure 2.5). At the same time in both France and the United Kingdom, dams were being built to satisfy the demand of local populations for water for irrigation and navigation.

Several embankment dams were built in the United States in the mid nineteenth century. These structures suffered various fates, as several were destroyed by floods. The first rockfill dam for irrigation dates from the late nineteenth century in California.

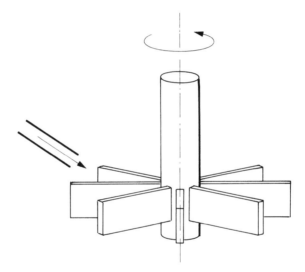

Figure 2.4 Example of a horizontal wheel with a vertical axis

2.1.3 The construction of dams in the twentieth century

Industrialization in the late nineteenth century and the growth and development of cities contributed to the increased construction of dams in the twentieth century. Figure 2.6 clearly demonstrates this growth in Switzerland, France, Canada, and the rest of the world.

At a global level, a regular increase in the number of dams can be observed over the first half of the century. From the end of the Second World War to 1970, the increase in the number of dams becomes extremely rapid in Switzerland, France, and Canada alike, as well as in the rest of the world. From 1970, a clear reduction is visible, which has continued until today. This stagnation, particularly obvious in Switzerland and France, can be explained firstly by the global economic downturn of the 1970s and secondly by the fact that the number of dams in industrialized countries began to reach or had reached the

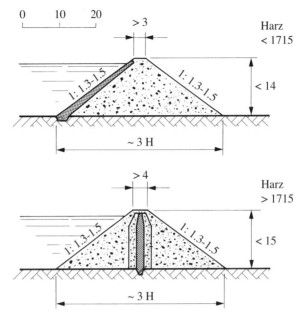

Figure 2.5 Embankment dams built in Harz, Germany, including the first to have a central core (from Schnitter, 1994)

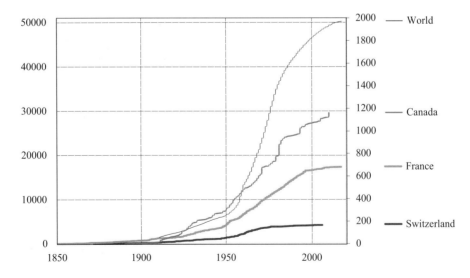

Figure 2.6 Increase in the number of dams (height > 15 m) in the world (vertical axis on the left) and in Switzerland, France, and Canada (vertical axis on the right). Only dams whose year of commissioning are known have been included.

maximum level economically possible. This limit is especially visible in Switzerland and France, where only a few, rare sites are still technically viable. Finally, the current economic situation and sensitivity to environmental issues make new dam projects difficult.

Figure 2.7 illustrates the increase across time of records in dam heights. A difference is drawn between concrete dams and embankment dams. This diagram highlights the role of Swiss engineers, who from 1962 held the record for the highest gravity dam in the world with the Grande Dixence (Figure 2.8). This concrete wall was only beaten in the mid-1980s by the Nurek embankment dam in Tajikistan, in 2010 by the Xiaowan gravity dam (294 m), and in 2014 by the Jinping I gravity dam (305 m). Once completed, the Rogun embankment dam in Tajikistan will become the highest dam in the world at 335 m.

The highest embankment dam in Switzerland is the 155–meter-high Göscheneralp dam (see Figure 2.21).

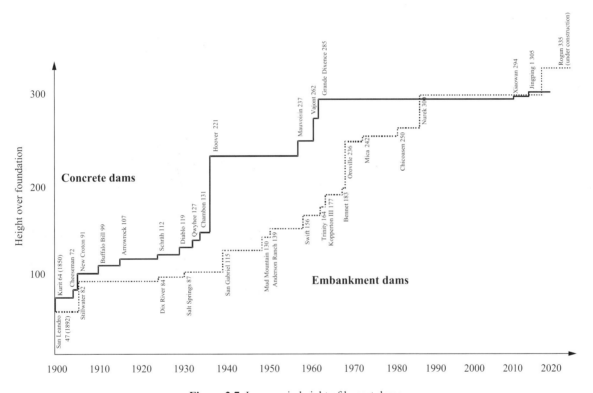

Figure 2.7 Increase in height of largest dams

Figure 2.8 Grande Dixence gravity dam (VS), H = 285 m

2.1.4 Dams today

According to the *World Register of Dams* (WRD) 2019 edition published by the ICOLD, more than 59,000 large dams are estimated to be in service (criteria: H > 15 m; 10 m < H < 15 m and volume of reservoir above 3 million m^3). It should be noted that their geographic distribution is very uneven. More than 40% of all dams higher than 15 m are found in China. One third are found in industrialized countries in Europe, North America, and Japan, while the remaining 25% are found in the rest of the world, primarily emerging and developing countries. This latter group, in particular South America, Africa, and Asia, with their increasing populations and growing need for irrigation and flood protection systems, represents a huge potential for dam construction.

According to the same source, 92% of the world's stock of large dams have a height lower than 60 m, 62% lower than 30 m, and only 2% higher than 100 m.

Today, the construction of large dams is the subject of considerable efforts on behalf of some countries (Table 2.9). The countries with the most activity in dam construction in 2021 were China, Iran, Turkey, and Indonesia. In Europe, most dams are built in Italy and Greece. These are also the countries that have the greatest number of dam structures under construction, given the need for controlling water supply for irrigation and drinking water.

Table 2.9 Dams under construction around the world, as at 2021
(from *Hydropower & Dams World Atlas*, 2021)

Height > 60 m			Height > 150 m		
1.	China	59	1.	China	16
2.	Iran	46	2.	Iran	5
3.	Turkey	37	3.	Pakistan, Ethiopia, Laos	3
4.	Indonesia	30	4.	Lesotho, India, Indonesia,	
5	Japan	15		Lebanon, Malaysia, Tajikistan,	
6.	Morocco	10		Turkey, Greece, Mexico, Colombia	1
7.	Laos	8			
8.	Vietnam, Greece, Italy	5			
World total		*301*	*World total*		*40*

Figure 2.10 Karun III Dam (Iran) commissioned in 2007 (Courtesy S. Emami)

2.1.5 Overview of the development of dam technology

Embankment dams

The first stability analyses on embankment dams were carried out in the early eighteenth century. Charles A. Coulomb (1773) developed his theory of earth pressure and introduced the ideas of cohesion and angle of friction. Rankine (1856) then made his own contribution to the same subject. We have Henry P. G. Darcy (1856) to thank for his simple experimental explanation of flow in porous media. He was able to demonstrate experimentally that the speed of water flow through a porous material is proportional to the gradient. However, it was Karl Terzaghi (1925) who gave a real boost to the development of soil mechanics, thus enabling a scientific and experimental approach for the study of earthfill dams. He notably was able to explain the consolidation of soil subjected to loading through the dissipation of pore water pressure. Arthur Casagrande also contributed to embankment dams and other hydraulic works. As for slope stability, Wolmar Fellenius (1932) published the slip circle method based on an idea of K. Petterson (1916) by introducing not only shear strength, but also cohesion. Taylor (1948) and Bishop (1955), who were later joined by Morgenstern (1960), refined this analytical method for stability (Figure 2.11). Ralph R. Proctor (1894–1962) defined the optimal saturation coefficient for compaction.

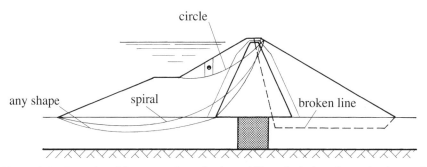

Figure 2.11 Examples of fracture surfaces for the stability calculation of an embankment dam

Due in particular to work by Terzaghi, the construction of zoned embankment dams continued to develop throughout the twentieth century (Figure 2.12). The construction of rockfill dams has also greatly progressed since 1950. Initially, quarried rockfill was dry compacted, but since then, the humidification of placed rockfill has become commonplace.

Mechanical means of compaction (rubber-tired compactor, vibratory roller, or padfoot roller) and quality control of fill have also greatly contributed to technological advances in the construction of embankment dams. The improvement of filters and drains has decreased the risks of internal erosion.

Although rare, rockfill dams that may be overtopped have been constructed in recent years.

Concrete dams

In 1840, Edouard H. T. Méry and Jean-Baptiste Bélanger formulated an initial theory for the calculation of gravity dams by proposing linear stress distribution. This method was still incomplete, however, as it did not allow for the effect of uplift. Thomas Hawksley (1807–1893) outlined the linear decrease in uplift in the direction of seepage flow, and George Deacon (1843–1909) recognized the effect of drainage on the

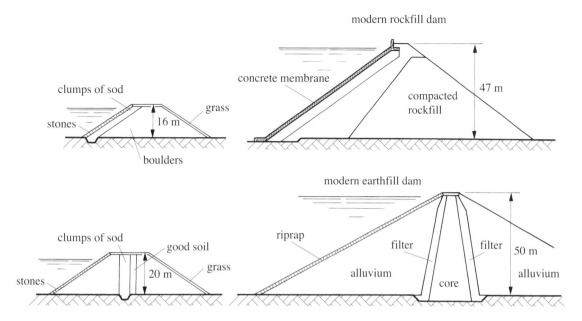

Figure 2.12 Development of embankment dams (from Sinniger, 1985)

distribution of uplift. Maurice Lévy (1843–1910) put forward his postulate that the vertical pressure in the upstream dam face, resulting from the dam's own weight and water load, should not be inferior to the water pressure at any point of the upstream face. Significant improvements to the safety of gravity dams were made throughout the twentieth century. They primarily concerned the widening of the profile, drainage, foundation choice, and an improved quality of binders.

François Zola (1795–1847) planned and designed the first masonry arch dam. He envisaged the arch dam as comprising independent horizontal arches bearing the load of water pressure equal to the depth of the arch below the water level; he was the first to propose the tube formula. Since then, of course, calculation methods have progressed. At the beginning of the twentieth century, the Buffalo Bill and Pathfinder dams (United States) were the first arch dams to use the trial-load method. Early arch dams with variable radius arches were designed by Lars R. Jorgensen (1876–1938) for the Salmon Creek dam in 1914 and by Heinrich E. Gruner (1873–1947) for the Montsalvens dam in 1920. For the design of the latter, Alfred Stucky (1892–1969) and Henri Gicot (1897–1982), who worked with H. Gruner, developed a method of analysis that was based on work published by Hugo F. L. Ritter (1883–1956).

New momentum was given to the construction of gravity dams thanks to the innovative process of roller compacted concrete (RCC). Early projects using this technique date from the 1960s, and it developed strongly throughout the 1980s.

Means of calculation
With advances in computer technology, today's engineers have access to highly effective numerical tools. Thanks to the availability of many types of software, the analysis of static and dynamic behavior of dams has made remarkable progress. Numerical modeling and simulation enable a precise approach when dealing with problems. The finite element method is an important example of this development (Figure 2.13).

Previously, only the dam itself was used as a basis for calculation; now these new methods allow the whole dam-foundation ensemble to be taken into account. Despite the increase in the availability of numerical tools, conventional analytical methods of calculation are nevertheless valuable for carrying out preliminary verifications and understanding the static and dynamic behavior of the structure.

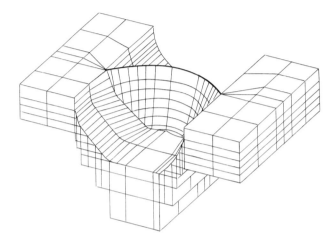

Figure 2.13 Example of the modeling of an arch dam and its foundations using the finite element method (from Jansen, 1988)

2.2 The Development of Dams in Switzerland

2.2.1 Introduction

The oldest dam structures still in use date from the nineteenth century. During the twentieth century, the country's economic development and ensuing energy needs had an influence on the rate of construction of dams associated with quite remarkable hydroelectric projects. While Switzerland today has many large dams, this is due to the driving force of eminent engineers who played pioneering roles. The most active period of dam building occurred between 1950 and 1970 (Figure 2.14). There are more than 220 dams in Switzerland under the jurisdiction of the Confederation (§ 2.2.7),[2] of which 87% are designed for the production of hydropower. Other uses include the storing of water for irrigation, supply of drinking water, and the production of artificial snow (3%), biotopes and leisure activities (3%), as well as protective structures for controlling flooding events and retaining sediment (7%). Among these dams, 56% are concrete dams (which can be further split into 54% gravity dams, 41% arch dams, and 5% multiple arch and buttress dams), 31% embankment dams (earth or rockfill), and 13% gated weirs. Twenty-five dams have a height greater than 100 m and four of these are taller than 200 m. The most impressive dams are located in the Alps. Finally, it is important to note the existence of several hundred dams and weirs of more modest dimensions and of various types and uses.

---

[2] WRFA, 2010; WRFO, 2012.

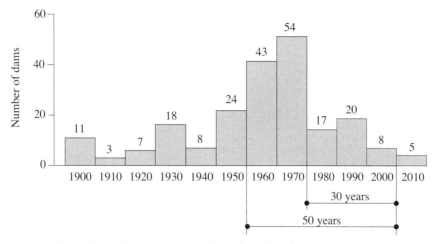

Figure 2.14 The age structure of large dams in Switzerland, by decade

2.2.2 Main stages in the construction of dams[3]

The growth in population and industrial development of the twelfth century led to the appearance of the first water storage structures in Switzerland. These installations were modestly sized and were used to produce hydropower. Only some of these structures still exist today.

Later, in 1695, the Joux-Verte dam (height 13 m) was built of dry stone to the north-east of Roche (VD) (Figure 2.15). Comprising an arched wall made of ashlar, this structure holds a special place in the development of arch dams. The water stored in the reservoir was regularly drained through a bottom outlet in order to float wood down into the Rhone Valley.

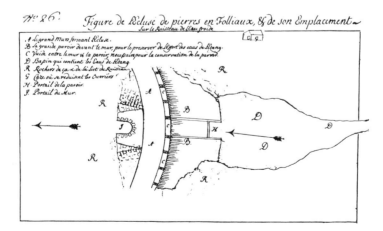

Figure 2.15 Map dating from the mid-18th century: the Joux-Verte dam (archives from the canton of Vaud)

---

[3] From Schnitter (1985) and Sinniger (1985).

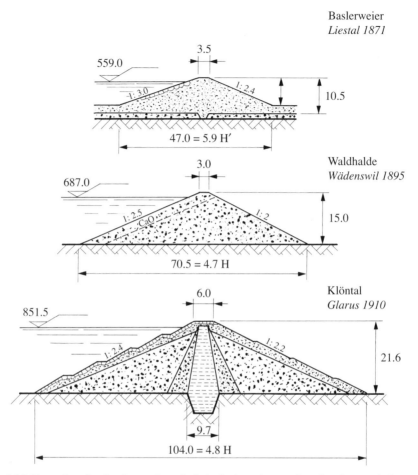

Figure 2.16 Examples of embankment dams built in the late nineteenth and early twentieth centuries (from Schnitter, 1985)

The industrial revolution in the eighteenth and nineteenth centuries brought about great economic growth, in particular due to the development of hydroelectric schemes in which turbines gradually replaced water wheels. From 1869 to 1872, Guillaume Ritter (1835–1912) built the Pérolles hydropower scheme on the Sarine River upstream of Fribourg. The slightly curved gravity dam (height 21 m, crest length 195 m) was at the time the largest dam in Switzerland. Its trapezoidal section includes two inclined faces. Its construction in concrete was also an innovation in Europe. Due to rapid sedimentation, the dam height was increased by 3 m in 1909. Interesting rehabilitation works across the whole dam structure were carried out from 2000 to 2004 (Figure 2.23, Figure 13.56).

In the late nineteenth and early twentieth centuries, many embankment dams were built (Figure 2.16). At that time, techniques for earthworks were based on empirical criteria far removed from the later science of soil mechanics. The failure of a small, five-meter dike in 1877 encouraged engineer Friedrich de Salis (1828–1901) to undertake a detailed study of the dam break wave, which was most probably the first calculation of its kind. Some of the more remarkable constructions include the Gübsen saddle embankment

(SG/1900/H = 19 m)[4] and the Klöntal embankment (GL/1910/H = 30 m), which was designed to raise the water level of a natural lake (Figure 2.16). During this same period, many gravity dams were also built, most often using traditional masonry techniques, including dams at Buchholz (SG/1892/H = 19 m), List (SG/1908/H = 29 m), Muslen (SG/1908/H = 29 m), and at the Bernina Pass (GR/1911/H = 15 and 26 m). Substantial rehabilitation works were carried out on the dams at the Bernina Pass in the late twentieth and early twenty-first centuries.

Due to increasing energy needs, from 1914 attention turned to the construction of water-storage reservoirs, in particular water issuing from summer snowmelt and glacier melt, and in order to guarantee the production of energy in winter when demand for electricity rises. This transfer of energy necessitated the availability of large reservoirs and therefore the construction of large dams.

The construction of the Montsalvens dam (FR/1920/H = 55 m) marks the beginning of a new period of dam development. It was the first double-curvature arch dam in Europe and was designed by Heinrich E. Gruner (1873–1947). As mentioned above, static calculations were based on a trial-load method devised by Hugo F. L. Ritter (1883–1956), which was then developed by Alfred Stucky (1892–1969) and Henri Gicot (1897–1982), who were both working with H. Gruner.

The Swiss Commission on Dams was founded on 2 October 1928 by six renowned scholars and practitioners in the construction of dams. Later, other important figures joined the commission (Table 2.17). According to its statutes, the aim of the commission was to "deal with problems related to dams and to collect information and knowledge about these constructions and their operation." In the beginning, the Commission dealt with issues brought up in conferences held by the International Commission on Large Dams (ICOLD), founded in 1928. It later went on to establish guidelines for the construction and maintenance of Swiss dams. In addition, it set itself the task of carefully examining the results of observations carried out on dams. In 1948 the Swiss Commission on Large Dams became the Swiss National Committee on Large Dams (SNCOLD) and in 1999, the Swiss Committee on Dams (SwissCoD).

In the 1930s, the global economic crisis slowed the rise in the consumption of electricity, as well as the need for new hydropower schemes, and, as a result, the construction of dams. However, it was at this time that the Dixence dam was built (VS/1935/H = 87 m), based on a project by Alfred Stucky. This buttress dam has a volume of 421,000 m^3 and until the end of the Second World War was the highest dam of its type in the world. Construction of the Verbois dam (GE/1943/H = 34 m) and the buttress dam at Lucendro (TI/1947/H = 73 m) (Figure 3.8) began during the Second World War. After the bombing of German gravity dams by the English air force in May 1943, the Federal authorities stipulated that horizontal buttress struts had to be added to the already completed Lucendro buttress dam, as well as major reinforcement to the Cleuson dam (VS/H = 87 m) built between 1947 and 1950. Thereafter, only very narrow hollows were to be admitted, as is the case for the dams at Räterichsboden (BE/1950/H = 94 m) and Oberaar (BE/1953/H = 100 m).

During the Second World War, a sharp increase in the consumption of electricity was observed, which continued in the post-war context. Hydroelectric schemes, of which large dams were the centerpiece, were constructed to respond to this demand for additional energy. In 1945 the construction of the Rossens arch dam began (FR/1947/H = 83 m). From 1950 and throughout the 1970s, dam construction boomed and more than one hundred dams were commissioned. The construction of the 237-meter-high Mauvoisin arch dam (Figure 2.18) and the 285-meter-high Grande Dixence gravity dam (Figure 2.8) began in 1951. Commissioned in 1957 and 1961 respectively, these two dams are still today among the highest operating dams in the world. Thanks to the talent of the above-mentioned people and engineers from top-level consultancy firms, many impressive arch dams were erected.

---

[4] (SG/1900/H = 19 m) = (The canton in which the dam is located/year commissioned/official height of the dam).

Table 2.17 Founding members of the Swiss Commission on Large Dams

Founding members of the Commission
- Dr H. Eggenberger (1875–1958), head engineer at the Swiss Federal Railways, Bern
- Dr H. E. Gruner (1873–1947), engineer, Basel
- Dr A. Kaech (1881–1965), head engineer at Kraftwerke Oberhasli AG, Innertkirchen
- Dr E. Meyer-Peter (1883–1969), professor in hydraulic works at the Swiss Federal Institute of Technology in Zurich (ETHZ)
- Dr M. Ritter (1884–1946), professor in statistics and reinforced concrete at the Federal Polytechnic School in Zurich
- Dr A. Stucky (1892–1969), professor in hydraulic works at the Ecole Polytechnique de l'Université de Lausanne (EPUL today the EPFL)
- A. Zwygart (1886–1972), director of the Nordostschweizerische Kraftwerke AG, Baden

In order to further constructive cooperation, in 1935 the Federal Department of the Interior seconded the following members to the Commission:
- M. W. Schurter (1889–1965), engineer, chief inspector of public works for the federal government

Over time, the following people joined the Commission:
- M. J. Bolomey (1879–1952), professor of stone materials at the Ecole Polytechnique de l'Université de Lausanne (EPUL today the EPFL)
- M. O. Frey-Bär (1909–1973), engineer, Motor-Columbus SA, Baden
- M. H. Gicot (1897–1982), consulting engineer, Fribourg
- M. H. Juillard (1896–1958), consulting engineer, Bern
- Dr M. Lugeon (1870–1953), geologist, professor at the University of Lausanne and Ecole Polytechnique de l'Université de Lausanne (EPUL today the EPFL)
- E. Martz (1879–1959), president of the Swiss Society of Manufacturers of Lime, Cements and Gypsum, Zurich
- Dr M. Roš (1879–1962), professor of materials technology at the Swiss Federal Institute of Technology in Zurich (ETHZ)
- Dr M. Roš jun. (1913–1968), consulting engineer, Zurich

In the design of arch dams, the traditional circular arches were replaced by parabolic or elliptical arches in order to ensure a better orientation of the pressure of the arches against the rock foundation and abutments. Among the structures with a height greater than 100 m, the following dams can be listed: Emosson (VS/1974/H = 180 m), Zeuzier (VS/1957/H = 156 m), Curnera (GR/1966/H = 155 m), Zervreilla (GR/1957/H = 151 m), Moiry (VS/1958/H = 148 m), Limmern (GR/1963/H = 146 m), Punt dal Gall (GR/1968/H = 130 m), Nalps (GR/1962/H = 127 m), and Gebidem (VS/1967/H = 122 m). The twin dams at Lake Hongrin (VD/1969/H = 125 and 90 m) should also be noted, whose double arches joined by a common, central abutment, give the structure a distinctive look. After the Grande Dixence and Mauvoisin dams, the 200 m barrier was broken, notably by the arch dams of Luzzone and Contra (Figure 2.20) in Ticino. The latter was designed by Giovanni Lombardi (1926–2017).

Utilizing primarily the core scientific and rational developments established by Karl Terzaghi, the founder of soil mechanics, the construction of several embankment dams was undertaken. In addition, on the initiative of Eugen Meyer-Peter (1883–1969), the ETHZ created an institute for foundation engineering and soil mechanics, whose research has proven to be of great use. The embankments at Marmorera (Castiletto) (GR/1954/H = 91 m), Göscheneralp (UR/1960/H = 155 m), and Mattmark (VS/1967/H = 120 m) are the largest embankment dams built in Switzerland (Figure 2.21).

Figure 2.18 The Mauvoisin arch dam

Figure 2.19 The double-arch Hongrin dam

Figure 2.20 The Contra arch dam

Figure 2.21 The largest embankment dams in Switzerland: (a) Mattmark, (b) Göscheneralp, and (c) Marmorera (Castiletto)

From 1980, new constructions integrated into hydroelectric schemes became less common. The largest are the structures in Solis (GR/1986/H = 61 m) and Pigniu (GR/1989/H = 53 m). However, to better operational conditions and ensure improved energy transfer, the arch dam at Mauvoisin (VS) was raised by 13.5 m to reach a height of 250 m in 1991 and the dam at Luzzone (TI) by 17 m to reach 225 m in 1997 (Figure 2.29). Two pumped-storage projects have been launched. The first is the scheme at Limmern (GL); here, the construction of the Muttsee gravity dam (GL/2010/H = 35) has raised the level of a natural lake, situated at an altitude of 2,500 meters above sea level, by 28 m. During construction of the Nant-de-Drance pumped-storage hydropower plant (VS), the Vieux-Emosson Dam, built in 1955, was raised by 20 m to reach a height of 65 m, which has doubled the storage capacity of the reservoir.

Furthermore, several structures with heights between 7 and 30 m have been built to protect against natural events such as floods and avalanches. In the late 1990s, reservoirs created to store water with the aim of producing artificial snow began to appear.

2.2.3 Strengthening and rehabilitation works

On a different note, from the early 1980s, attention turned toward old dam structures of all sizes whose safety had to be reassessed on the basis of modern standards and technological advances. Depending on the results obtained, it may be necessary to rehabilitate all or some of the structure elements in order to guarantee its safe operation for many more years. Of course, there are many reasons why the strengthening and rehabilitation of a dam are necessary. Often, the structure no longer meets the latest stability criteria. The accepted hypotheses for load on the structure from the initial project may have to be revised. These hypotheses may concern the weight itself, the distribution of uplift or the induced effects that may occur during an earthquake. New operational conditions, such as, for example, new flooding levels, a large accumulation of upstream sediment, or the installation of downstream rockfill may also have an impact on load. Several different types of rehabilitation works are possible and may sometimes be combined. Table 2.22 provides a list of the kinds of possible interventions.

Table 2.22 List of reinforcement and rehabilitation interventions

- Dam heightening
- Complete rehabilitation of all structures
- Treatment of facing
    - Laying membrane
    - Asphalt facing
    - Concrete cover, gunite
- Rehabilitation of dam body (concrete, embankments)
    - Grouting
    - Sealing
- Reinforcement of downstream toe
- Foundation treatment
    - Grouting
    - Drainage
- Flood safety
    - Modification of the spillway
    - Modification of the crest
    - Creation of a parapet wall
- Drawdown of the reservoir
    - Transformation of the bottom outlet
    - Implementation of a new bottom outlet

The following sections describe various examples of works carried out in Switzerland with a view to reinforcing and rehabilitating dams.

In the case of the Maigrauge dam, commissioned in 1872, the rehabilitation project was designed to improve safety in case of flooding by uprating the spillway, improving the dam's structural safety with the installation of prestressed rock anchors, and optimizing operating conditions with the modification of water intakes. Furthermore, the monitoring system was modernized. And last but not least, a ladder including a lift was constructed for fish to migrate upstream past the dam and a series of channels and pools were added for downstream migration. These works were carried out from 2000 to 2004 (Figures 2.23 and 13.56).

Figure 2.23 Maigrauge dam

When the stress and stability assessment demonstrates that safety conditions are not being met, drawing down the reservoir or reinforcing the dam becomes inevitable. For the Gübsensee dam, the chosen solution was to install post-stressed cables (Figure 3.17). As for the concrete dams at Muslen and List, their upstream and downstream faces were covered with a concrete shell while the height of the crest was also raised in order to increase the volume of the reservoir and thus optimize hydropower generation (Figure 2.24).

Remedial work may become necessary if the material at the core of the dam has suffered from major internal damage, due, for example, to swelling caused by an alkali-aggregate reaction (AAR), which can significantly impact concrete characteristics. To limit the development of this type of swelling, the upstream face of the Illsee dam (VS/1923–43/H = 25 m) was lined with a PCV geomembrane, and the Lago Bianco Sud dam (GR/1912–42/H = 26 m) was lined with a membrane comprising a synthetic liquid applied in successive layers. At the Illsee dam, the system put in place did not prevent water penetration via the foundation and thus did not slow the swelling phenomenon. Concrete drying measures were also taken without success. Vertical sawing cuts into the concrete are still to be done in order to relieve the stresses in the dam. After being commissioned in 1952, monitoring of the behavior of the Serra arch dam (VS/1952–2010/H = 25.7 m) was principally carried out through geodetic measurements and leveling. Concrete in the Serra dam, affected by an alkali-granulate reaction (AAR) leading to the swelling of the concrete, resulted in irreversible upstream deformation, accompanied by an uplift to the dam and diffuse cracking. Rehabilitation of the structure was necessary, as the gradual deterioration in its conditions of use and safety had been highlighted. The rehabilitation solution that was chosen consisted in building a new dam downstream of the original one. The partial and necessary demolition of the downstream toe of the old dam enabled a new and more favorable geometry to be determined. From a structural point of view, the Serra dam is close to a double-curved arch dam (SwissCoD, 2017a).

In cases of significant deterioration due mainly to frost, both faces must be treated. For example, around 1983, after damaged zones had been scored, the surface of the upstream face of the Schräh dam was covered with lightly reinforced shotcrete to a depth of 8 to 12 cm. In another example, wet shotcrete was applied across practically the whole surface of the upstream face of the Cleuson dam between 1995 and 1998. The affected area had previously been stripped by hydrodemolition. The bond between the base concrete and the sprayed concrete was guaranteed by a grid with mushroom-shaped anchoring bolts made of 12-mm-diameter reinforced steel placed at 4 bolts per m^2 (Figure 2.25).[5]

---

[5] From Rechsteiner (1994).

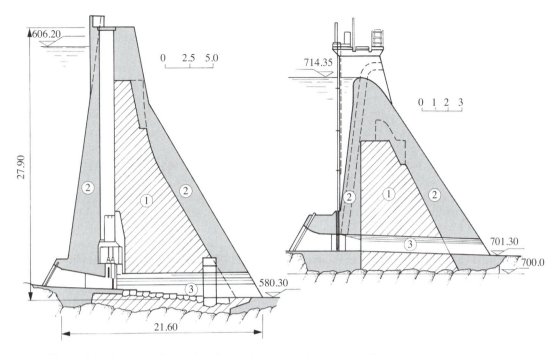

Figure 2.24 Strengthening project for the dams at Muslen and List: ① body of the original dam; ② new concrete shell; ③ bottom outlet

Figure 2.25 Cleuson dam: bonding system for shotcrete

Following a dam safety check for earthquake behavior implemented by the supervising authority in line with federal guidelines (SFOE, 2016), the Toules dam was strengthened with a mass concrete support built directly downstream of the dam on both the left and right sides of the dam. These braces were made of unreinforced concrete blocks. The joints were sealed with cement grout so as to ensure that the force was transmitted from one block to another and to form a monolithic structure.

After severe flooding in 1978 in Ticino and the blockage of the spillway channels at the Palagnedra dam (TI/1952/H = 72 m) by the massive piling up of driftwood (Figure 5.13) (SwissCoD, 2017b), the supervisory authority reviewed general safety criteria in cases of flood and asked for an investigation into the safety requirements to be carried out. It is important to note that in the Alps, floods can strike with devastating speed. For many dams, spillways or crests had to be modified, or a large parapet wall had to be added so as to provide increased retention capacity.

Studies have shown that during major flooding events, reservoirs attached to hydroelectric schemes contribute substantially to the reduction of peak flooding levels, due to their capacity for retaining water, even though that is not their primary function. In order to increase protective measures against flooding events downstream of a dam, without having to restrict production of hydroelectric power, one option is to transform a single operation into a multipurpose operation. The idea is to create an additional volume in the upper part of the reservoir that can be used hold a specific volume of water in cases of flood. Such a project was completed in 2001 for the Mattmark embankment dam by adapting the side weir spillway. Other proposals are also being considered.

In Switzerland, the general rule is that all dams must be equipped with a bottom outlet to empty the reservoir in cases of abnormal dam behavior or lower the water level for maintenance to be carried out. Due to insufficient capacity or obsolete equipment, some structures had to be completely transformed or a new bottom outlet had to be created. For example, a gallery was drilled into the foot of the Schräh dam (SZ/1924/H = 111 m). To ensure compliance, a bottom outlet was created in the Illsee dam (VS/1924–43/H = 25 m) by utilizing a gallery that had originally been used to lower the level of the natural lake at the time the dam was being built. In other cases, gates have simply been replaced or a new gate added.

Figure 2.26 Zeuzier arch dam (Courtesy M. Daneshvari)

Ground movement following the drainage effect of an investigation gallery for a tunnel project resulted in 1978 in abnormal deformations at the Zeuzier dams (VS/1957/H = 156 m) (Figure 2.26). Geodetic measurements brought to light settlement of approximately 10 cm and an upstream displacement of 9 cm at crest level, as well as a valley narrowing of about 5 cm between abutments at the level of the crest. Following these deformations, an opening of the vertical joints in the upper part of the upstream face and the development of cracks in the downstream face were observed. Works were carried out to check the condition of the grout curtain and to verify the contact point between concrete and rock. In order to re-establish the monolithic nature of the dam, grouting tests were carried out to determine the most appropriate grout product. An epoxy-based material was chosen (Rodur 510 and 520). Works were carried out with success and after a staged filling of the reservoir, it once again reached its normal level in the summer of 1978 (Pougatsch, 1990).

Geodetic measurements carried out between 1921 and 1937 had already highlighted weak plastic movement perpendicular to the bed of the downstream section of the left-bank abutment of the Montsalvens dam (FR/1920/H = 52 m). As this movement continued, additional monitoring devices (pendulum, extensometers) were installed in 1969 in order to ensure more systematic reporting of the behavior of the downstream zone. Analyses carried out later showed that the state of equilibrium was situated at the limit of elastic behavior. Strengthening works were decided on out of a fear that the deformations would only continue to intensify or that a rockfall would be caused by a seismic shock. Works on the left abutment were designed with two objectives in mind; firstly, bolts sealed with cement grout were applied to a concrete sprayed surface to protect the valley flanks against the risk of rockfall, and secondly, reinforced bars sealed completely into the rock with cement grout were designed to increase resistance to shearing along bedding planes. At the Pfaffensprung dam (UR/1921/H = 32 m), it was not known how long existing anchors would hold downstream of the left bank abutment, so it was decided that additional prestressed anchors would be installed with a system that enabled their tension to be controlled at all times.

2.2.4 Resolution of problems related to the reservoir

The area surrounding the reservoir

It is of the utmost importance to monitor the behavior of banks and slopes, as instabilities can occur, sometimes without any direct connection to activity at the reservoir. For example, upstream of the Mauvoisin dam, a crack was observed in a mountain road running alongside the reservoir. Snowmelt had saturated scree in the area, which had slipped in sections of differing depths. A monitoring system (geodetic measurements, inclinometric measurements from boreholes) was set up, and a limited water level was set while the zone was still unstable.

Similarly, experts inspected several glaciers in order to ensure that large sections of material did not break off and end up in the reservoir.

Sedimentation[6]

Due to climate change and its consequences (glacier retreat, the zero-degree line and permafrost levels rising in altitude, increased precipitation), an increase in the arrival of solid materials into alpine lakes must be expected. Solid materials issuing from soil erosion are transported toward water reservoirs in watercourses by bed load or suspended load. Whether this material settles in a particular place or spreads out into the reservoir depends on its size. This deposited sediment has a direct impact on the operation of

---

[6] See also, Part IX, Sedimentation.

the overall storage scheme, as well as on the safety of the dam. With regard to operation, this will above all be manifested in a loss of usable storage capacity. According to estimations, at a global level this reservoir volume loss is between 1 and 2% per year. Based on an analysis of 19 reservoirs (Beyer, Portner, and Schleiss, 2000), it is in the order of 0.2% for alpine storage schemes in Switzerland. The siltation of reservoirs can affect the lifetime of a dam. With regard to safety issues, there is a considerable risk of water intakes and bottom outlets in particular being obstructed. These situations must be avoided. For bottom outlets to remain operational at all times, a free space directly upstream of the discharge system must be guaranteed. For various reasons, it is vital that an appropriate amount of this deposited sediment be periodically removed. Several means for limiting the arrival of sediment into reservoirs already exist (settling basins, diversion galleries, sand traps, etc.). In many cases (Gebidem, Rempen, Palagnedra, Luzzone, etc.), a program of periodic flushing takes place in accordance with a predefined schedule. The legal basis regarding the protection of water establishes the terms relative to the flushing and emptying of reservoirs. Specifically, it is important to ensure that as far as is possible flora and fauna in the river downstream are not harmed during these operations. Furthermore, except in extraordinary events, permits are issued by the relevant cantonal authorities, some of which have established regulatory requirements. In the future, designers and operators will be required to take effective measures for preventing reservoir sedimentation.

In alpine reservoirs, turbidity currents that form during floods are responsible for the transportation of considerable amounts of fine particles of sediment along the reservoir. Turbidity currents, which are like underwater avalanches, also erode sediment that has already been deposited, bringing it nearer to the dam itself where it is more likely to block the entrance to bottom outlets or water intakes (Schleiss and Oehy, 2002; Oehy and Schleiss, 2003). The increase in reservoir sedimentation can force operators to undertake substantial work in order for these structures to retain their primary functions. For example, at the Mauvoisin dam, the water intake had to be raised by 38 m and the bottom outlet by 36 m by building new intakes and gate chambers. Many other cases also exist.

Fine sediment, principally transported along the bottom of the reservoir by turbidity currents, can contribute to more than 80% of sedimentation in alpine reservoirs. In addition to controlling turbidity currents in reservoirs through the use of obstacles (Oehy and Schleiss, 2003), the deposit of fine material in the vicinity of the dam can be avoided by venting it through bottom outlets. This approach is economically and environmentally beneficial. Artificial flood releases combined with sediment replenishment downstream of dams can work together with this venting of fine sediment and restore, as well as dynamize, bedload transport in the river downstream (Döring et al., 2018). Another promising option for the management of fine sediment is by using water jet installations in the reservoir near the dam to ensure its suspension before evacuating it at controlled concentrations through the powerhouse intake (Jenzer, Althaus et al., 2011).

2.2.5 The development of concrete pouring and placing

It is clear that concrete in its modern form has meant that stone masonry is no longer used in the construction of dams. In 1866, Boyds Corner was the first concrete gravity dam to be built in the United States. The first concrete arch dams were constructed in the early twentieth century in Australia. Techniques for manufacturing and laying concrete have continued to evolve throughout the years. For each project, however, a specific concrete mix has to be defined taking into account local constraints and the characteristics of materials used.

In Switzerland, the concrete in the Montsalvens dam, with a mix-weight of 250 kg/m^3, was compacted using pneumatic tampers that required only a modest set up. The transportation and laying of concrete using channel chutes began to appear, and poured concrete was used for the first time in the early 1920s at the Barberine dam (VS) with a height of 79 m. This method was then used for the construction of the

Figure 2.27 Laying concrete at the Grande Dixence dam (Courtesy H. Pougatsch)

dams at Schräh (SZ/1924/H = 112 m), which for many years was the world's highest dam, Rempen (SZ/1924/H = 32 m), three structures in the Oberhasli valley (BE), and the buttress dam at Dixence (VS/1935/H = 87 m). This method, which requires the addition of water and decreased resistance, was used for dams of a volume greater than 200,000 m^3; this was particularly the case for gravity dams, whose stresses are moderate. Signs of frost-damage started to appear, however, which led to a decrease in its use.

In the 1930s, immersion vibrators were used for the first time in the United States for the compaction of mass concrete. This method was adopted in Switzerland and is today the most common technique for concrete placing.

The construction of large dams with a high volume of concrete necessitates appropriate solutions for the production and laying of concrete. Concrete silos were employed to manufacture concrete, from where it was transported on more highly developed Blondin ropeways than in the time of poured concrete. Furthermore, for the laying and compaction of concrete, spreading was done by bulldozers and compaction by pervibrators attached to tracked vehicles (Figure 2.27). These techniques are still used.

Traditional concrete must today face up to a financially competitive rival, the roller-compacted concrete (RCC), which has a mix comprising little cement.

2.2.6 The development of monitoring systems[7]

To monitor the behavior of a dam and its foundations, as well as the surrounding environs, it is appropriate and recommended to measure a set of typical parameters. The first monitoring system designed to track behavior and obtain experimental data was set up at the Montsalvens dam (FR). Through triangulation, levelling, and inclinometers, as well as the installation of thermometers at various points, it became possible to measure deformation for different water levels and different thermal conditions. The aim of these measurements was to confirm the accuracy of the hypothesis used in the calculations.

In 1932 at the Spitallamm dam (BE), height 114 m, a structure that was part of the Oberhasli hydropower scheme, Henri Julliard developed and installed a pendulum as a mechanical reading system to measure deformation. This instrument became a vital part of dam monitoring systems in Switzerland and throughout the world.

Gradually, the monitoring of dams became widespread and the requirements for this type of surveillance have continued to evolve. With the successful development of measuring devices and methods, increased precision and simplified measurements were achieved. In addition, computer processing of data has enabled highly effective analyses. Given the importance attached to the monitoring system, various adaptations are regularly made with respect to new requirements and acquired knowledge, but also in response to new means and ways of measuring.

The updating of a monitoring system can include:

- The replacement of outdated or broken equipment
- The addition of devices providing complementary information (guaranteeing the monitoring of unequipped zones or following the observation of new phenomena)
- The installation of new types of measuring devices, such as fiber-optic sensors (measuring deformation and temperature, detecting seepage), measuring distances without reflection (laser scanning), spatial measurements by satellite (e.g., using the GPS network), 3D measurements of deformation in a drilling site, etc.

As a result of the development in electronics and computer processing, the importance and capabilities of automatized monitoring systems have only increased (Cottin, 1992). They enable a direct real-time connection between the dam and the team responsible for monitoring it. Such systems comprise means for measurement (measuring instruments), means for transmission of data, and automatic means for acquisition and storage of data (databases), as well as means of treatment (the timely evaluation and checking of the overall behavior of the structure) and presentation of data (analysis of the measured results, preparation of graphs, and writing of reports).

2.2.7 Legislative basis

As regards the safety of dams, the Swiss supervisory authority has two goals. Firstly, that of ensuring the safety of the dam and therefore that of the population, and secondly, of ensuring the safety of the operation itself. From a historical point of view, the June 22, 1877 Water Regulation Act amended in 1950 outlines the provisions relating to dam safety. It stipulates in one of its primary articles that concerning dams the Federal Council must take all necessary measures to prevent, insofar as is possible, any danger and damage that may result from their means of construction, their inadequate maintenance, or from acts of war. These details are regulated by means of an ordinance, and in 1957, the first regulation on dams was published.

---

[7] The design of monitoring systems is covered in Part VIII, Section 32.4.

It contained a number of minimum requirements and left sufficient leeway for innovation and progress. This regulation was modified in 1971 and again in 1985, primarily with regard to the flood alert system, before being replaced by the ordinance dated December 7, 1998 on the safety of water reservoirs (WRFA, 1998). Currently, the legislative basis includes an act on dams (WRFA, 2010) in effect since 2013, accompanied by an ordinance (WRFO, 2012). In addition to the terms relating to their safety, the act introduces the notion of civil liability due to the risks.

In terms of scope, the act (WRFA, 2010) specifies that the liability applies:

- To dams whose water level H above the low-water level of the watercourse or the thalweg (reservoir height) is at least 10 m, or
- If this water level is at least 5 m, for those whose reservoir capacity is higher than 50,000 m^3
- To dams of smaller dimensions, if they represent a specific potential risk for people and property; otherwise they are exempt.

Based on these size criteria, dams are divided into three categories (Figure 2.28). Dams in category I, which are subject to a five-yearly safety inspection, as well as those in category II are placed under the supervision of the Swiss Confederation; dams in category III fall under the supervision of the cantons. The legal provisions concern issues relating to structural safety, maintenance and monitoring, and emergency planning. These three elements constitute the key pillars upon which the safety of dams is based (see Part II, Chap. 5). Although not binding, certain guidelines assist in the implementation of legislation concerning dams (SFOE, 2014a, 2015a, 2015b, 2015c, 2016, 2017, 2018; Schwager et al. 2016).

Only dams with a potential for endangering human lives and causing serious material damages in the case of uncontrolled release or dam failure are subject to the legal conditions. This means that the requirements in terms of safety are the same for all dams, as everyone has the right to the same level of protection.

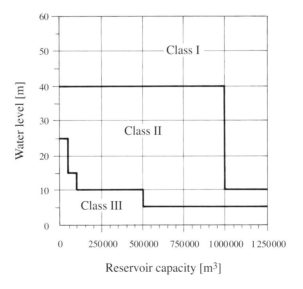

Figure 2.28 Illustration of dam categories

In countries that employ this type of classification, the levels are determined according to the potential for endangering human lives and the scale of damage to property and goods, as well as to the environment in cases of dam failure. Generally speaking, the following levels can be determined:

- Very high (many human lives are at risk, extensive damage)
- High (some human lives are at risk, significant damage)
- Low (no risk to human lives is expected, moderate damage)
- Remote (no risk to human lives, minor damage)

At a project level, the classification of a dam may affect the choice of frequency of occurrence for floods and earthquakes, and during operation on the monitoring and maintenance schedules. It may also play a role in the establishment of priorities relative to maintenance and rehabilitation works.

2.2.8 Measures to guarantee public safety[8]

Following the bombing of German gravity dams located in the Ruhr valley by the English air force during the nights of May 16 and 17, 1943, Swiss military and civil authorities were concerned about the vulnerability of dams to acts of war or sabotage. An initial measure introduced in June 1943 was to suspend a cable above dams as a means of protection against airplanes. In September 1943 the Federal Council, who had worked quickly on the legislation, announced an ordinance whose provisions covered active and passive dam protection measures against destruction in times of war, the use of reservoirs, and the lowering of their level, as well as the installation of an alarm system. It was decided that sirens would initially be installed in the near zone—the area subject to flooding 20 minutes after destruction of the dam. A list of dams that had to be equipped with the alarm system was published in late November 1943. In 1945 the Bannalp (NW) and Klöntal (GL) embankment dams were the first to be fitted with the flood wave alert system. The regulation concerning dams that came into effect in July 1957 gave a legal basis to the recommendation that alarm systems should be installed. However, this system still needed to be improved, and a technical committee was mandated to establish the terms of reference for a new system. A more concrete definition of the flood wave alert system was introduced in the 1957 version of the regulation concerning dams. For the first time, the near zone identified was extended to 2 hours (instead of 20 minutes previously) and a far zone was designated. The alarm systems for each zone are different. In addition, it was decided that the flood alert system would also be used in peace time and extended to all other possible hazards to dam safety. A new revision of the regulation concerning dams in 1985 brought further detail relative to the implementation of a complex system that could be engaged within one hour and that could function on its own. The introduction of degrees of preparation and the definition of criteria for the triggering of the flood alert system were also new elements. Later, a working group developed the "Flood Alert 2000" concept, as the renewal of the remote-control installation in particular had become necessary. The ordinance on the safety of dams came into effect on January 1, 1999 (WRFO, 1998), which stipulated that each dam with a reservoir capacity above 2 million m^3 had to be fitted with a flood alert system. This system also covers structures whose floodable area is at a high risk of danger, as the need to protect the population is not only related to the volume of the reservoir.

A guideline (SFOE, 2015d) establishes the means to be implemented and determines responsibilities relating to the organization and operation of the dam.

Depending on the situation, the alarm has two different sound tones. Firstly, the flood alert sirens sound in the 2-hour zone downstream of the dam. This alarm signals to the population that they must

---

[8] See also Part VIII, Section 3.3.

leave the zone. The general alarm sirens indicate to the population that they must listen to the radio for information.

The dam operator must also establish an emergency procedure that outlines its organizational structure, behavior, and responsibilities depending on the level of danger of the situation. This procedure also describes the emergency strategy, with its 5 levels of danger.

2.3 Looking to the Future

2.3.1 Introduction

In the future, dams will continue to play an essential role in the supply of drinking water, water for irrigation, and hydropower, especially in countries with a high potential for growth. Not to mention the protection that dams afford against flooding. Since 2001, the number of dams taller than 60 m being constructed in the world is most often between 320 and 370 (Figure 2.29).

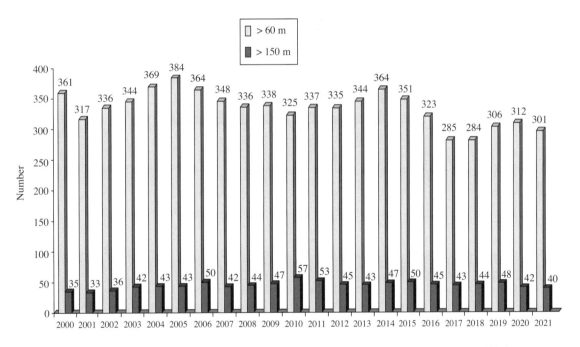

Figure 2.29 Number of dams under construction since 2000 (from Hydropower & Dams, 2021)

2.3.2 Future activities in Switzerland

The era of the construction of large dams in Switzerland is today practically over, as the sites with the most technical interest are for the most part already being used. The most recent large construction of a dam in Switzerland took place from 1995 to 1997 and involved the raising of the Luzzone arch dam, built in the early 1960s and whose height was increased from 208 to 225 m (Figure 2.30). Another major project concerns plans to raise the water level of the Grimsel reservoir by 23 m (the Spitallamm and Seeuferegg dams). This project is currently under study and is expected to go ahead.

Figure 2.30 Luzzone arch dam (TI), raised from 208 to 225 m

These examples demonstrate that dams are still attractive for design engineers, constructors, and operators. By way of their diverse uses, dams open up interesting prospects not only for projects aiming to transform or modernize existing structures, but also for new projects. It is also important to ensure the good health of existing dams. As structures and equipment age, certain problems need to be examined, such as the evolution of mass concrete (long term creep, alkali-aggregate reaction) or the stability of foundations (development of uplift, preserving drainage networks).

In the short term, monitoring and maintaining dams of all sizes are vital tasks for guaranteeing their safety. Although large and medium structures have been implementing these activities for many decades, the same cannot be said for small dams and barriers.

As for hydroelectric schemes, the future of hydropower falls under the general framework of sustainable development. Several socioeconomic and environmental parameters are set to influence operators in their future investments, that is, the deregulation of the market, changes in supply and demand, the price of electricity, the cost of construction, water-power royalties, and politics. First of all, many hydroelectric concessions are set to expire. When the time comes to renew them, the safety assessment of sometimes worn-out structures is a necessary phase that in most cases will necessitate consideration of major strengthening and rehabilitation works, or even the replacement of electrical and mechanical equipment in discharge systems. Other measures designed to modernize and optimize existing structures may be taken before the expiry of the concessions. The idea of heightening a dam in order to increase the capacity of its reservoir is a realistic option. Pumped-storage projects are once again coming to the fore. The advantage of these schemes is that they can store hydraulic energy and capitalize on the power from stations that cannot be regulated (thermal, solar, wind turbines) generated outside of high-demand periods. Some thirty potential sites were assessed during the 1970s. Today, projects aim to combine this type of construction with existing reservoir structures by equipping them with new water adduction systems and by increasing the reservoir capacity. For example, the Nant de Drance pumped-storage project, construction of which began in 2009, will use the difference in altitude of two reservoirs, Lac Emosson and the Lac du Vieux-Emosson (VS). This power plant is being constructed entirely underground, and it is due to be commissioned in autumn 2022 (see § 15.8.4).

Although most reservoirs were created with the aim of producing hydropower, the construction of reservoirs for storing water for the production of artificial snow developed considerably from the late 1990s. These constructions are not built on a river, but instead on a hillside or flat area; much of the artificial basin can be excavated. In view of local conditions and the availability of material, embankment dams are often used. There is no doubt that ski-lift operators will continue to use water reservoirs as a means to guaranteeing snow cover on ski slopes. Finally, hydraulic schemes can also lead to the creation of biotopes and areas for leisure activities.

Every year natural disasters (floods, avalanches, debris flows) cause substantial damage and result in considerable costs. The increased need for safety will lead to the design and construction of new protection systems including flood retention works and anti-avalanche dikes. Some protective measures will be revisited, improved, or completed. In Switzerland, reservoirs for existing hydroelectric schemes are generally speaking single-purpose reservoirs. They can be converted into multipurpose reservoirs by setting aside a clearly specified storage volume for retaining water during floods. This solution has already been implemented in certain cases and others may well follow. Various solutions are possible for maintaining the available capacity of reservoirs when necessary, including dam heightening, creating a supplementary connected reservoir, or by adding a seasonal pumped-storage system. It is possible to refer to a flood forecasting model so as to better manage reservoirs. The canton of Valais has taken this approach and developed the Minerve project, which simulates the overall hydraulic behavior of catchment areas and hydropower schemes in Valais. The model is designed to help cantonal officials make decisions (Garcia Hernandez et al., 2011, 2014; Raboud et al., 2001; Jordan et al., 2008).

Ongoing research and development over many years has led to technological advances. Various problems have been tackled such as safety under dynamic loading during earthquakes, extreme flood events, the long-term behavior of dams, and the behavior of foundations. Of course, these issues have not yet been fully resolved. Problems related to safety and the overall behavior of dams continue to occupy researchers in such domains as the behavior of overtopped dams, the long-term behavior of facing and drainage, and reservoir sedimentation. The development of data collection and methods for analysis and data measurement continue to be used in the monitoring of dams. And finally, new and increasingly applied construction methods, such as roller compacted concrete (RCC) and cemented soils, are the focus of specific studies.

Swiss expertise in the field of dams is recognized internationally and can thus be employed elsewhere in the world. Global demand for the construction of hydraulic structures and dams in particular is high and will continue to grow. Logically, the strategy we should be implementing is one that looks outward. To maintain this level of expertise, Switzerland must aim to develop the following vision: "In the twenty-first century, Switzerland will take on a lead role in the global production of environmentally friendly, multi-purpose hydraulic schemes" (Schleiss, 1999). Swiss industry and engineering are capable of achieving this vision thanks to more than one hundred years of experience in hydraulic construction and the international renown garnered since the 1960s by the construction of over 180 large dams outside of Switzerland. Succeeding in this challenge will only be possible by bringing together the skills of all of our many participants into a sort of hub for excellence. The wide competencies of industry, construction companies, engineering consultancy firms, electrical production and distribution companies, the financial world, and polytechnic schools must be gathered together (Schleiss, 1999).

To summarize, the prospects for a civil engineer in the field of dams are still positive, as Table 2.31 demonstrates.

Table 2.31 Prospects for engineers in the field of dams

	Switzerland	World
Maintenance of existing structures	✪✪✪	✪✪
Heightening of existing structures	✪✪	✪
Reconstruction Rehabilitation	✪	✪✪
New structures	✪	✪✪✪

✪ few, ✪✪ moderate, ✪✪✪ many

2.3.3 Dams and sustainable development in the twenty-first century

Figure 2.32 shows the location of more than 59,000 dams with a height > 15 m in use in 2019 according to the ICOLD World Register of Dams. The total reservoir volume of all dams registered by ICOLD is about 7,500 km^3 and thus satisfies the worldwide need for water, energy, and food. In view of climate change, reservoirs currently being used and those still to be constructed are becoming increasingly important.

According to Figure 2.29, the number of large dams under construction since the year 2000 has remained steady. Although a slight fluctuation can be observed, periods of economic crisis can be seen to have barely impacted the number of constructions. This demonstrates that dams and reservoirs are vital infrastructures that also influence the health of the global economy.

Dams and hydropower are strongly related (Schleiss, 2018). Hydropower is still the cheapest and most flexible renewable energy with a significant potential worldwide. So far, only about one quarter of the technically feasible hydro potential of 16,000 TWh has been developed. The potential for worldwide hydro development is currently estimated at almost 9,700 TWh. In South America, less than 30% of the technically feasible potential is in operation (*Hydropower & Dams World Atlas*, 2021). In Asia (including Russia and Turkey), the economy is rapidly growing and demand for power is skyrocketing, while also less than 30% of technically feasible hydropower potential is used today. In Africa only a tiny fraction of 13% of the economic and environmental feasible hydropower potential is used on average, with an approximate capacity of 17 GW under construction (2021). The remaining economically feasible hydropower potential in Africa corresponds to 190% of yearly generation of all hydropower plants operating in Europe (2021). This once again demonstrates the high potential for dam and reservoir development in Africa as a catalyst for increasing prosperity. Nevertheless, this development must be done in a sustainable way while protecting the continent's rich ecosystems, from humid zones with forest floodplains to wildlife reserves. "Green" electricity generation will be a selling point in the interconnected intercontinental electricity market in the future. In Europe, 70% of this potential is already used, in North America 75%. Some countries in Europe, for example, Switzerland (90%) and France (97%), are using most of their technically feasible hydropower potential. Nevertheless, considerable investment must be made in the near future to upgrade existing hydropower plants to increase flexibility and storage in view of the strong increase in new renewable energy sources such as sun and wind, whose production is highly volatile.

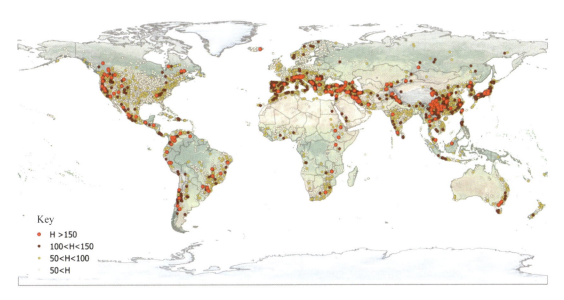

Figure 2.32 Location of the more than 59,000 dams with a height H > 15 m according to the ICOLD's World Register of Dams in 2019 (map prepared by Patrick Le Delliou)

The major challenges lying ahead for the world population in this century will be without doubt the safe supply of green renewable energy as well as the supply of water of sufficient quality and quantity to fight against famine, poverty, and disease in the world (Schleiss, 2000). Today water supply and sanitation services leave much to be desired; two thirds of the world's population suffer from a lack of safe water (insufficient quantity) or from the lack of safe sanitation (poor quality). Furthermore, an important part of the world population is threatened with famine. This risk could be considerably lessened by the irrigation of arid areas not cultivable today. Thus, in many countries, especially in Africa, there is still an urgent need for the increased development of water and energy resources as a basis for the economic prosperity and cultural wealth of local communities.

Regarding life cycle analysis, hydropower is one of the best options in view of sustainability. When looking at the so-called recovery factor or energy payback ratio of primary energy, which is obtained by calculating the total cost of non-renewable energy (direct and indirect) over a lifetime to operate an installation, hydropower is unbeatable (Schleiss, 1999, 2000). For hydropower plants with reservoirs created by dams the energy payback ratio is situated between 205 and 280, for run-off-river power plants between 170 and 270. In fact, these numbers clearly exceed those attained by other renewable energies such as, for example, solar photovoltaic (3 to 6) and wind (18 to 34). Their recovery factors are reasonably small today, but significant technological advances can be expected in the future. Furthermore, recent life-cycle analysis also confirms that hydropower can significantly reduce greenhouse gas emissions, even if only some of the remaining economically feasible hydro potential is developed. Recent studies in Switzerland have revealed that for hydropower the equivalent CO_2 emission is small compared to other energy generation types and is caused mainly by the acquisition of material required for construction and maintenance (Frischknecht et al., 2012; Bauer et al., 2017). Run-of-river power plants in Switzerland produce between 4 and 5 g CO_2–equivalent/kWh and storage power plants 6 to 7 g CO_2eq/kWh. For nuclear energy, these emissions are between 10 and 20 g CO_2/kWh depending on the type of reactors. For new renewable energies such as wind and solar (photovoltaic), values of 15 g CO_2/kWh and between 38 and 95 g CO_2/kWh respectively have been obtained. Biogas power plants in Switzerland with thermal-power coupling produce between 150 and 450 g CO_2eq/kWh. These values are still relatively small compared to the emission of gas power plants (480 to 640 g CO_2eq/kWh) and coal power plants (820 to 980 g CO_2eq/kWh).

The safe and sustainable development of dams and reservoirs can significantly contribute to the 16 sustainable development goals (SDG) set out in the United Nations Agenda 2030. The following goals are directly related to dam and reservoir development:

- Goal 2. End hunger, achieve food security and improved nutrition, and promote sustainable agriculture
- Goal 7. Ensure access to affordable, reliable, sustainable, and modern energy for all
- Goal 8. Promote sustained, inclusive, and sustainable economic growth, full and productive employment, and decent work for all
- Goal 9. Build resilient infrastructure, promote inclusive and sustainable industrialization, and foster innovation
- Goal 13. Take urgent action to combat climate change and its impacts

Hydropower and dam projects give rise to vigorous debate (WCD, 2000). In order to gain wide acceptance and obtain a win-win situation between all stakeholders, large water infrastructure project have to be designed as multipurpose projects by multidisciplinary teams with a complex systems approach. This needs excellence in engineering sciences and management.

Figure 2.33 shows where in the world large dams have been commissioned since 2000 (Schleiss, 2016a, 2018). Thanks to these new water infrastructures, a kind of a security belt around the world ensuring water, food, and energy can be identified. It ranges from Southern Europe over the Middle East to Central and East Asia. It covers areas with high water stress in arid and semi-arid regions as well as regions exposed to

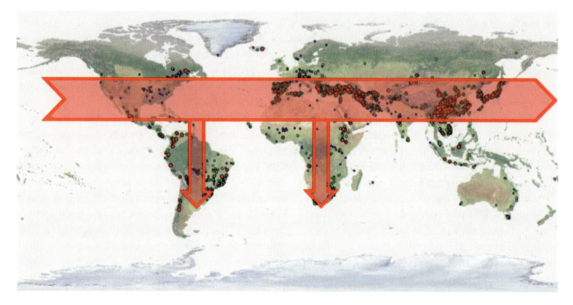

Figure 2.33 New dams and reservoirs commissioned since 2000 creating a security belt around the world to ensure water, food, and energy (NEXUS and SDGs) (map prepared by Patrick Le Delliou)

monsoon conditions with extremely high population density. The belt is less visible across North America, the world's most productive crop growing region, where only a few dams were built this century, but where significant development took place last century.

It should be noted that the regions around this belt are already noticeably affected by climate change, the effects of which will become even more dramatic in the future, according to predictions. Existing dams and reservoirs, as well as future projects will play a key role in the mitigation of the effects of climate change (Schleiss 2016a, b, c). The belt of dams and reservoirs which covers these threatened and vulnerable regions around the world will help ensure water security. This water-energy-food nexus can thus be called a "security belt."

Figure 2.33 highlights another worldwide problem, the huge economic gap between North and South, that is, between developed countries and emerging and developing countries. Overcoming this challenge is a millennium development goal. In South America, the shift from North to South of new dams is clearly visible. In Africa, unfortunately, a similar progression has yet to occur. To further extend the analogy, it could be said that the security belt should be attached to the South with suspenders, in order to ensure not only food, water, and energy but also equitable worldwide wealth for all countries. Dams and reservoirs are the vital infrastructure and primary element bringing strength to the security belt and its suspenders.

To conclude, this century's great challenge is to build *better dams for a better world* (Schleiss, 2017a).

3. Different Types of Dams

Sambuco arch-gravity dam in Switzerland, height 130 m, year commissioned 1956
(Courtesy © Swiss Air Force)

3.1 Introduction

Depending on the type of construction material used, dams can be grouped into two major categories:
- Concrete dams
- Embankment dams (dikes)

Some old dams, dating mostly from the nineteenth century, were built of masonry. Generally speaking, these are grouped along with concrete dams, as their form almost always categorizes them as gravity dams.

To this classification, hybrid or combined dams can be added. Some large dams include concrete and fill sections next to one another. The most frequent example is of a concrete construction including a spillway flanked by embankment sections, either on one or both sides. In other cases, the section of the structure is made up of several types of material. There are also some dams where an embankment is built against a masonry wall. In addition, the existence of dams designed to store waste products from industrial processes should also be mentioned. These tailings and mining dams comprise embankments made of layers of material of varying quality (excavated natural materials, waste material, rock fill).[1]

3.2 Concrete and Masonry Dams

3.2.1 Introduction

Apart from some exceptions (e.g., river dams), concrete dams are most often built on rock foundations with a high deformation modulus. As illustrated in Figure 3.1, concrete dams can be divided into three main groups, each with a certain number of subgroups.

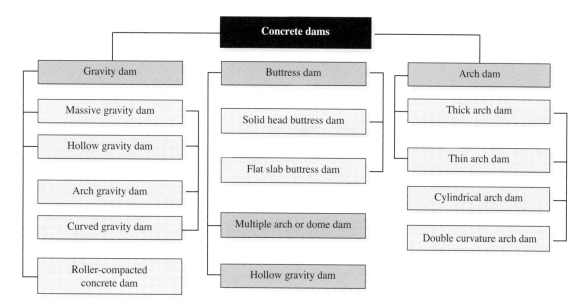

Figure 3.1 Main types of concrete dams

---

[1] Tailings and mining dams will not be covered in this book.

The three types of concrete dams can be categorized through their form, the nature of their static system, and how they resist water pressure.

- *Gravity dams*, as the name suggests, withstand the pressure of the water through their own weight. Gravity dams are made up of massive, juxtaposed elements called blocks. In order to save concrete, cavities are sometimes left between the blocks. These types of structures are known as hollow gravity dams. Another solution involves designing an arch-gravity dam. Dam stability here is guaranteed partly by its own weight and partly by the abutments against the valley flanks.
- *Buttress dams* also withstand water pressure through their own weight, but certain arrangements mean that the volume of concrete can be decreased compared to a gravity dam. This type of dam is made up of juxtaposed elements, known as buttresses, with complex geometry. Each buttress comprises a continuous facing on the upstream side and a web, and takes up the force exerted by the water pressure. The stress in the body of the dam and in contact with the foundation is higher than for a gravity dam of the same height. Multiple arch dams also fall into the category of buttress dams.
- *Arch dams* are tridimensional structures that act as a shell or a curved curtain. They are highly curved on the plane and transfer a large part of the force to the valley flanks. If all the required conditions are met, arch dams can save a considerable volume of concrete compared to the two previous dam types.

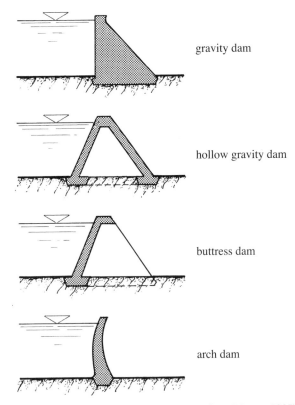

Figure 3.2 Cross-sections of concrete dams (from Mason, 1997)

Figure 3.2 illustrates the cross-sections of these types of concrete dams.

Concrete dams have certain shared elements. Firstly, the structure is made of mass, unreinforced[2] concrete that is rapidly placed and compacted by mostly mechanical means. Secondly, and generally speaking, the geometry is optimized to avoid tensile stresses occurring in the concrete at any point of the structure for normal operating conditions. However, tensile stress or the appearance of cracks that do not endanger structural integrity can be tolerated in cases of exceptional load, such as an earthquake.

Concrete dams, no matter their type, are constructed with individual blocks from 12 to 19 m in width, separated by contraction joints. The thickness of the dam determines the size of the block in the lengthwise direction (between 3 and 30 m). Each block is concreted in lifts of 1.5 to 3.5 m, resulting in concreting stages that can reach 1500 m^3 (Figure 3.3).

This method of construction:

- Facilitates concreting by adapting the volume of the stages to daily production
- Controls and facilitates the release of hydration heat
- Avoids cracking after concrete shrinking by allowing the joints to open

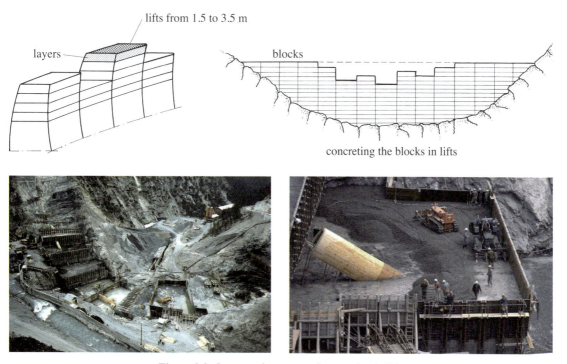

Figure 3.3 Concrete dams: staged concreting in lifts

---

[2] For concrete dams with a relatively "thin" section and with significant temperature effects (for example, multiple arch dams, water intake dams or spillways), the use of reinforcement may be necessary.

3.2.2 Gravity dams

Most gravity dams are massive and full with a triangular profile. The upstream face is vertical or slightly inclined (less than 5%). The downstream face is sloped at 75 to 80% (Figure 3.4). This geometry enables the dam, through its own weight, to withstand overturning and sliding under the action of external forces. Its foundations sit on rock.

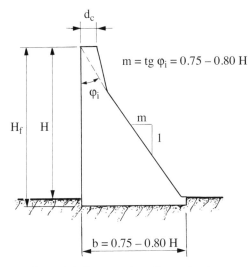

Figure 3.4 Cross-section of a gravity dam

Gravity dams can be any length and are particularly well suited to wide valleys. They can be straight, polygonal, or slightly curved, depending on the geology and topography of the site. Higher curvature influences the static behavior of the structure through a tridimensional effect. In this case, the term "curved gravity dam" is used.

Gravity dams are made from a series of 12- to 19-meter-wide blocks. These blocks are separated by joints (1 to 3 mm) that can open and close depending on the conditions. These expansion joints are in fact also shrinkage joints that open as the concrete cools and dries. They are protected by a sealing system on the upstream face (for example, using waterstop bars).

Through the simplicity of their shape, gravity dams used to be the most common type of concrete dam. However, concrete can be misused (particularly due to low stress), and this type of structure may not be the most economical (Raphael, 1970). As the cost of a gravity dam is directly related to the volume of concrete used, engineers have sought to eliminate it from areas where it is not well utilized. It is to this end that hollow gravity dams were designed, a solution that reduces uplift (Figure 3.5). The higher the dam, the more savings are made. The width of the cavity is determined by an economic cost/benefit assessment. To compensate for the lost weight of the concrete, the upstream face is inclined (up to 10%) and therefore benefits from a vertical component of water pressure. These cavities allow the behavior of the foundation and in particular the drainage system to be observed. When needed, additional grouting and drainage works can be carried out.

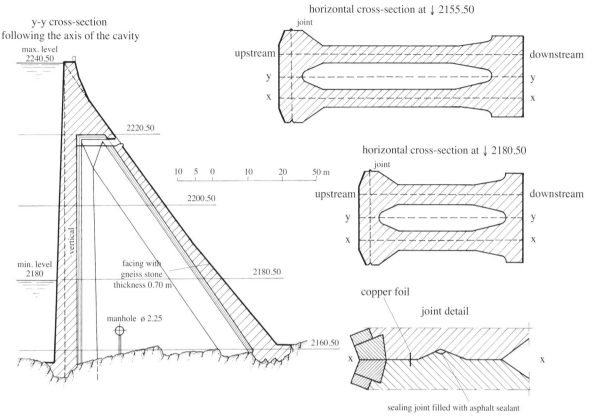

Figure 3.5 Transverse and horizontal cross-sections of the Dixence dam, the first hollow gravity dam built in Switzerland (from BTSR, 72nd year, 16 February 1946, no. 4, "Le barrage de Dixence," A. Stucky, p. 42)

3.2.3 Buttress dams

Buttress dams are always made from concrete. They have a triangular shape, and their upstream and downstream faces are sloped (Figure 3.6). The buttresses transfer the force through to the foundations, which must be able to withstand generally high contact pressures. In order to reduce the loading transmitted to the foundation, the base of the buttresses may be widened. As with gravity dams, buttresses are built side by side and are separated by a vertical joint. Due to the free space between buttresses, there is a substantially smaller volume of concrete needed than in an equivalent-sized gravity dam. However, there is more formwork, and its installation is more difficult.

There are various types of buttress dams, depending on the shape of the buttress itself. In the horizontal cross-section in Figure 3.6, the two buttressing zones of the buttress dam can be clearly identified:

- The *head*, which is 12 to 14 m wide. The buttress head in Figure 3.6 is known as a solid head or diamond head. This is the most common choice. However, other head shapes are sometimes chosen, and various options are illustrated in Figure 3.7. The heads have a gusset in their upstream end to transfer the force of water pressure to the web. The change in section is gradual to favor the transmission of the forces. Finally, a waterstop is located in the joint between two juxtaposed heads.

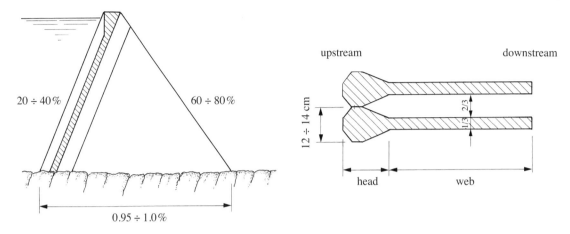

Figure 3.6 Buttress dam: profile cross-section and horizontal cross-section

- The *web*, which usually has a constant thickness and is approximately one third the width of the head. In some cases, the web is widened at the downstream end to reduce stress. The change in section must be gradual to limit the concentration of stresses.

The thickening of the downstream end of the web may in some cases reach the width of the head, so that on the downstream side of the dam there is a continuous face. This downstream facing may be intentional in order to make the buttresses more resistant to dynamic forces or to protect the web from the effects of high temperature variations (freezing, heating). The installation of horizontal struts between the buttress webs is also a way of taking up lateral forces during an earthquake.

To further limit the volume of concrete, some original options have been developed in specific cases:

- Flat slab buttresses (although these structures are particularly vulnerable to earthquakes)
- Multiple arch or dome dams are made of thin, often cylindrically shaped arches supported by the buttresses. In this type of construction, temperature effects can cause substantial tensile stresses in the arches, which must therefore be reinforced.

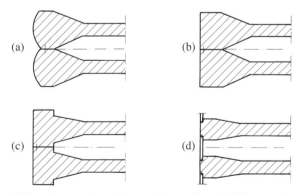

Figure 3.7 Buttress dam: different buttress head shapes: (a) round head; (b) hammer head; (c) tee head; (d) flat slab

Among the types of dams mentioned so far, these latter two types are most certainly the lightest. To ensure stability against buttress sliding, the lack of vertical load due to relatively low self weight must be compensated for by a large vertical component of water pressure. This stabilizing force is obtained by steeply inclining the upstream face of the dam, up to 100% (Figure 3.9). The Marécottes dam (VS/1925/H = 19 m) and the Oberems dam (VS/1927/H = 11 m) are two examples of multiple arch dams built in Switzerland (Figure 3.10).

Figure 3.8 Lucendro buttress dam (TI) H = 73 m

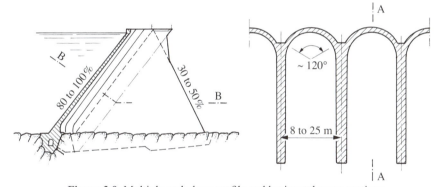

Figure 3.9 Multiple arch dam: profile and horizontal cross-sections

Figure 3.10 Examples of multiple arch dams: upstream view of the Oberems dam and downstream view of the Marécottes dam (Courtesy H. Pougatsch)

3.2.4 Arch dams

Arch dams are curved in plane. They are made mainly of concrete (RCC is now also used). On the face of it, arch dams require the placing of substantially less concrete than gravity dams. The concrete is also much better utilized. Due to the dam's curvature, much of the force from the water pressure is transferred over the abutments to the valley flanks.

To illustrate this tridimensional effect, arch dams can be modeled by a series of horizontal and vertical load-bearing elements, as in Figure 3.11:

- The horizontal load-bearing elements are curved beams with two abutments—the arches
- The vertical load-bearing elements are cantilever-type structures

In such a simplified model, water pressure applied at the intersecting point of the two elements is distributed according to their respective stiffnesses. It seems obvious in this context that the arches are much more rigid than the cantilevers (due to their hyperstatic behavior) and that the force due to water pressure is therefore advantageously guided toward the abutments at the valley flanks. It should be noted that the vertical force is also transferred from the central zone to the lateral abutments through the descending arch effect and the overall tilting or overhang of the dam. Furthermore, water pressure results in intense shear forces, which for the base of the cantilevers must be carefully considered during resistance analysis (§ 15.6.3). It is important to note that the behavior of rock abutments influences the deformation of the structure, which must be taken into account during calculations (Stucky et al., 1951).

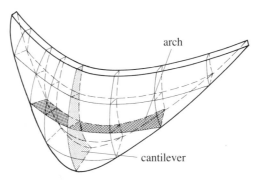

Figure 3.11 Arch dam: simplified static model

The thinner the arch the more the ratio of stiffness tends to direct the force towards the abutments of the valley banks. Arch dams can therefore be separated into thin arch dams, where the thickness at the base is in the order of 10 to 20% of the height, and thick arch dams, where the base thickness is greater than 25% of the height. A distinction can also be made between cylindrical arch dams (with horizontal curvature only, single curvature) and double curvature arch dams (horizontal and vertical) (Figure 3.12).

In the first half of the twentieth century, several cylindrical arch dams were built in particularly narrow valleys. These dams have a constant angle of curvature from the foundation to the crest. Some of these narrow cylindrical dams may have some light traction stresses, and due to this, reinforcement may be required (Serra[3] / VS / H = 22 m; St-Barthélémy C / VS / H = 51 m).

---

[3] Owing to the problems arising from alkali-aggregate reactions, the Serra dam was demolished; it was replaced by a new dam just downstream of it (see § 19.4.9) (SwissCoD, 2017a).

DIFFERENT TYPES OF DAMS

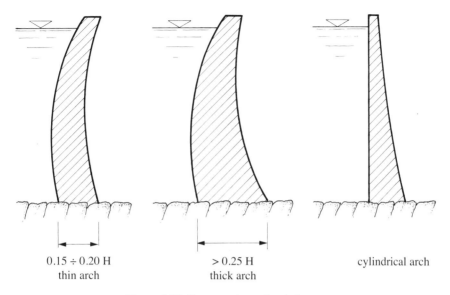

Figure 3.12 Cross-sections of arch dams

In modern arch dam design, traditional circular arches have been replaced by parabolic, elliptic, or logarithmic spiral arches in order to ensure a better orientation of the pressure of the arches against the foundation rock. Double curvature arch dams have become standard.

The arches can have a constant thickness with a gradual widening toward the abutments so as to form a gusset (fig 15.7). Another, less economical solution consists in designing arches whose thickness gradually increases.

As with gravity dams, arch dams are built with juxtaposed blocks, but one fundamental constructive element differentiates them. Although the joints separating the blocks in gravity dams are open, the joints in an arch dam are grouted with cement to render the arch monolithic and ensure transmission of horizontal force toward the abutments at the valley flanks. This joint grouting takes place before the reservoir is filled and in specific thermal conditions (in Switzerland, generally in winter), so that the resulting exterior load always creates compression in these joints (Figure 3.13).

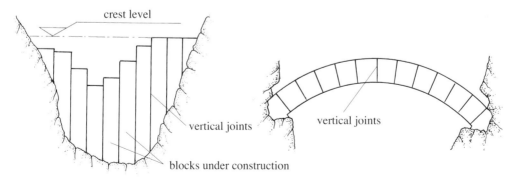

Figure 3.13 Arch dam: method of construction with vertical grouted joints

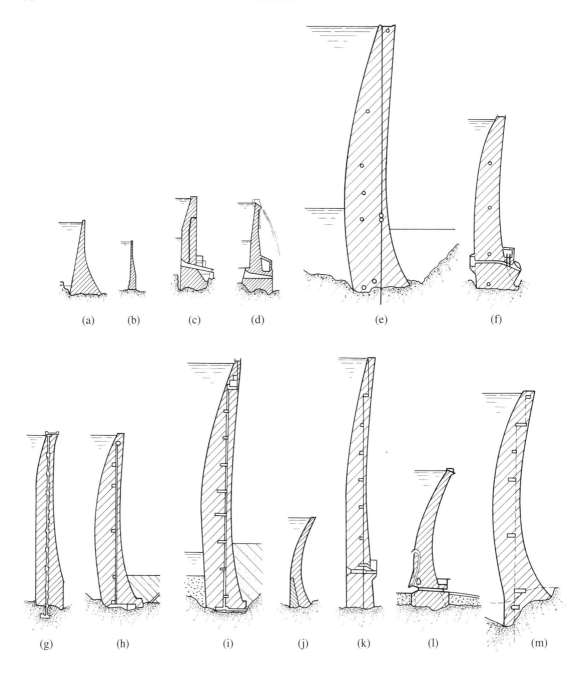

Figure 3.14 Examples of cross-sections of arch dams built in Switzerland between 1920 and 1974 ordered chronologically: (a) Montsalvens (1920; H = 55 m); (b) Pfaffensprung (1921; H = 32 m); (c) Rossens (1947; H = 83 m); (d) Chatelôt (1953; H = 74 m); (e) Mauvoisin (1957; H = 237 m); (f) Zervreilla (1957; H = 151 m); (g) Zeuzier (1957; H = 156 m); (h) Moiry (1958; H = 148 m); (i) Luzzone (1963; H = 208 m); (j) Les Toules (1963; H = 86 m); (k) Contra (1965; H = 220 m); (l) Gebidem (1967; H = 122 m); (m) Emosson (1974; H = 180 m)

Through its static system and relatively small thickness, the arch dam transmits significant load to the rock foundation of the abutments at the valley flanks. These must be resistant and not highly deformable. Although gravity dams can be adapted to any valley shape and any valley width, arch dams require specific topographical characteristics. Relatively narrow valleys are most favorable.

The slenderness of a dam is calculated as λ with the equation:

$$\lambda = \frac{\text{developed length of crest } (L_c)}{\text{dam height } (H_d)}$$

Generally speaking, it can be said that slenderness λ should not be greater than 4 to 5 (exceptionally 6). This limiting factor will depend in particular on the geology of the site. To illustrate the variety of shapes of arch dams, Figure 3.14 contains examples of cross-sections of dams built in Switzerland.

3.2.5 Dams built with roller-compacted concrete (RCC)

Since the late 1970s, a new technology has been developed to optimize the construction of gravity dams: roller-compacted concrete (RCC). Laying RCC concrete means that very dry concrete can be used with a low cement mix. The resulting concrete strengths are particularly low but are compatible with the requirements for gravity dams that resist water pressure through self weight. The properties of this concrete are used to their best effect when placing techniques borrowed from embankment dams are employed; this reduces workman requirements during construction (Figure 3.15).

Figure 3.15 Example of RCC Rialp dam in Spain (Courtesy A. Schleiss)

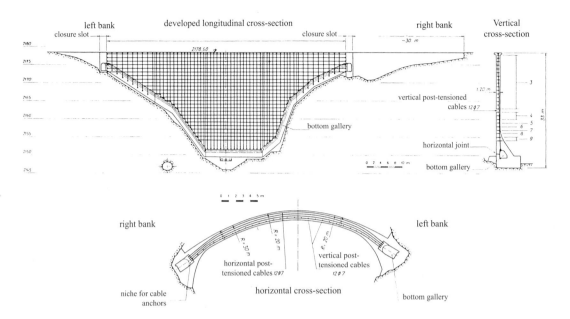

Figure 3.16 Tourtemagne dam: distribution of post-tensioned cables in a developed longitudinal cross-section and horizontal cross-section (from Panchaud, 1962)

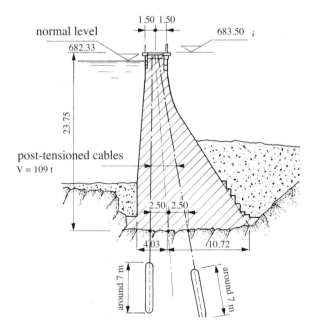

Figure 3.17 Strengthening of the Gübsensee dam by inserting post-tensioned cables (based on Ammann, 1987)

RCC dams were initially built with traditional gravity dam profiles. From 1990, French engineers (Londe and Lino, 1992) developed a new concept for RCC dams, called FSHD (Faced Symmetrical Hardfill Dam), (Figure 15.25, Figure 15.26) which proposed a thicker profile and used hard symmetric fill on the upstream and downstream faces. To guarantee the impermeability of the hard fill, corresponding to an RCC with a low cement content and aggregate up to 100 mm, an upstream facing made of conventional concrete is necessary. The triangular profile is the minimum volume profile that does not result in tensile stresses at the upstream heel of the dam, according to the concept of interstitial pore pressure (ICOLD, 2000).

In Japan, this concept for dams using cement materials (Lino et al., 2017; Jia et al., 2016) was taken up with a trapezoidal profile made of gravel and cemented sand—the Trapezoidal Cemented Sand-Gravel dam. In China, apart from cement, sand, and gravel, large blocks have been used to create a cement material called CSGR (Cemented Sand Gravel Rock Dam) (Jin et al., 2018a, 2018b).

3.2.6 Post-tensioned concrete dams

The Tourtemagne dam (VS/1958/H = 32 m) is a special case as it was built with prestressed concrete. The dam comprises a cylindrical vertical arch of a minimum thickness of 1.2 m. To reduce the risk of cracking due to temperature variation and to improve resistance to cooling and freezing, vertical and horizontal post-tensioned cables were placed. The post-tensioning was first applied vertically then horizontally after completion of the works, once the concrete temperature was at its minimum and shrinkage had finished (Figure 3.16).

Post-tensioning can also be used to strengthen an existing dam (Figure 3.17) and improve its stability and safety conditions or when heightening a dam. In the case of new dams, post-tensioning should not be the sole guarantor of the structure's stability.

3.3 Embankment Dams

Embankment dams (dikes) are made primarily of natural granular soil and rock material excavated near to the dam site. There are two types of embankment dams (Figure 3.18):

- Earthfill dam, made essentially from natural loose soil material taken from gravel pits
- Rockfill dam, most of which are made from crushed rock from quarries

As with other types of dams, embankment dams must fulfil two essential criteria: the static function, which involves transferring the pressure of the water stored upstream to the foundation, and that of providing an impermeable barrier.

Depending on its geotechnical characteristics, the loose material in earthfill dams may be sufficiently impermeable and perform both functions. Because of this there are a large number of dams (or dikes) made from homogeneous material.

When the permeability of the material from a borrow pit is too high, the solution is to create a zoned embankment dam, that is, with several materials used in different zones in the body of the dam.

Quarried rock materials, with which rockfill dams are made, are always permeable (to varying degrees). This rockfill is therefore always combined with another element that ensures the watertightness function.

Figure 3.19 shows a section diagram of the most common zones.

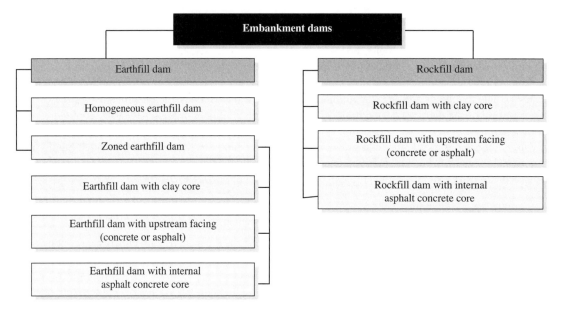

Figure 3.18 Types of embankment dam

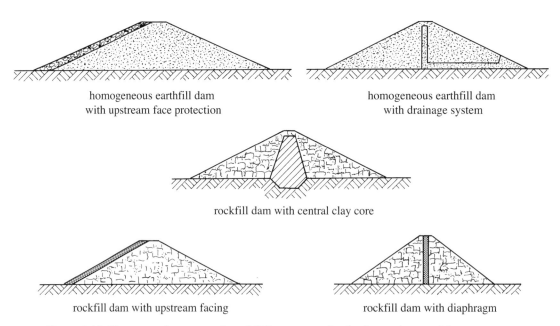

Figure 3.19 Diagrammatic cross-section of different types of embankment dams and the arrangement of watertightness elements

In comparison to concrete dams, embankment dams have the following advantages:

- The vast majority, if not all material comprising the body of the dam, comes from borrowing zones located in the near vicinity of the dam site
- The placement of material is highly mechanized and rapid, even though the volume of material to place is noticeably greater
- The loading of the foundation (stress) is to a much lesser degree
- Foundation settlement does not pose any major problems, as the material is sufficiently plastic and adapts

The two latter considerations are vital when choosing the dam type. Some types of embankment dam can be placed either on a rock foundation or on a foundation of loose ground, as long as continuous impermeability is guaranteed between the dam and its foundation.

The solution of the homogeneous embankment dam, made up of one single loose material that is sufficiently impermeable, has been employed for thousands of years. In order to maintain stability, the faces are gently sloped, which means that a large quantity of material is needed. The careful placement of drainage systems (including drainage blanket, chimney drains) can reduce the volume of the structure. The upstream slope is generally protected by rockfill to avoid erosion due to waves.

After the homogeneous embankment dam, the zoned embankment dam was a logical development. In the case of earthfill or rockfill embankment dams, the materials are selected depending their specific characteristics and availability on site. A relatively impermeable material (clay, silt, moraine) is used for the central core. The core is flanked on the upstream face by a supporting zone made with free-draining materials and a supporting zone on the downstream face. Filters ensure the transition between the core and supporting zone. This is particularly important downstream in order to prevent the migration of the core's fine particles due to seepage. Impermeability can also be assured with an internal diaphragm in asphalt concrete, concrete, or steel.

Finally, provision may be made for upstream facing to ensure watertightness in cases where it is difficult to find enough material with low permeability. This facing can be an asphalt coating, a geomembrane, or a concrete slab. The watertight facing must be able to adapt to deformation of the supporting dam material.

Embankment dams are constructed by the placement of successive layers of material covering the whole surface of the dam. The layers can be from 50 to 100 cm thick so as to allow appropriate compaction across the whole dam and regular settlement and consolidation (Figure 3.20).[4]

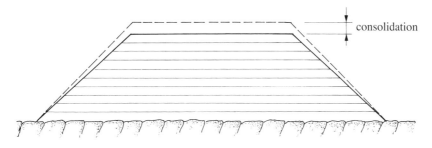

Figure 3.20 Embankment dam: construction and consolidation

---

[4] The thickness of the layers is generally established after on-site compaction tests.

3.4 Advantages and Specificities of Different Types of Dams

To summarize, the advantages and primary specificities of the main types of dams are listed below.

Gravity dam

Advantages:

- Weak stress in the concrete
- Weak stress transferred to the rock
- Variations in temperature only produce weak stress
- Low uplift gradient below the foundation
- Spillways can easily be integrated

Specificities:

- Large volume of excavated material
- Large volume of concrete
- Artificial cooling necessary for the curing of concrete
- Substantial uplift below the foundation
- Sensitivity to settlement
- Sensitivity to earthquakes

Buttress dam

Advantages:

- Smaller volume of concrete than for gravity dam
- Moderate stress transferred to the rock
- Weak uplift below the foundation
- Slight heating during curing of concrete
- Spillways can easily be integrated
- Limited risk of settlement

Specificities:

- Large volume of excavated material
- High localized uplift gradient below the foundation of the buttress heads
- Temperature stress can be high in the head
- High sensitivity to earthquakes

Hollow gravity dam

Advantages:

- Same as gravity dam
- Smaller volume of concrete
- Reduction in uplift at the concrete-rock contact point

Specificities:

- Same as gravity dam
- Local increase of seepage gradient near the foundation

Roller-compacted concrete (RCC) dam

Advantages:

- Low cement content
- Reduced water content
- Limited formwork surface
- Quick execution
- Reduced footprint compared to a gravity dam
- Implementation across a large surface in thin layers
- Lower cost of construction
- Good resistance to earthquakes and overtopping

Specificities:

- Requires a rock foundation
- Placement and compaction identical to embankment dam
- Limitation of seepage flow by the placement of a facing at the upstream face
- No system for the artificial cooling of concrete
- Possibility for including a surface spillway, monitoring galleries, and shafts

Multiple arch dam

Advantages:

- Low volume of concrete
- Relatively limited volume of excavated material
- Weak uplift below the foundation
- No thermal problems during concrete curing for thin structures

Specificities:

- High stress in the arches
- Need to reinforce arches to limit cracking
- Sensitivity to temperature gradients
- High localized uplift gradient below the foundation
- Sensitivity to differential settlement
- High sensitivity to earthquakes
- Integration of appurtenant structures is difficult
- Structure is very vulnerable and exposed to malicious acts

Arch dam

Advantages:

- Low volume of concrete
- Relatively limited volume of excavated material
- Weak uplift below the foundation
- High resistance to earthquakes

Specificities:

- High stress in the concrete
- High stress in the rock at the foundation

- Force transferred obliquely to lateral abutments
- Limited sensitivity to settlement (hyperstatic behavior)
- Heating during concrete curing may require particular pre- and post-cooling measures
- High uplift gradient below the foundation
- Drainage of cracks in the rock abutments must be rigorously dealt with
- Difficulties in integrating spillway into the dam

Embankment dam with central core

Advantages:
- Highly flexible dam body that adapts to the terrain
- Low sensitivity of structure to settlement and earthquakes
- Limited excavation of material
- Weak stress on foundation
- Low seepage gradient in the core and foundation

Specificities:
- Very large volume of material to be placed
- Availability of abundant clay material near the site
- Impossible to place the clay core during adverse weather conditions

Embankment dam with upstream facing

Advantages:
- Highly flexible dam body that adapts to the terrain
- Low sensitivity to overall settlement
- Low sensitivity to earthquakes, providing specific measures are taken
- Limited excavation of material
- No clay material to place
- No particular requirements regarding weather conditions
- Weak uplift on foundation

Specificities:
- High volume of material to be placed
- High seepage gradient under the plinth foundation
- Plinth foundation on low permeability rock
- Sensitivity to differential settlement of rigid facing (concrete) and fill (rockfill)

4. Impacts on the Environment

Cavagnoli arch dam in Switzerland, height 111 m, year commissioned 1967
(Courtesy © Swiss Air Force)

4.1 Introduction

It has long been recognized that the value of dams and reservoirs above all lies in their ability to generate energy, supply drinking water, and make water available for irrigation. It should be noted that hydropower, defined as a renewable energy source (and one that is considered particularly clean compared to other energy sources), generates 20% of the world's electricity. In addition, one fifth of the world's arable land is irrigated thanks to dams and reservoirs. Hydropower is suited to clean development mechanisms, which encourage investors in developing countries to support the energy transition to environmentally sound technologies and promote sustainable development. Not only that, but the ability to control and mitigate floods contributes to the safety of local populations. The fact nevertheless remains that the construction of such large infrastructures can have major effects on the environment and affect populated areas. These associated negative impacts are not always accepted, and in recent decades strong opposition to the construction of dams has emerged. Despite these setbacks, hydropower remains a major source of renewable energy that if appropriately planned and managed can be positive for the environment. To anticipate and avoid either environmental or socioeconomic issues, the requirement for an environmental impact assessment (EIA) has been introduced into many countries—some for several decades now—to accompany the planning, design, construction, and operation of any new project. An EIA implies the involvement of many experts whose job is to analyze the various consequences related to the project and suggest ways of resolving them, by mitigation if necessary. This multidisciplinary work requires mutual understanding from all groups and stakeholders involved. Any supporting measures can only be implemented if the experts consider they are technically practicable and economically acceptable.

In Switzerland, the EIA was introduced on 1 January 1985 as part of the Federal Act on the Protection of the Environment. Details concerning the EIA are outlined in an ordinance.[1] The environmental impact assessment is part of the decision-making process (approval process, acceptance process, and process for the granting of a concession) for the installation in question. Other specific legislation has also been passed with regard to the environment. It should be noted that in Switzerland and in many other countries, aspects relating to the environment were taken into account during the construction of dams well before any specific laws came into effect. For example, a hydropower storage scheme was even built in the Swiss National Park in the canton of Graubünden, a protected area if ever there was one.

4.2 Examples of Impacts

4.2.1 Introduction

The construction of a dam significantly affects the area surrounding the watercourse in which it is built. Dams constitute a barrier to the natural flow of water and consequently affect the migration of fish and other organisms, as well as the transport of sediment. By creating an artificial lake, a dam can have an impact on groundwater and also on the water quality. Furthermore, dams affect the flow regime downstream, which can lead to daily or seasonal fluctuations in river discharge, known as hydropeaking.

The aim of this chapter is to describe in a non-exhaustive way some examples of these impacts and their consequences. The environment is continually evolving, and it can be a useful exercise to take stock of the current situation. There may already be a pronounced human influence in the area. From the beginning of the project assessment, the engineer thus has the opportunity to learn about any possible negative impacts, before taking measures to limit these impacts and include the risk-benefit ratio into the final analysis.

---

[1] The specific elements of an environmental impact assessment study are covered in Part III, Section 12.1.

An analysis of environmental impacts includes physical, biological, human, and socioeconomic aspects (Figure 4.1). Physical aspects involve the atmosphere (climate issues), hydrosphere (issues related to surface and groundwater), and the lithosphere (issues related to geology such as the stability of dam foundations and the reservoir slopes, as well as the impermeability of the reservoir basin). Biological aspects are related to the flora and fauna, as well as the vegetation generally. And finally, human and socioeconomic aspects relate to the presence, property, and activity of communities living in the zone around the dam and the reservoir. Communities are said to be completely affected if inhabitants lose their houses, land, and other resources, including even their means of subsistence; they are partially affected if they lose some of their resources (this usually involves the loss of land) and retain their house (as located outside of the flooded zone).

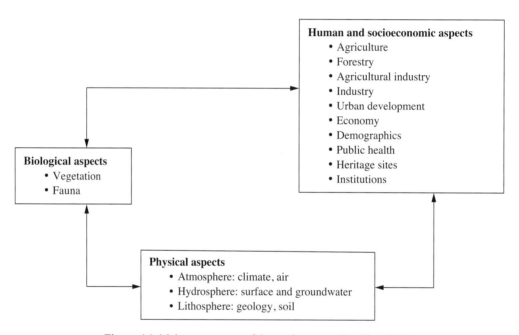

Figure 4.1 Main components of the environment (Zwahlen, 2003)

4.2.2 Impacts related to the atmosphere

Impact of the climate

It is recognized that large natural bodies of water (seas) can have a direct impact on their surrounding environment, particularly on temperatures and humidity, while small bodies (lakes) have an influence on their local microclimate. However, it is difficult to access reliable data relating to the effect of artificial lakes on the climate. This point is the subject of much discussion, in particular regarding large reservoirs created in tropical regions (the Aswan dam on the Nile in Egypt, the Akosombo dam on the Volta river in Ghana).

Greenhouse effect

The greenhouse effect is a natural phenomenon. When the sun's rays reach the Earth, they are either absorbed or reflected back toward the atmosphere. The atmosphere contains gases and molecules that capture the rays and reflect them back toward the Earth, thus maintaining an average temperature of 15°C. The main greenhouse gases are carbon dioxide (CO_2), methane (CH_4), nitrous oxide (N_2O), ozone (O_3), and water vapor.

It is acknowledged that hydropower is a clean energy compared to thermal power stations. For example, for a hydroelectric plant with a large reservoir in a cold climate, carbon dioxide (CO_2) and methane (CH_4) emissions are 30 to 60 times lower than for thermal power stations (Gagnon, 2002).

Studies carried out since the early 1990s have shown that reservoirs in hydropower schemes emit greenhouse gases such as those listed above. These emissions have been recorded in both temperate and tropical regions. It should be noted that the nature of the gases differs depending on the region and that the phenomenon is more significant in tropical regions. Initially, it was believed that the emissions were only due to the degradation of the flooded biomass and that they would stop once the organic material had decomposed. However, it has been observed that this process continues throughout the life of the dam, as it is fed by the carbon brought down by the upstream watercourse. According to measurements taken in Canada, it appears that there are substantial emissions during a period of ten years or so. Decomposition of organic material initially concerns the flooded vegetation and ground. One third of all carbon is stored in the form of cellulose in vegetation, while the other two thirds are in the ground in the form of humus and dead organic matter. CO_2 emissions are also given off by vegetation around the reservoir and plant debris brought down from the waterway upstream. Methane is generated from the organic decomposition that occurs in anaerobic conditions and as the water passes through the turbines and discharge structures.

The level of emissions depends on several factors, namely the type of ecosystem affected, the shape of the reservoir, the local climate, etc. It varies dramatically from one reservoir to another, and even from one part of a reservoir to another. The assessment of the impact of reservoirs must take into account the net level of emissions, that is, the emissions present before the dam was built. Gas emissions can also originate from the construction period, but these are not significant across a reservoir's lifetime. Finally, it has been noted that natural lakes and rivers emit similar quantities of greenhouse gases compared to reservoirs (Scherrer, 2007).

As there are still several discrepancies and unanswered questions around emissions from reservoirs, particularly in regard to the quantification of these emissions, it is clear that further research is necessary to gain a better idea of the actual impact of reservoirs on climate warming. In this respect, it must be stressed that each project has its own characteristics, and it is therefore difficult to highlight conclusions of a general nature.

Air quality

This can be locally affected during the construction phase by the excavation and transportation of material, but without causing any particular issues, providing that personal protection systems are made available.

4.2.3 Impacts related to the creation of a barrier to water flow

Fish migration

If specific measures are not installed (fish ladders), the migration of fish upstream and downstream is interrupted (Figure 4.2). However, the colonization of the reservoir by other species is often observed.

Figure 4.2 View of part of a fish ladder installed at the Verbois dam (near Geneva)

Barrier to the transportation of sediment

Reservoirs created by dams act as a giant decanter, which leads to the sedimentation of the reservoir. As a result, the water discharged into the downstream watercourse lacks sediment. This is illustrated by the notion of "hungry rivers." The equilibrium of solid transport is disrupted, and an eroding of the riverbed can often be observed downstream. Furthermore, water taken for irrigation has been emptied of fine suspended particles, thus reducing the natural input of nutrients and leading to the use of chemical fertilizers. Downstream from large cities that have no sewerage treatment, reservoir sediments are highly contaminated, which can represent a threat, especially for groundwater.

Some studies have shown that hydropower schemes impact the carbon cycle by holding back sediment that should end up in oceans. This sediment plays a not inconsiderable role by absorbing a substantial amount of CO_2 from industrial activities. Atmospheric CO_2 is absorbed by phytoplankton through the process of photosynthesis, before sinking to the depths once the organisms die, thus sequestering the carbon for a period of up to 1,000 years (Scherrer, 2007). This suspended sediment can, however, be found downstream from a dam after transiting through turbines and water release structures.

4.2.4 Impacts related to the creation of a reservoir

Flora and fauna

The flooding of land due to a reservoir has many consequences, including upon flora, fauna, and the overall biota. Establishing an inventory of the different species of flora and fauna across several seasons is recommended. This work will eventually highlight the presence of rare species that may become threatened, as well as those already endangered.

Effect on fish

Several scenarios may play out: the number of fish can increase, decrease, or they can disappear entirely. An increase in fish fauna can be observed in early years of the reservoir, as fish will feed on and prosper from the food originating from the biomass and soil. Later, the fish population may stabilize as this food source disappears. It is not uncommon for reservoirs to be stocked with fish.

Habitat destruction of terrestrial animals

A similar situation exists in terms of loss of habitat for wild animals. Following the destruction of small animals and a myriad of microorganisms caused by the filling of a reservoir, a new equilibrium establishes itself around the reservoir.

Forest destruction

Compared to the clearing of forest regions or forest and logging operations for the production of luxury wood items, the flooding of forests by a reservoir represents a tiny percentage of affected areas.

Water quality

In deep reservoirs in arid countries, considerable thermal stratification has been observed. This phenomenon prevents water from circulating within the reservoir. As it biodegrades, organic matter brought into the reservoir by the river consumes a large quantity of oxygen, thus leading to a deficit in dissolved oxygen. This phenomenon increases if the basin area of the reservoir was not fully cleared of trees before filling. In some cases, an acceleration of the modification of the aquatic environment can be observed (eutrophication).

Risk of landslides

The raising or lowering of the water level in a reservoir can result in the movement of unstable land on banks and valley slopes. The largest of the landslides at the Vajont dam (Italy, 1963) took place as the reservoir was being filled for the first time, when a landslide of about 250 million m^3 collapsed into the reservoir; the resulting overtopping impact wave led to the death of approximately 1,900 people. The valley flanks may comprise some naturally unstable zones situated outside of the reservoir itself, which may in turn lead to falling rocks.

Reservoir triggered earthquakes

The impoundment of water behind a dam can cause induced earthquakes. This phenomenon has been observed over several decades and has been the subject of research. Under certain geological conditions, the overload caused by a reservoir may provoke earthquakes if the first filling is not carried out in rigorously controlled stages. It must be noted, however, that certain tectonic conditions may make induced earthquakes inevitable for reservoirs taller than 100 m (see also § 10.4.3).

4.2.5 Impacts related to the modification of a downstream flow regime

Runoff of residual flow

In cases of diversion, only the minimum required environmental flow is returned to the river downstream. The riverbed is thus modified, which may affect the ability of the river to discharge floods. Furthermore, small and medium flood events are eliminated or heavily reduced by storage in the reservoir. The periodic flooding of the surface of spawning beds, swamps and ponds, small islands and riverbanks, as well as other areas of high environmental value is thus interrupted. Swiss legislation makes provisions for the maintenance of an appropriate fixed compensation flow based on a flow rate that is reached or exceeded for 347 days a year.[2]

---

[2] Title 2, Chapter 2 in the Waters Protection Act of 24 January 1991.

Another problem may arise during the initial impoundment of large reservoirs in cases where the filling may last months or even years. In this situation, total downstream de-watering must be avoided.

Fluctuation in flow

Modifications in flow due to daily and seasonal fluctuations may be observed. Rapid fluctuations may result from the operation of a hydropower plant, through the stopping or starting of its turbines. This is known as hydropeaking. Operating in this manner can have an effect on the aquatic life and breeding grounds of fish or other organisms living in the water. It can be remedied by setting up ramping programs for the start-up and shut-down of turbines or by the creation of a compensation basin, among other things.

The discharge of turbined water, depending on its source, may also affect the temperature of river water downstream. In addition to turbine operations, other activities related to dam operation may lead to a sudden increase in flow (opening of gates, flushing), which may even endanger users of the river downstream. Generally, the public should be informed or warned about these operations.

Transportation of sediment

Modification in the transportation of sediment can also be observed, because the sediment brought down confluents can no longer be mobilized. In these cases, flushing combined with artificial flood release can be carried out in a controlled manner.

4.2.6 Human and sociological aspects

Land flooded by the reservoir

This can have serious consequences and may result in the displacement of a large number of people. One example of this is the Three Gorges dam in China. Property and arable land are the most important resources for rural populations, and they are often to be found in the submerged zone. It should be noted that good soil is often found in the proximity of rivers.

When flooding of this land occurs, communities affected by the project must receive fair compensation. They must be rehoused and indemnified for their losses to ensure they are not disadvantaged. This is an important stage in the project and requires meticulous planning and follow-up.

Impact on public health

In hot and humid regions, the creation of reservoirs can have an effect on major parasitic diseases (for example, malaria, schistosomiasis, onchocerciasis). Preventive measure may be taken under the framework of a public health program relating to water supply and hygiene measures.

Safety of communities living downstream

Dam failure occurs to only a very small percentage of constructed works. This percentage falls even further for recent constructions. It is nevertheless useful to develop an emergency strategy to inform the public of evacuation procedures in the event of an uncontrolled release of a large quantity of water following partial or total dam failure or following a flood wave due to a large landslide into the reservoir.

Cultural impact and heritage protection

Ancient remains or monuments of a historical nature may be flooded once a reservoir has been filled. Action must be taken to conserve heritage sites, as was the case for Abu Simbel in Egypt. The temples of Ramses II and Hathor were to have been submerged by the waters of the River Nile following the construction of the Aswan high dam. On the initiative of UNESCO, dismantling of the temples began in 1965, and they were reassembled at the summit of a spur overlooking the same point in the river. The final cost for the works, which were finished in 1968, was 42 million dollars.

4.2.7 Impacts outside of the reservoir

Creation of new lines of communication
Once work at the construction site starts, there will be a labor force for the duration of construction and a need for the transport of equipment and materials. In order to access the construction site of the whole dam and reservoir complex, new roads often need to be built or the existing network may need upgrading. The creation of these access routes may also bring new settlers to a region. The presence of the dam and reservoir may make access to existing sites difficult, or even impossible, which may also imply the construction of new lines of communication including bridges.

Site installations
Accommodation for employees, traditional construction site installations, as well as the opening of borrow pits for material in the vicinity of the dam will noticeably modify the surrounding area. Once construction is complete, the affected, non-submerged areas will have to be restored and rehabilitated.

4.3 Description of Impacts

The following terms are generally used to describe the impacts (Zwahlen, 2003):

- *Direct-indirect*
  A direct impact is caused by the project itself (for example, the submersion of a forest) and an indirect impact is a consequence of a change caused by the project (for example, the presence of mosquitoes in the reservoir zone, which may lead to a rise in malaria cases).

- *Total-partial*
  This describes the magnitude of an impact in regard to a specific element (for example, a human settlement can be totally or partially submerged by the reservoir).

- *Permanent-transient*
  This describes the duration of an impact (for example, the dam interrupts the river permanently, whereas the noise and dust emissions from construction only last for a limited time).

- *Important-negligible*
  This describes the significance of an impact (for example, the submersion of historic monuments is to be considered important, while the submersion of non-arable land is considered negligible).

- *Positive-negative* (sometimes *desirable-undesirable*)
  These terms are used to characterize impacts of human activity but depend on the perception of the impacts and their consequences (for example, the submersion of a village is a negative impact, while the opportunity for fishing in the lake is a positive impact).

- *Acceptable-unacceptable*
  Acceptable means "according to a legislative basis"; in the absence of such a basis, consensus must be sought. The consequence of an unacceptable impact may lead to a reassessment of the project.

Although the categories "direct-indirect," "total-partial," and "permanent-transient" are clearly identifiable, this is not the case for the others. Defining an impact as "important," "positive," or "acceptable" is a question of personal appreciation or of convention in a given socioeconomic context.

4.4 Technical Measures

When discussing the mitigation of negative impacts, there are three types of measures that can be taken (Zwahlen, 2003).

1. Avoidance measures (abandonment, conservation)

These are measures (project modifications, choice of an alternative site, etc.) that will prevent a certain impact from actually happening. In many cases they lead to a "yes-or-no" type decision, as some impacts can only be avoided by abandoning a project. This must be considered as an option.

2. Minimization measures

These reduce the magnitude of an impact (for example, reducing dam height and therefore reservoir size to limit the area to be submerged or limit clearing and reduce impact on water quality).

3. Compensation measures

These may be implemented in cases where an impact cannot be avoided, and adequate mitigation is not possible. With this solution, the value that is lost is replaced by something of equal or higher value located elsewhere. Such a compensation would, for example, provide agricultural land of the same size and value to a farmer or replace a loss in forested area by reforesting a similar area elsewhere.

Obviously, it is not enough to simply outline mitigation measures. They must be implemented if the project goes ahead. The following additional conditions must be met: (1) careful planning of the measures, including identification of suitable and available land in the case of compensation measures; (2) identification of the costs of the measures and their integration into the overall project budget; (3) a firm commitment by project developers to actually carry out the measures; (4) strict conditions formulated by the competent authorities that stipulate that the measures are part of the project; and (5) follow-up and monitoring that the measures are implemented (this latter point is usually part of the monitoring program).

4.5 Stance of International Organizations

In 1997 the World Commission on Dams (WCD) was set up on the initiative of the World Bank and the World Conservation Union. It received financial support from many international organizations. In principle, the WCD was to have been dissolved once it had achieved it aims.

Its mandate was to examine the impact of large dams in terms of development and assess the available options for the elaboration of water and energy resources, but also to develop criteria, guidelines, and acceptable standards at an international scale, for the planning, design, assessment, construction, operation, and monitoring of large dams, as well as their decommissioning. In November 2000 the WCD published a report that presented the results of its findings. It wanted to demonstrate the reasons for which dams serve or fail to serve development objectives, by establishing five core values—equity, feasibility, efficiency, shared decision-making, and responsibility—values that are also accepted by other professional organizations directly related to the dam industry.

It proposed an approach based on the evaluation of rights (identifying legitimate claims and rights) and risks in order to guide planning and decision making. This was achieved by identifying all the groups who could legitimately participate in the negotiation of choices and agreements in terms of development.

The WCD then issued a set of strategic priorities and principles of ancillary actions for the development of hydraulic and energy resources, that is, gaining public acceptance, providing an exhaustive assessment of the various options, optimizing existing dams, conserving waterways and livelihoods, recognizing rights and the sharing of benefits, checking compliance with standards, sharing waterways for peace, and ensuring development and security.

Finally, it set criteria and guidelines related to good practice with regard to strategic priorities. The report was accompanied by 26 recommendations.

The WCD believed that its arguments and recommendations offered possibilities for progress through a comprehensive approach to integrating social, environmental, and economic dimensions of development, guaranteeing heightened transparency, and giving greater assurance to all parties concerned and increasing trust in the ability of nations and communities to satisfy their future needs in water and energy (WCD, 2000, Summary Report).

This report did not receive full endorsement on the part of large central organizations such as the International Commission on Large Dams (ICOLD), the International Hydropower Association (IHA), and the International Commission on Irrigation and Drainage (ICID). In their position statements, they raised the point that evaluating 125 large dams out of the 45,000 works identified would mean that most dams would not be reviewed. It is true that any infrastructure development will have an impact. Denying this goes against the notion of progress and can prevent some individuals from accessing a minimum level of prosperity. It is therefore important to ensure the development of water resources to satisfy the needs of all.

It is worth remembering here the role of the International Commission of Large Dams (ICOLD), created in 1928. One of its goals is to improve techniques in the dam industry in every respect and at every stage throughout the conception, design, construction, and operation of large dams and their ancillary structures.

The ICOLD guides the profession toward safe, efficient, and economic dam construction and operation with a minimum impact on the environment. For several decades now, the ICOLD has focused on raising engineers' awareness of natural and social environmental issues and on widening their perspectives so that these aspects can be afforded the same attention as technical issues. From 1973, this concern was reflected in the following terms: "The real problem to solve is to find out whether dams are useful or harmful, whether overall they improve our environment and human well-being or whether they deteriorate it, and to find out, in each case, whether they should be built, and with which characteristics."

In a charter published in 1997, the ICOLD indicated that in order for a project to be acceptable today, the determining factor, on a par with safety, is protection of the environment. It is aware of the need to protect and conserve the natural environment. In the technical bulletins published by the Commission, it deals with a large number of issues related to dams, concerning the socioeconomic, environmental, and geophysical context as well as water quality.

In the future, the ICOLD will be called on to intensify its activities in order to harmonize the development of water resources with protection of the environment, all the while taking into account the communities that are affected by these projects. It will have to improve its understanding of environmental interactions and progress made in regard to controlling these, through the collection, evaluation, and publication of real experiences, and by making recommendations based on these experiences. It will then be able to recommend the application of criteria and goals favorable to the environment and the establishment of a legal and administrative framework adapted to each country depending on its specific conditions and needs (ICOLD, 1997).

With the help of the Electricity Production Sustainability Index (EPSI), electric power plants can be compared on the basis of environmental, societal, technical, and political criteria (Lafitte and De Cesare, 2005). Storage projects, even those with large reservoirs, have a higher EPSI index than thermal power stations.

4.6 Education

Dams currently make and will continue to make a major contribution to the management of water resources. This will be even more important in developing countries where the construction of reservoirs and the utilization of hydropower can have a huge impact. For example, for a country whose subsoil does not contain any natural resources such as coal or oil, the possibility of developing hydropower (a clean and renewable energy) is an opportunity to be seized. This is a high-value energy, and the use of hydroelectric schemes allows for considerable flexibility in supply and an ideal way of meeting demand.

To be able to continue building dams, the study of impacts on the natural and social environment must now accompany a project from its inception and subsequently throughout its design and construction, as well as during operation. This environmental impact assessment has in fact been compulsory since 1971 in several of the Commission's member states. It is also important to mention and summarize some of the recommendations in the ICOLD charter, in the spirit of ensuring appropriate water management while conserving the natural environment:

- In the preliminary project stage, assess solutions other than the dam that may meet the same objectives
- Appoint a multidisciplinary team of experts to collect the necessary information
- Assess the project based on the latest technical knowledge and modern criteria for the protection of the environment; study and implement means for reducing negative impacts
- Carry out realistic economic analyses and avoid underestimating the costs or overestimating the benefits
- Consider the issue of human displacement with care and sensitivity; the relocation of these communities must lead to an improvement in their standard of living
- Analyze the impact on local communities before and after the works; do not ignore the information
- After the works, assess the level of satisfaction with the measures taken
- Assess the impact of working installations at regular intervals
- Expand environmental research on installations that have been operating for many years

The only lesson that can be drawn from imperfections and errors highlighted by past water resource development projects with dams is not "no dam" but, as always, "better projects" (Pircher, 1993).

Projects for large dams and reservoirs around the world can undoubtedly cause public controversy. But in most cases, the "no dam" option is not the best solution, as it means no water, no food, no energy, no development in the region, and no future for the next generation. The great challenge, and often the only option, is to build "better dams for a better world" (Schleiss, 2016).

Finally, to close this vital chapter, let us recall the wise words of Nelson Mandela, former South African president, who declared that "the problem is not the dams. It is the hunger. It is the thirst. It is the darkness in a township. It is the townships and rural huts without running water or sanitation. It is the time wasted gathering water by hand. There is a real, pressing need for power in every sense of the word."

II. DAM SAFETY

Should they fail, dam structures and reservoirs can cause serious loss of human lives and extensive damage to property (dwellings, lines of communication, infrastructures). There can also be a huge impact on the environment: flora and fauna destroyed, farming operations devastated, or a risk of pollution due to damaged industrial sites. Safety programs must be extremely stringent, which means following a rigorous approach right from the start of the project and during its construction. Once the dam is in operation, particular emphasis must be placed on prevention, and consequently, a clearly defined monitoring system must be put in place that applies the same safety standards across all of the structures and their operation. As there is no such thing as zero risk, the monitoring system must be able to detect any anomaly concerning the dam's behavior, foundations, appurtenant structures, and surrounding area.

Authorities under whose jurisdiction dams fall are responsible for establishing laws and regulations outlining the measures that must be taken to safeguard against danger and damage resulting from the construction method, operation, insufficient maintenance, or even an act of war or sabotage (in the case of Swiss legislation). The responsibilities and tasks of dam operators and other stakeholders must also be defined. Currently, thanks to the safety management system that has been in place in Switzerland for many years, no failure of a dam under the surveillance of the Confederation has yet occurred.

Chapter 5 sets out the essential elements of the dam safety concept implemented in Switzerland, which is built on three key pillars: structural safety, surveillance and maintenance, and emergency planning. Chapters 6, 7, and 8 cover the dangers and hazards that may affect a dam and its foundations, examples of types of failure, and possible protection measures. And finally, chapter 9 outlines some of the concepts of risk analysis applied to dams. Expert appraisal and a wide experience in the specific field of dams have an important role to play in the analysis of risk. However, it should be remembered that unlike products issuing from factories, each dam must be regarded as a unique system within its own distinctive environment.

5. Dam Safety Concept

Curnera arch dam in Switzerland, height 153 m, year commissioned 1966
(Courtesy © Swiss Air Force)

5.1 Introduction

According to the definition taken from code SIA 160 of the Swiss Society of Engineers and Architects (SIA), relating to the action of load-bearing structures, "safety in the face of risk must be maintained even when this risk is kept under control by appropriate measures or remains below an acceptable minimum. Absolute safety can never be guaranteed."

To ensure the safety of the structure and to counter possible hazards, the supervisory authority of the Swiss Confederation has developed and implemented a comprehensive safety approach for dams, based on the three following pillars (Figure 5.1):

- *Structural safety* implies the appropriate concept, design, and construction of dam structures
- *Surveillance and maintenance* imply the strict implementation of an organized system for the monitoring of the behavior and condition of dam structures and their foundations
- *Emergency planning* implies appropriate preparation in case of a crisis situation

The first pillar enables risk minimization, while the two others allow the residual risk to be managed. It can be accepted that this safety concept, which prioritizes a deterministic approach, introduces principles of risk management.

Several European countries (including France, Italy, Austria, and Spain, among others) also apply this safety approach based on three pillars.

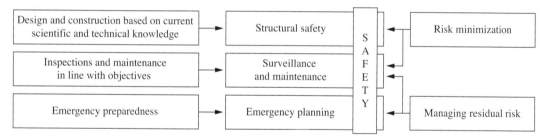

Figure 5.1 Diagram of the safety concept in effect for dam structures in Switzerland based on the three pillars (from SFOE, 2015a)

5.2 Structural Safety

5.2.1 Introduction

Structural safety concerns the dam, its foundations, its surrounding area, and the appurtenant structures. It is important to again emphasize that dam and reservoir structures are built in such a way as to guarantee safety during construction, at the time of first impoundment, and during their operation.

The concept and design of a dam draw on current scientific and technical knowledge. The goal of structural safety is to ensure that the structure can withstand all possible and foreseeable loads. The design process also factors in all the parameters that might act on the structure, which implies the dam's ability to withstand various types of individual loads. The combination of individual actions allows the structure to be checked under even the most severe stress. Three types of combinations are defined: normal (type 1), exceptional (type 2), and extreme (type 3). A distinction can be drawn between permanent actions, such as self weight; variable actions, for example, water and/or sediment pressure and climatic conditions; and accidental actions due to natural events such as floods or earthquakes. Structural safety also implies taking all appropriate actions to guarantee not only the correct behavior of the structure, but especially the safety of people and property. Protective structural arrangements should also be included in order to deal with potentially dangerous situations (Figure 5.8).

5.2.2 Service criteria agreement and basis of design

The SwissCode SIA 260 (1999) introduces, among other things, the notions of "service criteria agreement" and "basis of design."[1] The service criteria agreement is the description of protection and service criteria objectives, as well as conditions, requirements, and instructions. The basis of design is the technical transcription of the service criteria agreement. These two documents are also usefully employed in the field of dam structures, both for new constructions and existing dams (SwissCoD, 2000b; SFOE, 2017).

The service criteria agreement, drawn up in agreement with the project owner, allows the project's baseline data to be set (function and characteristics of the installation, conditions of operation) and the specific structural provisions to be taken into account during the project's concept and design phases, as well as to outline the requirements of the supervisory authority. It is also possible at this stage to specify the general hypotheses concerning design.

The basis of design is an important document, as experts are required to undertake research into all possible hazardous situations and determine the structural dispositions or other appropriate measures. Acceptable risks (nature of tolerated damage) will have a direct influence on the actions to be taken.

Once these elements have been gathered together, it becomes possible to set out all the hypotheses for the design and verification of the dam and its appurtenant structures (spillway, bottom outlet, etc.). Finally, the basis of design must also ensure that the specific elements to be included in the service and surveillance instructions can be determined. It also sets out the protection measures.

Figure 5.2 Ferden dam: water release system integrated into a concrete dam: initial spillway (top) and bottom outlet (bottom) (Courtesy H. Pougatsch)

---

[1] Previously, "service plan" and "safety plan," introduced in the code SIA 160 (1989).

5.2.3 Structural provisions

Water release structures are vital so that the water level can be managed when dam safety is threatened. With this in mind, the installation of a bottom outlet (and possibly also a mid-level outlet) and spillway must be provided for as a structural safety provision (Figure 5.2). The intake structure draws water from the reservoir for various uses (turbine operation, irrigation, drinking water, industrial water, etc.). These appurtenant structures may be independent or, depending on the type of construction, integrated into the dam structure.

Bottom outlet

The bottom outlet is primarily a safety system. Its main functions are to:

- Control the rise in water level during the first impoundment
- Enable the lowering or even total emptying of the reservoir in cases of imminent danger of landslide or military threat, as well as during monitoring and maintenance works
- Maintain a low water level when required (exceptional event, rehabilitation works)
- Release floods

Furthermore, it allows

- Sedimentary deposits to be evacuated (providing this activity has already been authorized)
- The reservoir to be emptied
- A discharge to be maintained into the downstream river in exceptional cases (for example, routine overhaul of the hydropower plant)

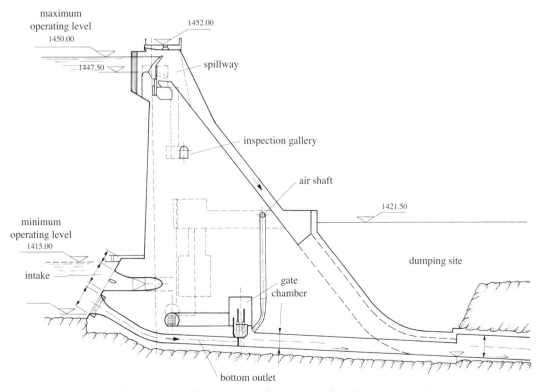

Figure 5.3 Panix dam (GR): example of a bottom outlet with two consecutive gates

The water release structure must be built in such a way that the gates do not get obstructed or blocked. When there is risk of blockage by sediment in the reservoir, appropriate and constructive preventive measures or regular flushing must be planned, with the aim of ensuring that the inlet entrance is clear of sediment.

For reasons of operational safety, the bottom outlet is usually equipped with two gates, one as a safety element (maintenance gate) and the other as an operating or regulating element (service gate) (Figure 5.3). In some cases, an upstream stoplog is designed to close the bottom outlet entirely in cases of maintenance works and routine overhaul. For gate operation, backup control systems should be installed for cases where the main system fails. For example, provision should be made for a backup power supply for the gate hoist motors. These must also be able to be hand operated. For remote-controlled gates, the opening must take place in stages so that an involuntary total opening does not occur. In addition, access to on-site controls must always be guaranteed.

A plan or manual describing how to operate the gates during floods must be developed and incorporated into the surveillance regulations. Safe gate operation must be checked at least once a year, if gates have not already been actioned during the year (Section 32.3).

Operational checks for the gates of water release structures and the bottom outlets must occur under the same conditions as those that might exist in an exceptional situation requiring gate operation. The inspection must occur with a discharge ("test with water release") and for a high water level.

Spillway

A spillway is also a safety structure that prevents the uncontrolled rising of water in a reservoir and protects the dam and its appurtenant structures against overtopping, which could lead to damage due to erosion and instabilities. Flow conditions are outlined in Section 10.6.3.

Spillways can be weirs controlled by gates (flat gates, flap gates, drum gates, sectors, rubber gates) or not (Figure 5.4). Fuse plug systems (fuse gates, erodible dikes) are complementary solutions.

Among the possible types are:

- Front spillway with or without a chute
- Side channel spillway
- Bell-mouth spillway (morning glory)
- Labyrinth spillway

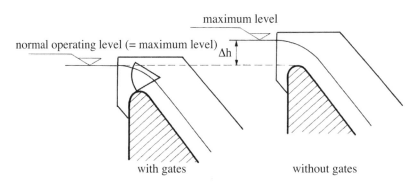

Figure 5.4 Illustrations of spillways with and without gates

The width and height of weir sections and freeboards must be sufficient to avoid obstruction by tree trunks or other floating debris in cases of a flood (Figures 5.13 and 10.38, Table 10.40). The same provisions that apply for bottom outlets in regard to the operation of moving structures are also applicable for gate-controlled spillways. In all cases, access to on-site controls must always be guaranteed, and the operating station, electrical source, and cable routing must not be installed in areas that may be submerged during flood events. Depending on winter conditions, the installation of an air bubble system prevents ice from attaching to the surface of gate structures; systems for heating slots of regulating gates and guide walls also exist.

Water intake
The water intake is an appurtenant structure, whose size and withdrawal rate depend on the purpose of the dam scheme (hydropower, irrigation, drinking water, artificial snow, etc.). The characteristics of the water intake are generally settled with the client following a technical and financial study. The intake is equipped with a protective trash rack. Depending on requirements, the water intake may also be used to partially lower the reservoir water level.

5.2.4 Operating dams

Commissioning
The first complete filling of the reservoir created by a dam is a delicate operation. It is important in that it allows the dam's operational ability to be assessed. This procedure must be meticulously carried out alongside close monitoring. Dam failure is most likely to occur during this impoundment.

Generally speaking (except for run-off river dams and retention basins), the first impoundment of a reservoir occurs in two phases and according to a predefined schedule that outlines the filling process, the way in which monitoring will be organized, and the measurements and observations required to assess dam behavior. Filling can only begin if the water release structures and monitoring systems are operational. Authorization for dam operation will only be granted if the results obtained during this test phase are conclusive.

Periodic safety review
It is vital to maintain and improve, if necessary, the safety of existing structures. The aim of these periodic reviews and analyses is to reveal whether the structure meets the most recent and state of the art safety criteria. The structures must therefore be maintained in a good state of repair and be periodically upgraded to meet the standard required for new structures. A special analysis of structural safety may lead to strengthening or rehabilitation works (§ 2.2.3). Structural safety must be reviewed following any modification in operation, new loads, or a deterioration in any of the structures.

Concept of service life
A limit to service life does not mean the dismantlement or destruction of the structure or of one of its elements, but rather the need to invest substantial means to ensure its rehabilitation with the aim of extending its working life or delaying its replacement. It must also be noted that good maintenance results in an extended service life.

Experience shows that service lives of materials, installations, and identical or comparable equipment can vary greatly. Mention should firstly be made that reservoirs in Switzerland are often integrated with hydropower schemes for which a concession has been granted for varying periods of time, but never exceeding 80 years. They may, however, be renewed. Obviously, some parts of the structure (grout curtain, facing, anchors, etc.) have a shorter service life than the body of the dam, and they must be remediated while the dam is operating.

The same applies for mechanical equipment, which may, for example, need parts replaced during their service life. Such equipment should be able to function for 40 to 50 years without being replaced. Steel lining and gates should be able to last over 50 years (SwissCoD, 2000b).

Particular attention should be paid, however, to the monitoring system (§ 2.2.6), as technological advances (Sect. 32.4) can have a substantial impact in this area.

Sedimentation and flushing

The entrance area to the bottom outlet must remain clear so that flow capacity through this water release structure is guaranteed. One area, generally conically shaped, needs to be kept free of material. If upstream sediment deposits hinder flow, measures must be taken to eliminate these deposits. Flushing or other environmentally friendly methods can be used. It is recommended that these operations should be repeated as often as necessary, which prevents sediment deposits building up in large quantities. It is important to ensure that as far as possible flora and fauna are not harmed during flushing operations.

Decommissioning

If, for various reasons, a dam must be decommissioned, it should be transformed in such a way that no water can be stored on a long-term basis and, apart from usual work maintaining the watercourse, no maintenance of the structure is required. One solution consists in creating a breach or orifice in the retaining structure that is large enough to let a flood (for example, a 100-year flood) transit through without danger (SFOE (SFWG), 2002b).

5.3 Surveillance and Maintenance[2]

5.3.1 Introduction

As risk cannot be entirely eliminated, it is vital that strict and regular surveillance and maintenance of the structure is undertaken to ensure satisfactory behavior and condition in all circumstances. It should be noted that during the lifetime of the dam, phenomena that are understood to a greater or lesser degree, or that are unpredictable, may occur.

Surveillance must ensure that any irregular behavior of the dam and its foundations, any specific damage, and any external risk (such as a landslide into the reservoir) can be detected rapidly and adequately. In the case of such an event, all actions necessary to mitigate the possible harm to the structure's safety must be undertaken without delay (see Section 5.4, Emergency Planning).

Maintenance is carried out primarily to prevent any malfunction and to repair any eventual damage. In this way, necessary refurbishment, repair, rehabilitation may be performed.

5.3.2 Main activities

In order to assess the behavior and condition of the dam, three principal tasks must be carried out (Figure 5.5):
- Visual checks
- Direct measurements through the monitoring system
- Operational checks of gate elements, water release structures, instruments, and means of communication

---

[2] See also, Part VIII, Chapter 32.

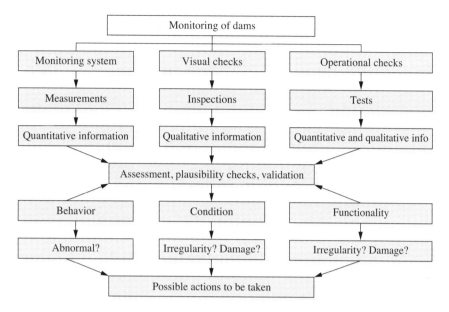

Figure 5.5 Surveillance process for a reservoir

The importance of *visual checks* must be emphasized, as they allow not only the condition of the reservoir and its appurtenant structures to be checked (deterioration of materials, cracks, etc.), but also the visible parts of the foundations and the stability of the reservoir slopes. Worldwide, almost 70% of specific events concerning a reservoir have been brought to light through visual checks.

With regard to *direct measurements*, these must be carried out regularly and the results rapidly analyzed. Processing results of direct measurements concerning displacement, deformation, seepage, and pressure allows the general behavior of dams, foundations, and their surroundings to be evaluated.

The importance of *operational checks* should not be overlooked. For example, during exceptional flood events, it may not be possible to operate gates due to failures in the gate operating systems, which may result in considerable damage to the dam and downstream zones. These systems may need to be activated frequently throughout the year (for spillways), or they may have limited or even no use (in the case of the bottom outlet). In this latter case, the test becomes all the more important, as the fact of carrying out a gate operation with water release at least once a year allows the whole control system to be tested and, if necessary, a faulty part can be replaced. In addition, staff members can be trained in the set procedures for these operations (which also comes under the umbrella of flood management).

5.3.3 Organizational structure

In order to ensure efficient monitoring, it is a good idea to base it on a four-level check system (Figure 5.6). According to the model in use in Switzerland, surveillance and maintenance are primarily the responsibility of the operator (level 1), who must organize regular visits and monitoring activities. After the readings from the dam monitoring system have been reviewed, the results are rapidly transferred to an experienced engineer for an initial appraisal. The operator carries out all the planned operating checks. This is a key role, as through

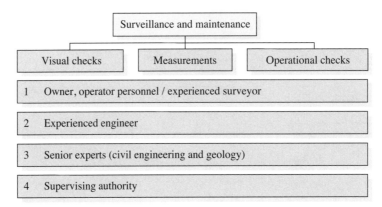

Figure 5.6 Organization of surveillance and monitoring

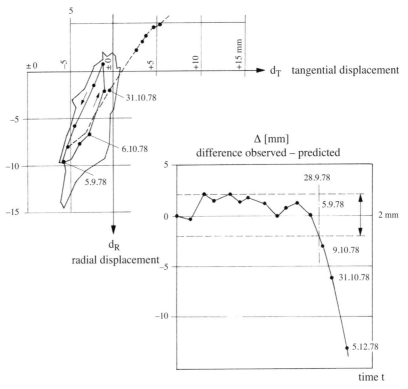

Figure 5.7 Abnormal deformations at the Zeuzier dam: displacement from the point of suspension of the median pendulum (CNSGB, 1982)

their frequent presence at the dam, the operator will be the first to detect any irregularity. Furthermore, the operator has to carry out specific checks following the detection of a danger that might jeopardize the safety of the structure or of an earthquake (for example, from a magnitude larger than 3 on-site), as well as during and after a major flood event. These checks, which must be adapted to each specific situation, can include measurements, visual inspections, and operational checks. Depending on the event, a framework program may be set up.

The work carried out by the operator is checked by an experienced engineer (level 2), who may or may not be employed by the operator. The engineer is required to continually analyze the measurements that are directly transferred by the operator, and, if necessary, to raise the alarm. It is thanks to this system of continuous analysis that the exceptional deformations at the Zeuzier dam arising from a rock mass drainage problem during the drilling of a tunnel were able to be identified so quickly (Figure 5.7). This experienced engineer must also carry out periodic on-site inspections and draw up annual reports related to the behavior and condition of the dam and reservoir.

Level 3 is assigned to senior experts (civil engineers, geological engineers, geologists), who are required to intervene every five years. They carry out periodic safety reviews (five-year review) that focus in particular on long-term behavior. They are also called upon in cases of specific behaviors. These senior experts may be mandated by the supervisory authority to carry out specific analyses (analysis of old structures, safety in regard to natural hazards).

And finally, the entire organizational structure is overseen by a supervisory authority (level 4), which acts as an independent body and carries out its own general assessment of the condition and behavior of dams and reservoirs.

5.4 Emergency Planning[3]

5.4.1 General information

One of the aims of the dam safety concept is to ensure the protection of local communities from the consequences of an uncontrolled water release from the dam. Given that the risk of a dangerous situation cannot be excluded, it is vital to have an emergency plan. To best deal with a danger that may impact the safety of the dam, several important measures must be taken in order to manage the situation, and the means must be in place to sound the alarm, inform, and possibly evacuate at-risk populations. It is therefore necessary to plan these measures and establish an emergency plan that outlines the strategy for managing dangers, indicates the various actions that must be taken and the preparatory work required, and gives the instructions for warning and evacuating the population at risk, as well as instructions on what to do.

The various types of hazards that can create a dangerous situation are described below (Biedermann, 1987b). Measures for managing these hazards are also indicated. Items relating to the planning and implementation, as well as the elaboration of a set of emergency procedures are covered in Chapter 33.

5.4.2 Potential hazards

There are many hazards that may lead to a dangerous situation and to the implementation of response measures. These include:

---

[3] See also, Part VIII, Chapter 33.

- Behavioral anomalies (dam, foundations) and operating failure (appurtenant safety structures)
- Landslides or rock/ice falls (the possible or proven arrival of a large volume of loose or rocky material, snow, or ice into the reservoir, generating an impulse wave)
- Floods
- Extreme earthquakes
- Sabotage (infrawarrior threat)
- Armed conflict (act of war)

The first three hazards are identifiable, while the last three are unpredictable events. Should an exceptional event be identified, technical interventions and operational measures can be envisaged. Evacuation is the final measure possible, particularly when the risk of an uncontrolled water release is probable (Figure 5.8). In the case of floods, earthquakes, sabotage, or an act of war, evacuation is the only active prevention measure.

Hazards \\ Measures	Abnormal behavior	Landslide, rock/ice fall	Floods	Earthquake	Sabotage	Act of war
Rehabilitation	(I)	Poss.				
Partial drawdown		(I)				Poss.
Complete drawdown	(II)					
Precautionary evacuation	(III)	(II)	(I)			
Post-event evacuation		(II)		(I)	(I)	(I)

Figure 5.8 Matrix of hazards and available measures in cases of emergency
Key: I First possible measure II Measure to be taken should the first be insufficient
III Measure in cases where the situation worsens

Abnormal behavior and operational failure
A multitude of specific events directly affecting the dam and/or its foundations, as well as the appurtenant structures (spillways, bottom outlet) may occur.

The abnormal behavior of a dam and/or its foundations is a major issue, which is notably expressed through the exceptional deformation of the entire dam or parts of the dam, including its foundations, increases in pressure or flow, and failures in watertightness. In the beginning, these events may develop slowly, but can also accelerate (for example, internal erosion in an embankment dam). In many cases, technical interventions are often possible in the form of drainage works (grouting works, creating a drain, sealing a crack in the foundation—Figure 5.9) or remedial works (sealing cracks). If too much time is required to carry out these works and restore dam safety, the reservoir water level can be lowered, and, in an emergency, precautionary evacuation of the population can be carried out.

An operational failure of the gated water release structures is a possibility that must be taken into account. In this case, overhaul work, replacement or even transformation may be needed.

Finally, the upstream deposit of sediment can lead to the obstruction of the entrance to gate systems and prevent them from being used when necessary. If possible, regular flushing should be carried out to free the passageway.

Figure 5.9 Placing a waterproof membrane directly on the surface of the rock—downstream toe of the Albigna dam

Landslide, rock/ice fall, avalanche, serac

After a landslide, rock fall, or avalanche, the arrival of a large volume of loose or rocky material, snow, or ice into a reservoir may create an impulse wave, which, depending on the amount of material and the water level of the reservoir, can surge over the crest of the dam (§ 11.5.3; § 27.1.2.4) (Huber, 1980; Müller and Huber, 1992; Heller, 2008). This event may take place very rapidly, but some warning signs may be able to be detected (Figure 5.10). When a risk (the triggering or reactivation of a landslide, instability in a section of rock) has been identified, a partial drawdown or operating restrictions can be imposed in order to prevent any water from overtopping the dam. In this case too, a precautionary evacuation of the population may be envisaged.

Figure 5.10 Landslide at the Bornes du Diable in the area surrounding the Mauvoisin dam and reservoir

After a large snow avalanche into the Räterichsboden dam and reservoir (BE), an avalanche freeboard for the winter period from early December (possibly November) to late April was put in place for several dams in the Alps. The movement of glaciers overlooking reservoirs can be observed using various methods, including theodolite measurements based on the position of ranging rods set into the snow, GPS, drilling sites fitted with inclinometers, aerial and terrestrial photographs, photogrammetry, and laser-scanning. In some cases, a safety margin is included when filling the reservoir. The normal operating level can only be reached if the glacier overlooking the reservoir does not pose any risk of ice fall.

Figure 5.11 Ferden dam hit by the Faldumbach avalanche in February 1999 (Bremen et al., 1999)

Cases of floods

Floods are phenomena that can be violent and develop extremely rapidly. However, they can also be detected very early on. Many dam structures have been severely damaged by overtopping (damage to the crest, erosion of the abutments, scouring of the downstream toe), and some have even been destroyed (Figure 5.12). As a result, a high level of safety in regard to floods must be adhered to. Spillway

Figure 5.12 Embankment dam destroyed by overtopping during a flood (Glashütte (D) flood protection basin; H = 9 m, reservoir vol. = 60,000 m^3) seen from upstream on the left and downstream on the right (Courtesy R. Pohl)

Figure 5.13 Blocked passes at the Palagnedra dam (TI) during the flood on August 7, 1978

structures must be sized to take into account the largest possible events at the site. Their flow capacity must be such that the passage of an extreme flood can be handled without critical damage, that is, the damage must not endanger the stability of the dam, even with a full reservoir. Furthermore, spillway openings must be sufficiently large to ensure they are not obstructed by floating debris (Figure 5.13). Transformation works for water release structures may be necessary after a safety reassessment in cases of flood. Conditions for the passage of floods are outlined in Section 10.6.3. Finally, it should not be overlooked that during flood events, various incidents external to the dam may occur: access road to the dam cut off, power failure, blocked gate systems, flooding in gate control rooms, etc.

When a flood endangers the dam as a whole, precautionary evacuation is the only possible means of protection. Finally, experience has shown that the presence of on-site staff is strongly recommended to observe how the situation evolves and then to take any necessary on-site measures. In some countries, cameras are installed to monitor remote dams; however, a disruption in the transmission of images due to local meteorological conditions cannot be excluded. A regulation containing information on the organizational structure of the operator, the roles of staff members, and gate operation should be established.[4]

An extreme flood, possibly combined with the failure of a water release structure (gate operation impossible or limited, obstruction of the opening) may lead to an uncontrollable increase in water level, even when following the instructions for gate operation. In this case, the moment at which the operator has to take action or even set off danger alerts must be established (SFOE, 2015c).

For dams situated near glaciers, the bursting of subglacial lakes can suddenly cause increased water flow.

Earthquake

Although earthquake-prone regions are listed, this kind of natural event occurs suddenly and without warning. So far, the literature makes no mention of the failure of a large dam due to an earthquake, provided it was designed and constructed according to modern specifications. However, an earthquake can nonetheless cause various issues ranging from small to substantial (Section 6.1). For example, several embankment dams were damaged by the Bhuj earthquake in the state of Gujarat, India, in January 2001. However, no uncontrolled water release occurred due to the low water levels in the reservoirs (Figure 5.14). In addition, in areas surrounding a dam, earthquakes can provoke landslides, modify hydrogeological conditions,

---

[4] See Part VIII, Section 31.1.

Figure 5.14 Embankment dam damaged by the Bhuj earthquake (Gujarat, India) on January 26, 2001 (Courtesy Wieland, 2006)

Figure 5.15 Damage to the Shih Kang dam (Taiwan) caused by movement in a fault during the Chi-Chi earthquake (September 21, 1999) (Courtesy P. F. Foster; Wieland, 2002)

or create a seiche (generation of waves that can overtop the dam). The possibility of a reservoir induced earthquake should also be considered (see § 10.4.3). This phenomenon has been observed in the Kariba, Koyna, Hsinfengkiang, Nurek, and Assouan lakes (Jansen, 1988).

Given the catastrophic consequences of an eventual dam failure, dams must meet an especially high level of safety criteria for earthquakes. The requirement that the dam must not experience critical damage means that it must be designed and dimensioned so as remain stable during even the most violent earthquake possible at the site, even if the reservoir is full. For old structures, an earthquake safety reassessment is strongly recommended.

After an earthquake and depending on access conditions, the first task is to carry out a detailed check of the dam in accordance with the earthquake intensity recorded at the site. Generally speaking, the objective is to highlight any changes in the dam and its surrounding area. Firstly, detailed visual checks are made of the dam, its abutments, and immediate surroundings (especially banks and valley slopes of the reservoir), and the appurtenant structures. The available results of behavior measurements are then analyzed (deformations, uplift, percolation rate). Once these elements have been analyzed, the necessary steps to take can be established.

Due to the unpredictable nature of earthquakes, evacuation is the only protective measure that can be taken once the event has happened.

Sabotage (infrawarrior threat)
An act of sabotage is generally a planned operation undertaken by experts. It may result in limited or sometimes substantial damage, without, however, completely destroying the dam. A list should be drawn up of all the installation's vulnerable elements, and no operation should be undertaken that could cause a major outflow. To guard against possible acts of sabotage, the operator should remain vigilant, particularly in regard to monitoring access to the site. This surveillance may take the form of video cameras or an infrared alarm system. Only authorized persons should have access to the site.

Armed conflict
From 1937 to 1993, some twenty acts of war concerning dams have been identified (Schnitter, 1993). Damage may range from limited to major (formation of a breach and uncontrolled release of water). One well-known case is the bombing of the three dams in the Ruhr valley in May 1943 during the Second World War. Two of the dams (Möhnetalsperre and Edertalsperre) were severely damaged. The attack on Möhnetalsperre caused a breach 60 m long and 22 m high, and the flood wave reached a height of 8 m (Figure 5.16). In January 1993, troops occupying the Peruča dam in Croatia placed explosive in five places (on the spillway and in control galleries in the embankment dam). The explosion formed a crater on each bank and resulted in damage to the gallery entry at the toe of the dam. Despite the extent of the damage, the dam was not destroyed (no breach and no uncontrolled release) and was later remediated (Figure 5.17).

In the case of armed conflict, attacks may be either terrestrial or aerial and should be guarded against by strengthening security measures (for example, by pre-emptively lowering the water level) and by ensuring continual monitoring of the dam. A good alarm system and procedure for evacuating local communities in case of an uncontrolled water release following partial or total dam failure must also be in place (§ 2.2.8 and Chap. 33).

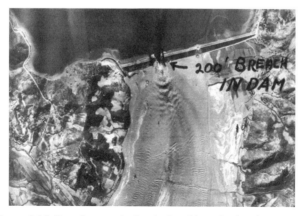

Figure 5.16 Breach created after the bombing of a dam in Germany

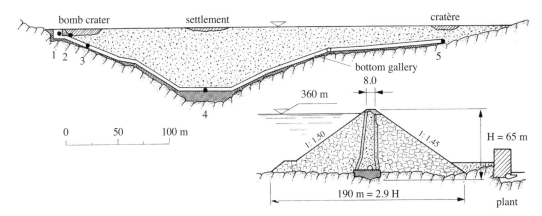

Figure 5.17 Explosion damage to the Peruča dam in Croatia (Schnitter, 1993)

Table 5.18 Hazards for a dam and their consequences (based on Biedermann, 1987b)

Hazards	Observations	Assessment	Possible consequences
Abnormal behavior of the dam and/or its foundations	Exceptional deformations Increase in uplift, interstitial pressure Increase in seepage flow	Detection by official inspections Slow development in general	Damage Partial or total failure
Malfunction in water release structures	Operation impossible Blockage of inlets	Detection by official inspections	Uncontrolled rise in water level Overtopping
Natural hazards:			
Landslide, rockfall, avalanche, ice fall	Initiation or reactivation of a landslide Instability of reservoir slopes Significant snowfall Instability of a glacier	Possible early detection Sudden activation	Overtopping of a wave over dam crest depending on water level
Flood	Rise in water level	Rapid development Timely detection	Overtopping Partial or total failure
Earthquake		Unexpected	Damage Partial or total failure
Sabotage	Damage (explosive or other)	Unexpected	Relatively low damage
Acts of war	Bombing	Unexpected, only in case of armed conflict	Partial failure

The protocol additional to the Geneva conventions of August 12, 1949 and relating to the protection of victims of international armed conflicts (Protocol 1), adopted on June 8, 1977 must also be mentioned. It stipulates in article 56, paragraph 1 that *"works or installations containing dangerous forces, namely dams, dikes, and nuclear electrical generating stations, shall not be made the object of attack, even where these objects are military objectives, if such attack may cause the release of dangerous forces and consequent severe losses among the civilian population."*

Table 5.18 summarizes hazards and their consequences that must be taken into account in the emergency plan.

5.4.3 Management of proven hazards

The management of proven hazards in an emergency plan may include several levels of danger depending on the progression of the detected or ongoing hazard and on actions taken (SFOE, 2015c). Table 5.19 provides an example.

Table 5.19 Definitions of different levels in an emergency strategy

Level of danger	Situation	Short description
1	No or low danger	The behavior and condition of the dam allow for safe operation.
2	Limited danger	An abnormality regarding safety has been observed in the behavior or condition of the dam. Further investigation or the taking of constructive or operational measures is necessary. No uncontrolled release of large quantities of water will take place.
3	Significant danger	The situation can be controlled. An uncontrolled release of large quantities of water is not expected.
4	Serious danger	The danger can be controlled for the moment. A deterioration in the situation with ensuing uncontrolled release of large quantities of water might occur.
5	Extreme danger	The situation can no longer be controlled. An uncontrolled release of large quantities of water is about to occur or has occurred.
	End of danger	Return to normal level (possibly return to enhanced monitoring level or lower danger level).

As long as the situation is normal, the operator applies the usual monitoring and operating program (regular visual checks, measurements, and operating tests).

Then, depending on the severity of the detected situation, various interventions can be undertaken at the first sign of danger or in view of its development. The following interventions can be considered to manage the emergency:

- Increased monitoring
- Technical measures
- Operational measures
- Precautionary drawdown (except in case of flood)
- Evacuation of local communities

To go into more detail, as soon as a specific condition arises, for example, a clear abnormality in the behavior of the dam or a worrying observation, it is a good idea in the first instance to increase surveillance by carrying out increased analyses of the situation. It is up to experts and experienced engineers, possibly in conjunction with other specialists (glaciologists, geometricians), to assess the level of danger and give advice on the technical and operational measures that should be taken. Assessment criteria for operating staff can also be established.

If the situation can be controlled, various technical measures (increased monitoring, carrying out works to prevent further damage, etc.) and operational measures (limited operating level or precautionary partial drawdown) may be taken.

As soon as the situation becomes alarming, that is, control of the situation is no longer guaranteed, developments must be followed on-site by operating staff. A precautionary partial drawdown (except in cases of flood) or minimum operating level is one protective measure. Thought can be given to a precautionary and partial evacuation of local communities.

When the level of danger is extreme, because an uncontrolled release of water is possible from the dam or is already occurring, the alarm must be set off and local communities must at all costs be evacuated. This release may be due to partial or total dam failure, or to a landslide or rockfall that brings a large volume of material into the reservoir and causes an impulse wave to flow over the crest of the dam.

Finally, if analyses demonstrate that there is no danger for communities or for the safety of the dam, the level of danger can lower or return to normal, but this may not necessarily mean that the increased level of monitoring should come to an end.

5.4.4 Planning evacuation measures

The first stage involves determining the area that could potentially be submerged in the case of total and sudden dam failure with a full reservoir (Pougatsch et al., 1998; Müller, 2001). Guideline documents from the Swiss Federal Office of Energy SFOE (SFOE, 2014b and 2014c), based on the Beffa (2000) and CTGREF (1978) publications, outline simple procedures for estimating the flood wave. Based on the results of these calculations, a flood map can be established that allows the relevant authorities (cantons, communes) to draw up an evacuation plan. With the help of this document, the alarms sirens, observation stations, and the water alarm center can be located, for which the National Emergency Operations Centre (NEOC) sets the requirements (OCENAL, 2007).

It should be noted that in Switzerland flood wave calculations had already been made by the end of the Second World War following the bombing of the Ruhr dams (Figure 5.16). These early calculations were made by hand using methods published in 1945 by G. De Marchi. Later, the possibilities provided by electronic calculations enabled the Laboratory of Hydraulics, Hydrology, and Glaciology at the ETH Zurich to develop more sophisticated models with 1 or 2 dimensions. Today, there is a wide range of software available, developed in Europe and the United States. The ICOLD bulletin 111 "Dam Break Flood Analysis" (International Commission of Large Dams, Paris, 1998) provides a summary of the various programs available.

To evaluate the flood wave, the hypothesis of a sudden (that is, occurring at once) dam failure is assumed. The gradual formation of a breach, with either a gradual failure or partial fault, is not assumed. As for the formation of the breach, a distinction is made between concrete and embankment dams (Table 5.20, Figure 5.21).

Table 5.20 Assumed failure mechanisms for calculating the flood wave following dam failure

Type of dam	Assumed failure mechanism
Arch dam	Total dam failure
Gravity dam	Total dam failure (possibly over a width of several blocks)
Embankment dam	Formation of a breach, as a rule, of trapezoidal shape, with a base equal to 2 times the hydrostatic height and with a 1:1 gradient of the lateral slopes; the width must not be larger than the dam crest itself

As for the calculations, these can be performed at 1D or 2D depending on the topography of the potential floodplain downstream of the dam. If the shape of the affected area is clearly defined and the water release advances in one specific direction, the use of a 1D model may be appropriate. The relevant stretch of river is modeled with a series of cross-sections along the axis of the drainage area. However, if within the area of flow, the cross-sections are irregular and the direction of flow is not totally identical, 2D modeling is preferable, as it can be applied to an area of any shape; it is particularly well-suited for lowland areas, where the flow can go in several directions. The region in question is discretized into several cells within which flow is determined in two directions. The spatial and temporal distribution of the flood wave can be established for any terrain elevation.

The results from flood wave calculations allow a map to be drawn up that indicates the limits of the flood zone, which are defined by the energy line (water height h_w + kinetic energy $v^2/2g$) (Figure 5.22). If the flow of the flood wave remains confined to the riverbed, only the water level is considered.

New models allow the effects of erosion and sediment transport and deposit to be taken into account, which can be helpful for calculations for a dike created by a landslide or a sediment trap. With the existing calculation possibilities, digital terrain models can serve as a topographical basis. However, these models sometimes require specialized work to define certain areas by refining the data. These models also allow flood maps to be drawn up for planning the evacuation of downstream populations.

Figure 5.21 Example of a trapezoidal breach following the failure of an embankment dam (Courtesy Müller, 2001)

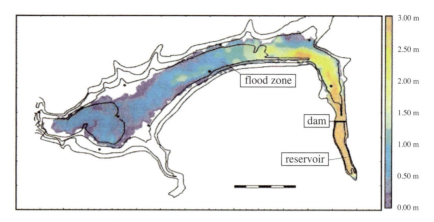

Figure 5.22 Illustration of the results of flood wave calculations (Courtesy Müller, 2001)

5.4.5 Alarm systems

To alert local communities, the use of sirens is recommended. The use of both stationary and mobile sirens increases the range of the alarm across a wider area. The sounding of an alarm indicates to the public that certain instructions are to be followed. In Switzerland, there are two types of sirens; general alarm sirens and flood alert sirens have different tones and signals (Figure 5.23). General alarm sirens (installed across the entire country) indicate to people that they must listen to the radio and follow the instructions; flood alert sirens are installed in a so-called "near zone" and indicate that people must leave the flood zone immediately due to the threat or arrival of an uncontrolled release of a large quantity of water (OProP, 2020; OAL, 2018). New "combined" sirens transmit both the general alert and the flood alert.

Figure 5.23 Flood alert sirens (Courtesy Federal Office for Civil Protection)

All alarm systems must allow those involved in emergency management to communicate among themselves and pass on information. As such, voice communication is vital and must be maintained at all times.

Alarms are worthless if the public does not know how to behave in an emergency situation. Pre-established evacuation plans and the distribution of memory aids provide general guidance. Radio stations can be an additional means for broadcasting instructions and information to the public about how the situation is developing at the dam site.

5.4.6 General organization

In Switzerland, the Federal Office for Civil Protection has installed alarm sirens to warn members of the public. Sirens are tested once a year on a set date. The National Emergency Operations Center (NEOC) is the specific group responsible for managing tasks during an emergency. It provides the link between the cantons and the Confederation. The NEOC has flood maps and emergency response files from operators. Other departments of the Confederation are also informed if the level of danger is high (ABCN, 2010).

It is up to cantons and communes to prepare evacuation plans, memory aids and instructions for the public, shelters, traffic management, broadcasting material, and any other necessary equipment, as well as to ensure that the sirens for which they are responsible work (general alarm). When necessary, they must sound the alarm, and inform the public about how the situation is developing, while reminding them of the meaning of the different tones and what to do.

The operator must draw up a set of emergency regulations. It must also organize and train its staff to deal with a possible emergency, by providing the necessary instructions (SFOE, 2015c; SFOE, 2015d). The instructions are to be specified in a special file (§ 33.4.1).

6. Hazards and Damage Affecting Dams and Their Foundations

Illsee gravity dam, Switzerland, height 25 m, 1927, year commissioned (1st stage/2nd stage) 1927/1943
(Courtesy Argessa AG)

6.1 Description of Weak Points

First of all, it is important to note that the foundations are a vital element as they serve as the base for any dam structure. They must firstly be able to take up the force transferred by the dam and secondly serve as a barrier to the water against the valley flanks and underneath the dam. The rock mass is not homogeneous and instead constitutes a discontinuous environment characterized by the fragmentation and anisotropy of its structure. It can be differentiated by its cracking (stratification joints, diaclase, faults), deformation, and permeability. In the natural environment, the direction and strike and dip of layers, the orientation of fractures, the type of material between the layers, and the cracks and faults should all be taken into account. The foundations may have potential slip surfaces due to insufficient shear strength or may be the path of internal erosion, which can cause intense seepage and particle transport. During operation, a large crack may form at the upstream heel of concrete dams, which results either in an increase in flow collected by the drainage system or by an increase in piezometric levels. Finally, when abutments are being consolidated, the strength and holding force over time of rock anchors must be monitored, as they can corrode and suffer from loss of tension.

With regard to concrete dams, exceptional deformations may appear following foundation movement, temperatures above or below seasonal norms, shrinkage phenomena, creep, swelling (alkali-aggregate reaction, AAR), or exceptional rises in water level. The stability of gravity dams may be affected by an increase in uplift due to a poor drainage system or in cases of cracking on the upstream face due to tensile stress. The stability of abutments is a vulnerable point for arch dams. Following a major earthquake, a gravity dam may experience more or less severe cracking in the zone of the crest, more particularly along planes of weak resistance such as the surfaces of horizontal lifts; in addition, openings in joints and movement relative to blocks may appear. It must be noted that no gravity dam has suffered major damage from a seismic shock. The appearance of cracks and/or opening of joints may occur, however. For RCC dams, cracks can form along horizontal lift joints (Wieland, 2003).

The body of embankment dams may be endangered by internal erosion depending on the quality and condition of placed material (poor zoning, cracking, hydraulic fractures). Internal erosion occurs when particles are detached from their original location and transported by seepage through the infill and its foundation. The transport of materials is either concentrated along a pipe (piping) or diffused throughout the intergranular space (suffusion). Piping is dangerous and quickly leads to dam failure. Although suffusion develops slowly, it modifies the permeability of the surrounding environment and may also lead to the failure of the dam structure. This erosion may also appear along the contact surface. The clogging of drains may cause an increase in pore pressure. The surface of slopes may be the site of swelling, collapse, or local slope instability (FRCOLD, 1997; ICOLD, 2017).

After a strong earthquake, the following effects have been noted in embankment dams: longitudinal and transverse cracks of greater or lesser depth, crest settlement, upstream movement, slope instability due to a loss in shear resistance, and an increase in pore pressure (phenomenon of the liquefaction of saturated soil without cohesion). Settlement can diminish the freeboard and lead to overtopping of the embankment dam in cases of extreme floods.

For embankment dams with upstream sealing, monitoring must be carried out on the behavior and condition of asphalt facing (the appearance of fine cracks, faiençage, creep, or blisters in a double-layer lining) and in concrete (cracking, leakage at damaged joints).

The earth/rockfill-concrete structure interface is also a vulnerable zone, which may experience differential settlement and localized seepage. The development of internal erosion can also be of concern.

In the case of a flood, a failure in electronic and mechanical equipment, power supply (or even an interruption in telecommunications) may result in the loss of gate operation control and an uncontrolled increase in water level, leading to overtopping.

Table 6.1 summarizes the principal hazards and damage that may affect dams and their foundations.

Table 6.1 Examples of hazards and damage affecting dams and their foundations

Concrete dam	Embankment dam	Foundations	Reservoir
Exceptional increase in water level	Exceptional increase in water level	Exceptional increase in water level	
Exceptional deformation Irreversible deformation Differential displacement of blocks Displacement of abutments	Exceptional deformation Settlement of the dam, foundations (decrease in freeboard) Differential settlement	Exceptional deformation Deformation of abutments Differential displacement of abutments Settlement	
Damage and deterioration of dam body Cracking Loss of section Swelling	Damage and deterioration of dam body Wave action Formation of burrows, animal shelters Development of vegetation (trees, plants)	Defective sealing system Degradation of grout cut-off	Reservoir leakage
Increased seepage, water circulation Ineffective or defective drainage system (drain clogging) Increase in uplift	Increased seepage, water circulation Ineffective or defective drainage system (drain clogging) Increase in pore pressure (loss of shear strength)	Increased seepage, water circulation Ineffective or defective drainage system (drain clogging) Increase in uplift and pore pressure	Emergence of springs downstream
	Internal erosion in dam body (piping)	Internal erosion (piping)	
Overtopping External erosion (downstream face)	Overtopping Wave action External erosion (damage to faces and toe of dam)	Scouring	Landslides Rockfalls Ice falls Avalanches
Instability (risk of sliding, overturning)	Slope instability Local subsidence	Potential sliding surface	Landslides
Cracking at the concrete-rock contact point of the upstream heel		Cracking at the concrete-rock contact point of the upstream heel	
Blockage of a discharge structure (bottom outlet, spillway) Malfunction/blockage of release structure	Blockage of a discharge structure (bottom outlet, spillway) Malfunction/blockage of release structure		Sedimentation Sediment deposits Arrival of floating debris, trees Ice run

6.2 Description of Causes

The causes of hazards and damage (Table 6.2) may be varied and connected to the design of the dam itself. Weaknesses should be identified during exploratory works. Geological issues that were not identified or underestimated can lead to an unfavorable setting out of the works and an inefficient treatment of foundations (defective execution of consolidation grouting or grout curtain, insufficient drainage, etc.). Undercalculation may be the result of incorrect assumptions, for example regarding load (distribution of uplift under a gravity dam), flood flow (risk of overtopping with a deterioration in the downstream face or scouring at the downstream toe). Finally, in terms of execution, an inappropriate choice or incorrect use of materials may be the cause of shortcomings.

Table 6.2 List of causes of hazards and damage affecting dams and their foundations (from SwissCoD, 2000)

Initiating events	Special events
Natural events	Power supply failure
Floods	Mechanical equipment failure
Earthquakes	Communication failure
Avalanches	**Project characteristics**
Debris flow	Shortcomings in the project
Landslides	Insufficient exploratory works
Rockfalls	Undersizing
Ice falls	Incorrect assumptions
Sediment accumulation	Execution errors
Caused events (following a third-party intervention)	Incorrect use of material
Underground work	Inadequate checks during construction
Sabotage	Dam condition
Explosion	Poor concrete quality
Bombing	Major cracking
Aircraft crash	**Operation**
Other effects	Operating error
Scouring	Rapid drawdown
Lightning (effect on electronic equipment)	Floating debris
	Abundant vegetation
	Digging of shelters by animals

While the dam is operating, the conditions of static, dynamic, or even thermal loading and the shrinkage phenomenon will lead to more or less major cracks, which must be detected and monitored in some cases. Over time, the dam condition may become a source of concern, due to the decrease in material quality, the appearance of swelling in concrete, and a deterioration in the drainage system or the grout curtain. Wear and tear of electrical and mechanical parts and the increase in sediment and floating debris may make water release structures (at the top or the bottom of the dam structure) inoperative.

Landslides, rock or ice falls, and avalanches into the reservoir can create, depending on the water level, an impulse wave capable of overtopping and causing additional damage (VAW-ETHZ, 2009). The case of the Vajont dam (Italy) is described below.

And finally, dam deterioration can be due to external interventions (works taking place in the surrounding environs, sabotage and vandalism, explosion and bombing).

It should be mentioned that some of these listed causes can be attributed to human errors or intervention (for example, design error, poor management of water level, sabotage).

Natural events such as floods and earthquakes, as well as huge landslides that displace material into the reservoir should not be forgotten either, as previously described.

Landslide on the left bank of the Vajont dam (Italy)
The Vajont arch dam is 261.6 m high from the riverbed and 190.15 m wide along the dam crest. The crest is 3.4 m thick at the top and 22.11 m at the bottom. The dam sits on a "pulvino" joint. The dam is built on a foundation of middle and late Jurassic rock formations in a narrow gorge in the Dolomites.

*The impoundment of the reservoir began in March 1960. In fall 1960, creep across a large area of the left bank near the dam was observed, and a superficial rock fall of 700,000 m^3 occurred. The level of the reservoir was reduced and a few weeks later, the movement stopped. In the summer of 1962, filling resumed after the completion of a by-pass tunnel around the area of the landslide. The movement was reactivated once the water in the reservoir reached a level 25 m higher than that previously reached. The water level was again reduced, and the landslides stopped some weeks later. In the summer of 1963, impoundment and the movement began again. In October 1963, the slippage transformed into the sudden slide of a mass of rock of some 250 million m^3. It displaced 40 million m^3 of water, which overtopped the dam in a 260-meter-high wave that plunged into the valley below and destroyed four villages. The dam did not burst (*Water Power & Dam Construction, *1985).*

7. Types of Failure

Räterichsboden gravity dam in Switzerland, height 92 m, year commissioned 1950
(Courtesy Kraftwerke Oberhasli-KWO)

7.1 General Information

The International Commission on Large Dams (ICOLD) defines failure as the *collapse or movement of part of a dam or its foundation, so that the dam cannot retain water. The result is generally the release of a large volume of water, leading to risks for people and property downstream.*

The conclusions of the ICOLD Bulletin 99 (ICOLD, 1995)[1] state that the percentage of failure has decreased since 1951. The failure rate from this year onward is at 0.5% whereas previously it had been at 2.2%, which suggests that the progress made in construction techniques is making modern dams more reliable. It has also been observed that failures occur in the first 10 years of dam operation, and primarily in the first year. One in two failures occurs during the first impoundment of the dam, the risk being not as great for concrete dams as for rockfill dams. Additionally, failures mostly affect small works, which are the majority of constructed dams.

The risk of the unexpected and sudden failure of a dam is low. Possible failure is related to the continuing deterioration of the dam, which should be rapidly detected through permanent and vigilant monitoring exercises.

The failure phenomenon depends on the type of dam. For embankment dams, it may be gradual, through retrogressive erosion. For concrete dams, the overturning or sliding of one or several blocks may occur suddenly; however, there are generally warning signs. Failures may emerge out of issues related to the project (faulty design and construction, insufficient preliminary studies and execution checks), technical difficulties (major damage to the body of the dam, lack of dam and foundation stability, failure of water release structures), natural hazards (floods, earthquakes, landslide into the reservoir), and specific operating conditions (uncontrolled rise in water level, insufficient surveillance and maintenance).

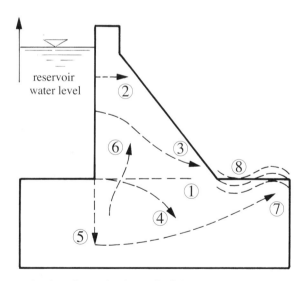

Figure 7.1 Failure mechanisms in gravity dams (①, ② horizontal cracks; ③, ④ curvilinear cracks; ⑤ vertical foundation cracks; ⑥ extension of existing foundation discontinuity into dam body; ⑦ sliding plane in the foundation; ⑧ buckling failure of thin bedded strata. Source: *Manuel d'utilisation "CADAM"* (Léger et al., 1997).

---

[1] This Bulletin has since been updated (Final Draft December 2019).

7.2 Types of Failure in Concrete Dams

From an analysis of failures in concrete dams, it is apparent that issues relating to the foundation are the most frequent cause of failure. The collapse of the dam is usually caused by internal erosion and insufficient shear strength in the foundation. It can also be caused by excessive deformation in the foundation.

The failure of a gravity dam (Figure 7.1) may be related to insufficient stability following a degeneration in properties of some materials, irreversible displacements, pressure, and excessive stress, or a reduction in section through the presence of cracks. Sliding may also occur at the dam-foundation interface. Uplift also plays a considerable role, as it can act upon upstream cracks provoked by tensile stress and become excessive in cases of inadequate or nonexistent drainage. A rise in water level well above permissible values modifies the forces due to water pressure and uplift, therefore threatening the overall stability of the structure.

For arch dams, failure may be due to problems in the foundation and abutments (insufficient shear strength, uplift, internal and external erosion), cracks in the structure, and deterioration of the concrete. The most well-known case is the failure of the Malpasset arch dam (France) in 1959.

Finally, major external erosion at the downstream toe, such as scouring caused by impacting jets, may be the cause of instability.

Failure of the Malpasset arch dam (France)

The Malpasset arch dam was 66 m high and had a crest length of 222 m. The crest was 1.5 m thick at the top and 6.78 m at the bottom.

On the evening of December 2, 1959, the Malpasset dam was swept away as the water level neared the spillway weir. This was during the first impoundment of the dam. Practically the whole arch collapsed at once; only a part of the dam on the right bank and the base of the central section remained. The dam was completely destroyed on the left bank, and a large cavity in a dihedral form appeared in the rock. The abutment shifted 2 meters horizontally. The remaining part rotated as a whole around the edge of the right bank with a maximum displacement of 80 cm. Enormous blocks of concrete, to which the rock was still attached, were found in the valley below.

The dam failure was the result of an unexpected combination of reasons, including some that had not previously been envisaged. The Malpasset site was characterized by a 45°-angle downstream fault directly under the dam and by a large number of possibly undetectable slip surfaces descending downstream. The uplift caused by the reservoir water acting against a large surface of the deep rock foundation developed sufficient force to uplift the dam and the wedge of rock supporting it. The uplift developed due to the weakness of the rock and the behavior of moderately impermeable gneiss in normal conditions. Gneiss becomes impermeable when it is compressed, acting here like a watertight cut-off in the foundation against which water infiltration from the reservoir accumulated. Under the effect of hydrostatic pressure from the low cohesion between the potential slip surfaces, the foundation shifted slightly upstream, causing a crack in which uplift was able to occur due to the impermeability of the downstream fault. This fault and the upstream crack delineated a rocky wedge in the foundation; the uplift raised the block and dam so much, it were as if the foundation had exploded (Source: Water Power & Dam Construction, *1985).*

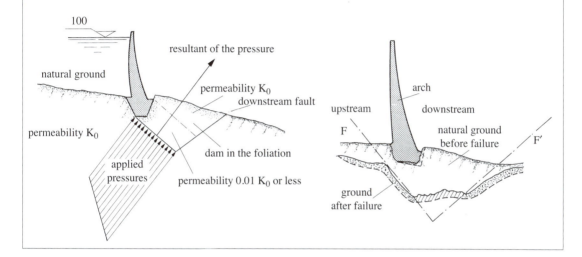

Failure of the Shih Kang gravity buttress dam (Taiwan)

The dam is located over a fault, but the full extent of the fault had not been detected during the design phase. This active fault was the cause of the Chi-Chi earthquake on September 21, 1999, which had a magnitude of 7.3. A vertical differential displacement of 9 m was measured. The part of the dam that was destroyed was located right on the fault, and only two blocks were affected.

7.3 Types of Failure in Embankment Dams

According to statistics from the ICOLD, overtopping is the main cause of failure for embankment dams. This overflowing via the crest may result from an uncontrolled increase in water level, poor management of water inflow into the reservoir (Zeyzoun embankment dam, Syria), or a wave caused by the arrival of a large volume of material into the reservoir. A decrease in freeboard through settlement following an earthquake may also favor this flow over the crest.

The second case is related to internal erosion in the earthfill and its foundations, resulting in many incidents and cases of failure. Four paths to failure can be identified: through the earthfill, through the foundation, along a buried pipe, and at the interface of an adjoining structure (ICOLD, 2017). Erosion occurs through the migration of particles along preferential paths in the body of the dam or in the foundation. Internal erosion is caused by rapid seepage or poorly designed filters. The channel created by the leak grows (this phenomenon is called piping) and can eventually lead to the destruction of the dam. Internal erosion[2] can occur during the first impoundment or the first five years of operation (Teton embankment dam, USA).

The third case deals with external erosion causing a deterioration in the crest and slope surfaces due to overtopping. At the crest, the failure mechanism begins at a downstream edge and gradually forms a breach. This flow can also cause damage at the toe of the slope.

A final case of failure concerns the mass sliding of a large volume of the body of the dam and/or its foundation, which shifts along a failure surface. This mass movement is generally initiated by the excessive development of pore pressure, which creates a decrease in effective stress leading in extreme cases to shearing displacement of the fill and/or the foundation (Lower San Fernando embankment dam, USA, under seismic stress).

The causes of cases involving internal and external erosion, as well as mass sliding, are related to dam geotechnics and can occur in a combination.

Failure of the Teton embankment dam (USA)
The 95.1-meter-high Teton embankment dam was made of earthfill and had a core of impermeable clay. Dam failure occurred on June 5, 1976 during the first impoundment of the reservoir.

On June 3, springs were observed at the level of the riverbed approximately 450 m downstream of the dam. On the following day, other springs appeared at the downstream toe of the dam. On June 5, seepage at an initial rate of 0.5 to 0.8 m³/s began to develop on the downstream face of the dam, about 40 m below the crest. The water, flowing at a rate of 0.05 m³/s, emerged from the right side of the dam near the contact point between the dam and the abutment. Dirty water was then observed leaking from the slope near the toe of the dam at the rate of 0.7 m³/s. The flow in the upper part continued to increase. Later a large sinkhole formed and eroded the dam, and finally a breach appeared at the level of the crest. The dam completely failed.

---

[2] Section 26.8 covers the process of internal erosion and its consequences.

Failure of the Lower San Fernando embankment dam (Van Norman) USA

On the morning of February 9, 1971, a severe earthquake measuring 6.6 magnitude generated by a 20-kilometer fault shook the San Fernando Valley, where the Upper and Lower San Fernando embankment dams were built. The upper dam showed signs of cracking and irreversible deformation, while the upstream shoulder of the lower dam slipped 46 m into the reservoir. A residual freeboard of 1.4 m of material, with cracks in some places, remained. Investigations revealed that the liquefaction of the lower part of the upstream shoulder was responsible for the slip. The maximum displacement of the upstream heel was 61 m. Analysis of the seismographic data showed a maximum acceleration of between 0.55 and 0.6 g. The slip appeared 20 to 30 seconds after the shocks had stopped. The upstream shoulder slid for 40 s at a speed of 1.5 m/s; the reservoir was not full (Source: Pathologie des barrages: de l'analyse au diagnostic. *FRCOLD Bulletin, 1997).*

Case of the Zeyzoun embankment dam (Syria)

The Zeyzoun embankment dam, 43 m high and with a length of nearly 5 km, was an earth-rock dam with a central core.

Its reservoir had a capacity of 71 million m³ and the water, used for irrigation, was pumped in. On June 4, 2002, an 80-meter-wide breach was caused by overtopping due to an overfilling of the reservoir as the inflow was not stopped. There was a total stored volume of 82 million m³ of water at the time of the failure.

8. Protection Measures

Robiei gravity dam in Switzerland, height 68 m, year commissioned 1967 (Courtesy © Swiss Air Force)

To avoid any issues occurring during dam operation, preventive actions can be taken during the project design stage and while construction is being carried out. Interventions to improve the structural safety of the dam can be carried out during operation.

At the project stage, in-depth geological studies must be undertaken. Borehole surveys and tests must be sufficient to provide comprehensive knowledge of the foundation, as well as of its geological and geotechnical characteristics (strength, permeability, dissolution, etc.). On the basis of these results, the dam can be set out and subsoil treatment defined (grout, drainage). Care must be taken over upstream watertightness and downstream permeability to avoid underground water circulation that could scour the foundations or allow seepage. The general stability of the abutments must be examined and any eventual strengthening measures, such as grouting or anchors, must be implemented.

When designing the structures, the loads applied to the dam must be clearly defined, in particular resulting from uplift due to seepage as well as from the effects of natural hazards (earthquakes, floods). The stability of the overall structures must be checked against previously determined criteria.

For concrete dams, tests are required to identify the mechanical characteristics of materials (concrete-rock interface, lift joints, and discontinuities in the rock mass) and to establish allowable stresses. For embankment dams, geotechnical tests must be carried out, as well as placement and compaction trials for earth- and rockfill.

During the design phase of the dam, geometric data such as crest width, freeboard height, and width of spillway openings must be established. Specific and constructive measures should be planned for the construction and maintenance of water release structures, drainage and sealing works, as well as the selected approach for joints (joint grouting, shear boxes, keyways). This is also the time to decide on the use of special materials. For embankment dams, the slope protection system must not hinder visual observation (evidence of settlement, instability, cracks, or seepage). Steps should also be taken to prevent animals from digging holes and runs, to make sure drainage systems are not blocked by roots, and to avoid deterioration of sealing systems.[1] Finally, the monitoring system is also a key part of the design project.

During construction, quality control of material (aggregate and concrete) and rockfill placement tests must be carried out. It is also important to do a systematic survey of the foundation geology, as well as updating plans and drawings so that all information concerning the dam and its structures is reliable and correct in cases of unexpected events.

During operation, maintenance, measurements, and checks needs to be organized and carried out.[2] Periodic safety analyses are strongly recommended to ensure that the dam meets the most recent safety requirements. If necessary, strengthening measures may be undertaken, such as reinforcing the stability with anchors, creating or improving the drainage system, strengthening concrete or foundations by grouting, improving the grout curtain, constructing a diaphragm wall (cut-off wall) or installing a membrane. After the 1978 floods in Switzerland, new criteria for flood safety were established. Following an assessment on the basis of these new assumptions, rehabilitation works had to be carried out on some twenty dams. Modifying the spillway or dam crest and building a parapet were the principal actions taken.

An earthquake safety assessment of more than 1,200 dams carried out in the United States showed that strengthening measures were necessary for more than 100 of them (primarily embankment dams) (Babbitt, 2003). Operating restrictions were imposed as an initial measure until the upgrading works had been completed. The improvements included structural strengthening (construction of berms, buttresses,

---

[1] These various elements are covered in more detail in the relevant chapters.
[2] These tasks are described in detail in Part VIII, Operation.

foundation treatment), rehabilitating water release structures, increasing freeboard, widening the crest, flattening slopes, and replacing inappropriate foundation materials.

A verification campaign for earthquake safety in dams falling under the jurisdiction of the Swiss Confederation was launched in 2003. Checks were made on the basis of guidelines relating to dam safety, particularly with regard to earthquakes (SFOE (SFWG), 2003; SFOE, 2016). The whole dam structure and reservoir had to be verified (dam, appurtenant structures, and reservoir area). The operator had to demonstrate that no failure leading to an uncontrolled release of water from the reservoir could occur. If that was not the case, constructive or operational measures had to be taken. Following these verification studies, strengthening works were carried out on various dams (Darbre et al., 2018).

9. Risk Analysis

Panix gravity dam in Switzerland, height 53 m, year commissioned 1989 (Courtesy K. Maj Nigg)

9.1 Introduction

Risk analysis was first employed in industrial and complex mechanical systems sectors, particularly aeronautics and the nuclear industry. In the field of risk management, it plays an important part in reducing potentially unfavorable effects following an accident, without, however, entirely eliminating them. This approach provides tools for assessing and preventing the risks inherent in the use of a specific system (means of production, product, or structural components), therefore improving its level of safety. Furthermore, identifying weaknesses in these elements enriches databases that are useful for the statistical analysis required in feasibility studies.

Many countries have issued guidelines and recommendations for regulating such risk analyses.[1] In Switzerland, the Environmental Protection Act (LPE) and the Major Accidents Ordinance (OPAM) require any person who operates or intends to operate installations that, in exceptional circumstances, could seriously damage people or the environment to take the measures required to protect members of the public and the environment. The operator is responsible for technical safety measures, as well as monitoring the installation and organizing an alarm system. The OPAM goes on to define risk, as well as general and specific safety measures. Although dams are not expressly affected by these legal documents, it should be noted that the concept of dam safety and how it is organized also includes the abovementioned elements.

In the civil engineering domain, and in particular that of dams, a risk-based approach for assessing safety is relatively recent. It must be emphasized that each dam is a unique structure and can be considered a prototype, unlike elements produced in a factory. Certain countries, including Canada, South Africa, Australia, and some Scandinavian countries, have played a leading role in this risk-based approach. Misgivings concerning such risk analyses are due to the fact that deterministic concepts have already proven their efficacy. However, risk analysis can be a complementary tool to support traditional methods, without actually replacing them. It can make a valuable contribution to the diagnosis of events and their consequences (cause and consequence studies). For its proponents, this risk-based approach helps in making decisions to protect the public from the effects of dam failure. It can set priorities, provide a breakdown of necessary investment, and justify measures aimed at reducing risk (Darbre, 1998a, b, 2000).

The experience and judgement of experts from the field of dams in particular is essential for evaluating risk when introducing a probability analysis into the assessment of failure processes. Throughout the various stages of analysis, it is important to have in-depth knowledge of the operation and behavior of dams and reservoirs and their surrounding environment.

The aim of this chapter is to present the core elements upon which risk analysis is founded.[2]

9.2 Fields of Application for Dams

Risk analysis can be carried out at any stage of the design (concept design, detailed design), construction, or operation of dams. It should be noted that over time, the condition, impacts, dangers, and criteria may evolve, which can lead to further analyses and a revision in the implemented programs.

---

[1] For example, ANCOLD, Guidelines on Risk Assessment, 1994; Canadian Standard Association, Risk Analysis Requirements and Guidelines, 1991; USBR, Safety Risk Analysis Methodology, 1999.

[2] The information in this chapter is primarily based on the internal report from the "Risk Analysis" working group from the Swiss Dam Committee and the General Report on Question 76 from the 20th Congress on Large Dams, Beijing 2000 (Kreuzer, 2000).

Indeed, risk analysis goes beyond simply assessing dam safety. In addition to the classical fields of geotechnics, floods, and seismology, there are other domains where this analysis can be employed in dam engineering (Kreuzer, 2000), in particular:
- Helping decide on the cost/benefit ratio for new dams or refurbishment work on existing dams
- Classifying hazards for existing dams
- Preparing alert and emergency plans
- Defining the public's risk acceptance criteria
- Drafting recommendations and regulations

The reason for these diverse applications is clear, as dams are high-risk structures. Risk assessment is a useful process for taking uncertainty into account and evaluating impacts on society and the environment.

9.3 Definition of Risk

Risk combines the likelihood and severity of a hazardous event in regard to human life, property, and the environment. Risk is assessed based on the mathematical prediction of the consequences of such an event occurring. Risk R can be expressed as the probability P of an event occurring by measuring the magnitude of consequences C (threat to human lives; social, economic, and environmental loss; damage), or

$$R = P \times C.$$

This formula can also be written as $R = P \times C^\alpha$, where α can be equal to 1 (or higher if it is considered a wise decision to increase the risk, such as in cases where the event may have particularly serious consequences).

The total risk of a system is equal to the sum of all individual risks related to primary causes (initiating events).

Some different concepts have also been put forward to describe risk. For example, by combining risk with specific conditions of load (such as floods, earthquakes, or static loading), risk can be expressed as:

$$\text{Risk} = (p_1 \times p_2) \times (\text{consequences}),$$

where

p_1 = the probability of the initiating event occurring and p_2 = probability of failure if the initiating event occurs.

As part of risk management and given that no intervention is possible to modify the probability of a natural event occurring, the likelihood of negative effects occurring (for example, dam failure due to an extreme flood), and well as their consequences, should be minimized by implementing appropriate measures that mitigate the damaging effects (for example, emergency planning).

9.4 Theoretical Basis for Risk Analysis

9.4.1 General information

There is no single process for carrying out a risk analysis. However, the selected approach must enable the risks to be identified, as well as related uncertainties. The latter concern hydrology, seismology, geology, loads, construction quality, construction materials, aging, and human factors. Be that as it may, it is a question of determining which probable disastrous events may occur, estimating their likelihood of occurrence (admitting the possibility of total or partial failure), and evaluating the consequences (magnitude of damage). The necessary conclusions may then be drawn and a decision made on the means to implement for minimizing risk (Figure 9.1).

Figure 9.2 and Table 9.3 outline and define the terminology used in risk management (Kreuzer, 2000).

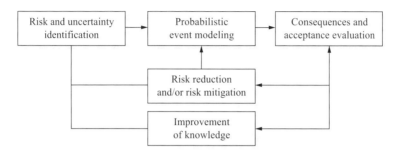

Figure 9.1 General framework for risk management (from Faber and Stewart, 2001)

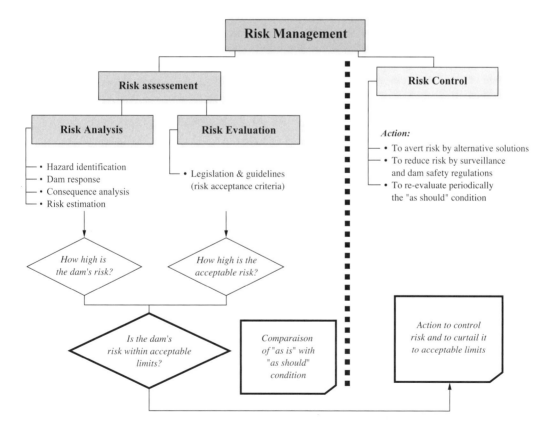

Figure 9.2 Example of terminology used in risk management (from Kreuzer, 2000)

Table 9.3 Definitions of terms used in risk analysis (Kreuzer, 2000)

Acceptable risk	Description of level of risk deemed acceptable or tolerable
Bayes theorem	A formal framework in probability analysis to combine intuitive, subjective judgement with observed data to obtain a balanced estimation
Bounded distribution	Probability density functions (pdf) with finite upper and/or lower tails
Conditional probability	A probability given a prior event has occurred
Consequence	The outcome or result of a risk being realized; impacts in the downstream and upstream areas of a dam resulting from failure or incident
Dam failure	Failure or movement in a dam, part of a dam, or its foundation following a static or dynamic load such that the dam is no longer able to retain the water
Deterministic approach	An approach that leads to reasonably clear-cut solutions based on prescriptive rules, without considering uncertainties in the analytical process; the factor of safety is a typical result of this
Event tree analysis	A technique that describes the possible outcomes (range and sequence) that may arise from an initiating event
Extreme-value statistics	Statistical inference for estimating the min./max. of data sets with extreme value distributions
Fault tree analysis	A graphic illustration of a system (scenario), which provides a way to follow the logical flow of events resulting from a system failure; the fault tree commences with the failure event
Follow up	Process of evaluating, at regular intervals, the impact of decisions made in relation to risk
Hazard	That which has the potential for creating adverse consequences
Hazard identification	Recognizing the existence of a hazard and defining its characteristics
Maintenance	Action to ensure that dams are in good condition and that electrical and mechanical installations operate as they should, allowing any anomaly to be detected and all work to be carried out appropriately
Monte Carlo simulation	A procedure that seeks to simulate stochastic processes by the random selection of values in proportion to known probability density functions
Options analysis	Searching for solutions for the management of risk
Portfolio risk analysis	Risk assessment for a group of dams belonging to a given operator
Probability density function (pdf)	Shows the values of intensity of probability of a continuous random variable; this function is not a probability, the area below the pdf is the probability
Quantitative risk assessment	Technique of assessing the probability of an unwanted event and its measurable consequences in terms of loss of life and socio-economic damage
Reliability analysis	Analytical process: loads and resistance are modeled as random variables in order to consider uncertainty in the outcome; a reliability index is the typical output

Table 9.3 Definitions of terms used in risk analysis (Kreuzer, 2000) (cont.)

Risk	(1) Measure of the probability and severity of an adverse effect to human life, property, or the environment; risk is estimated by the mathematical expectation of the consequences of the adverse event occurring (i.e., the product of the probability of occurrence and the event multiplied by the consequences) (2) The likelihood or probability of adverse consequences; the downside of a gamble
Risk analysis (RA) based on indices	Estimating numerical values for the weighting of risk parameters, to be incorporated into a comparative decision-making process (for example, estimated values for dam condition indices, the critical level of hydrological or geological conditions of a dam site, reliability of gate operation, etc.)
Risk estimation	Process used to determine the level of risk
Risk evaluation	Step during which values and judgements enter the decision-making process by considering the importance of estimated risk and its social, economic, and environmental consequences, with the aim of identifying solutions for the management of risk
Risk management	A comprehensive dynamic strategy for the evaluation, treatment, and administration of risks that threaten well-being (Figure 9.1)
Socially acceptable risk	The socially acceptable level of risk in terms of events that impact on society at a community, regional, or national level
Statistical uncertainty (epistemic, random)	Uncertainty within the sample-to-sample variation of experimental observations and testing

9.4.2 Risk analysis outline

The process of analyzing risk includes several stages for determining if the risk associated with a certain activity is acceptable. It can also be used as a basis for decision making. As an example, Table 9.4 lists the main points generally covered in a risk analysis.

Table 9.4 Example of the flow of risk-based decision analysis (from Faber and Stewart, 2001)

Define the framework of risk analysis
- Define the context
- Define the system

Analyze risk
- Identify hazards and their sources (scenarios)
- Estimate risk
  - Analyze consequences and probability
  - Identify risk scenarios (sensitivity study)

Risk evaluation
- Risk acceptability
- Decision making/Options analysis

Risk control
- Risk treatment (risk management)
- Surveillance and monitoring

9.4.3 Defining a framework for analysis

Firstly the specific context within which the risk analysis will be carried out needs to be clearly defined. This occurs through the identification of issues and the determining of objectives. The audience for the results of the analysis needs to be determined, and the political, social, financial, and economic aspects that may influence the analysis process need to be evaluated. Defining acceptable risks based on human safety (injury, illness, or death), economic issues (destruction, loss of productivity), and environmental criteria (damage to the environment, impact on flora and fauna) is a vital phase in the risk analysis.

It is important to clearly outline the system, its characteristics (technical, environmental surroundings), and its organization in order to correctly understand how it operates and behaves. How the system is described will affect the level of detail of the analysis (Figure 9.5).

Portfolio level				
Dam A	Dam B	Dam C	Dam D	

Dam systems level				
Main dam	Counter dam	Spillway	Bottom outlet	Water intake
Penstock				Reservoir

Dam structure level		
Foundation	Concrete structure	Access
Power supply system		

Dam substructure level	
Drainage system	Rock/concrete interface
Dam specificities	Grout curtain	Monitoring system

Figure 9.5 Components of the dam system (from SwissCoD, 2003a)

9.4.4 Risk analysis

Identifying hazards and their sources

There is a specific identification stage that corresponds to the question "what might happen?" It outlines all the potential hazards and their sources, the consequences of which will probably be harmful for local communities the economy and the environment. The sources are connected to technical, natural, or human causes.[3] It should be noted that in civil engineering, human error may occur in the any of the stages of the dam: design, construction, operation, or maintenance. A hazard may lead to the total or partial destruction of the overall system, or of one of its components, or even a subsystem. It is advisable to draw up a list of all the plausible types of failure. The results of on-site inspections, which enable weak points to be identified, should not be overlooked. They should also be considered when selecting the primary causes. This approach means that appropriate conclusions can be applied for the rest of the analysis.

---

[3] The causes and types of failure are described in Chapter 7.

Initially, a preliminary analysis of the dangers should be envisaged, identifying in a simple manner the events leading to critical situations, their possible causes, and finally their effects. Early measurements could also be considered (Table 9.6).

Table 9.6 Example of preliminary hazard analysis (from Jenssen, 1997)

Hazard	Causes	Consequences	Actions
Water level too high	Defect in gate operation	Access to gate chamber blocked Overtopping Dam failure	Backup mechanisms for gate operation
	Blocked spillway opening(s)		Remove floating debris Widen opening(s)
	Water inflow greater than spillway capacity		Increase discharge capacity of spillway

The following stage involves researching and analyzing various scenarios using a logical system. Several techniques for identifying types of faults and their effects have been developed, which also provide useful information for the creation of logic trees.[4] Finally, at this stage of the study, it is important to utilize past experience, particularly by calling on the judgement of experts and by bringing together documents and information related to events that led to accidents, or even failure.

Using logic trees to analyze faults and events
A fundamental element of risk analysis concerns the establishment of logic trees that relate initiating events (causes) to their consequences (events leading to a critical situation). This sequential approach makes it possible to identify a series of individual events that culminate in undesirable damage, or even failure. A likelihood of occurrence is assigned to each branch of the tree. These likelihoods are obtained by statistical analysis (objective probability) or by the judgement of an expert (subjective probability), or even a combination of the two. This evaluation work, both qualitative and quantitative, relies greatly on judgement (Salmon and Hartford, 1995).

Traditional risk analysis includes the following logic trees:
- Fault tree (causes)
- Event tree (consequences)
- Cause/consequence charts

The fault tree is a logical, one-directional flow chart (without loops or returns) that enables the user to go back to the root cause of events leading to dangerous situations (deductive conclusion). Departing from an undesirable event (*top event*), the process advances retroactively toward the causes (initiating events). While establishing a fault tree, symbols are used to describe events and logic gates. As seen in Figure 9.7, probabilities (p_{11}, p_{12}, and p_{13}) are first assigned to base events E_{11}, E_{12}, and E_{13} (initiating events), and then the likelihood of other events occurring is calculated sequentially depending on the logic gates.

---

[4] For example: *Failure Mode and Effect Analysis* (FMEA), *Failure Mode Effect and Criticality Analysis* (FMECA), *Hazard and Operability Studies* (HAZOP).

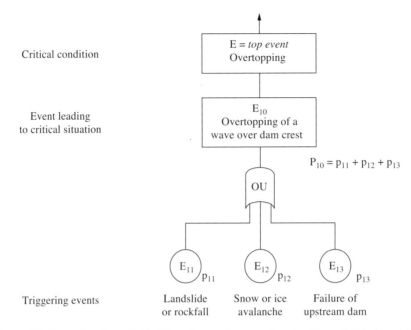

Figure 9.7 Example of a partial fault tree for overtopping (from Grütter and Schnitter, 1982)

Event trees enable all the possible consequences of precursor events to be established, which must then be identified (inductive approach). An initiating event may lead to a first series of events, which may themselves lead to a new series of events. This progression continues until the last possible level is reached. An induced event can only occur if the preceding event occurs. The sum of probabilities in one node is always equal to 1. The amount of detail in logic trees can have a considerable impact on the results of the risk analysis. Finally, the probability of the various types of failure considered is calculated, and the final result is obtained by multiplying the probabilities of all the events present in the series (Figure 9.8 and Figure 9.9).

The risk connected to an extraordinary event may also be estimated from the sequential flow of a cause-damage-consequence analysis. This is a combination of fault trees and event trees. The example in Figure 9.10 is based on a six-stage concept (Bury and Kreuzer, 1986).

For dams, events that may lead to exceptional situations include floods, earthquakes, and unstable slopes of a valley, but also errors in design, construction, or operation. These events lead to dangerous conditions. Grouped together under types of failure, the probabilities of failure (here P_1 to P_4) are obtained from event trees. The total probability is equal to the sum. An impact assessment gives the risk for each scenario (R_1 to R_4).

The flow in Figure 9.10 is also represented in Figure 9.11 in the form of a block diagram for all stages in the cause-consequence analysis. The blocks can be replaced by a more detailed and interconnected event diagram, for example, by event logic trees.

RISK ANALYSIS 133

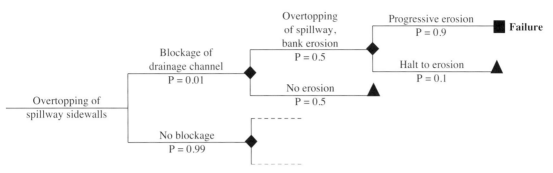

Figure 9.8 Example of an event tree for overtopping (from Le Delliou, 1998)

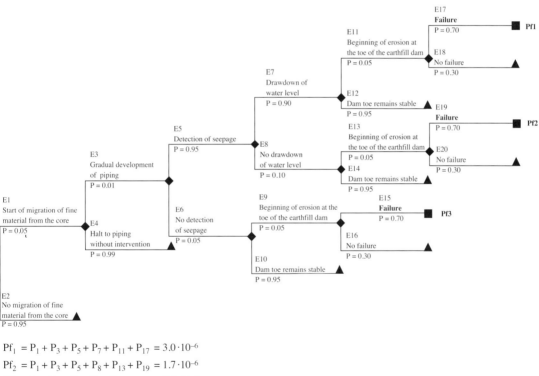

$Pf_1 = P_1 + P_3 + P_5 + P_7 + P_{11} + P_{17} = 3.0 \cdot 10^{-6}$
$Pf_2 = P_1 + P_3 + P_5 + P_8 + P_{13} + P_{19} = 1.7 \cdot 10^{-6}$
$Pf_3 = P_1 + P_3 + P_6 + P_9 + P_{15} \qquad = 8.8 \cdot 10^{-7}$
$\Sigma\, Pfi = Pf_1 + Pf_2 + Pf_3 \qquad = 5.6 \cdot 10^{-6}$

Figure 9.9 Example of an event tree for internal erosion (piping) (from Johansen et al., 1997)

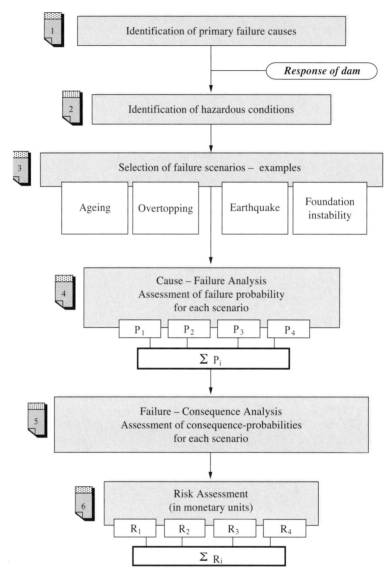

Figure 9.10 The six stages of sequential, cause-consequence analysis to assess the safety of a dam (from Bury and Kreuzer, 1986)

Taking uncertainty into account

Uncertainty is a complex notion that is omnipresent in dam engineering. One task in risk assessment is to examine the source of this uncertainty. The most common sources are outlined in Table 9.12 (Kreuzer, 2000).

RISK ANALYSIS

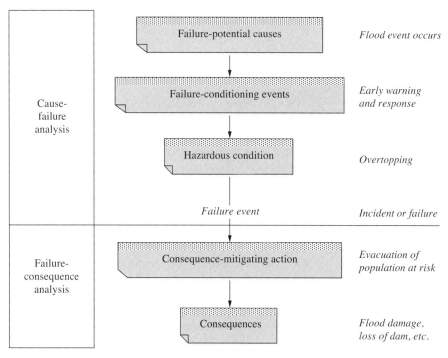

Figure 9.11 Generalized block diagram for a cause-consequence analysis

Table 9.12 Sources of uncertainty (Kreuzer, 2000)

Source/Types	Example of origin	Analytical treatment
Physical	Limited understanding of geology, hydrology, structural behavior	Bounded pdf's Monte Carlo simulation
Statistical	Sample-to-sample scatter	Standard deviation, mean error, confidence limits
Epistemic	Simplifications of mathematical models	Monte Carlo simulation Bayes' theorem, event trees
Decision	Subjective human views of a hidden state	Event trees Bayes' theorem
Prediction	Uncertain future event	Event trees Bayes' theorem
Public reaction	Lack of trust or impossibility of conveying it	Communication
Human error		Gross error theory

The treatment of uncertainty is varied and may be addressed in three ways: determination through additional information, avoidance by modifying the design or even taking actions, and implementation of preventive measures and additional training (Mouvet and Darbre, 2000).

9.4.5 Risk estimation

Analysis of consequences

Analyzing consequences involves evaluating downstream damage following an uncontrolled release due to total or partial dam failure or overtopping.[5] The repercussions of such a catastrophe will directly affect local communities and the environment.

To get an idea of the impact, the types of failure need to be defined and the extent of the flood wave needs to be calculated. The consequences of a failure result firstly in direct loss for everything in the flood zone and secondly in indirect losses, such as effects on the economy, as well as harm to the physical and mental well-being of local populations and an ensuing lack of trust. Consequences are measured in terms of monetary units for social, economic, and environmental damage that has been caused and in terms of the threat to human lives. Attempts have been made to determine the economic value of human lives, which poses its own ethical dilemmas.

Assessment of probability

Data used to analyze risk are obtained from statistical databases on failures and incidents, the opinions of expert engineers, and analytical treatment (Kreuzer, 2000).

Estimating the likelihood of occurrence (which is required at various stages during the analysis) depends on the quality and quantity of available data, which necessitates careful research. Examining and checking that the level of data is in line with the set objective is recommended. In addition, the amount of data necessary will differ depending on whether the decision to be made is at a preliminary or final level (Figure 9.13).

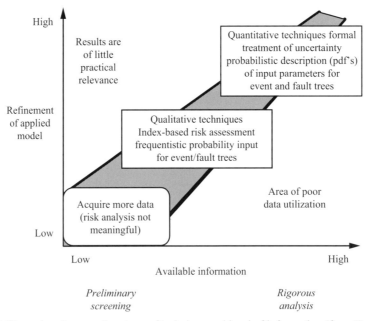

Figure 9.13 Hierarchy of approaches: type of technique and level of information (from Kreuzer, 2000)

---

[5] The causes are outlined in preceding chapters.

Statistics are useful for assessing fault and event trees probabilities, carrying out cost-benefit analyses for medium dams, and calibrating judgment. Much data is available in the literature; however, further progress will be made through improved quality and better information sharing.

Judgement is useful for interpreting statistical data. It helps to validate or refute analytical results. Judgement by an expert improves reliability and reduces the number of options at the decision-making stage. Indeed, a response can be formulated verbally; to express this as a probability, experts have established conversion tables (Table 9.14).

Table 9.14 Converting verbal descriptions into probability

Verbal description	Probability		
	Vick (1997)	Hoeg	Cyganiewicz and Smart
Virtually impossible	0.01	0.01 (0.001)	0.001
Very unlikely	0.1	0.1	0.01
Unlikely			0.1
Neutral	0.5	0.5	0.5
Likely			0.9
Very likely	0.9	0.9	0.99
Virtually certain	0.99	0.99 (0.999)	0.999

9.4.6 Risk evaluation

Calculating risk

Risk is defined as resulting from the probability of a specific consequence with the impact of this consequence expressed, for example, in monetary terms. The total risk is therefore the sum of all possible consequences given in monetary terms:[6]

$$R = \Sigma_i \, (P_i \times C_i).$$

An example of a risk calculation for the overtopping of a dam is given in Table 9.15; the values assigned to the various envisaged consequences are drawn from the article "Analytical Risk Assessment for Dams" (Grütter and Schnitter, 1982).

---

[6] This calculation is generally made separately for the various types of consequences (material consequences and human consequences).

Table 9.15 Example of a risk calculation in the case of dam overtopping (from Grütter and Schnitter, 1982)

Case		Specific consequences	Probability p.a.	Consequences (in monetary terms) 10^6 \$ US p.a.	Risk \$ US p.a.
Temporary overtopping due to arrival of large mass of rock into the reservoir	A	Damage to dam and/or installations	$1.06 \cdot 10^{-5}$	3	30
	B	Damage to dam and/or installations, damage due to flooding and loss of productivity	$5.00 \cdot 10^{-6}$	100	500
	C	Damage to dam and/or installations, damage due to flooding, loss of productivity, and victims	$2.15 \cdot 10^{-6}$	250	540
	D	Damage due to flooding, loss of dam, and loss of productivity	$6.30 \cdot 10^{-7}$	600	380
	E	Damage due to flooding, loss of dam, loss of productivity, and victims	$2.70 \cdot 10^{-7}$	850	230
Permanent overtopping due to flood	A	Damage to dam and/or installations	$1.94 \cdot 10^{-4}$	5	970
	D	Damage due to flooding, loss of dam, and loss of productivity	$1.45 \cdot 10^{-5}$	900	13,000
	E	Damage due to flooding, loss of dam, loss of productivity, and victims	$2.55 \cdot 10^{-6}$	1,200	3,060
Total					18,710

Risk acceptance

Reducing risk to a level that society considers acceptable (or tolerates) is desirable. Risk acceptance generally depends on the value attributed to things, on trust, and on societal attitudes.

Finding out whether risk is tolerated implies an external approach, one that is outside of the scope of usual quantification. The human dimension here is a determining factor; estimating risk only takes on its true meaning once the people involved adopt a position in regard to it. In order to determine society's tolerance to the risk posed by dams, the attitude that a population might have about the negative consequences of an event outside the bounds of generally accepted phenomena must be established. This attitude in turn depends on a variety of factors: sociocultural position of people involved, value system of the society affected by the event, general development of safety needs, development projects for a territory, etc. (SwissCoD, 2003d).

The domains affected by this range of factors clearly goes beyond the scope of the engineer. Irrationality plays a major role in a population's perception of a poorly defined danger. The disciplines able to characterize this tolerance to risk can be found in the social sciences.

Furthermore, through its very approach, risk analysis weighs up the loss of human life and the probability of it occurring. Implicitly, it admits that some lives may potentially be sacrificed should the feasibility of a project justify it.

With regard to consequences for people, two criteria can be established:

- The individual risk that a person is prepared to accept on their own behalf and that may differ depending on whether the individual is in a critical situation of their own choosing
- The societal risk that expresses the fact that the more victims an event causes, the more it becomes collectively intolerable

Assessing risk brings together calculated risk (the likelihood of events × consequences) and levels of risk considered acceptable by society or the operator. Developing risk acceptance criteria involves (Faber and Stewart, 2001):

- Perception of risk (ensuring an acceptable or tolerable level of risk to the system)
- Formal decision analysis (analytical techniques balancing or comparing risks against benefits)
- Regulatory safety goals (legislative and statutory framework developing and enforcing risk acceptance criteria)

The risk acceptance criteria generally adopted by authorities is that risk should be As Low as Reasonably Practicable, ALARP, or As Low as Reasonably Achievable, ALARA. This means that measures for reducing risk may be carried out until a level is reached where any additional reduction brings about disproportionate cost in relation to that already agreed to. As the terms "low, reasonably, possible, and achievable" are subjective, attempts have been made to define these criteria in a more tangible way (Figure 9.16).

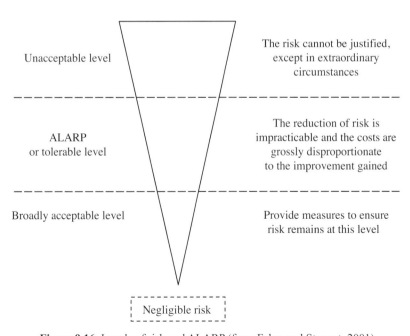

Figure 9.16 Levels of risk and ALARP (from Faber and Stewart, 2001)

It can further be mentioned that societal limits of tolerance and acceptance of risk are generally presented as an F/N diagram, with F generally the cumulative failure probabilities of scenarios with loss of life and N the number of lives lost (Figure 9.17). These socially acceptable risk charts are based on the analysis of statistics relative to accidents and the perception of risk in each country. According to the various recommendations, the accepted limits are situated between 10^{-4} and 10^{-5} or 10^{-6}/year/person. However, some experts doubt the usefulness of such diagrams or have pointed out their subjectivity. These diagrams are available in Australia, the United States, South Africa, the Netherlands, and Canada, among others. There are no limits for hydropower structures in Switzerland. However, limits for societal risk have been established through the Major Accidents Ordinance (OPAM).

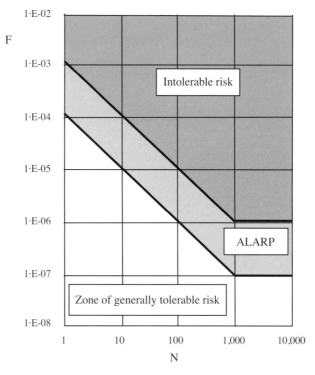

Figure 9.17 Example of F/N diagram: F = cumulative failure probabilities; N = number of lives lost (from ANCOLD) (ALARP = "as low as reasonably practicable")

Decision making

The introduction to this chapter outlined the value of risk analysis in the context of decision making and options analysis. For experienced engineers, it can be particularly useful in regard to the cost-benefit aspect, as it relates the search for a reduction in risk to that of costs incurred.

During the design phase, this method is helpful for comparing various designs and the factors that may have an impact on failure. It makes it possible to assess the risk at a given moment in the life of a dam and reservoir (design, construction, impoundment, operation) and, when necessary, to determine the need for carrying out additional studies or undertaking actions to improve safety. Furthermore, risk analysis can prove to be a basis for setting out a works program and investment strategy.

Finally, risk analysis can also play a role in the field of public liability insurance. In some countries, it is mandatory for the owner to take out insurance. Risk analyses are sometimes recommended for setting premiums instead of referring to simplified parameters such as volume of reservoir. Establishing an insurance portfolio is one solution for reducing premiums. A neat solution is to charge the same premium for all dams, as is the case in Sweden and Norway. Categories based on indices assessing the risk of specific dams can also be established (Lafitte, 1993, 1996).

9.4.7 Risk management

Treatment of risk

Risk cannot be entirely eliminated. Wise risk management must therefore aim to identify risks that can be minimized and those that must be managed. Should a risk analysis demonstrate that the calculated risks are not acceptable, appropriate measures will need to be taken. Various solutions are available in such cases.

An initial measure consists in minimizing the risk by modifying the system. One example of this would be to transform a spillway in order to increase its capacity or adapt a gate command system to avoid an untimely operation.

Risk can be reduced by seeking to limit consequences, for example, by constructing structures downstream that protect against floods.

Transferring risk is a common approach among risk management strategies. This can be achieved by taking out insurance that covers the consequences of an event or by finding other financial arrangements with third parties, after consultation and agreement between the relevant parties. In principle, risks that do not have financial implications are not transferable.

And finally, intervening in the acceptance of risk may be an option, especially if the cost for the above solutions is disproportionate to those incurred due to the risk, should the event occur.

Surveillance and monitoring

Risk analysis can be seen as an evolving process fed with continual information from the system in question. Through monitoring, it is possible to regularly assess the impact of decisions made in relation to risk.

Should new or specific elements occur, the risk analysis may be updated and the results used to improve the system. For example, structural modifications may be planned (widening spillway openings), surveillance and monitoring systems can be stepped up, modifications or restrictions to operation can be made, and finally, emergency plans can be reviewed.

III. Basis for the Design and Construction of a Project

A project cannot be designed and become reality without all the parameters being available, as well as essential information relating to the chosen site. Chapter 10 identifies the key points that must be addressed. Of these points, full knowledge of the nature of the foundations is critically important. It therefore goes without saying that geotechnical and geological surveys are vital stages, to which must be added the search for material suitable for construction of the works. The basis for the project must also include local conditions such as hydrology (controlling floods), seismicity (dynamic stress), and climate (temperature effects).

Chapter 11 outlines the different types of load supported by the dam during construction and operation. The engineer must define and combine the total loading to successfully complete the structural calculations. These loads can be permanent, variable (i.e., non-permanent), and exceptional.

Chapter 12 covers some of the diverse administrative requirements (requests for concessions, construction permits, impact studies) to which a project must adhere. These requirements vary between countries.

10. Identifying Conditions Related to the Site

Zayandehrood arch dam in Iran, height 100 m, year commissioned 1970 (Courtesy A. Schleiss)

Choosing the type of dam is a complex task that requires a particularly large number of factors and information to be taken into account. The aim is to offer the most economical solution while guaranteeing the highest level of safety and minimizing the impacts caused by the dam itself, the construction site, and dam operation.

When identifying sites, the following points are to be studied:

- Valley shape (morphology)
- Geology
- Availability of construction material
- Seismicity
- Climatic conditions
- Flood control

10.1 Topographical Conditions

10.1.1 Topographical survey

There is no question that topographical information is indispensable for carrying out geological surveys and setting out the dam and its appurtenant structures, as well as for determining the volume of the reservoir depending on the water level. Land surveys must cover not only the site of the dam but also the area situated downstream, the entire reservoir zone, and areas for possible borrow pits. They must also include the catchment area overlooking the reservoir, up to a sufficient height.

The spatial representation of the land in the form of maps may be carried out by aerial and terrestrial photogrammetry. In the case of aerial photogrammetry, shots must be taken so that each object is covered by at least two photos (overlapping). With the help of instruments and software, a photogrammetric model is produced. If this model is integrated within a coordinate system with points visible on the photos and known coordinates, it can be reproduced to scale. Complementary detailed mapping can take place on the ground (for example, contour lines, profiles). This topographical output serves to establish a network of fixed points that will in turn be used to set out the works, but also identify sites for boreholes, shafts, and exploratory trenches, and other specific sites on the ground. Depending on the stage of the project and construction, topographical maps can become more detailed, and the scale can be adjusted. Scales of 1:500, 1:200, or even 1:100, with contour intervals of 5, 2 or 1 m (possibly 0.5 m) are commonly used in the design of dams and their appurtenant structures. For the reservoir zone and depending on its surface area, scales can range from 1:1,000, 1:2,000, 1:5,000, or even 1:10,000 with contour intervals of 25 m, 10 m, or even 5 m. Qualified surveyors (geometricians) are responsible for the creation of these topographical maps.

10.1.2 Topographical criteria

Subject to geological conditions, it is important to search for the most favorable topography in order to develop a project that is both realistic and economically viable. In general, a narrow topography can prove to be an advantage. The morphology of the valley (Figure 10.1), narrow or wide, has a considerable influence on the possible choices for the type of dam. In some cases, the geometry of the valley is not suitable for certain types of dams, which can then be excluded from the outset.

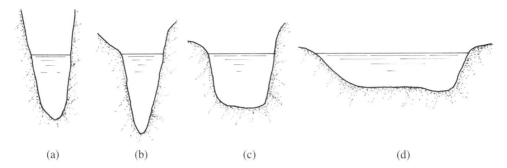

Figure 10.1 Diagram of different valley shapes: (a) gorge or canyon; (b) V-shaped valley; (c) U-shaped valley; (d) wide valley

A *narrow valley* can be shaped like a gorge or canyon (steep-walled valley with near vertical flanks), "V"-shaped, or "U"-shaped. Narrow valleys are especially suitable for the construction of arch dams and gravity dams (Figure 10.2).

Figure 10.2 Different valley shapes: Les Toules and Sta Maria arch dams (U-shaped valley) (top), Zeuzier arch dam (V-shaped valley) (bottom left), Zimapan cylindrical arch dam, Mexico (gorge) (bottom right)

In geologically favorable areas, erosion through waterways generally forms a narrowing of the valley that is suitable for the construction of an arch dam. The dam should be located just upstream of this natural narrowing, so that the load is transferred to the foundation in the zone of the highest geo-mechanical quality (Figure 10.3).

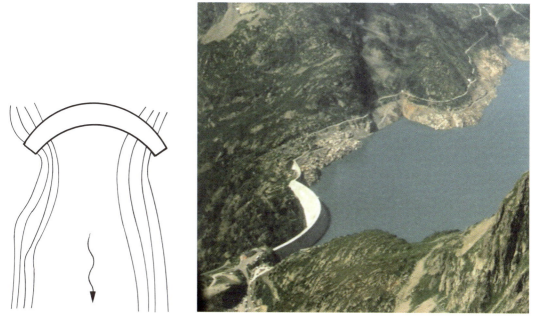

Figure 10.3 Site location directly upstream of a rocky narrowing of the valley: Emosson arch dam (Courtesy H. Pougatsch)

For arch dams, slenderness λ (see § 3.2.4), which is the ratio between the developed length of the crown L_c and the height of the dam H_c, is also important. Normally the slenderness ratio is limited to 5 (or 6) in a V-shaped valley and 4 (or 5) in a U-shaped valley (Figure 10.4). These limits may, however, be exceeded for some structures when the nature of the foundation rock is particularly supple, that is, with a relatively high ratio between the elasticity modulus of the concrete and the foundation rock ($E_B/E_R > 3$).

As for slenderness λ, if it is lower than 2, the dam resists like an arch; if it is between 2 and 4, part of the water pressure is taken up through the arch effect and the other through bending effect; finally, if it is higher than 5, water pressure is primarily taken up by bending.

In addition, the valley needs to be symmetrical to ensure the static function of the arch dam. A lack of symmetry can be corrected by creating a massive abutment. Most often, this abutment is formed by the gradual thickening of the longer wing, so as to shift from a narrow profile to a gravity dam profile without discontinuity (Figure 10.5).

In a gorge, if the width is almost constant along the whole height, a cylindrical arch dam may be envisaged (Figure 10.6).

Figure 10.7 portrays an arch dam set out in a topography where the contour levels of the supporting rock are almost parallel to the axis of the valley.

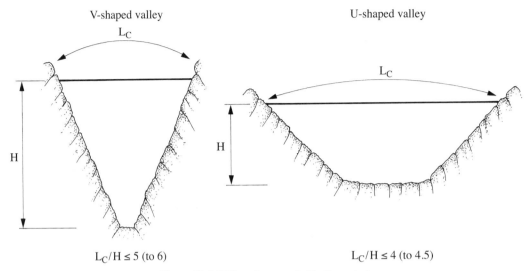

Figure 10.4 Valley shapes suitable for arch dams

V-shaped valley: $L_C/H \leq 5$ (to 6)

U-shaped valley: $L_C/H \leq 4$ (to 4.5)

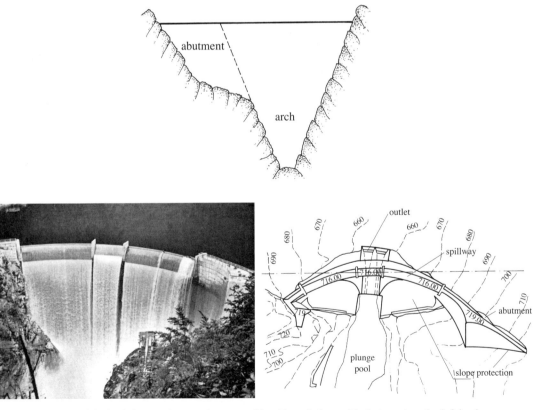

Figure 10.5 Arch dam against an abutment: Chatelôt arch dam with abutment on the left bank (from SNCOLD (SwissCoD), 1964)

Figure 10.6 Example of a cylindrical arch dam: St-Barthélémy C dam under construction and once completed (Courtesy H. Pougatsch)

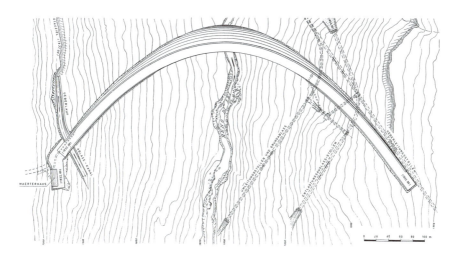

Figure 10.7 Example of the set out of an arch dam (Valle di Lei dam)

For "V"- or "U"-shaped valleys, gravity dams and rockfill dams with concrete facing are also possible. Rockfill dams with a central core are not to be considered, due to the steep valley flanks (risk of differential settlement and cracking of the core).

A *wide valley* is one that has a U-shaped section or a trapezoidal shape; it is suitable for embankment dams (Figure 10.8), buttress dams, hollow gravity dams, and roller-compacted concrete dams (RCC). A dam with a need for a high slenderness ratio λ of the dam may tend to rule out arch dams. Topographical considerations are less of a determining factor for embankment dams.

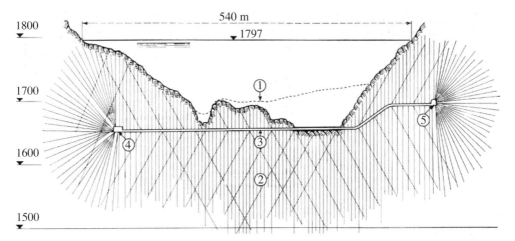

Figure 10.8 Longitudinal cross-section of the Göscheneralp embankment dam: ① natural ground; ② grout curtain; ③ grouting gallery; ④ bottom outlet gate chamber; ⑤ headrace tunnel gate chamber

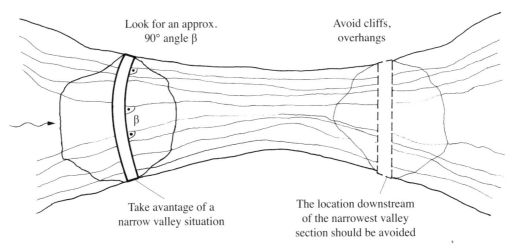

Figure 10.9 Topographical criteria for the setting out of an embankment dam in a valley

Embankment dams can adapt to practically any topographical form. However, ideally a narrowing of the valley should be taken advantage of, while avoiding the downstream extremity of the narrowing. Overhangs should also be avoided. Figure 10.9 illustrates the rules that are intended to manage the deformation effects to which the embankment dam will be subject.

10.2 Geological and Geotechnical Surveys

10.2.1 The importance of geology

Geology is a fundamental issue as it is one of, if not the essential criterion for determining the choice of dam type. A detailed analysis of all the geological parameters must be undertaken before making any decision about the feasibility of various types of dams. It is also the reason why, no matter the type of structure and its dimensions, a site investigation program is vitally important to ensure in-depth knowledge of the ground foundations. The nature, quality, and characteristics of the foundations will play a prominent role in the choice of dam.

The feasibility study that must be undertaken is a joint task between the engineer and the geologist. The geologist must give his or her opinion, in particular on the following questions:

- Catchment area:
  – regional geology
  – hydrogeological conditions
- Reservoir:
  – reservoir impermeability
  – stability of valley flanks and banks
  – transport and inflow of solid material
- Site:
  – abutment impermeability
  – stability of abutments
  – stability of surface and underground excavations
  – groundwater flow
  – degree and depth of surface weathering
- Materials:
  – inventory of usable rock and loose materials
  – nature of quarries
  – approximate volume of rocky and loose material available for construction
  – seismicity

The answers to these questions will become increasingly detailed as the study progresses.

10.2.2 Description of field investigations

Field investigations form part of a gradual process that enables the permeability, hydrogeological conditions (particularly the level of the water table), deformation characteristics, and strength characteristics of soil and rocks to be assessed, as well as possibilities for excavation. To find out the general surface conditions, a preliminary geological site investigation is first carried out, that is, topographical surveys of outcrops, overlapping materials, and unstable sections of the future reservoir. These preliminary in situ observations will then allow an initial work program for more in-depth geological study to be established. Surface investigations include trenches and test pits, while boreholes and tunnels are used for deeper study. In situ and laboratory studies are carried out in order to acquire more knowledge of the mechanical and hydraulic characteristics of the soil and rocks encountered.

Reservoir study

The first, vital stage is to ensure that the basin is watertight. This means checking for the existence of permeable and non-permeable zones, tectonic accidents (faults, cracking), an epigenetic valley, or karstic zones where water can infiltrate and gradually cause an outburst (Figure 10.10). Should this occur, possible seepage losses must be assessed in relation to the inflow that is stored, and a decision must be made on whether grouting is required. Tunnels planned for in the project stage were excavated into the banks of the Linth-Limmern (GL/1963/H = 146 m) and Sanetsch dams (VS/1965/H = 42 m), as well as in the Salanfe dam (VS/1957/H = 52 m), during operation, when a grout curtain was installed to cut off preferential seepage paths and avoid water losses. It should be noted that karstic rock formations are particularly

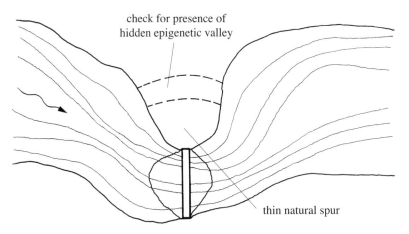

Figure 10.10 Danger represented by an epigenetic valley

dangerous. They are made up of limestone rocks, parts of which are soluble and gradually dissolve in the presence of water. The seepage paths through the rock may grow over time, and eventually it becomes impossible to seal them.

The second stage involves identifying potentially unstable zones that under diverse conditions could be at the origin of landslides or rockfalls, and therefore pose a risk to dam operation. Furthermore, in view of modifications brought about by permafrost, it is advisable to check whether there is a potential risk of rockfall. Fluctuations in water level or a rapid drawdown, without an accompanying dissipation of pore pressure, may jeopardize stability in zones with loose overburden material, which may result in the need for strengthening works or specific operating modes. For example, an embankment slope protection was constructed along the right bank of the Rossinière dam (VD/1972/H = 30 m) (Figure 10.11).

The third stage concerns sediment yield into the reservoir. Wind and rain are the main factors behind soil erosion. On average, erosion may be in the order of a millimeter per year in alpine regions. The sediment is then transported by watercourses as bed load or as suspended solids toward dam reservoirs. Whether this material settles in a particular place or spreads out into the reservoir will depend on its size. These deposits have a direct impact on the long-term use of the reservoir (sedimentation and loss of storage capacity) and on the safe operation of water release structures such as intakes and the bottom outlet. There are, of course, ways in which the arrival of sediment into reservoirs can be reduced (settling basins, bypass tunnels, sand traps, etc.). Where necessary, a regular flushing program should be organized once the dam is in operation.[1]

---

[1] See Part I, § 2.2.4 and Part IX, § 34.4.3.

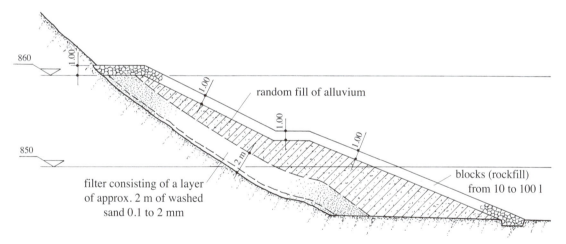

Figure 10.11 Rossinière dam, slope protection by embankment on the right bank (Courtesy H. Pougatsch)

Dam site selection

Directly underneath the dam site, unweathered rock may be found with a partially weathered surface and/or loosely marked by superficial deposits that should be removed before construction.

The rock may also be covered by a greater or lesser quantity of loose ground. Some valleys are covered in a substantial layer of more or less homogeneous and permeable fluvial deposits (depending on the nature of its composition), which are suitable for the construction of an embankment dam. This formation may present, for example, in sandy-gravel (permeable, good mechanical strength) or silty-clay (more or less impermeable) forms.

Mud and peat are in principle unfit to support dam structures (in exceptional cases, very small dams may be built with the usual precautions).

When the dam must be built on loose ground, studies rely on information from the excavation of trenches and/or test pits, as well as on a drilling program capable of reaching the existing rock, or even geophysical explorations designed to identify the individual layers and the rock surface. The extent of the exploratory drilling program will depend on the size of the dam structure. Carrying out in situ tests and collecting samples for laboratory tests allows useful information to be gathered, relating in particular to:

- The identification of soil types
- The nature of the heterogeneity of the soil (layers, lens)
- Hydraulic conditions (horizontal and vertical permeability of the various foundation materials, position of the groundwater table)
- Mechanical parameters (deformability, shear strength)

During the design phase and in view of loading from the self weight of the dam and water pressure, the stability of the foundation and settlement should be assessed. In addition, adequate measures must be taken to ensure sufficient watertightness in the foundation.

For rock foundations, particular care is required in establishing the following points:

- Stratigraphy, tectonics, and lithology:
  - cracks and fissures
  - faults recurring throughout the foundation (active or not)
  - orientation and nature of discontinuities (opening, persistence, fill material, etc.)
- Seepage flow regime (foundation permeability, position of groundwater table)
- Long-term behavior under the effect of seepage (internal erosion, dissolution)
- Overall mechanical characteristics (deformability, shear strength, and compressive strength)
- Classification (overall quality assessment)

During the geological study, a survey of the orientation of stratification joints and schistosity is to be carried out, as well as the demonstration of any geological accidents (faults, fractures, diaclases, fissures), while addressing the nature of filling materials issuing from weathered rocks. Care must be taken with preferential slip surfaces that may constitute discontinuities. A statistical survey of cracks enables specific seepage zones to be identified, which indicate a risk of leaks or uplift.

In-depth rock investigations are carried out by way of boreholes and exploratory galleries, or even using geophysics. Exploration galleries mean that specific uneven passages can be identified and that the quality of the rock present at greater depths can be observed, as well as the filling of cracks and faults. In galleries and test pits, in situ tests can be carried out (jacking and plate-bearing tests). Mechanical strength tests ascertain whether the load transferred by the dam is greater than the shear strength in the dam foundation.

On the basis of these in situ observations and the results of all investigations, the geologist can comment on the stability of abutments and suggest possible strengthening measures, as well as on foundation rock permeability by suggesting necessary grouting to ensure watertightness in the upstream section and for the drainage system in the downstream zone. Exploration galleries may, depending on their location, be used as grouting galleries for installing the grout curtain and/or be integrated into the drainage system.

The geologist must also decide how much of the surface weathered part should be stripped and excavated in order to achieve an appropriate bedrock for the dam.

10.2.3 Description of types of geomechanical and geotechnical investigations and tests

In situ and laboratory tests on intact or disturbed samples allow the appropriate mechanical and hydraulic characteristics to be estimated in order to establish the type of dam required for the local conditions. Samples are collected in test pits or trenches, as well as from core drilling. Table 10.12 gives an overview of the main geomechanical and geotechnical tests carried out in situ and in laboratories.

Table 10.12 Overview of main geomechanical and geotechnical tests for loose and rocky ground

	Loose ground	Rocky ground
Type of investigation	Geophysical methods Trenches, test pits Exploratory boreholes	Geophysical methods Galleries Exploratory boreholes
In situ tests	Mechanical characteristics • Dilatometer test • Penetration test (penetrometer) • Shear vane test Permeability (water test) • Lefranc test • Pumping test	Mechanical characteristics • Shear test • Load-bearing test (plate-bearing, jacking) • Deformation test • Rock elasticity Permeability (water test) • Lugeon test
Determined in the laboratory	• Identification of soil type • Shear strength • Internal angle of friction • Cohesion • Settlement • Consolidation • Permeability	Strength and deformability • Direct shear test • Simple compression test • Triaxial compression test • Dynamic test (deformation modulus measurement) • Hydraulic characteristic • Permeability test

Description of modes of exploration

Geophysical investigation is a way of indirectly observing the subsurface with the aim of detecting the various layers in the foundation and identifying discontinuities. There are several geophysical methods including seismic and geoelectric tests. Interpreting these measurements is the job of an experienced geophysician, who can also rely on calibration from core drills.

The underlying principle of seismic methods is based on the wave propagation speed caused by a small explosion or artificial vibrations. These waves move at the surface, along layer boundaries, or inside rocky bodies (pressure waves, known as P, and shear waves, known as S), and their travel time, speed, and path are then analyzed. There are two main seismic methods: reflection and refraction. Seismic reflection is used to explore the ground generally at a depth of more than 20 m. This method identifies surfaces that reflect the waves due to an abrupt change in physical properties (density, elasticity). Seismic refraction is based on the detection of the first waves to arrive (P waves, or pressure waves). It enables discontinuity surfaces separating the rocks or soils to be identified, due to the difference in velocity of seismic waves. The obtained results are complementary to those from seismic reflection. This method is principally used to determine the depth of the rock substratum and fractured zones at a shallow depth.

The geoelectric method measures the electrical resistivity of the ground and enables the various materials with a resistivity contrast to be identified. This method is used to detect and localize water seepage and the level of the groundwater table, as well as for permeability measurements.

These methods are especially suitable in regular terrain such as alluvial plains. However, care must be taken in their interpretation. In a seismic test, low speeds may indicate either silt or highly broken up rock. High speeds generally indicate healthy rock with few fractures.

Carrying out *investigations with trenches and galleries* is an effective solution as it provides the opportunity to view the rock on site. However, it is expensive. The construction of some galleries may be necessary to adequately study the dam foundation.

Drilling investigations play a major role in dam construction. They can reach great depths (usually a depth equal to the height of the dam) and give generally satisfactory results. Drilling boreholes can be supplemented with water injection testing that determines the degree of ground permeability. Much care must be taken when drilling boreholes, and the professional skill of drillers plays an important role. It is a good idea to entrust this task to experienced professionals.

In fluvial deposits, drilling is usually vertical and tubed. In rock, it is not usually necessary to tube the boreholes, which may be vertical or sloped in any direction. They are drilled from the ground surface or from within tunnels. The usual diameter of exploratory boreholes ranges from 86 to 101 mm (the diameter should not be less than 60 mm). More or less intact cores can be extracted with a diameter of 40 to 60 mm. To execute these boreholes, a simple core drill is used that turns with the drill bit (which results in more wear on the core) and/or, usually in poor quality rock, a double tube core barrel where, as the inner tube does not rotate, there is less wear (see Figure 29.1). Substantial damage to the core due to the pieces rubbing against each other indicates friable rock. When the core is less damaged, this instead indicates good rock. The core samples are collected, photographed, and stored in boxes labelled with their depth.

Based on the analysis of these drill cores, a rock quality designation (RQD) index can be defined. One method involves counting the number of fractures by linear meter (lm) of drilling and then measuring the coring rate t_c expressed as a %, that is, the total length of the collected cores divided by the total length of the borehole. An RQD coefficient is thus defined as:

$$RQD(\%) = 100 \frac{\Sigma \text{ length of the collected cores} > 10 \text{ cm}}{\text{length of the borehole}}$$

A grading scale was developed by Deere:

Rock quality	RQD
Excellent	between 90 and 100
Good	between 75 and 90
Fair	between 50 and 75
Poor	between 25 and 50
Very poor	between 0 and 25

Description of in situ tests

Dilatometer tests using penetrometers and shear vanes are mostly employed in instances of poor foundations. ***Dilatometer tests*** measure deformation of the borehole wall depending on the pressure exerted by the dilation of the probe. The deformation modulus can then be calculated in situ. ***Penetrometer*** tests involve forcing a rod with a conical end into the rock. The rod is forced down by a hammer dropping from a given height (dynamic test) or by pressure with a hydraulic press (static test). The results are illustrated on a diagram showing resistance to penetration depending on depth, which gives an indication of the nature of the various rock types encountered. ***Shear vane tests*** enable the undrained cohesion value of soft clay to be calculated. In these tests, a rod with vanes at the end is inserted into the ground and rotated. It measures the torque required to shear a cylinder of soil limited to the vanes attached to the device.

In situ mechanical tests on rock (Einstein and Descoeudres, 1972)

Deformability measurements of a rocky mass can be undertaken by tests under static loading (triaxial test, plate-load or jacking test, loading test in the borehole with a dilatometer or in a test room) or by tests under dynamic loading (seismic refraction measurements). The static load test using the plate-load or jacking method involves running loading cycles – the unloading of a rock surface by pushing against the tunnel or gallery wall, or, for an open-air test, by using a rock anchor as a support reaction (Figure 10.13). Deformation is measured under the plate and in the borehole behind the loaded surfaces. The Boussinesq equation is used to interpret these measurements.

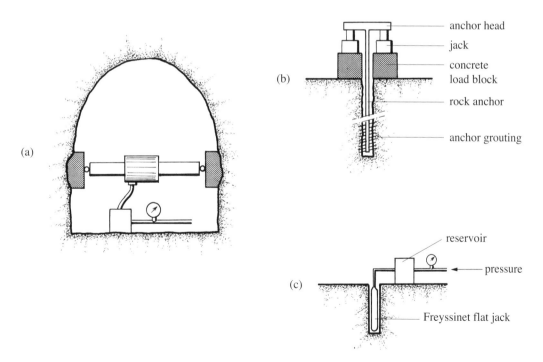

Figure 10.13 Rock load tests: (a) plate-load test in a gallery; (b) open-air plate test (reaction using anchors); (c) hydraulic jacking test

Dilatometer tests enable rock deformability to be calculated, at various depths, by measuring borehole wall displacement. And finally, radial jacking tests in a gallery involve placing a circular chamber under uniform radial pressure using jacks or by filling a chamber with water under pressure.

Strength tests measure the shear strength of rock test samples with a limited volume, which may or may not exhibit a discontinuity. These tests include direct shear tests and triaxial or torsion shear tests.

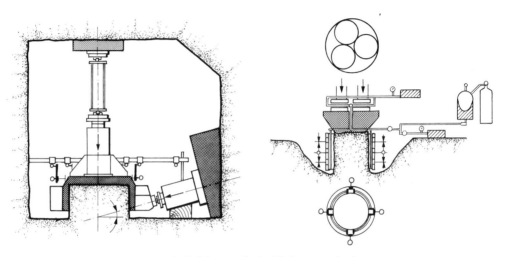

Figure 10.14 Direct and triaxial shear tests in situ

Water tests

Swiss geologist Maurice Lugeon (1870–1953), from Lausanne, developed a ***water test*** that has since become universal (Figure 10.15). As the drilling proceeds, water injection tests are carried out every 5 m, under a 1 MPa pressure (generally measured at the entrance to the borehole) for 10 minutes (1.5 MPa or more for large dams; generally 10 bar for a zone between 0 and 100 m deep under the level of the future reservoir, and so on). The quantity of water that is lost into the ground is measured. This unit of measure (1 l/m × min under 10 bars) is called the *Lugeon* (Lugeon, 1933/1979), which corresponds to a Darcy permeability of $1.3 \cdot 10^{-7}$ m/s. Experience has shown that the ground is watertight (from the point of view of the engineer) if the loss is inferior to 1 liter per minute of drilling for 10 minutes, that is, below 1 Lugeon. It is rare for natural rock to be that watertight. In general, the scale is as follows:

Rock quality	Lugeon unit
Excellent	0 to 1 l/m × min
Good	1 to 5 l/m × min
Fair	5 to 10 l/m × min
Poor	more than 10 l/m × min

The ***Lefranc test*** involves injecting or pumping water into an open cavity in the ground below the water table. Measurements are carried out at a constant level (injection or pumping of water until the level has stabilized) in ground with a permeability greater than 10^{-4} m/s, or with variable levels (sampling or injection of a given volume) for less permeable ground.

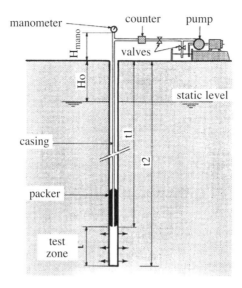

Figure 10.15 Arrangement for carrying out a Lugeon test

Pumping tests involve measuring the variation in water level with piezometers set out around a borehole or test pit, in which the water level has been lowered by pumping it out.

Classification of rock masses
Without going into detail, there are various methods for classifying rock masses. The most widespread are those by Z. Bieniawski (RMR classification) and N. Barton (Q classification).

10.2.4 Types of possible dams depending on the nature of the foundations

A *foundation set on loose soil or rock* is generally suited to embankment dams, which are able to tolerate the deformability of such ground. It is important to ensure that the watertightness of the reservoir continues into the foundation under the dam and to manage seepage through an appropriate drainage system. Sometimes, the construction of low concrete dams can be envisaged on this type of foundation, providing, of course, that suitable precautions regarding seepage and the risk of differential settlement are taken.

A *rock foundation* is suited to the construction of any type of dam, particularly concrete dams. Concrete dams require sound rock foundations. It is vitally important that the foundation bedrock has adequate shear strength and sufficient bearing capacity to respond to the required stability criteria. The deformability of the rock also plays a role. The strike and dip and orientation of the layers must be recorded. Depending on the cracking and permeability of the rock, it must be grouted on the surface (contact and consolidation grouting) and depthwise to ensure watertightness directly under the dam foundation. The presence of geological accidents (fault, shear zone) necessitates specific provisions in the design to anticipate possible movement. To ensure good concrete-rock contact, all weathered material and overburden must be removed from the foundation zone. In addition, excavation should be adjusted to avoid an abrupt change in the slope of the abutment, subvertical abutments, and especially overhangs, and a regular bedrock surface must be ensured.

An upward-sloping foundation is preferable; should the original ground surface be downward sloping, benched excavation is required.

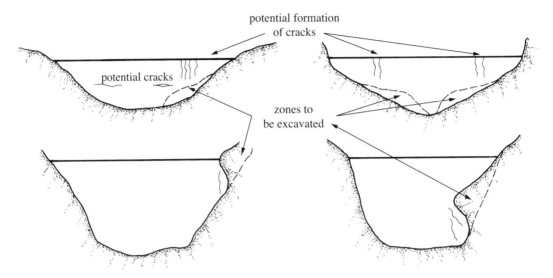

Figure 10.16 Rectification of excavation: eliminating overhangs and irregular slopes

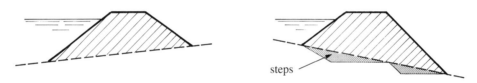

Figure 10.17 Foundation direction

With regard to the quality of original ground and subject to several reservations, the rock can be roughly categorized into the following groups:

- *Good quality rock*

Good quality rock, whose deformation modulus E_R is greater than 8000 MPa, is favorable for all types of dams. Exploration galleries can be excavated in this type of rock without the need for rock support.

- *Medium quality rock*

Medium quality rock, whose deformation modulus E_R ranges from 4000 to 8000 MPa, is suitable for all types of dams except arch dams for which the stresses at the level of the foundation are too great. The bearing capacity of the rock must be meticulously studied to check that the dam can tolerate deformations in the foundation. Homogeneous behavior in the foundation is very important. With this type of rock, exploration galleries can be excavated almost without any rock support measures besides local use of rock bolts and shotcrete.

- *Poor quality rock*

If the deformation modulus E_R is lower than 4000 MPa, the rock deformability is too great to place a rigid structure like a concrete dam on it. Preference should therefore be given to an embankment dam with a core or possibly with upstream facing. In this type of rock, the excavation of exploration galleries requires specific rock support measures such as steel girders and shotcrete.

10.2.5 Interventions by a geologist during the works

The involvement of a geologist (or geological engineer) and, depending on the situation, a geotechnical engineer during construction and later operation is essential.

Unexpected geological situations may arise during construction, and expert monitoring ensures that any intervention and all appropriate measures can be taken as necessary.

Prior to being covered, all tunnels, galleries, and excavations must be inspected by the geologist, who assesses the rock quality. He or she draws up geological surveys that indicate cracking, fissuring, the strike and dip of layers, schistosity, and any points of water inflow, and compiles a file of photographs. After the inspection and if required, the geologist may put forward recommendations for any necessary adaptations. At the same time, a topographical plan of the excavation line must be drawn up by a surveyor.

After having given instructions concerning the grouting program, the geologist oversees its implementation and, depending on the type of geology encountered and the results of grout take tests, suggests possible modifications.

During impoundment tests, he or she assesses the efficacy of drainage systems and proposes possible additions of grouting and drainage.

The geologist is involved in assessing the stability of excavations and natural slopes, and gives instructions on the implementation of any strengthening judged necessary. He or she identifies any facts or events that may have occurred during the works so that all the geological information that may be useful for the future are available to draw on should there be any specific events during first impoundment and/or operation.

Figure 10.18 Excavation footprint of an arch dam

The geologist puts together a geological reference file, which, depending on the circumstances, may be completed over the life of the dam.

10.3 Searching for Material

No matter the type of dam, construction requires the placing of large quantities of material. Construction costs are heavily influenced by the excavation, transport, and placement of material. Borrowing pits must therefore be located in the immediate vicinity of the site, and costs for the processing of materials (crushing, washing, selection) must be optimized.

The choice of dam may depend on the quantity and quality of available material. This material must be present in sufficient quantity to meet the required needs. One of the major difficulties in preliminary dam studies is accurately guaranteeing sufficient and homogeneous quality across the total amount required.

In regard to construction, there is a key difference between concrete dams and embankment dams:

- For a concrete dam, the construction material is industrially manufactured and so is more or less of consistent quality. In theory, only extreme temperatures (especially low temperatures) can prevent placement.
- For embankment dams, the construction material is a natural product whose properties may vary from one point to another and also depending on the season. Low temperatures and precipitation (rain, snow) may affect the placement of some materials, especially cohesive materials.

10.3.1 Aggregates for concrete dams

Aggregates are inert materials that, in order to be suitable for manufacturing concrete, must meet certain quality criteria.[2] The shape, physical and chemical constitution, and cleanliness of particles are important characteristics that affect concrete properties. Granulometric analyses enable aggregate to be graded according to the largest particle size: filler, sand, chippings, gravel, stones (river stones), or broken up stones (crushed material). Managing particle size is extremely important. Aggregate should not react with the cement and should remain stable in air, water, and frost. Porous materials are not recommended. Various other conditions also come into play for the manufacture of concrete and are covered in Part V.

With regard to shape, round smooth aggregate from alluvial gravel pits are almost sphere-shaped, which allows the particles to fit closely against each other. Another category includes crushed aggregate from quarries, which is instead irregular and less spherelike. Particles of flat and elongated elements can influence the mixing and placing of concrete. However, crushed aggregate can offer good resistance to cracking and sliding along a weak plane. Its workability can be improved by adding fine sand and rounded aggregate. To prevent the inclusion of inadequately shaped aggregate, a volumetric coefficient C can be used, which is the ratio between volume v of the particle in question and the diameter of the circumscribed sphere of the largest size of particle d:

$$C = \frac{v}{(\pi/6) \cdot d^3}$$

---

[2] There are several standards that establish criteria: the SIA in Switzerland, AFNOR in France, ASTM (American Society for Testing and Materials) in the US.

For a family of particles, the formula becomes:

$$C = \frac{\Sigma v}{(\pi/6) \cdot \Sigma d^3}$$

The lower limits range from 0.20 for gravel with a size of 12.5 mm to 15 mm to 0.15 for sizes greater than 25 mm (Faury, 1958).

The characteristics of aggregate are related to cleanliness, surface condition, frost resistance, mineralogical composition, hardness, apparent density, and the presence of schistosity. Cleanliness is a vitally important quality. Aggregate must be mechanically washed so that there are no traces of harmful impurities such as humus, plant debris, gypsum, mica, dross, or clay nodules. Only a small proportion of soluble and extrafine material (silt, mud, clay) are acceptable (2%):

$$\text{percentage of impurity} = 100 \cdot \frac{P1 - P2}{P2}$$

where

$P1$ = mass of dry aggregate before washing
$P2$ = mass of dry aggregate after washing

Surface condition also plays a role. The rough surface of a crushed material offers better adherence to the cured binder (cement) but requires more wetting water. The possible frost resistance of aggregate must be considered. It depends largely on the presence of cracks or cleavage. From the point of view of mineralogical composition, aggregate must not contain any harmful elements such as gypsum or anhydrite. The presence of amorphous silica may lead to issues with swelling, resulting from reactions with cement's alkaline component. There are various types of alkali-aggregate reactions (AAR) including alkali-silica, alkali-silicate, and alkali-carbonate (seldom observed). To avoid damage from these reactions, a petrographic analysis is recommended. This indicates whether a mix, a fraction, or some components may be considered potentially reactive. Reliable indications on the reactivity of aggregate can be obtained by measuring the variations in length of concrete or mortar samples. The "Microbar" test (as per AFNOR P 18-588) has proven its worth. Table 10.19 lists the main reactive rocks and minerals (Hammerschlag and Merz, 2000). During the construction of a dam in Switzerland, the limestone rock aggregate that was used contained nodules of amorphous silica. Swelling due to a reaction with the alkali in the cement was treated by adding 30% pozzolana to the Portland cement.

The behavior of aggregate during mixing and curing of the binder (cement) must also be assessed. Hardness is usually measured using the Los Angeles or Deval test. The aim is to subject the material to abrasion through rotation. The weight of the tested material and the duration of rotation differ depending on the type of test used. At the end of the test, the materials are passed through a 1.6 mm sieve. The Los Angeles coefficient is the ratio between the weight of the particles < 1.6 mm and the total initial weight (5 kg). The Deval coefficient is $A = 400/U$, where U is the weight in grams of sand produced during the rotation that pass through a 1.6 mm sieve and calculated per kilo of aggregate.

Preliminary concrete tests are vital.

Table 10.19 Main reactive minerals and rocks (from Hammerschlag and Merz, 2000)

Rock family	Rocks in which reactive mineral phases can occur	Reactive mineral phase	Types of reaction
Crystalline rocks	Granite, granodiorite, diorite, etc.	Porous microfiber quartz	Alkali-silicate
Volcanic rocks	Rhyolite, dacite, andesite, basalt, obsidian, tuff	Unstable high-temperature forms of quartz: trydimite, cristobalite	Alkali-silica
		Crypto-crystalline silicic acid: chalcedony	Alkali-silica
		Amorphous, hydrated silicic acid: opal	Alkali-silica
Metamorphic rocks	Gneiss, schist, mylonite, quartzite, hornblende	Deformed, cracked quartz	Alkali-silicate
		Deformed, porous, weathered feldspars	Alkali-silicate
		Fine crystalline micas	Alkali-silicate
		Crypto- and microcrystalline quartz	Alkali-silica
Sedimentary rocks	Sandstone, greywacke, siltite, flint, siliceous limestone	Deformed, porous, weathered feldspars	Alkali-silicate
		Fine crystalline clays, mica	Alkali-silicate
		Crypto- and microcrystalline quartz	Alkali-silica
		Crypto-crystalline silicic acid: chalcedony	Alkali-silica
		Amorphous, hydrated silicic acid: opal	Alkali-silica

10.3.2 Materials for embankment dams

When planning an embankment dam, the material chosen for earth/rockfill is a key element. It must meet the following conditions:

- Quality
    - nonorganic
    - unweathered
    - extraction, transport, and placement are possible
    - compaction is possible
    - high shear strength j′ and cohesion c′ (for the core)

- Available in sufficient quantity and quality near to the site
  For example, by utilizing coarse material or rockfill and only a limited amount of impermeable material, a zoned embankment dam may be considered. However, if rockfill is all that is available, a rockfill dam with upstream facing may instead be selected.
- Cost effective

In theory, all nonorganic material can be used to construct embankments and dikes, providing that placement (taking into account climatic conditions: precipitation, frost in cold regions or alpine conditions), compaction, and resistance to weathering are guaranteed.

A wide range of materials may be used, from clay to large blocks. Effort should be made to select material with a low sensitivity to climatological effects, firstly for placing, and secondly for long-term durability. At the same time, the highest possible shear strength φ' and cohesion c' are sought. The art of constructing embankment dams (dikes) resides in the ability of the engineer to use the natural materials available near the site of the dam, while also meeting safety requirements and being cost effective.

To construct watertight elements such as the core, cohesive materials must be used as they have low levels of permeability. Clayey silt, with its low permeability ($k < 10^{-7}$ m/s), is used, as well as moraine with a high fines content.

However, the body of the abutment and filters must be made of non-cohesive material (gravel, stones, blocks). The requirements for the physical properties of material in the various zones of the dam are outlined in Figure 10.20.

Of course, before selecting a material, it must undergo the usual geotechnical tests in order for its properties to be defined. These tests measure water content, identify the soil (granulometric analysis, Atterberg limits), and measure compaction characteristics (Proctor test) and their behavior (triaxial CU and UU tests, shear strength, compressibility, permeability).

The most important characteristics of borrow materials are listed in Table 10.21.

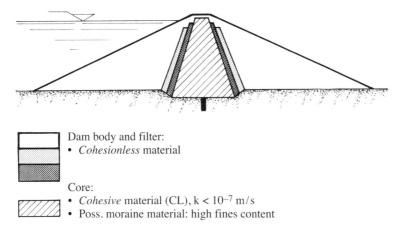

Figure 10.20 Physical properties of materials in the various zones of a rockfill dam with a central core

Table 10.21 Characteristics required for borrow material

Earth/rockfill for shells	
• Alluvial material for earthfill dams; low fines content (possibly after washing)	
• Quarry material for body of a rockfill dam	
Core	
• Clay material with very low permeability	
Filter material	
• Alluvial or possibly quarry material, washed; very specific granulometric requirements	
Rip-rap	
• Large blocks of rock resistant to the dynamic weathering of waves, often hard to find; possibly concrete blocks (tetrapods)	

10.4 Seismicity

10.4.1 Seismic elements[3]

The origins of earthquakes

In his general report relating to Question 51 (Dam resistance to earthquakes) covered in the 13th ICOLD Congress in New Delhi, 1979, R. G. T. Lane noted "that no region in the world is immune to the threat of earthquakes" (Wieland, 2003). This observation is still the case and will continue to be so in the future. History has shown that regions with some seismic activity (including small earthquakes) can also experience violent earthquakes.

The plates that make up the Earth's crust are constantly moving and shifting against one another. Stress builds up in these contact zones, and sudden displacements result in an earthquake. Worldwide distribution of earthquakes confirms this connection with plate boundaries.

The causes of earthquakes can be tectonic (internal fractures in the Earth's crust), but also volcanic or due to the presence of a reservoir. Most large earthquakes occur on existing faults. Different types of slow movement produce deformations, which in turn lead to an increase in stress along the fault. As frictional resistance is overcome, a rupture occurs, resulting in an earthquake (Figure 10.22).

How earthquakes occur

An earthquake occurs when there is a sudden and violent release due to the disturbance of the balance of mass and the transformation of potential energy into kinetic energy. Two main types of waves are generated: longitudinal or primary waves (P) and transversal or secondary waves (S). They propagate to the surface of the Earth, where they form waves known as long waves (L). These can be vertical or lateral and are known as ***shocks***. The initial point of rupture is called the ***hypocenter*** or focus. The size of the event depends on the depth of the hypocenter. The ***epicenter*** is the point at ground level directly above the hypocenter. Depending on the size of the earthquake, the shocks can last for some seconds (magnitude 3) to 20 to 30 seconds (magnitude 5). Longer durations are also possible. An earthquake is generally followed by aftershocks of a shorter duration.

---

[3] The elements present in this chapter refer in part to the book by Markus Weidmann, *Tremblements de terre en Suisse* (Verlag Desertina, Chur).

IDENTIFYING CONDITIONS RELATED TO THE SITE 169

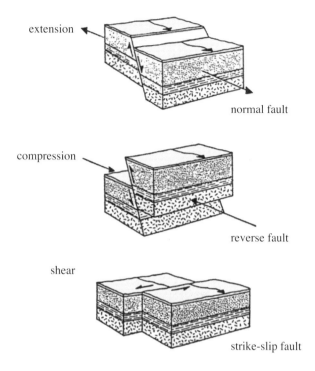

Figure 10.22 Main types of faults

Earthquake characteristics
The parameters for describing the violence of an earthquake are Richter's ***magnitude*** and the scale of ***intensity.***

The term magnitude was introduced by Richter in 1935. Magnitude, which describes the amount of energy released at the earthquake's hypocenter, is a logarithmic value presented in Arabic numerals (M = 6.4; M = 7.2). An increase of magnitude 1 denotes a release of energy approximately 30 times greater (Table 10.23). This increase can be calculated with a mathematical formula using values defined by a seismogram (distance from the seismological station to the earthquake hypocenter and the maximum amplitude of earth movement).

In principle, the greater the magnitude, the greater the rupture surface, and the greater the amplitude of movement. Table 10.24 demonstrates the relationship between magnitude, the size of the rupture area, and slip.

There are several scales that can describe the intensity of an earthquake (Rossi-Forel – RF, modified Mercalli – MM, etc.). Although the modified Mercalli scale is used in the US, the most commonly used scale of intensity in Europe is the MSK 64 (Medvedev-Sponheuer-Karnik, 1964), which establishes 12 degrees (from I to XII, in Roman numerals), describing the effects felt by people and observed damage, mainly to buildings. Table 10.25 presents an extract from the MSK scale. An estimation of associated

Table 10.23 Relationship between magnitude, earth movement, and released energy (from Weidmann, 2003)

Difference in magnitude	Difference in earth movement (factor)	Difference in energy released (factor)
0.1	1.26	approx. 1.4
0.2	1.6	approx. 2
0.3	2	approx. 2.8
0.4	2.5	approx. 4
0.5	3.16	approx. 5.7
0.6	4	approx. 8
0.7	5	approx. 11.3
0.8	6.3	approx. 16
0.9	7.9	approx. 22.6
1	10	approx. 32
2	100	approx. 1,000
3	1,000	approx. 33,000
4	10,000	approx. 1 million
5	100,000	approx. 33 million
6	1,000,000	approx. 1 billion
7	10,000,000	approx. 34 billion
8	100,000,000	approx. 1 trillion

Table 10.24 Relationship between magnitude, size of the rupture area, and energy released (from Weidmann, 2003)

Magnitude	Diameter of the front face of the rupture	Size of the rupture area	Slip
2	100 m	0.008 km^2	4 mm
3	300 m	0.017 km^2	1 cm
4	1 km	0.8 km^2	4 cm
5	3 km	7 km^2	10 cm
6	10 km	80 km^2	40 cm
7	60 km	2,800 km^2	1.2 m
8	250 km	49,000 km^2	10 m

acceleration is given, expressed as a fraction of gravitational acceleration g. The table refers primarily to damage inflicted on three types of constructions:

- "A" constructions: Adobe constructions (blocks of clay dried in the sun) or uncut and poorly bound fieldstones
- "B" constructions: Brick houses, made of prefabricated concrete element walls, wood and brick, cut stone
- "C" constructions: Houses in reinforced concrete, well-constructed wooden chalets

Figure 10.26 provides empirical relationships between magnitude and various physical values.

Table 10.25 MSK 64 scale, description of damage to constructions

MSK degree	Description	Effects	Acceleration, expressed in g
I	Not felt	Only recorded by seismographs	
II	Scarcely felt	Only felt by people at rest in houses, especially on upper floors	
III	Weak	Felt indoors by a few people (like the passing of a truck)	
IV	Largely observed	Felt by many people, rattling, hanging objects swing	
V	Strong	Many sleeping people awaken, hanging objects swing considerably	0.012-0.025
VI	Slight damage	People are frightened, small cracks in adobe walls and plaster	0.025-0.05
VII	Damage	Large cracks in A constructions, slight damage in B constructions, collapse of chimneys, variation in water level in wells, lake water becomes muddy	0.05-0.1
VIII	Heavy damage	Partial collapse of B constructions, cracks in C constructions, cracks in the ground, statues and monuments move	0.1-0.2
IX	Destructive	Destruction of some B constructions, heavy damage to C constructions, landslides	0.2-0.4
X	Very destructive	Partial collapse of C constructions, large cracks in the ground (open up to 1 m), damage to roads, rail lines, underground pipelines	0.4-0.8
XI	Devastating	Major damage to even the most resistant constructions: bridges, dams	0.8-1.6
XII	Completely devastating	Total upheaval of the ground surface, all human constructions are destroyed	> 1.6

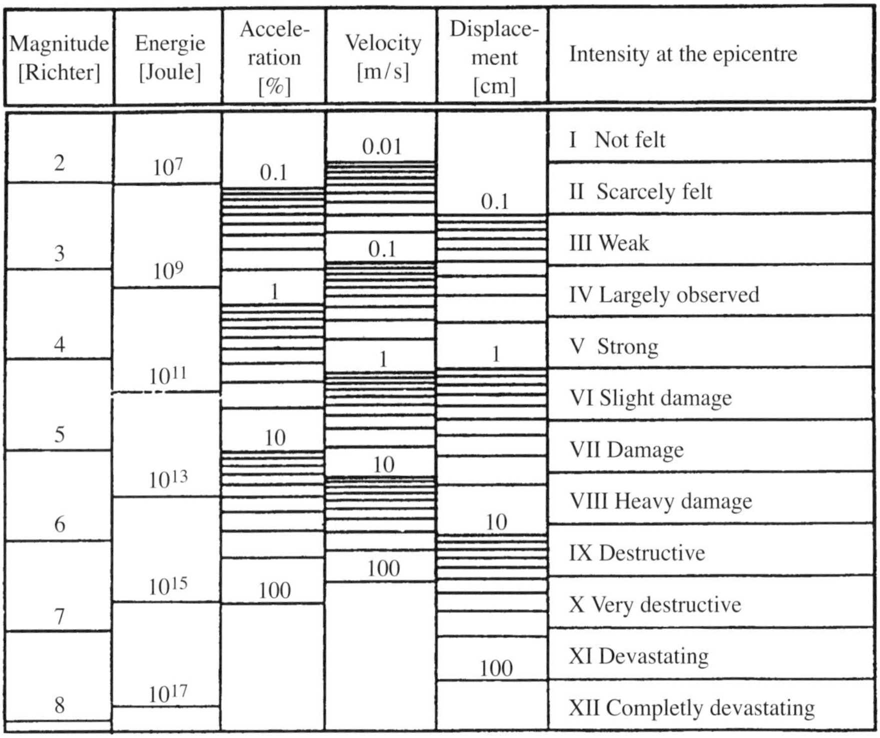

Magnitude [Richter]	Energie [Joule]	Acceleration [%]	Velocity [m/s]	Displacement [cm]	Intensity at the epicentre
2	10^7	0.1	0.01	0.1	I Not felt
					II Scarcely felt
3	10^9		0.1		III Weak
		1			IV Largely observed
4	10^{11}		1	1	V Strong
					VI Slight damage
5		10	10		VII Damage
	10^{13}			10	VIII Heavy damage
6			100		IX Destructive
7	10^{15}	100			X Very destructive
				100	XI Devastating
8	10^{17}				XII Completly devastating

Figure 10.26 Characterization of earthquakes and magnitude according to Richter

Recording earthquakes

Seismographs and accelerographs are instruments that, depending on their design, record the oscillating movement of the base of the equipment with respect to time in two perpendicular horizontal directions and one vertical direction (three components). Some instruments have only one component. The instrument has a substantial mass whose center of inertia remains relatively stationary while the rest of the instrument moves during an earthquake. The output from the recording is called a seismogram or an accelerogram. Modern seismological stations no longer draw the ground motions onto paper or film, but instead transmit the data to a computer, which transforms it into a digital seismogram. An earthquake epicenter can be located using seismograms from several stations. The information they contain is also used to calculate magnitude and the depth of the hypocenter.

The accelerogram in Figure 10.27 illustrates the recording of acceleration a (m/s^2) in a given direction of an actual earthquake (Taft, 1952, M = 7.7). It should be noted that acceleration is often expressed in terms of gravitational acceleration g.

Seismographs do not directly measure acceleration, but the amplitude of the displacements caused by the earthquake. Gravitational acceleration is then calculated.

Assuming a stress that causes a shock to oscillate sinusoidally, as per the following equation

$$\delta = A_{TT} \sin(t/T),$$

where δ is the displacement, A_{TT} is the maximum movement amplitude, and T is the period of movement (in reality, this may vary between 0.1 second and several seconds over the course of one event).

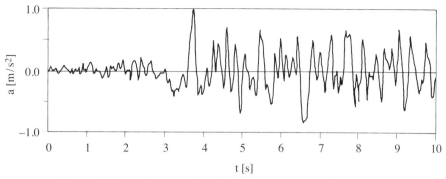

Figure 10.27 Accelerogram from the Taft earthquake, California, 1952

Maximum acceleration is expressed as

$$a = \frac{4\pi^2 A_{TT}}{T^2}$$

And so, for the period T = 0.1 s, acceleration becomes:
- for A_{TT} = 1 mm, a ≅ 0.4 g
- for A_{TT} = 0.5 mm, a ≅ 0.2 g

For a period T = 1 s, a maximum amplitude of movement of 0.1 m equals an acceleration of 0.4 g.

Readings from seismographs demonstrate that an acceleration of 0.4 g corresponds to a devastating earthquake. In Switzerland, and broadly speaking in Europe (except for particularly sensitive zones such as in Greece and southern Italy), a horizontal acceleration of 0.1 g to 0.3 g is accepted as a common value, but one that must be justified in dam design, depending on the region (MSK intensity 7.0 to 8.8). The maximum horizontal accelerations of the Izmit earthquake (August 17, 1999) in Turkey did not go beyond 0.36 g.

Seismic hazard and risk

Seismic hazard is the probability with which an earthquake of a given strength will occur in a given region within a given period. The earthquake force can be expressed by its intensity and also by the maximum ground acceleration. The more intense an earthquake, the more the probability of occurrence (return period) in a given region is low. Seismic hazard is determined on the basis of historical information on earthquakes, and past and recent seismic activity. Modern seismology is capable of predicting the probability of occurrence of the most severe earthquakes on the basis of small, recorded earthquakes. Thanks to wave attenuation models and laws (Sägesser and Mayer-Rosa, 1978; Smit, 1996), it is possible to calculate intensity at a given location (see § 13.7.2.1).

In the case of dams, ***seismic risk*** depends on the hazard (expressed, for example, by the probability that a given peak ground acceleration value will be exceeded), on its resistance to earthquakes (vulnerability), and the consequences of damage caused by the earthquake on the dam (for example, loss of storage capacity, uncontrolled releases of water, damage to a dam and its appurtenant structures). It should be noted that the vulnerability of a dam can be noticeably reduced by taking physical actions to rehabilitate, if necessary, existing structures and by applying the most recent earthquake standards.

10.4.2 Seismicity in Switzerland

Within the European geographical context, Switzerland is one of the countries that has weak to moderate seismic activity. In Europe, zones of strong seismic activity are mostly found in Italy, the Balkans, and Turkey, to which can be added some limited regions in central Europe.

Seismicity in Switzerland's alpine region is directly related to the movement of the African plate toward the Eurasian plate. Switzerland is located at the northern edge of the Adriatic, or Apulian microplate, which is a finger-like extension to the African plate (Figure 10.28). The collision of these plates also leads to the uplifting of the Alps. The Alps are a kind of boundary zone between the African and Eurasian plates. Earthquakes in the Alps are caused by the collision of continental lithospheres in the zone of the Apulian microplate. Earthquakes in the region of Greece and islands in the Aegean Sea are due to the subduction of the oceanic lithosphere of the African plate under the Eurasian lithosphere. Earthquakes have also been recorded north of the Alps in relation to the formation of the southern end of the Upper Rhine Plain, which explains the seismicity in the Basel region.

The Swiss Seismological Service (SSS) at the ETHZ is responsible for monitoring seismic activity in Switzerland and interpreting the data. It initially operated a network of highly sensitive seismic stations capable of recording very weak shocks that was set up from the 1970s onward (SSMNet). From 1998 to 2001, these early stations were replaced by digital broadband stations (Figure 10.29 and Figure 10.30). The recordings were directly transmitted to on-call personnel at the SSS. From 1980, the SSS began developing a network of accelerometers (SDSNet) along with the then Swiss Federal Office for Water and Geology (SFWG) and the Principal Nuclear Safety Division (HSK), which is today the Federal Nuclear Safety Inspectorate (IFSN). The aim was to obtain improved source data to study the hazard and evaluate the safety of dams and nuclear plants.

This second network, put in place between 1991 and 1995 included instruments capable of measuring large earthquakes (Strong Motion stations). Field stations were installed in the country's most active regions and later, five dams were equipped with these instruments (see text box p. 179).

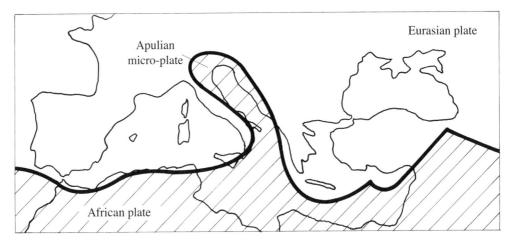

Fig. 10.28 Boundary between the African and Eurasian Plates (from Weidmann, 2003)

A project designed to modernize and increase the density of the accelerometric network (bringing the total to 100 modern stations) was carried out in two phases from 2009. In the first phase, 30 stations were installed. During the second phase, which began in 2013, a further 70 stations were set up. In addition, 43 field stations designed to investigate specific local conditions were also installed. Figure 10.29 illustrates the location of seismic monitoring stations.

The recording of data by seismic instruments is the only way to improve and validate seismic hazard models for Switzerland. This occurs through the analysis of data from small tremors and larger events, which are nevertheless rare. In cases of strong earthquakes, real-time recording will enable the effects on buildings and infrastructure to be rapidly evaluated (Figure 10.30).

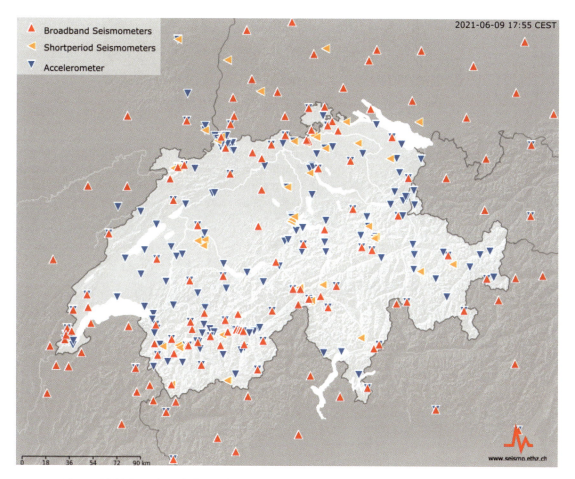

Figure 10.29 Location of all real-time stations monitored by the SED in Switzerland, as at 2021 (Courtesy SSS)

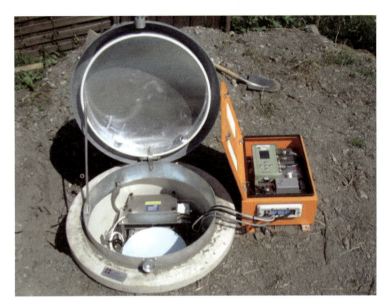

Figure 10.30 Installation of a field seismic station (Courtesy SSS)

With regard to historical statistics from the year 1000 to today, some 640 earthquakes with an intensity equal to or greater than V have been recorded: 125 with an intensity equal to or greater than VI, 40 with an intensity equal to or greater than VII, 12 with an intensity equal to or greater than VIII, and 1 with an intensity equal to IX (Basel, 1356). Table 10.31 lists the strongest earthquakes ever to occur in Switzerland.

Table 10.31 List of the strongest earthquakes recorded in Switzerland (with strong aftershocks) (Courtesy SSS)

Date	Location	Magnitude	Intensity
18.10.1356	Basel (BS) with strong aftershocks	6.6	IX
03.09.1295	Churwalden (GR)	6.2	VIII
25.07.1855	Stalden-Visp (VS) with strong aftershocks	6.2	VIII
11.03.1584	Aigle (VD) with strong aftershocks	5.9	VIII
18.09.1601	Unterwalden (NW)	5.9	VIII
??.04.1524	Ardon (VS)	5.8	VII
25.01.1946	Sierre (VS) with strong aftershocks	5.8	VIII
09.12.1755	Brigue-Naters (VS)	5.7	VIII
10.09.1774	Altdorf (UR)	5.7	VII
03.08.1622	Ftan (GR)	5.4	VII

The map in Figure 10.32 shows the epicenter location for the earthquakes with a minimum magnitude of 2 recorded by the network of seismic stations at the Swiss Seismological Service (SSS) from 1975 to 2018. During this period some 10,000 earthquakes were felt. Figure 10.33 demonstrates the annual repartition of earthquakes with a magnitude of 2.5 or more.

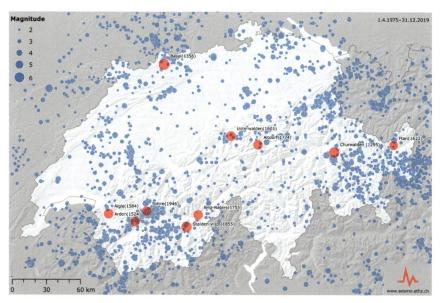

Figure 10.32 Location of the 10 strongest earthquakes of the last 1,000 years in Switzerland (without strong aftershocks) and earthquake epicenters with a minimum magnitude of 2 recorded by the SSS from 1975 to 2019 (Courtesy SSS)

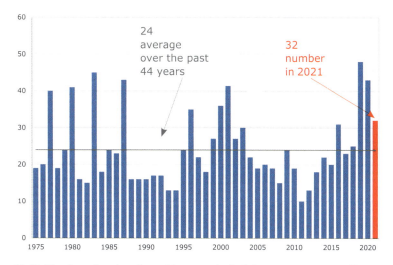

Figure 10.33 Number of earthquakes with a magnitude 2.5 or greater per year (Courtesy SSS)

According to the 2015 seismic hazard model for Switzerland (that replaced the 2004 model), the evaluation of regional distribution has not changed greatly over the past ten years. Valais is still the region with the highest risk of seismic danger, followed by Basel, Grisons, St Gallen Rhine Valley, Central Switzerland, and the rest of Switzerland. Grisons has a slightly higher level of seismicity compared to 2004. These are the regions where there is an increased risk of earthquakes, as demonstrated by the seismic hazard map (Figure 10.34). There is a danger of strong acceleration and damage in zones with high seismic risk (dark areas on the map).

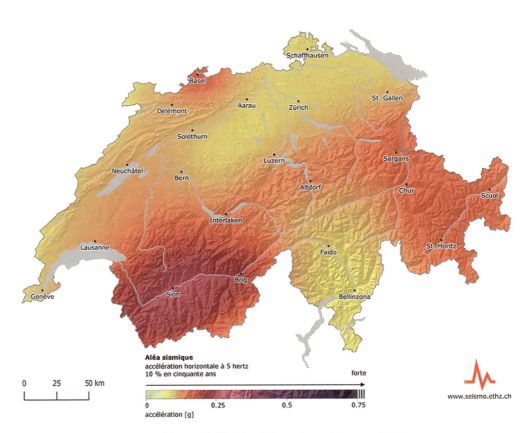

Figure 10.34 Seismic hazard map (Courtesy SSS)

Installation of accelerographs in dams in Switzerland

Some dam owners participated financially in a research project by paying for accelerographs to be installed in dams. The aim of the study was to obtain information about dam behavior under considerable dynamic forces. The decision was made to equip a representative selection of dams constructed in Switzerland (that is, arch dams, gravity dams, and dikes) in the most seismically active areas. The following 5 dams were chosen: the Mauvoisin (VS/H = 250 m), Emosson (VS/H = 180 m), and Punt dal Gall (GR/H = 130 m) arch dams, the Grande Dixence gravity dam (VS/H = 284 m) and the Mattmark dike (VS/H = 120 m).

The instruments were installed based on the type of dam and agreed goals. Generally speaking, the primary elements that were studied were the effective loading, the propagation of waves along the abutments, the dynamic properties of the dams, and possible nonlinear, inelastic behavior. In addition, in each location, an exterior device was installed at a certain distance from the dam in order to pick up movements not influenced by the dam and its reservoir. The below figure illustrates the layout of 11 accelerographs in the Mauvoisin arch dam, which was reported as being the best equipped dam in the world (Chopra, 2020). Five accelerographs were installed in the Grande Dixence gravity dam, 3 in the Mattmark dike, 5 at the Punt dal Gall arch dam, and 5 in the Emosson arch dam. The instruments were arranged in a star layout and connected to a control center that recorded acceleration values (Darbre, 1993a; Darbre et Pougatsch, 1993; Darbre, 2004).

This layout proved to be effective, and the results obtained were helpful for modelling the foundation and for preparing guidelines.

(a)
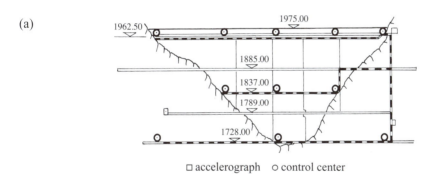

(b)

Mauvoisin dam: (a) layout of accelerographs; (b) installation of an accelerograph and the control center for the dam network (Courtesy P. Smit)

10.4.3 Earthquakes and dams

First of all, it should be noted that earthquakes have an impact on every part of a water-retaining facility, that is, the dam, appurtenant structures, the foundation, uplift, and hydromechanical and electromechanical equipment. Rock subjected to an earthquake has good to moderate behavior, which means that it barely amplifies or prolongs the ground motion. However, if the foundation includes loose rock or soil of a height greater than 5 m, surface movements in particular can be amplified.

Dams are designed to take up horizontal forces, as hydrostatic pressure has a major horizontal component. Due to this, dams are able to resist horizontal seismic forces issuing from ground motion in this plane.

The dams that are the most resistant to dynamic forces are:

- Arch dams and arch gravity dams, owing to their hyperstatic behavior
- Rockfill dams with a central clay core, through their ability to tolerate large deformation

The transversal joints of gravity dams are usually open, which may constitute a sensitive zone during an earthquake. This situation can be substantially improved if the joints are filled with cement grout and present a surface that can tolerate shearing (dam joint with shear box, see Fig. 13.44). The Lower Cristal Springs gravity dam, near the San Andrea fault, is an example of this. The rubbing of the joints and the interlocking of the blocks prevented movement perpendicular to the joints during the San Francisco earthquake in 1906.

Through the shape of their structure, buttress dams can tolerate very little transversal stress. Widening the buttress webs at the downstream end so that they join up is one way of solving this (Part IV, Section 14.4).

Rockfill dams with upstream facing can suffer from vulnerability in the impermeable layer. During the Wenchuan earthquake (Sichuan province, China) on May 12, 2008 (magnitude 7.9), the upstream facing of the Zipingpu concrete dam (H = 159m) was damaged. However, the highly permeable shell body was a positive factor.

Indeed, in the course of the Wenchuan earthquake, some 1800 dams and reservoirs were damaged; most of these were small embankment dams in addition to a further 400 hydropower plants (Wieland and Chen, 2009).

In the past, earthquake resistance to dams was achieved with a pseudo-static approach, a technique that was developed in the 1930s. For concrete dams, this simple method takes into account the dam's inertia force and the hydrodynamic pressure exerted by the reservoir. For embankment dams, the static inertia force is applied to the sliding mass. A value of 0.1 g was usually employed as a horizontal seismic coefficient. In Switzerland, the case of load due to an earthquake, while taking into account the static force of the replacement, was also used in the design of all dams built after the 1950s. The development of dynamic analysis tools has led to further progress in the evaluation of the seismic safety of dams.

From the 1980s, modern design parameters for the resistance of dams to earthquakes were developed by the ICOLD technical committee on "Seismic Aspects of Dam Design" (Bulletin 148, 2014). Analysis must also cover appurtenant safety structures such as spillways and bottom outlets (SFOE, 2018). Current practice has been summarized by Wieland (2014).

The case of induced earthquakes during impoundment and dam operation must also be raised (see § 4.2.4). The result of findings on this subject by an ICOLD working group revealed that this phenomenon only occurs if tectonic conditions are favorable, for example, the presence of a fault near to the rupture point. Rapid variations in reservoir water level can also be the cause of an increase in this type of seismic activity. It should be noted, however, that a dam designed and constructed in accordance with professional standards can resist an induced earthquake without difficulty, as this kind of seismic activity can never be greater than the *maximum credible earthquake* (MCE). Finally, the implementation of an appropriate monitoring system is recommended to track induced seismic activity during and after first impoundment; five seismographic stations are considered to be the minimum (ICOLD, 2011).

Dam inspection after earthquakes is an important part of a dam's overall safety concept (ICOLD, 2016).

10.5 Climatic Conditions

Climatic conditions significantly affect the construction of a dam, and therefore the time frame for delivery of the project. The same can apply with regard to the durability of the dam. Some specific cases are described below as examples:

- Clay core of embankment dams (conditions during construction)
  Water content is the most important criteria for the optimum placement and compaction of the core. In regions where the rainy season is long and intense (tropical rains), placement is often interrupted due to excessive saturation levels in the materials.

- Buttress dams (conditions during operation)
  The difference in temperature between the head, including the upstream face in contact with the cold reservoir water, and the web, subject to the sun's rays, results in major thermal gradients that can lead to concrete cracking.

- Asphalt concrete of upstream facing
  This material is particularly sensitive to the effects of extreme temperatures:
  – plastic deformation under high temperatures
  – accelerated ageing under the effect of ice and exposure to the sun's rays

10.6 Hydrological Study

10.6.1 Flood control

The design flood is a particularly complex element in dam design, due to the uncertain and probabilistic nature of the selected values. The hydrological study must be carried out by experts (engineers, hydrologists, meteorologists).

The issue of flood control and the integration of appurtenant structures are key criteria when selecting the dam type. Concrete dams can tolerate a possible exceeding of the project flood and therefore overtopping without excessive damage, as long as stability is guaranteed during and after the overtopping, tensile and compression stress remain within the limits, and conditions at the downstream toe tolerates it (for example, no risk of scouring). However, overtopping in embankment dams is catastrophic and can result in the total collapse of the dam and considerable damage downstream.

Flood control is dependent on:

- The characteristics of the catchment and its hydrology: to better understand the behavior of a catchment, it is helpful to find out more about its topographical, geological, and pedological characteristics and the climatic conditions of the region (rainfall, air temperature); flow measured in watercourses also provides vital information for hydrological studies
- The effect of flood attenuation related to the reservoir (surface area, freeboard) and spillway structures
- The type of dam

Gravity dams and buttress dams
Broad weirs can be installed on the dam, which enable high flow rates with very favorable conditions. Existing dams with a spillway capacity of more than 3000 m³/s can often be seen.

Arch dams
In narrow valleys, weirs located on the crest have a limited capacity. Submerged spillway structures (orifice spillways) can easily be integrated, or the spillway must be designed at the sides of the dam abutments.

Embankment dams
Spillways cannot be integrated into the dam, given the incompatibility in deformation between the infill body and the reinforced concrete structure of the spillway. A solution along the abutments of the valley flanks at the sides of the dam must be found. On particularly long dams, in some cases the combination of a section of gravity or buttress dam including a spillway component with long sections of embankment on both sides may be seen.

10.6.2 Definition of a flood

A flood is defined by its hydrograph, which is primarily characterized by its peak or peaks and volume, as well as its duration, rate of water level rise, time of water level rise, and time of water level fall (SFOE (SFWG), 2002b) (Figure 10.35).

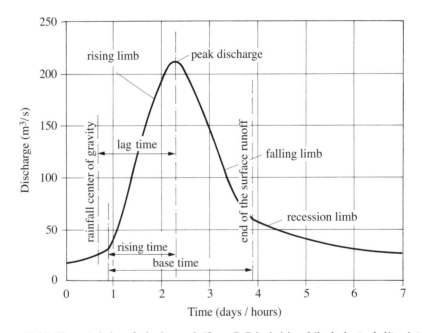

Figure 10.35 Characteristics of a hydrograph (from G. Réménérias, *L'hydrologie de l'ingénieur*)

Floods include natural and indirect inflow resulting from operation (SFOE, 2018). Inflow into the reservoir can be defined as follows:

$Q_D(t)$ natural inflow from the direct catchment area
$Q_I(t)$ quantity of water fed in from an indirect catchment area (feed-in capacity)
$Q_T(t)$ quantity of water fed in from turbines from a hydropower plant further upstream (turbine capacity)
$Q_P(t)$ quantity of water pumped in from a hydropower plant further downstream (pump capacity)
$Q_R(t)$ quantity of water flowing back from the surge tank of a hydropower plant further downstream

When analyzing flood safety, hydrographs must be calculated for the most extreme levels affecting the safety of the dam and reservoir (see § 10.6.3), while taking into account retention in the reservoir and release capacities. Different scenarios can be envisaged depending on the operating conditions involving the above elements for extraordinary and extreme situations. If the reservoir capacity is negligible, only the peak can be taken into account.

On the basis of the flood hydrograph entering the reservoir and initial water level (assuming that the water level is at the reservoir's normal level, or the initial water level elevation that corresponds to the maximum operating level), the calculations describe

- Peak discharge (Figure 10.35)
- Maximum rise in water level in the reservoir during the passage of a project flood (Figure 10.36)
- Flood storage volume (Figure 10.36)

In Figure 10.36, total freeboard and safety freeboard can be identified. By definition, freeboard corresponds to the vertical distance between a water level and the crest of the dam.

It should be noted that a flood is often accompanied by additional processes such as erosion, bed load, deposit formation, and the arrival of wood and floating debris (SwissCoD, 2017b).

10.6.3 Floods considered

Definitions

The size and exceedance probability of floods must be selected so as to guarantee dam safety. To protect against dam failure and its consequences, flood control must be in line with exceptional and extreme events (Table 10.37). The analysis of exceptional and extreme floods that have occurred in Switzerland has shown that the ground was saturated at the time of the event. Due to this, a flow coefficient equal to or near 1 may be assumed.

In Switzerland (SFOE (SFWG), 2002b; SFOE 2018), flood events taken into account for the safety of water-retaining facilities in cases of floods are defined in the following manner:

- Design flood (this corresponds to an exceptional situation)
- Safety flood (this corresponds to an extreme situation; it used to also be called "Deluge")

The ***design flood*** corresponds to an exceptional event. It must be released in usual flow conditions, without creating any damage (either to the dam structure itself or to the water release structures) and with a safety margin set by the final freeboard, known as the safety freeboard. In cases of water-retaining facilities with medium to high storage heights, the design flood must be equivalent to an exceedance probability of $1/1000$ (1000-year flood Q_{1000}).

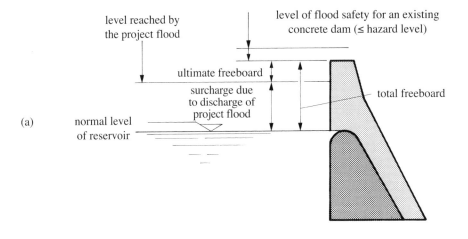

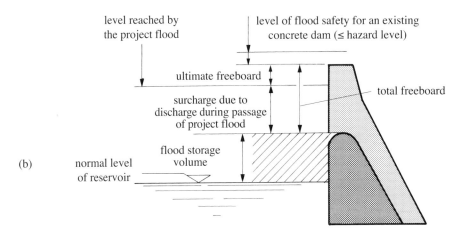

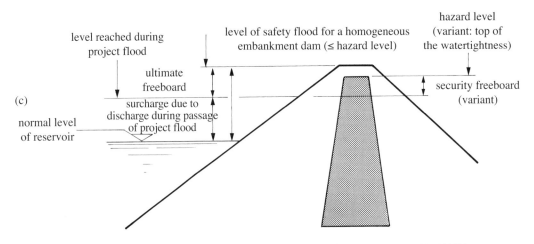

Figure 10.36 Definitions of levels and reservoir storage (from SFOE (SFWG), 2002b)

Table 10.37 Correlation between various flood events and their return period

Event	Frequent	Rare	Exceptional	Extreme
Type of flood			Design flood	Safety flood
Return period (years)	30	100	1000	n * 1000 PMF [4]

Furthermore, the safety freeboard between the level of the crest and the maximum water level reached in a design flood must be sufficient for the passage of the safety flood and avoid overtopping due to waves[5] (Figure 10.36). Experienced engineers are responsible for establishing the value of the safety freeboard on a case-by-case basis. Table 10.38 gives indicative values for the safety freeboard (see also Section 27.1).

Table 10.38 Indicative values of safety freeboard in relation to the crest

Type of dam \ Height of dam	H < 10 m	10 m ≤ H > 40 m	H ≥ 40 m
Concrete dam	0.50 m	1.00 m	1.00 m
Embankment dam			
• homogeneous	1.00 m	2.00 m	3.00 m
• with core	1.00 m	2.00 m	3.00 m
• with upstream facing	1.00 m	1.50 m	2.50 m

The *safety flood* corresponds to an extreme event that the dam must be capable of tolerating and releasing. Its arrival into the reservoir results in the maximum permissible level, which corresponds to the *danger level*. The safety flood is the PMF (*Probable Maximum Flood*), which is based on PMP (*Probable Maximum Precipitation*) or is contingent on the 1000-year flood (Q_{1000}). In the latter case, the safety flood allowed for in new projects or when evaluating the safety of existing dams is assumed to be $1.5 \times Q_{1000}$. The safety flood must be released without exceeding the danger level, which is defined as the water level above which the water-retaining facility is endangered due to damage to the crest, erosion of abutments, scouring at the dam toe, internal erosion, and an increase in uplift. For concrete dams, the danger level can be situated above the level of the crest (or a parapet that resists water pressure). However, the danger level of embankment dams must not pass the crest (for homogeneous earthfill dams) or the upper level of the sealing system (core, membrane, etc.). Be that as it may, it is up to an experienced engineer to establish the danger level.

---

[4] *Probable Maximum Flood* (PMF)
[5] It is important to remember that for embankment dams, the freeboard must also allow for possible settlement during earthquakes. Total freeboard must also take this into account.

Conditions during floods

As is often the case, a flood is accompanied by numerous other factors: difficult or impossible to access the dam, power cut, or the arrival of floating wood and other debris. As such, specific situations have to be allowed for when verifying safety during a flood. For example, it may not be possible to operate all mobile water release structures. Table 10.39 presents the conditions for adequate verification. Analyzing the effects of an obstruction or the disabling of other water release systems necessary for flood evacuation is also recommended.

Table 10.39 Conditions around the operation of mobile water release structures during flood safety verification (SFOE, 2018)

Concrete dam	Design flood	The most efficient structure among "n" gated discharge and outlet structures is out of service ("n – 1" rule).
		No water can be diverted through intake works, except if the power plant is protected against flooding and the passive release of water (for example, via bulb turbines in open position) or the continued operation of turbines (for example, if two separate high-voltage power lines are available to release energy). In any case, no more than "n – 1" turbines may be taken into account during verification.
	Safety flood	All discharge and outlet structures can be employed.
		No water can transit via intake works.
Embankment dam	Design flood	The most efficient structure among "n" gated discharge and outlet structures is out of service ("n – 1" rule).
		No water can be diverted through intake works, except if the power plant is protected against flooding and the passive release of water (for example, via bulb turbines in open position) or the continued operation of turbines (for example, if two separate high-voltage power lines are available to release energy). In any case, no more than "n – 1" turbines may be taken into account during verification.
	Safety flood	The most efficient structure among "n" gated discharge and emptying structures is disabled ("n – 1" rule).
		No water can transit via intake works.

Incidents from floods

a) Floating wood and debris
In terms of the structural conditions, the arrival of floating wood and debris has an impact on the size of the passes and the clearance zone of a structure.

Large floating wood debris can be picked up and shifted by flood waters. This includes trunks and large roots of natural trees (dead or fresh wood), timber from forestry operations and deforestation, or wood from bridges and/or riverbanks (SwissCoD, 2018b).

There are three options for dealing with floating debris upstream of dams: adopting measures in the intake zone to reduce the accumulation of large trunks as much as possible, allowing debris to pass through the spillway, and capturing debris and removing it from the reservoir. According to observations made following large floods (in particular in 1978), tree trunks in mountain rivers and streams transported by a flood are soon broken down into maximum lengths of approximately 10 m. For dams on large rivers and in plains, the width of the spillway passes must be greater than 10 m.

The width of spillways passes and the vertical freeboard to superstructures, such as fully open gates and bridges, must be sufficiently wide and open to avoid trees and other floating debris building up and creating an obstacle.

In addition, to ensure that tree trunks and other floating objects transit through the dam, water release structures must ideally be designed as a free overflow spillway. As far as is possible, the construction of superstructures and piers should be avoided.

When passes are unavoidable, the width of openings should be in the order of 80% of the possible length of trunks. The table below gives minimum indicative values for openings (CFBR, 2013).

Minimum width for openings of water release structures based on the elevation of water level z

Reservoir altitude	Z <= 600 m.a.s.l.	600 m.a.s.l. < z < 1800 m.a.s.l.	Z <= 1800 m.a.s.l.
Minimum width of the outlet opening	15 m	Linear interpolation	4 m

When there is a superstructure (for example, a bridge), there is a risk of obstruction if the free distance between the water level and a superstructure is lower than the value indicated in the figure below.

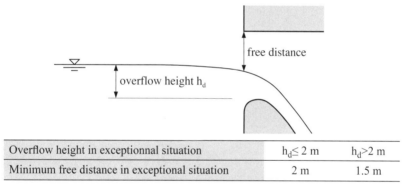

Overflow height in exceptionnal situation	$h_d \leq 2$ m	$h_d > 2$ m
Minimum free distance in exceptional situation	2 m	1.5 m

Definition of the free distance between the water surface and a superstructure

After the spillway at the Palagnedra dam (TI) became blocked (Figure 5.13), a new bridge was built upstream of the dam (SwissCoD, 2017b; Pougatsch et al., 2011). A central pier of the spillway at the Ferden dam (VS) was also removed during dam operation (Figure 5.2). Other instances have been described in the SwissCoD report "Floating Debris at Dam Spillways" (2017b).

b) Scouring

Scouring at the downstream toe of a dam can threaten its stability (Schleiss, 2002). To limit this phenomenon, the energy of the falling jets can be dissipated in a plunge pool or, if topographical and geological conditions are suitable, released into the downstream watercourse by a ski jump. If no plunge pool has been designed, the geometry of the predicted scouring must be established, and if required, dam stability in this zone must be checked.

In cases of downstream scouring, carrying out a periodic topographical study in the zone is recommended.

10.6.4 Methods for flood estimation

Rare floods should not be calculated by applying one single formula or method. Instead, all appropriate methods should be used and the available data specified. Calculations must be adapted on the basis of prior experience, and the results must be regularly reexamined. Measurements from a broad range of sources, new information, the latest developments, and the most recent events must be taken into account.

Different methods enable a flood to be characterized. The hydrologist is responsible for defining the methods to apply in each specific case. These include:

- Historical methods
- Empirical formulas
- Probabilistic methods
- Deterministic methods (PMP-PMF)

Historical methods and empirical formulas should be used with care. They can be useful for establishing an initial approximation, but the other methods are more appropriate.

10.6.5 Historical methods

The aim is to evaluate flood peaks that occurred at a time when no direct hydrological measurements were available. Historical data is only available for large catchment areas. This method is therefore not appropriate for dams with a small catchment. In any case, the design flood should not be smaller than a historical flood.

10.6.6 Empirical formulas

Empirical formulas were developed to calculate the "maximum" peak flow value of the flood depending on the catchment area only or by taking into account other parameters specific to the catchment such as

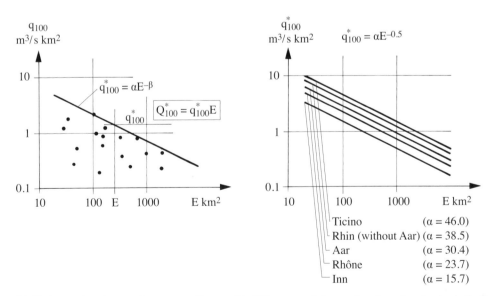

Figure 10.40 Approximative envelope curves for specific 100-year peak flows of various river basins in Switzerland (from Biedermann et al., 1996)

the runoff coefficient and soil characteristics. Other formulas consider the flood return period, rain, or mean flow. The time of concentration can also be evaluated using empirical formulas. The maximum flood values observed and calculated can be illustrated in a graph depending on the area of the catchment. The envelope curve of the corresponding points can be plotted; the applicable values for other catchment area surfaces can then be calculated (Figure 10.40). These formulas must be used carefully. Generally, the area of the catchment should not be greater than 100 to 200 km^2.

10.6.7 Probabilistic methods

The values are grouped according to their exceedance probability. Low exceedance probabilities are obtained using distribution laws. The Gumbel, log-Pearson type III, and log-normal are the most commonly used probability laws for determining extreme values. The measurements must range at least from 30 to 50 years, so that a 100-year flood can be extrapolated (Figures 10.41 and 10.42).

Direct methods
Peak flood flow with a low probability of occurrence can be calculated through measured flows.

Indirect methods
The flow is calculated on the basis of rainfall. These methods can be used in a region where there is a dense network of rainfall stations, as well as measurements that span a long period of time (more than 30 years). When analyzing these elements, rain intensity over a given time period is used. The runoff coefficient value must always be near to 1, but greater than 0.8 (for catchment areas up to 200 km^2). Other values are only accepted if they have been verified. If possible, verification on the basis of real events must be carried out.

Hypotheses for precipitation-discharge model
A precipitation-discharge model can be used to assign a time-dependent inflow into the reservoir to a precipitation event (event-based modelling) or a precipitation time series (long-term simulation). Such a model must be able to depict the hydrological behavior of the catchment area during extraordinary and extreme events.

Aggregated Gradex
The Gradex method is not only based on the statistical treatment of observed inflow, but also incorporates the measurement of precipitation, for which observation periods are generally longer (CFGB, 1994). This introduces the hypothesis whereby the frequency of rainfall over a given time period and in a given location follows a simple exponential function, for which the exponent or Gradex (gradient of extreme values) is deduced from a series of measurements. Another hypothesis postulates that the ground is practically saturated for flood flows with a high return period, and consequently, rainfall runoff is total. An increase in rainfall corresponds to an equivalent increase in flow. The rain-flow relationship becomes almost linear. The intensities of rainfall, as well as flow, are extrapolated onto a Gumbel probability graph (Figure 10.43). Once the flood volume for a given period of time has been calculated, peak flow is deduced from mean flow. This method can be applied to catchment areas from ten or so square kilometers to several thousand, providing the rainfall distribution is relatively homogeneous.

Analysis on a regional basis
For a hydrologically homogeneous region, this method involves combining all the related observations in order to calculate a flood flow with a given return period. A typical relationship for the whole must be

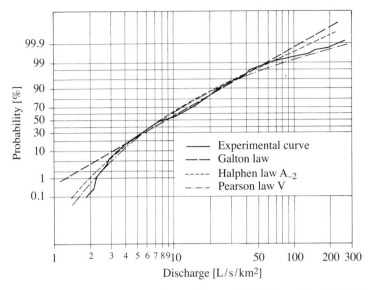

Figure 10.41 Example of adjustment using various probability laws (from G. Réménérias, *L'hydrologie de l'ingénieur*)

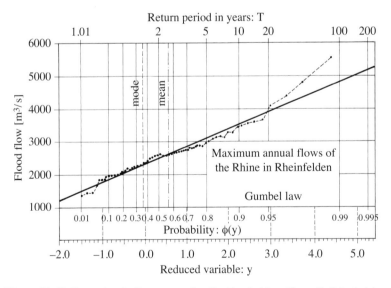

Figure 10.42 Example of adjustment using the Gumbel law (from G. Réménérias, *L'hydrologie de l'ingénieur*)

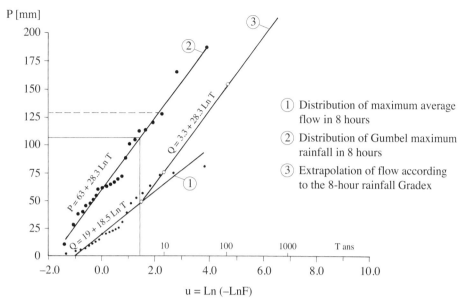

Figure 10.43 Example of the Gradex method (from CFGB, 1994)

established (for example, between the mean flow from various catchments and their surface areas) and then the available various measurements must be homogenized (for example, flow measurements). Thanks to this procedure, dimensionless values are obtained and constitute one single series, which can be used as a basis for a frequential analysis.

10.6.8 Hypotheses for synthetic hydrographs

If synthetic hydrographs are established on the basis of precipitation, the assumption has to be made that the volume of water flowing into the reservoir is equal to the volume of precipitation, as the runoff coefficient is assumed to be equal to 1.

Should the contribution from snow and glaciers toward flood events be significant, this has to be taken into account; in this case, preference should be given to a precipitation-discharge model.

If no other specific clarifications have been carried out, in an initial approximation the synthetic flood hydrograph in accordance with Maxwell (Sinniger and Hager, 1984)

$$Q(t) = \left(\frac{t}{t_{max}} e^{\left(1 - \frac{t}{t_{max}}\right)} \right)^n Q_{max}$$

where the corresponding flood volume

$$V = Q_{max} t_{max} \frac{e^n n!}{n^{n+1}}$$

may be assumed. Time t_{max} corresponds to the duration up until the flood peak; it may be assumed that this is equivalent to the duration of precipitation.

Exponent n is assumed to have the value 6. Deviating values between 1 and 6 may be used if these have been determined through studies of the specific characteristics of the catchment area.

10.6.9 Procedure for calculating the flood safety level based on the design flood

With this procedure, the hydrograph of the natural inflow of flood safety level $Q_{D,S}(t)$ can be estimated based on the hydrograph of the corresponding part of the design flood $Q_{D,B}(t)$.

For existing dams: by increasing the inflow by 50% (Figure 10.44) (Biedermann et al., 1988):

$$Q_{D,S}(t) = 1.5\, Q_{D,B}(t)$$

For new or altered dams: by increasing the inflow and event duration by 50% each (Figure 10.44) (Biedermann et al., 1988; SFOE, 2018):

$$Q_{D,S}(t) = 1.5\, Q_{D,B}(\tfrac{2}{3}t)$$

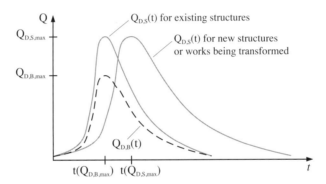

Figure 10.44 Safety flood schematically derived from the project flood

10.6.10 Deterministic methods (PMP-PMF)

These involve the simulation of a runoff process. The probable maximum flood (PMF) is determined from the heaviest precipitation physically possible (PMP). The PMF is derived from the most severe combination physically possible of meteorological and hydrological conditions in a specific region in a given season and for a given timeframe.

The transformation of rain into runoff usually takes place in two stages:
- Calculation of net rain
- Use of a transfer function that enables runoff from net rain to be calculated while conserving the total flow volume.

PMP maps for Switzerland are presented in Hertig et al. (2007). Up until today, little experience has been had in applying these PMP maps. Calculated precipitation (PMP) and flow (PMF) should therefore be compared and assessed alongside the results from other methods.

One PMP-PMF approach, developed by several EPFL institutes and other partners under the name of CRUEX, and implemented as CRUEX++, is described below in the text box.

CRUEX Project

The CRUEX research project, which is aimed at calculating extreme floods, was funded by the Office fédéral des eaux et de la géologie (SFWG). The interest and originality of the developed method resides primarily in:

- The interaction of meteorological, hydrological, and hydraulic phenomena
- The consideration of spatial variation and the temporal curve of these phenomena
- The calculation of the rainfall duration leading to a "critical" PMF for the reservoir

The three academic partners for the project were:

- the Laboratoire de Constructions Hydrauliques (LCH-IHE): for hydraulic modeling and questions relative to the operation of dam facilities
- the Laboratoire de Systèmes Energétiques (LASEN-IHE): for modeling of precipitation and meteorological questions
- the Hydrologie et Aménagements unit (HYDRAM-IATE): for modeling hydrological functions of production and transfer

The PMP-PMF issue

Areas situated in zones influenced by watercourses are not only subject to the vagaries of weather, they are also tributary to anthropogenic interventions that modify the behavior of their catchment area. Large dams and their reservoirs thus produce a flood attenuation effect that enables the extent and frequency of floods downstream to be reduced. In order to completely master this beneficial effect, the structural and functional safety of these dams must be guaranteed, particularly in regard to their ability to withstand extreme floods. This conceptual entity known as a *Probable Maximum Flood* (PMF) (and also called a *deluge*) derives from the assumption that there is a physical limit to both precipitation and the resulting floods. Models based on physical laws are used to investigate the probable maximum precipitation and "critical" PMF, which, in the case of a dam, may be defined as follows:

The flood that results from the most severe combination of meteorological conditions physically possible in a given region and whose inflow into the reservoir, initially full but with water release systems operating, will lead to the maximum water level.

The CRUEX methodology

The aim of the CRUEX project was to develop a PMP-PMF type methodology capable of integrating the hydrological specificities of the alpine environment, where orographic effects play a major role in the generation and abatement of meteorological conditions. This research was directly related to the safety of dams located in this geographical context.

The proposed analytical progression went through four distinct stages, which are described below.

First stage: Formulation, calibration, and validation of rainfall-runoff models
The first stage consists in defining rainfall generation models, hydrological transfer and hydraulic routing, which are used for the simulation and then its calibration.

An initial model must be able to reproduce the precipitation in the catchment area after a specific meteorological situation. Ideally, the parameters to be considered are depth of precipitation and its spatiotemporal distribution. For the small catchment areas of alpine reservoirs, the volume and duration of precipitation are generally sufficient, as the hypothesis of uniform rainfall distribution may be assumed and its temporal distribution subjected to a sensitivity study.

The second model is designed to simulate the functions of production and transfer, which enables the precipitation in a catchment area to be transformed into a flood entering the reservoir. In alpine catchment areas characterized by a rock geology with a low level of permeability and for extreme events, it can be assumed that the total amount of rain will be transformed into inflow. This conservative hypothesis compensates for possible flow resulting from high-altitude snowmelt. It is important for the model to allow continual simulation and a sectorization of the catchment area that is based on physical reality.

The third model is required to reproduce the flood attenuation effect of the reservoir during a flood. It is simply a question of solving the storage equation that, in addition to the incoming flood, has an effect of the reservoir's "level-volume" and "level-outgoing flow" relationships, as well as the initial impoundment condition. During an extreme flood, the reservoir is initially considered as full, and the safety instructions relating to the operation of water release structures are applied, according to the relevant flood safety guidelines.

Second stage: Maximizing orographic precipitation (PMP)
The PMP is defined as the maximum amount of water that can fall over a given duration, across a given surface, in a specific location, and at a given time of the year.

In highly uneven terrain, such as in the Alps, the spatial distribution of probable maximum precipitation is essentially calculated by the original orographic component, for which relief plays a major role. Non-orographical processes, such as those associated with summer storms, should not be overlooked when establishing extreme precipitation maps.

At each point in the domain, the probable maximum precipitation ultimately corresponds to the maximum stable orographic contribution or to convective contribution.

Third stage: Maximizing flood hydrographs (PMF)
Generally speaking, a hydrograph is characterized by the volume of water, maximum flood peak, and temporal curve. The PMF can therefore adopt various forms depending on the studied characteristic or effect. For a reservoir, "critical PMF" is the one that will cause the maximum increase in water level, depending on spillway operation.

Fourth stage: Retention calculation to find critical PMF
The final step consists in giving a temporal curve to the PMP for different durations of precipitation. The minimal duration to consider generally corresponds to that which will allow the entire catchment area to contribute to the formation of water flow at the outlets.

Different approaches can be adopted to define the precipitation hyetograph, for example, by the homothetic weighting of historic episodes, the use of stochastic models, or the definition of a design flood for the catchment area. At the end, the PMPs generated for various durations of precipitation must undergo flood retention calculations for the reservoir in order to obtain the maximum outflow. Under that condition the reservoir level will be the highest and the required safety condition must be satisfied.

(from Boillat and Schleiss, 2002)[6]

---

[6] See also Communication 5 from the LCH, 1996.

The CRUEX++ research project

The CRUEX++ project (2012–2017) is the continuation and culmination of the CRUEX project (Zeimetz et al., 2017). CRUEX++ developed a new methodology for estimating extreme floods. A computer program has been created that makes application of the CRUEX++ method accessible.

This research work develops an approach combining statistics and hydrological modeling techniques to propose an innovative ready-to-use method to address the complex issue of extreme flood estimations. In order for this to be successful, some gaps in scientific knowledge had to be addressed before focusing on the combination of statistics and simulations. Concerning the temporal rainfall distribution, it can be shown that a unique rainfall mass curve is assumed for the entire territory of Switzerland. Regarding a coherent combination of temperature and extreme precipitation, linear relations between the duration of the precipitation event and the zero-degree isothermal altitude can be determined. In the context of hydrological modeling, the influence of initial conditions for extreme flood simulations have been assessed and methodological recommendations for the choice of initial conditions for extreme flood simulations can be formulated. Furthermore, the maximum assumed spatial expansion of the PMP events derived from the Swiss PMP maps, elaborated during the CRUEX project, can be estimated to be 230 km². Finally, the combination of the simulation results with the approach of upper bounded statistical extrapolations can be shown to be advantageous: sample sensitivity is reduced, and the plausibility of the extrapolations is enhanced compared to conventional statistical distributions.

Ultimately, a holistic methodology for extreme flood estimations can be formulated. The main advantage of the methodology is that it allows extreme flood hydrographs to be estimated using hydrological simulation and a return period to the simulated peak discharge to be plausibly attributed, taking a deterministically determined upper discharge limit into account by referring to the PMP-PMF approach. The methodology thus combines the possibility of flood routing estimations with the knowledge of the occurrence probability of the peak discharge of the event, which is normally a reference quantity in flood safety guidelines.

In Zeimetz (2017), the CRUEX++ methodology process was illustrated through applied examples (the Limmernboden, Contra, and Mattmark dams in Switzerland). The results were compared with estimations based on conventional statistical methods. In this way, the advantages of the new CRUEX++ methodology were able to be demonstrated.

(from Zeimetz, 2017).

11. Actions and Forces

Chatelôt arch dam in Switzerland, height 74 m, year commissioned 1953 (Courtesy ENSA)

11.1 Types of Loading

During its construction and operation, a dam is subject to loading and actions that lead to deformation and stresses. These loads are part of verification calculations for a dam and can be divided into the three following groups, depending on how they act on the dam:

- Permanent loads
- Variable loads
- Exceptional loads

Permanent loads are those that are always present. However, they may appear over time and continue without undergoing any modifications.

Variable loads vary depending on operating conditions or the natural environment.

Exceptional loads generally occur after sometimes violent natural events whose effects may be sudden and may only last for a limited time.

Table 11.1 and Figure 11.2 list the types of loads in each group.

Table 11.1 Loading considered for the verification of dams

Permanent loads	Variable loads	Exceptional loads
• Self weight (structure, gates)	• Water pressure	• Floods
• Earth pressure (earth/rockfill)	• Sediment pressure	• Earthquakes
• Anchoring forces	• Uplift	• Avalanches
	• Concrete temperature	• Debris flows
Under the permanent presence of water:	• Seepage pressure	• (Aircraft crash)
• Water pressure	• Pore pressure	
• Uplift	• Snow	
• Pore pressure	• Ice load	
	• Bearing loads	

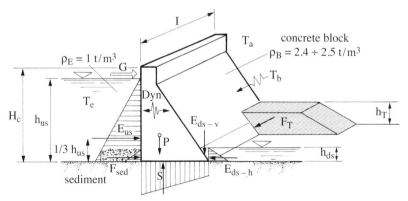

Figure 11.2 Distribution of forces and actions to be considered. Key: P = self weight; E_{us} = upstream horizontal water pressure; E_{ds-h} = downstream horizontal water pressure; E_{ds-v} = downstream vertical water pressure; F_T = earth pressure; F_{sed} = sediment pressure; S = uplift; T = thermal effects; T_b = dam body temperature; T_a = air temperature; T_e = water temperature; Dyn = dynamic loading; G = ice load; H_c = dam height from foundations; h_{us} = upstream hydrostatic head; h_{ds} = downstream hydrostatic head; h_T = height of a downstream embankment.

11.2 Combining Loads

According to the Swiss legislation relating to the safety of water-retaining structures (WRFO, 1998; WRFA, 2010), the safety of these structures must be guaranteed in all foreseeable cases of loading and operation. The engineer is responsible for establishing the loads in question as well as the possible combinations of loads depending on the type of dam (concrete dam, embankment dam) and its purpose (water storage, protective structure). Load scenarios for an empty or full reservoir, while taking flood water levels into account, must also be integrated, depending on the operating mode and type of dam. Construction phases (particularly for arch dams) and the end of construction (particularly for embankment dams) also need to be included.

Possible operating modes that may affect hydrostatic pressure include drawdown water levels, deactivation times for the maintenance of hydromechanical equipment, operating faults in water release structures, inability to action turbines, and the consequences of possible break downs (for example, interruptions to power supply).

The most unfavorable load combinations to be verified must be identified – while remaining realistic. Various operating modes can have an effect on the water levels that set hydrostatic loading. The combining of extreme cases should be avoided (for example, an extreme flood and an earthquake at the same time). When combining loads for an operating dam, the 3 following situations are generally identified:

- Type 1   normal loads
- Type 2   exceptional loads
- Type 3   extreme loads

Normal loads (type 1) are those that regularly act on a dam during the usual operation of the facility. This combines the effects of self weight, water pressure (generally with a full reservoir), uplift, and thermal loads (due to seasonal variations and thermal gradients in the dam and reservoir), to which can be added upstream sediment pressure for concrete dams, ice pressure in cold regions, and downstream earth pressure.

Exceptional loads (type 2) may occur during the lifetime of the water-retaining facility, but not necessarily. In this case, slight damage is tolerated. Situations of exceptional load correspond to cases of normal loads with the addition of either a design flood, an avalanche, the rapid drawdown of the reservoir level (embankment dam), or clogged drains.

Extreme loads (type 3) involves the most unfavorable loads to which a water-retaining facility may be subjected. Some are hypothetical and constitute physical limits. In these cases, major damage may occur without however threatening the dam. As previously mentioned, an extreme situation should not combine a flood with an earthquake. However, after an earthquake, a flood with a return period of at least 10 years must be able to be tolerated.

Verification of structural safety is necessary for all water-retaining facilities, especially:

- In the case of a new construction or transformations as part of the planning approval process
- For existing dams that have not undergone earthquake safety verification
- When this is necessary to implement advances in scientific and technical knowledge
- When this is necessary to consider modifications in hypotheses from a previous safety verification

11.3 Permanent Loads

11.3.1 Self weight

Whether for concrete, earthfill, or rockfill, laboratory tests take into account the quality of material used and provide the most appropriate unit weight values. Its placement and compaction also have an impact on these values.

Besides its compaction, the unit weight of concrete also depends on its maximum aggregate size. The exact value of the unit weight of concrete is of particular interest in gravity dams.

For reference, the unit weight values provided below are generally assumed (SFOE (SFWG), 2002c):

Concrete
- Masonry dam   22 to 23 kN/m³
- Poured concrete   23 kN/m³
- Vibrated mass concrete[1]   24 to 24.5 kN/m³ (poss. 25 kN/m³)
- Reinforced concrete 25 kN/m³

Embankment dam
- Soil self weight generally varies between 16 and 22 kN/m³
- The self weight of rockfill is highly dependent on the volume of voids and may vary between 17.5 and 21.5 kN/m³

11.3.2 Earth pressure (downstream backfill) (Figure 11.3)

Firstly, it should be noted that earth pressure from downstream backfill may vary across time and does not necessarily act as a stabilizing force (SFOE, (SFWG), 2002c).

Earth pressure depends on the relative movement between a wall and the ground, among other things. When no movement occurs, earth pressure at rest is used in the verification. The force created by earth pressure at rest lies between that of active and passive earth pressure. Active pressure develops if the wall shifts by 0.1% of its height; passive pressure occurs with a movement of 1% of wall height. Earth pressure at rest is greater than active earth pressure. It is therefore wise to proceed carefully when determining earth pressure.

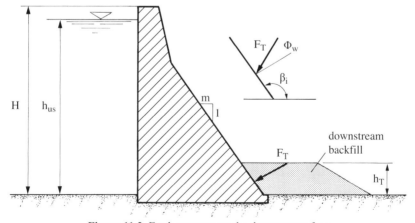

Figure 11.3 Earth pressure on the downstream face

---

[1] For preliminary calculations, a conservative value of 24 kN/m³ should be adopted.

Earth pressure at rest that acts perpendicularly on a vertical wall is given by the equation:

$$F_{TH} = \tfrac{1}{2} \gamma_T h_t^2 K_0 = \tfrac{1}{2} \gamma_T h_t^2 (1 - \sin\Phi),$$

where γ_T = unit weight and Φ = angle of internal friction.

11.3.3 Anchoring forces

Anchors provide the additional support necessary to reinforce the stability of an existing dam if the usual criteria are not met. Anchors are also installed when heightening concrete dams. Attention should be paid to ensuring that stability safety factors are not smaller than 1.0 without anchors. Current standards should be applied (for example, SIA 267). The ability to check anchors during dam operation is recommended (SFOE, (SFWG), 2002c; SFOE 2017). Section 13.12, which deals with the strengthening of gravity dams, provides some helpful information concerning anchors.

11.4 Variable Loads

11.4.1 Water pressure

Water pressure exerts a force perpendicular to the surface of the dam face (Figure 11.4). The unit weight of clean water is equal to 10 kN/m³; however, water laden with suspended sediment may range between 10.5 and 11 kN/m³, or even more.

To calculate water pressure, the highest water level relative to the type of load combination is chosen. In a normal situation, the normal operating water level is applied. If a flood is being considered, the level to be applied is that reached during a project flood or a safety flood. The upstream level is important for small dams whose stability is affected by any increase in water level. Counter-pressure may also be exerted by downstream water level.

The values of the various pressures that act on the center of gravity of their representative surface are the following:

- Upstream horizontal pressure (without overtopping): $E_{us-h} = \tfrac{1}{2} \cdot \rho_E \cdot g \cdot h_{us}^2$
- Upstream horizontal pressure (with overtopping h_i): $E_{us-h} = \tfrac{1}{2} \cdot \rho_E \cdot g \cdot (h_{us} + h_i) \cdot h_{us}$
- Upstream vertical pressure: $E_{us-v} = \tfrac{1}{2} \cdot \rho_E \cdot g \cdot m_1 \cdot h_{us}^2$
- Downstream horizontal pressure: $E_{ds-h} = \tfrac{1}{2} \cdot \rho_E \cdot g \cdot h_{ds}^2$
- Downstream vertical pressure: $E_{ds-v} = \tfrac{1}{2} \cdot \rho_E \cdot g \cdot m_2 \cdot h_{ds}^2$

The width of the base depends on the angle of the faces (relative to the vertical axis):

$$m_1 = \operatorname{tg} \varphi_1 \qquad m_2 = \operatorname{tg} \varphi_2,$$

where ρ_E = unit weight of water; h_{us} = upstream hydrostatic head; h_{ds} = downstream hydrostatic head; h_i = height of the headwater.

For dam sections with overtopping, the force exerted by the water against the upstream face can be included, depending on the quantity of flow determined as part of a stability analysis.

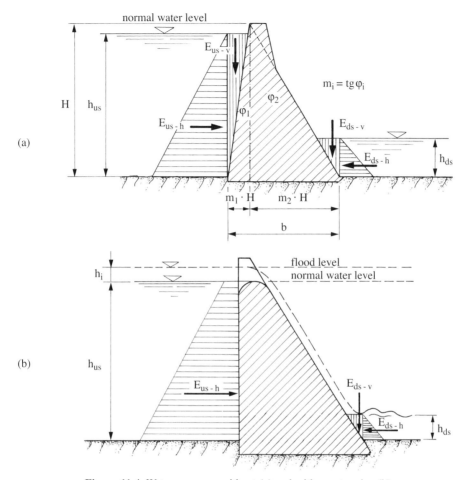

Figure 11.4 Water pressure: without (a) and with overtopping (b)

11.4.2 Sediment pressure

Sediment can accumulate at the upstream face of a dam and reach a considerable depth or may even grow to be practically the same height as the dam (Figure 11.5). This sediment exerts horizontal pressure on the upstream face, which is added to hydrostatic pressure. Like water pressure, sediment pressure has a triangular distribution expressed by:

$$F_{sed} = \tfrac{1}{2} \cdot \gamma_i \cdot h^2_{sed} \cdot K,$$

where

K = earth pressure coefficient:

- Fluid $\qquad\qquad\qquad K = 1$
- Pressure at rest $\qquad K = 1 - \sin\phi$
- Active pressure $\qquad K = (1 - \sin\phi)/(1 + \sin\phi)$
- Passive pressure $\qquad K = (1 + \sin\phi)/(1 - \sin\phi)$

where γ_i = unit weight of submerged sediment in kN/m³ (in 1st approximation, 10 kN/m³)²; h_{sed} = depth of sediment layer; ϕ = angle of internal friction of sediment (generally accepted as being between 15° and 30°).

In standard practice, the effect of sediment pressure can generally be accounted for, under static loading, by a 4 kN/m³ increase in unit weight of the reservoir water (SFOE, 2017).

For earthquakes, it should be noted that liquefaction can affect silt, which may be important when carrying out verification during or after an event.

In North American standard practice, it is more commonly assumed that the layer of sediment in contact with the upstream face of the dam that oscillates is liquefied. Sediment is then considered as a liquid whose density is greater than that of water, which modifies the Westergaard pressure.

Mononobe-Okabe put forward an approach deduced from Coulomb's theory that considers inertia forces acting on the ground (Matsuzawa et al., 1985; USACE, 2005). Sediment pressure is expressed by:

$$F_T = \tfrac{1}{2} \cdot K_{AE} \cdot \gamma \cdot (1 - a_v) \cdot h^2.$$

The active pressure coefficient K_{AE}, by assuming pressure against a vertical wall, vertical acceleration a_v, and a zero angle of friction, as well as a horizontal sediment slope, is defined by:

$$\left[K_{AE} = \frac{\cos^2(\phi - \psi)}{\cos^2\psi \left[1 + \sqrt{\dfrac{\sin\phi \sin(\phi - \psi)}{\cos\beta \cos\psi}} \right]^2} \right]$$

where $\psi = \mathrm{tg}^{-1}(a_h)$ = angle of seismic inertia; a_h = horizontal gravitational acceleration.

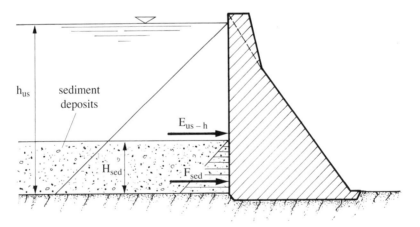

Figure 11.5 Sediment pressure

---

² The unit weight of the mud or silt can reach values of 16 to 19 kN/m³.

11.4.3 Uplift

Statement

Uplift exerts a major active hydrostatic force. It acts upon both the inside of a concrete dam and its foundation. It can sometimes be difficult to assess, which is why there are several hypotheses regarding its distribution, which in turn implies several different possible values of intensity.

Due to the difference in pressure between the upstream and downstream faces, water gradually penetrates the concrete and rock, as in any porous, cracked, or fractured media. A seepage flow pattern gradually forms inside the dam and the rock mass. Uplift is almost the same as hydrostatic pressure (100%) upstream and is almost zero downstream, as long as there is no counter-pressure. In addition, uplift follows variations in water level – however, with a certain lag in some cases.

If a concrete prism is carefully dried and weighed in a tub of water, after a while the water level decreases and the weight of the prism increases. This demonstrates that concrete is porous. Volumetric porosity is in the order of 5 to 10%. The pores in the concrete in a dam are therefore subject to pore pressure equal to upstream hydrostatic pressure, which decreases in a downstream direction. The application surface of this pressure is large and varies between 93 and 97% of a plane section. Indeed, a concrete dam is constructed by placing a series of lifts. If the lift joints are poorly executed, seepage will be visible downstream.

No matter the quality of the foundation rock, water will infiltrate and occupy the pores, thus applying pressure to the walls. This pressure plays a major role in dam stability at the surface of the concrete-rock interface. Its role is not only restricted to the dam structure, but also concerns the stability of the foundation itself or a rock abutment. Uplift can occur along a potential sliding surface or perpendicular to a geological formation change (Figure 11.6).

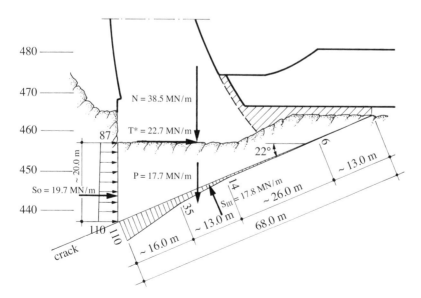

Figure 11.6 Measurements of uplift and forces along a crack for the calculation of sliding safety (SwissCoD, 1992)

To summarize, pore pressure exists in the foundation, along horizontal concrete lift surfaces and in some cracks, but also within concrete, which is porous. The challenge lies in establishing the most appropriate uplift distribution, in view of local conditions and construction layout.

Uplift coefficient λ

If an uplift coefficient $\lambda = 1.0$ and triangular distribution is admitted, this assumes that total uplift is active under the foundation and that permeability distribution is homogeneous across the whole width of the foundation.

From the time the uplift phenomenon was clearly defined under dam foundations until the 1970s, it was generally admitted that the uplift coefficient depended on the quality of the rock foundation and grouting of the concrete-rock interface. The most commonly assumed values were:

- $\lambda = 0.75$ à 0.8 Sound rock, well grouted
- $\lambda = 0.75$ à 1.0 Medium quality rock, but well grouted
- $\lambda = 1.0$ In cases of doubt

This approach was justified, it was believed, by the decrease in surface against which uplift acts, as demonstrated in Figure 11.7.

Today we know that only cracks larger than 100 to 200 µm can be filled with cement grout. The appropriate model is instead illustrated in Figure 11.7 (b), as in fractured rock a continuous surface can always be found against which uplift can act.

As a rule, $\lambda = 1.0$ should always be chosen for the concrete-rock interface and for the rock itself.

Within a block in a dam, it is difficult to see how uplift could act on the whole surface, and it is accepted that uplift only acts on a fraction λ of the surface in question. If the concrete is not cracked and the lift surfaces and concrete joints are of good quality, an uplift coefficient $\lambda = 0.9$ can be assumed within the dam body.

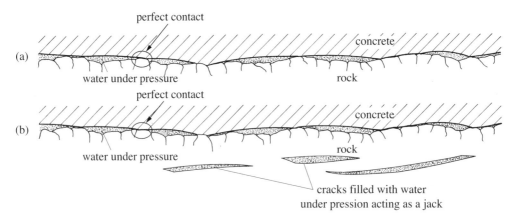

Figure 11.7 Uplift application surface: (a) according to outdated principles; (b) according to current scientific knowledge

Force exerted by uplift (Lévy's rule)[3]

In the case of a gravity dam, assuming a triangular uplift diagram with maximum head upstream ($\lambda \cdot \rho_E \cdot g \cdot h_{us}$) and zero head downstream (Figure 11.8), the resultant of uplift force S acts vertically and is situated in the upstream third of the base. Supposing that under the effect of water pressure E_{us} and despite the stabilizing effect of self weight P, the dam will heave slightly upstream. In this case, uplift will start to act on the whole surface and the diagram will change. The force of the uplift becomes greater and stability conditions are no longer met. This reasoning leads to ***Lévy's rule***, which postulates that at no point on the upstream face should the stress resulting from self weight and water pressure E_{us} be lower than hydrostatic pressure at that point (that is, $\rho_E \cdot g \cdot h_{us}$) or at that pressure multiplied by λ:

σ upstream ($P + E_{us}$) must be greater or equal to $\rho_E \cdot g \cdot h_{us}$ or $\lambda \cdot \rho_E \cdot g \cdot h_{us}$.

With regard to arch dams, uplift is not usually taken into consideration. An arch dam is instead a hyperstatic structure: the strengthening effect of the arches comes into play should the force of uplift increase because of upstream heaving. However, uplift is introduced in cases where the rock is of a dubious quality.

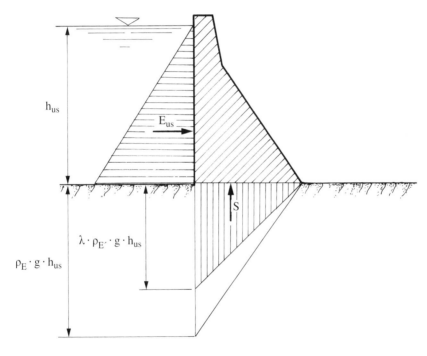

Figure 11.8 Force exerted by uplift (Lévy's rule)

Uplift distribution

Uplift distribution under the foundation depends on the gradient of seepage flow, and this gradient in turn depends upon permeability conditions. Several typical cases can be demonstrated (Figure 11.9).

---
[3] See also Part IV, Section 13.3.

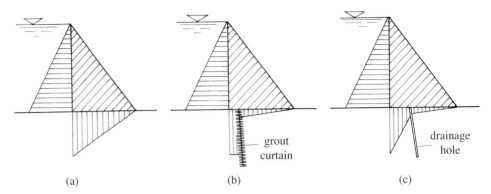

Figure 11.9 Typical forms of uplift distribution: (a) homogeneous and isotropic permeability; (b) a curtain with weak permeability (grout); and (c) drained zone

Cases (b) and (c) in Figure 11.9 have the direct effect (if they are correctly implemented) of reducing uplift, which tends to lift the dam. In practice, these two effects are combined by installing a grout curtain placed in the upstream part of the dam and drainage holes placed immediately downstream. Figure 11.10 shows the arrangement of these measures and their effect on uplift distribution.

Installing a grout curtain into the foundation is a long and complex task. Grouting relatively cracked rock in fact creates a rather well-defined zone with reduced permeability. Grout holes should be arranged so that their respective zones of influence join up and overlap. This cannot be entirely guaranteed. A heterogeneous curtain with variable permeability is therefore the result, through which seepage flow present in the foundation may develop a preferential path (see Part VII). For this reason, it is important to connect the drainage holes downstream of the grout curtain, thereby avoiding possible loading, and manage the development of seepage under the foundation. By observing the turbidity of seepage water, the tightening efficiency of the grout curtain can be verified over time.

Drainage holes set a limit to the development of uplift, which theoretically cannot go beyond the level of the drain outlet (usually the tunnel at the toe of the dam). Furthermore, significant seepage gradient appears in the grouted zone comprising the grout curtain. Upstream of the curtain, pressure is practically constant and almost equal to hydrostatic pressure. Downstream of the drain, the decrease in pressure is influenced by permeability distribution in the rock (linear if the rock is homogeneous and isotropic). Safety in regard to uplift can in fact be guaranteed to a large extent by drainage. In extreme cases, a drainage network can have the same effect whether the ground is highly permeable or highly impermeable. Only outflow from drains will vary (Sabarly, 1968).

In practice, the heterogeneity and anisotropy of the foundation rock mean that the true effects of grout and drainage will not exactly match the theoretical model. It is therefore important to regularly measure pressures under the foundation across many points to check to what extent the hypotheses are verified. If they are not verified, it is always possible to drill new drains or strengthen the grouting. This is why concrete dams almost always have a gallery in the toe that is large enough for a drilling machine and that runs along the foundation a few meters from the upstream face.

Other solutions for reducing uplift may also be applied: implementation of a drainage shaft (Figure 11.11), opting for a hollow gravity dam (Figure 11.12).

11. ACTIONS AND FORCES

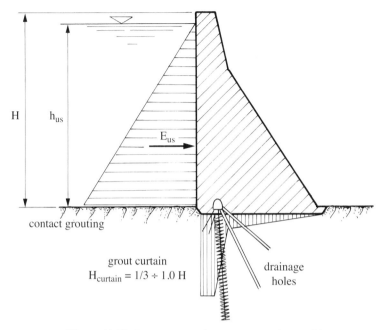

Figure 11.10 Arrangement of measures to reduce uplift

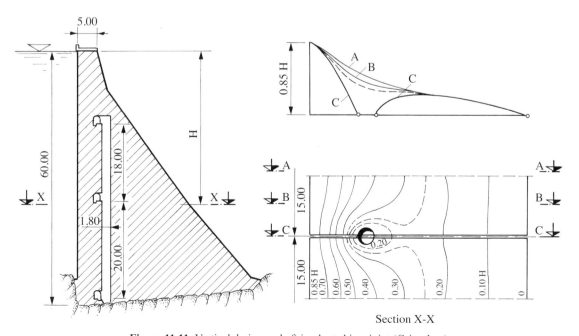

Figure 11.11 Vertical drainage shaft implanted in a joint (Gries dam)

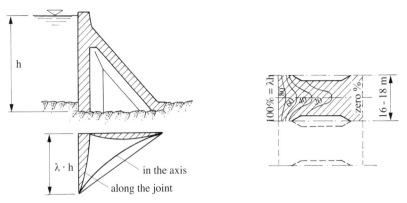

Figure 11.12 Drainage using joints with cavities

Distribution modes for accepted uplift

Previously, given the uncertainty around the final uplift distribution, different rules and recommendations were implemented in various countries. Examples of often assumed distribution modes are presented in Figure 11.13.

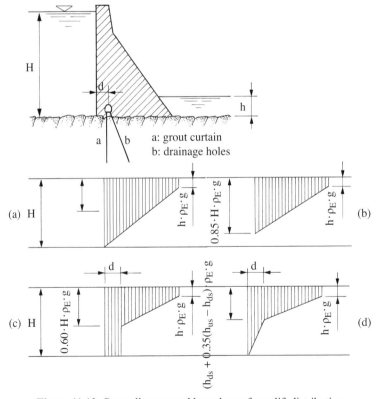

Figure 11.13 Generally accepted hypotheses for uplift distribution

Cases (a) and (b) (Figure 11.13) accept a trapezoidal uplift distribution. In case (a), uplift pressure is equal to upstream and downstream hydrostatic pressure respectively. In case (b), upstream uplift can be assumed to be at 85% of total hydrostatic pressure to allow for the effect of a grout curtain and drains (Schnitter, 1956). As for case (c), the grout curtain may affect uplift distribution. And finally, in case (d), the reduction in uplift is due to the installation of drainage systems (galleries, drillholes). Other hypotheses on uplift distribution are also possible; however, they must be justified (measured values, presence of drains). For a new project, following conservative and cautious hypotheses is recommended.

The uplift diagram can also be deduced from the modeling of a seepage flow pattern. In the case of an existing dam, the measured values can be introduced during a review of dam safety. Finally, it should be noted that the uplift diagram may change following an earthquake. If, for example, a crack forms upstream, maximum uplift must be considered along the length of the crack (Figure 11.14).

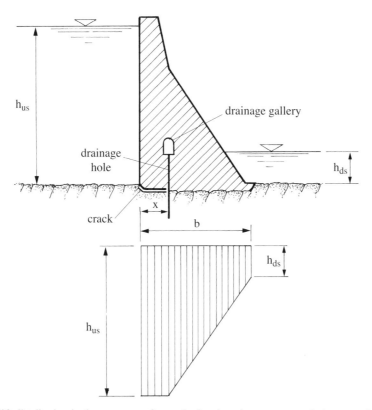

Figure 11.14 Uplift distribution in the presence of a newly developed upstream crack that goes right to the drainage (from USACE)

Conventional forms of uplift diagrams adopted by the SFOE (SFOE, 2017)

A — Grout curtain
B — Drainage curtain
X — Measurement point at the interface
p_j — Pressure without the drainage effect at the interface
p_i — Pressure under the effect of the drainage at the interface
$p_i = p_j - k(p_j - p_m)$, where
p_m — the greatest pressure value in $\gamma_w h_w$ and $\gamma_w h_d$ (according to the elevation of the gallery)
k — reduction/drawdown coefficient (drainage curtain)

Alternatively, the following distributions may be used in the absence of adequate uplift measurements or in the design phase:

- In the absence of a grout curtain and drainage system: distribution that is triangular (without the presence of tailwater) or trapezoidal (in the presence of tailwater), with upstream and downstream uplift pressure equal to the respective hydrostatic pressures
- In the presence of an upstream grout curtain: a reduction in pressure at the level of the grout curtain is only acceptable if it is functioning, otherwise, no reduction may be introduced
- In the presence of a drainage system (drainage gallery, draining boreholes): a reduction in uplift pressure to a maximum of 50% perpendicular to the drainage, providing the drains are perfectly functional

(Obernhuber, 2014; USACE, 2000; SFOE, 2017)

Effectively measured uplift

On the basis of uplift measured across more than 20 gravity dams and arch dams in Switzerland, Figure 11.15 provides a representation of the various uplift diagrams, assuming a normalized base length.

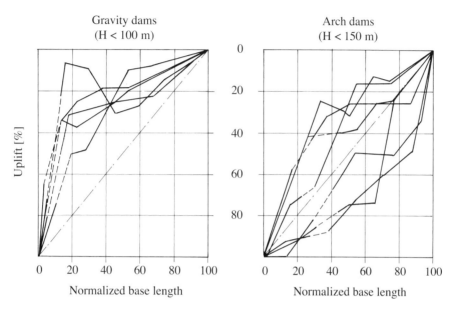

Figure 11.15 Normalized uplift diagrams for various concrete dams in Switzerland (from CNSGB (SwissCoD), 1992)

11.4.4 Concrete temperature

First, the temperature from the heat generated during cement reaction is added to that of fresh concrete. Thereafter, a natural cooling of the concrete is observed. This is a very slow process that takes place mainly through the upstream and downstream dam faces and, providing they are in contact with the air, through the lateral and upper faces of the blocks. In some cases, it can be necessary to accelerate this cooling using appropriate methods. For a thick dam, temporary gaps can be left between the blocks. One artificial cooling process involves embedding metal or plastic pipes (cooling pipes) between the concrete lifts, through which cold water circulates (Figure 11.16).

When the dam reaches its thermal equilibrium, variations in temperature arise due to exchanges with the direct environment: exposure to solar radiation, reservoir water temperature, air temperature. Heat penetrates the dam through the faces. Inside, the more the temperatures drop below those on the faces (phase shift), the more they are attenuated. Daily variations in exterior temperature are only felt near the faces at a depth of around 1 meter. Solar radiation on the faces also produces significant seasonal heating.

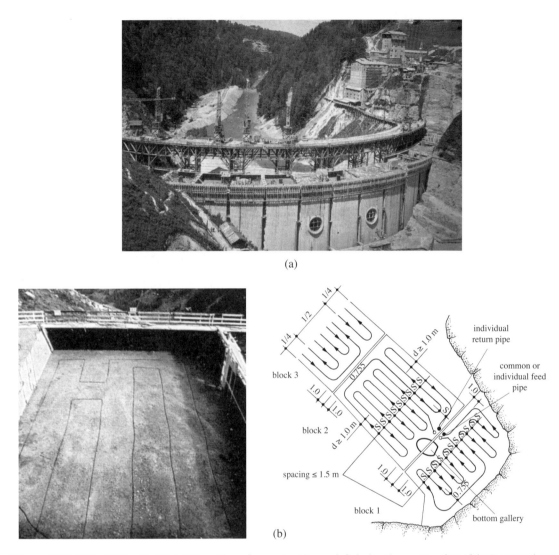

Figure 11.16 Natural (a) and artificial (b) cooling of concrete: (a) gaps left during the construction of the Rossens dam; (b) illustration of a network of cooling pipes

Temperatures affect the stresses and deformation of the dam structure, particularly for arch dams. Variations in temperature must be considered when designing arch dams by analyzing several different states, for example, full reservoir, empty reservoir, and summer/winter seasons (Figure 11.17).

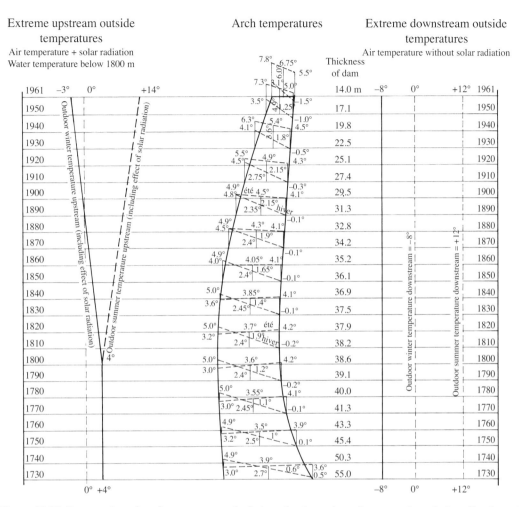

Figure 11.17 Bases and results of temperature calculations for the arches of a mountain arch dam (Stucky and Derron, 1957)

Each state can be either favorable (compression) or unfavorable (tensile stress) for different zones in the dam. Across the width of the cross-section, a linear relation is assumed between the two faces (trapezoidal shape), which includes a uniform variation in temperature compared to a reference temperature and a temperature gradient between the two faces (Figure 11.18).

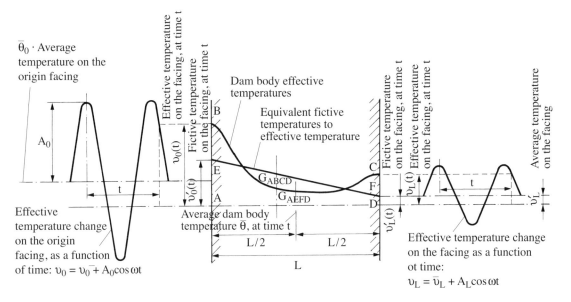

Figure 11.18 Definition of real and fictive temperatures (Stucky and Derron, 1957)

11.4.5 Flow pressure from percolation water

Seepage through an embankment dam can wash out particles of material and lead, through the process of internal erosion, to the creation of pipes that gradually become bigger. This erosion phenomenon is known as *piping*, and it must be carefully analyzed. In addition, investigations should be made to ensure that under the effect of seepage pressure (or uplift), there is no risk of parts of the dam lifting up (SFOE (SFWG), 2002c).

11.4.6 Pore pressure

Pore pressure can vary depending on time, permeability, hydrostatic loading, and drainage conditions. A network of seepage flow lines and equipotential lines indicates the intensity and distribution of pore pressure (Figure 11.19). With regard to stability, it should be noted that pore pressure can reduce shear strength (SFOE (SFWG), 2002c).

11.4.7 Snow

Snow, which is considered as a variable fixed action, can act as a load for embankment dams with a small reservoir height. This load is affected by the location of the dam and local conditions. Snow load value must be taken from current standards.

As an example, load due to snow can reach the following values:

- Altitude H_{alt} below 400 m $p_s = 90 \text{ kg/cm}^2$
- From 400 to 800 m $p_s = 40 + (H_{alt}/55)^2 \text{ kg/cm}^2$
- Above 800 m $p_s = 40 + (H_{alt}/100)^2 \text{ kg/cm}^2$

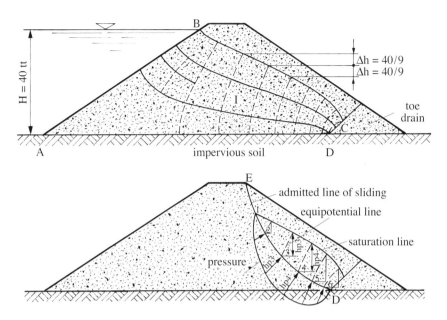

Figure 11.19 Network of flow and equipotential lines (from Lambe and Whitmann, 1969)

11.4.8 Ice load

Ice load is due to thermal expansion but also depends on the shape of the body of water, variations in water level, and wind. It is not a factor during major fluctuations in water level, as the layer of ice is likely to break up. It also remains detached from the face, as the dam itself is a thermal reservoir that emits heat, melting the ice to a distance of 10 to 20 cm. However, ice does have a considerable effect on concrete dams with low or moderate water levels in the reservoir.

The intensity of ice load is a value that has not yet been clearly defined. According to the literature and test results (in particular, Comfort, 2000), it varies between 20 and 300 kN/ml. The USACE (1982) suggests a value in the order of 150 kN/ml for ice to a depth of 60 cm.

Other sources indicate that ice is around 30 cm thick up to an altitude of 500 m.a.s.l. and 80 cm from an altitude of 2,300 m.a.s.l. A linear variation between these values can be assumed. As for total ice load, suggested pressure is 200 kN/m^2; values of 150 or 250 kN/m^2 are also found.

In cold regions, there is a possible risk of blocks of floating ice knocking parts of the dam (for example, spillway gates) during the spring run off.

11.4.9 Bearing loads

When bearing loads are to be considered, their values must be taken from current structural standards (SIA 261, 2014).

11.5 Exceptional Loads

11.5.1 Floods

As described in Section 10.6.2, water release conditions must be verified for a design flood (exceptional loading, type 2) and a safety flood (extreme loading, type 3). To establish appropriate flood safety criteria, calculations must be based on the water level demonstrated in the design flood and safety flood hydrographs (the maximum rise in water level).

11.5.2 Earthquakes

History

In the past, earthquake resistance to dams was achieved with a pseudo-static approach, a technique that was developed in the 1930s. For concrete dams, this simple method considers the dam's inertia forces and the hydrodynamic pressure exerted by the reservoir (Figure 11.20). For embankment dams, the static inertia force is applied to the sliding mass. It was common practice to use a horizontal seismic acceleration coefficient α with a value equal to 0.1 g, even though it was understood that this simplified calculation did not reflect the actual behavior of a dam and reservoir during an earthquake. A real earthquake is characterized by transient accelerations and oscillations in three directions. A pseudo-static analysis only assesses the effect of horizontal acceleration oriented in the least favorable direction. In Switzerland, the case of load due to an earthquake, taking into account replacement static forces, was already considered in the design of all dams built in the 1950s.

With the development of computers and numerical analysis methods, much progress has been made in assessing seismic safety for concrete and rockfill dams under dynamic forces. These models are used in Switzerland for studying new projects, transforming dams (for example, heightening a dam), and assessing the safety of existing dams (Wieland, 2006; Schwager et al., 2016).

Today, provided that earthquake acceleration is low (a < 0.15 g), a simplified pseudo-static analysis can give results suitable for preliminary design.

Hydrodynamic pressure according to Westergaard

During upstream displacement caused by an earthquake, the dam pushes against the water in the reservoir (Figure 11.20), and the water's inertia force creates a hydrodynamic pressure on the upstream face of the dam. Westergaard formulated a parabolic relationship describing this phenomenon based on depth z:

$$p_E = K_e\, C_e\, \alpha\, \rho_E\, \sqrt{Hz}\ ,$$

where K_e = the coefficient depending on the slope of the upstream face:
- vertical: $K_e = 1.0$
- sloped: K_e varies linearly with the tilt angle of the face
  (as long as $\delta \leq 20°$): as in Figure 11.21
  for $\delta = 10°$, $K_e = 0.88$ for a = 10 m and h = 100 m.

$$C_e = \frac{0.817}{\sqrt{1 - 7.75 \left(\dfrac{H}{1000\, T}\right)^2}}$$

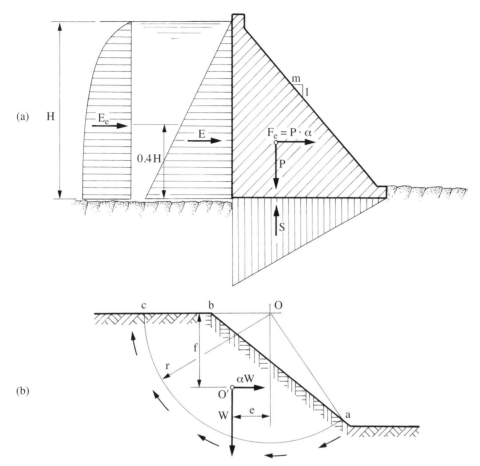

Figure 11.20 Seismic overloading based on a pseudo-static analysis: (a) concrete dam; (b) embankment dam. Key: F_e: inertia force of the dam; E_e: inertia force of the water.

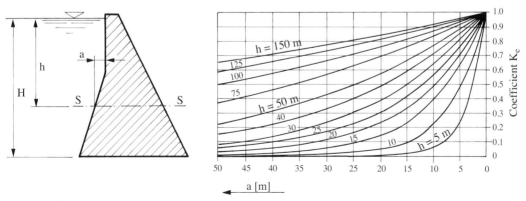

Figure 11.21 Hydrodynamic pressure according to Westergaard – coefficient K_e based on the slope of the upstream face

where C_e: Westergaard coefficient, H: dam height, in m, T: vibration period generally assumed between 0.5 and 1 s ($C_e \approx 7/8$).

Total hydrodynamic pressure equals:

$$E_e = \int_0^H p_E(z)\, dz = \int_0^H K_e\, C_e\, \alpha\, \rho_E \sqrt{h_{us}\, z}\, dz = K_e\, C_e\, \alpha\, \rho_E\, \tfrac{2}{3} h_{us}^2$$

and

$$M_e = \int_0^H p_E(z)\, (h_{us} - z)\, dz = K_e C_e \alpha\, \rho_E\, \tfrac{4}{15} h_{us}^3,$$

which enables the application point and resulting force E_e ($h_e = 0.4\, h_{us}$) to be determined.

Design earthquakes

For verification calculations, different types of design earthquakes are recommended. Firstly, the Operating Basis Earthquake (OBE) assumes that the dam remains operational following the event, and only minor, easily repaired instances of damage are acceptable. The proposed return periods are 145, 200, or 500 years. Large dams must be able to resist the impact of the strongest earthquakes expected at the site. A Maximum Credible Earthquake (MCE) is also defined and applied. In theory, the MCE is an earthquake that generates the largest ground movements at the site of the dam. In practice, the MCE is usually determined on a statistical basis with a return period of 10,000 years for countries with low to moderate seismicity. Note that the terms MDE (Maximum Design Earthquake) and SEE (Safety Evaluation Earthquake) are used instead of MCE (Wieland, 2003). Dam stability must be guaranteed in even the most violent cases of ground movement at the site of the dam, and no uncontrolled release of water must occur, even if there is major damage. If the dam becomes damaged, it should still be possible to drawdown the water level in the reservoir.

Ground movement due to an earthquake is represented by Peak Ground Acceleration, which is a very important parameter. This peak acceleration, in the order of a few thousandths of g during a small tremor, can reach 2 g during a violent earthquake. Due to the solidarity of the dam with its foundation, ground acceleration is transmitted to the dam structure. Ground movement can be heavily amplified at the level of the crest. Acceleration greater than 2 g in an upstream and downstream direction were observed at the crest of the Pacoima arch dam (H = 111) during an earthquake in 1994.

Earthquake safety analysis recommended in the Swiss Directive[4]

Verification objectives

In Switzerland, where seismic activity is low to moderate, analysis is made on the basis of a standardized verification comprising probabilistic considerations. No dam failure with an uncontrolled release of water is accepted for a dam under loading from a verification earthquake, nor is any damage to ancillary structures that may threaten the safety of the water-retaining facility. Local damage that does not affect the physical integrity of the dam structures is, however, tolerated. In addition, the directive advises operators to assess the behavior of appurtenant structures and the reservoir zone following seismic loading. With regard to appurtenant structures, operators must check that (a) water release and bottom outlet structures have not been damaged; (b) a flood can still be released; and (c) monitoring instruments designed to highlight a weakness still work. In the reservoir zone, the operator must ensure that potential ground movement

---

[4] From SFOE (SFWG), 2003; SFOE, 2016.

following an earthquake (such as landslides, bank instability, rock falls, falling boulders, or seracs) does not compromise the safety of the water-retaining facility. Thought must be given to the possibility of an event leading to overtopping or dam overloading. The information at the end of Section 11.2 outlines the situations in which the verification of structural safety must be carried out.

Initial conditions
Following the criterion of potential damage to people and property in the case of dam failure, dams are divided, depending on their official height and reservoir capacity, into three categories with different requirements (Figure 11.22).

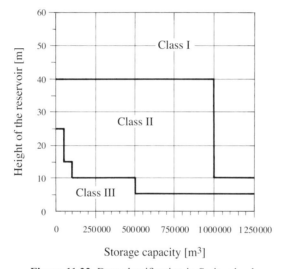

Figure 11.22 Dam classification in Switzerland

As an extreme stress, seismic activity is combined with cases of normal loading. The safety assessment is generally based on the case of a full reservoir. The accepted level corresponds to the level that determines the height of the reservoir (see SFOE, 2015a).

Material characteristics are usually determined through laboratory and in situ tests; representative tests for the parameters of each material must be chosen. The results of tests carried out during construction may be used. In the absence of such tests or should these tests be insufficient, new tests should be carried out. Material parameters may also be calculated from observed measurements at the dam, if it can be demonstrated that a back calculation can determine the necessary parameters. For dam structures in category III, the values may be obtained from the literature. As a reference, values are listed in Section 13.7.2.3.

Baseline data
The probability of occurrence of a verification earthquake is established based on the dam category. The exceedance probability is given for an interval of 100 years, expressed as a return period (Table 11.23).

Table 11.23 Return period for verification earthquake for the three dam categories

Dam category	Time interval considered	Average exceedance probability	Average return period
I	100 years	1%	10,000 years
II	100 years	2%	5,000 years
III	100 years	10%	1,000 years

Seismic loading depends on peak ground acceleration, the response spectrum, and the development over time of the corresponding acceleration.

The values of peak horizontal acceleration a_h that correspond to return periods of 1,000 and 10,000 years are determined on the basis of the following transformation:

$$\log a_h = 0.26 \cdot I_{MSK} + 0.19 \ (cm/s^2)$$

The I_{MSK} intensities for a return period of 1,000 and 10,000 years are taken from intensity maps (Figure 11.24 and 11.25).[5] Values for a return period of 5,000 years are interpolated as follows:

$$I_{5000} = 0{,}3 \cdot I_{1000} + 0{,}7 \cdot I_{10\,000}$$

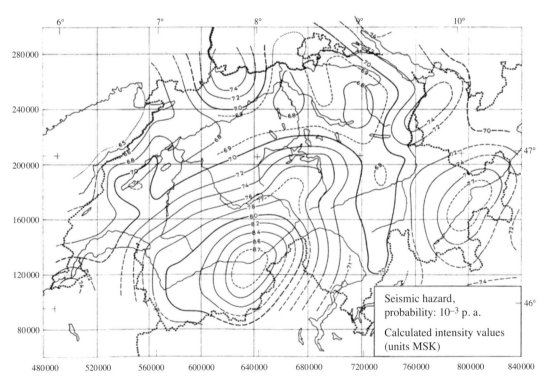

Figure 11.24 Intensity values for an exceedance probability of 10^{-3} p. a. based on the Swiss earthquake hazard map – *Determination of Earthquake Hazard*, 1977

---

[5] Sägesser and Mayer-Rosa, 1978.

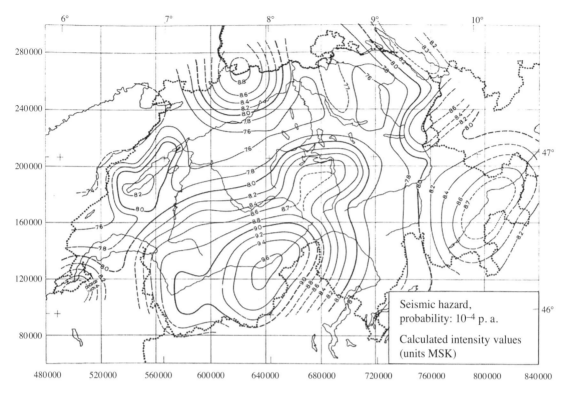

Figure 11.25 Intensity values for an exceedance probability of 10^{-4} p. a. based on the Swiss earthquake hazard map – *Determination of Earthquake Hazard*, 1977

Peak horizontal acceleration a_h is independent of direction. As for peak vertical acceleration a_v, it can be calculated from the horizontal component a_h by reducing it by a third ($a_v = 2/3 \cdot a_h$).

The response spectrum is the maximum dynamic response of an oscillating mass excited around its attachment point by an earthquake. The spectrum is depicted based on the natural frequency and absorption of the oscillating mass.

Three categories of foundation can be identified depending on the profile of the foundation layers:

- Category A foundation: Rock and rigid deposits of sand, gravel, or well-compacted clay. Shear wave velocity greater than 400 m/s.
- Category B foundation: Deep deposits of sand or gravel with moderate compaction or moderately rigid clay. Shear wave velocity between 200 and 400 m/s.
- Category C foundation: Deposits of loose, non-cohesive ground with layers of only slightly cohesive material, as well as deposits formed primarily of cohesive, soft to moderately rigid soil. Shear wave velocity lower than 200 m/s.

For the verification of dams on rock foundations (Category A foundation), the standardized response spectra for acceleration in Figure 11.26 are applicable (the value of the amplification of acceleration is multiplied by peak ground acceleration). Figures 11.27 and 11.28 give the standardized response spectra for Category B and C foundations. The response spectra are valid for both horizontal and vertical directions.

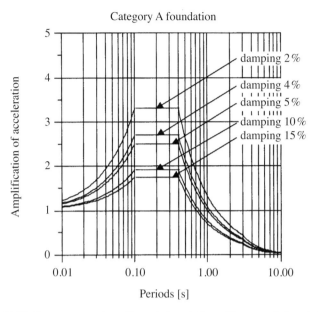

Figure 11.26 Response spectrum for rock foundations (Category A foundation)

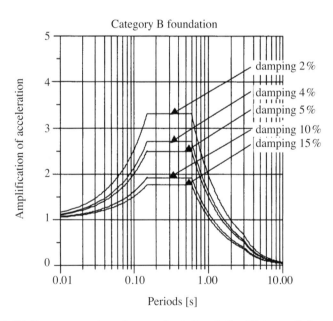

Figure 11.27 Response spectrum for a moderate foundation (Category B foundation)

11. ACTIONS AND FORCES

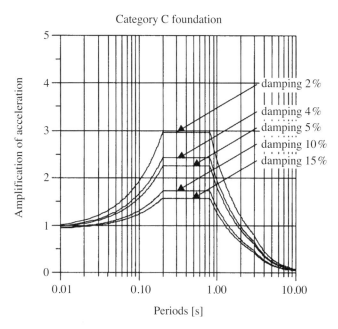

Figure 11.28 Response spectrum for a soft foundation (Category C foundation)

Finally, appropriate recordings or artificially generated traces may be used. The recordings must be compatible with the given spectra, otherwise they must be adapted. The stationary duration required for artificially generated traces is given by the formula:

$$T_s = 10 + 50\,(a_h/g - 0{,}1)\;[s]$$

T_s must be at least 10 s.

Verification procedure
The verification procedure involves 4 stages which are described below.

Stage 1) <u>Modeling</u>
Modeling must consider foundation ground, the dam structure, and the reservoir basin.
Geological and geotechnical configuration of the foundation, as well as the mechanical properties of the dam and foundation, must be determined.

Stage 2) <u>Analysis</u>
The behavior of the water-retaining facility during and after an earthquake must be analyzed. When analyzing behavior during an earthquake, deformation and permanent damage must be determined in relation to the seismic stress.
When analyzing behavior after an earthquake, the resulting consequences must be examined. Note must be made of possible deformation, damage, high pore pressure, and any permanent modification in uplift conditions.

The type of earthquake behavioral analysis employed is chosen based on the category of the water-retaining facility; the following minimum requirements generally apply:
- For Category I water-retaining facilities: dynamic calculation with evolution over time, with a finite element model
- For Category II water-retaining facilities: response spectrum method (also for dikes: simplified response spectrum method taking into account several modal forms)
- For Category III water-retaining facilities: quasi-static procedure (also taking into account the first modal form)

Methods specific to dynamic analysis are described in Section 13.6.5.5.

Stage 3) Interpretation and assessment
Dam behavior during and after the earthquake must be assessed.

If any measures are planned to be implemented after an earthquake, the safety level for the resulting conditions must also be assessed as a fundamental part of earthquake safety verification, particularly, for example, in the case of rapid reservoir drawdown.

In the long term, and after any necessary measures, the safety of the water-retaining facility after seismic stress must correspond to the pre-earthquake level of safety. In the short term, that is, immediately after the seismic stress and until the completion of any necessary measures, safety must not fall below 80% of the safety coefficient accepted in the case of normal and extraordinary loads.

Stage 4) Analysis process or the planning of measures
If the earthquake safety of a Category II or III water-retaining facility cannot be verified with the analysis procedure corresponding to the dam category, it can be verified by following the procedure from a lower level. If this is not possible either, structural measures or measures relating to operation are necessary.

Minimum requirements for the verification of earthquake safety (SFOE, 2016)

Category	Category of water-retaining facilities		
	I	II	III
Earthquake return period	10,000 years	5,000 years	1,000 years
Vertical acceleration	Yes	Yes	No
Material parameters	Laboratory and in situ tests (also tests carried out during construction), possibly back-calculation from behavior measurements	Laboratory and in situ tests (also tests carried out during construction), possibly back-calculation from behavior measurements	Laboratory and in situ tests (also tests carried out during construction), possibly back-calculation from behavior measurements; for existing works also, based on the literature or comparable dams
Concrete resistance to static tensile stresses	Tests specific to the dam	Tests specific to the dam	Tests specific to the dam or on the basis of the formula $(MPa) f_{ts} = 3/8 \cdot f_{cs}^{2/3}$, f_{ts} max. 3 MPa
Characteristic dynamic ground values	Tests specific to the dam	Literature or comparable dams	--
Model for gravity dams	2D finite element model incl. foundation	2D FEM incl. foundation or springs	2D FEM, rigid foundation
Model for gate weirs	3D finite element model incl. foundation	3D FEM incl. foundation or springs	3D FEM, rigid foundation
Model for arch dams	3D finite element model incl. foundation	3D FEM incl. foundation or springs	3D FEM incl. foundation or springs
Model for embankment dams	2D finite element model	2D FEM	2D FEM
General analysis process	Dynamic calculation with time history analysis	Response spectrum method (taking into account several eigenvalues)	Quasi-static procedure (taking into account the first eigenvalue
Acceleration in determinant slip circles of an embankment dam	Calculation from finite elements with time history analysis	With simplified procedure, with the help of response spectra	Simplified by amplification of maximum ground acceleration by a factor of 1.5
Verification of stresses for gravity dams and gated weirs	Yes	Yes	No

11.5.3 Avalanches

A distinction is drawn between the direct impact of an avalanche hitting a structure specifically designed for this reason (avalanche protection structure) and an impulse wave caused by an avalanche descending into a reservoir.

Avalanche hitting a protective structure (Figure 11.29)
This case concerns low-height protective structures that block the descent of an avalanche. Their return period is usually accepted at 300 years (SFOE (SFWG), 2002c). Pressure q_f exerted by an avalanche is given by the following equation[6]:

$$q_f = 0.5 \cdot c_d \cdot \rho_f \cdot v_f^2 \; [kN/m^2],$$

where v_f = avalanche velocity [m/s], c_d = 2 to 3 (resistance coefficient) and ρ_f = 0.3 t/m^3.

Furthermore, a single force q_e due to the impact of an object (tree, large block, etc.) may potentially be considered.

Impulse wave
When there is a possibility of an avalanche descending into the reservoir, the impulse wave generated by an avalanche with a return period of 300 years must be estimated and the risk of overtopping assessed (VAW-ETHZ, 2009). Construction or operational measures must be taken if there is a possibility of the wave going above the danger level (for example, by constructing a parapet or permanently or temporarily increasing the freeboard).

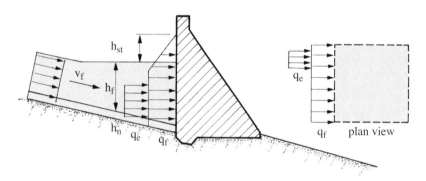

Figure 11.29 Distribution of load due to an avalanche

11.5.4 Debris flows (Figure 11.30)

This case is to be considered if the total force on the dam structure is greater than the hydrostatic pressure. It concerns low-height dams (SFOE (SFWG), 2002c).

---

[6] Indications relating to the characteristics of avalanches (§ 11.5.3) and debris flows (§ 11.5.4) are taken from *Richtlinie, Objektschutz gegen Naturgefahren*, published by Gebäudeversicherungsanstalt des Kanton St Gallen.

The pressure exerted by a debris flow is given by the following equation:[7]

$$q_f = 0.5 \cdot c_d \cdot \rho_f \cdot v^2_f \ [kN/m^2],$$

where v_f = velocity [m/s], c_d = 1.5 to 2.0 and ρ_f = 1.8 t/m³.

Furthermore, a single force q_e due to the impact of an object (tree, large block, etc.) may potentially be considered.

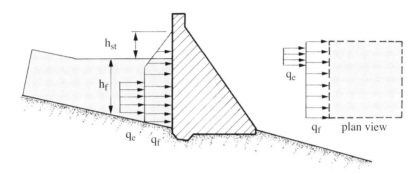

Figure 11.30 Distribution of load due to a debris flow

11.5.5 Other individual actions

Creep and concrete shrinkage
Delayed impacts such as creep and the shrinkage and relaxation of concrete develop gradually over time, which leads to a modification in elasticity laws. These impacts must be considered, as they can affect the stress state in the dam (see also Section 19.3).

Concrete swelling of a chemical nature
In the proven case of concrete swelling of a chemical nature in an existing concrete dam (notably related to an alkali-aggregate reaction), the associated internal stress due to the hyperstatic behavior of the static system (arch dam), as well as the possible damage to the concrete (advanced stage of the reaction) are to be determined on a case-by-case basis. This also applies to their impact on safety (see also Section 19.3).

Earth- and rockfill settlement
For embankments, a camber along the crest must be included during the design phase to guarantee sufficient freeboard following settlement (see Figure 20.11).

Aircraft crash into a water-retaining facility
This case falls under residual risk and requires no verification.

Superstructures
Superstructures such as communication masts are not generally considered to be appurtenant safety structures. However, operators must ensure that they do not create local instabilities at their foundation level.

---

[7] Ditto 6.

12. Administrative Procedures and Requirements

Schiffenen arch dam in Switzerland, height 47 m, year commissioned 1963
(Courtesy © Swiss Air Force)

12.1 Granting of a Concession and Construction Permit

As in other domains, the Swiss Confederation has set out the provisions that hydropower schemes, including dams, must meet. In particular, the Confederation has established and enacted laws and ordinances that relate to the rational use of hydraulic power, the protection of water resources, the safety of water-retaining facilities, protection of the environment, the drawing up of an environmental impact assessment study, and the restoration of watercourses for the purposes of flood and water protection. On the basis of Swiss legislation, the cantons have also passed laws, ordinances, and regulations to define their responsibilities and the services they are to provide, as well as the obligations of the applicants.

Construction of a dam cannot go ahead without a construction permit. Furthermore, a concession is required for all hydropower schemes.

12.1.1 Request for concession[1]

In agreement with the Swiss Constitution, the right to use the water from a river falls under the jurisdiction of the cantons. Some cantons have delegated this right to communes. Ultimately, it is up to the cantonal authority to decide on the granting of a concession and issue a construction permit. However, in collaboration with the relevant cantons, the Confederation is responsible for granting concessions for international projects.

In order to obtain a concession for water rights (Figure 12.1), the applicant (owner) files a request with the competent cantonal authority, with an accompanying application file. This file includes a technical section (project description and general drawings of planned installations), a financial section (cost estimate and funding possibilities), as well as a report on the environmental impact assessment (EIA). Generally, the competent cantonal department organizes internal consultation with all the various departments affected by the project. In some cases, the canton transfers the file to the competent federal departments, which must also assess whether the project complies with regulations.

After official notification, the cantonal authority opens up the file for consultation by members of the public and associations who are directly involved. Any person affected by the project may, usually within a period of 30 days, submit any comments and reservations, or even file an opposition, which must then be addressed. Opposition is discussed by the applicant and the party granting the concession in order to find an amicable solution. Should this not occur, private law opposition is sent to the civil courts. Public law opposition is a matter for the cantonal authorities or for the Confederation. If the opposing party is not satisfied with the judgement from the court of first instance, it may appeal to the Federal Court, whose judgement is final.

The concession contains information regarding the concession holder, scope of usage rights, concessionary obligations (especially financial requirements such as fees), and duration. A concession for a hydropower scheme is generally granted for a duration of 80 years starting from the date of operation and renewable at the end of its term.

It must be noted that expropriation proceedings are possible should the interests of the public carry greater weight, according to the law.

---

[1] This chapter is based on legislation concerning the use of hydraulic power from the Canton of Valais. Authorization procedures and construction approval may vary from one canton to another.

REQUEST FOR CONCESSION Technical, financial, environmental reports	APPLICANT
ASSESSMENT OF THE REQUEST	COMPETENT AUTHORITY
PUBLIC INQUIRY	MEMBERS OF THE PUBLIC, ORGANIZATIONS
INQUIRY AND PROCESSING OF OPPOSITIONS	COMPETENT AUTHORITY
CANTONAL JUDGEMENT	CANTONAL ADMINISTRATIVE COURT
FEDERAL JUDGEMENT	FEDERAL COURT
EXPROPRIATION	COMPETENT AUTHORITY

Figure 12.1 Process for granting a water rights concession (based on an oral presentation, see Bischof et al., 2000)

12.1.2 Construction permit

Before beginning construction of a new dam, or transformation and repair works, the owner must seek approval for the plans and specifications from the competent cantonal authority. If the dam meets the federal criteria, the supervisory authority must sign off the project for the dam and its appurtenant structures from a technical safety viewpoint (WRFA, 2010; WRFO, 2012). This approval is required for works to begin and forms part of the construction permit issued by the competent authority (canton, possibly supervisory authority). In order for a dam project to be assessed, the application must demonstrate that the planned water storage facility has been designed and will be constructed in accordance with current scientific and technical knowledge so that its safety is guaranteed in all foreseeable cases of loads and operation. The technical file must contain explanatory notes, general project plans (location, elevation, cross-sections, monitoring systems, ancillary structures), results from geological and geotechnical investigations, a hydrological study, static, dynamic, and stability calculations, as well as hydraulic calculations for water release structures.

The project is also submitted for public consultation. The opposition procedure and its treatment follow the same path as that of concessions. It is useful to note that the concession can be accompanied by a deadline that corresponds to the beginning of construction works.

Once works are completed, the canton and supervisory authority verify that the project was constructed according to the approved plans and all relevant laws. The owner must submit a file containing a report on construction works (including results from verification tests carried out) and drawings of the executed construction.

The supervising and approval of plans by cantonal and federal authorities does not relieve the owner of its responsibility with respect to third parties for damage caused during construction and operation of the dam and reservoir.

12.2 Environmental Impact Assessment (EIA)

Environmental aspects were taken into account long before the implementation of a specific law. In Switzerland, a legislative basis was developed in the 1980s. It defined the criteria for an environmental impact assessment (EIA) as well as mitigation and compensation measures. The EIA relates to the construction or modification of installations that may have a significant impact on the environment. The EIA carried out in either the preliminary phase or the construction phase enables officials to assess whether the planned installation respects legal provisions and to verify whether the impacts of the project are acceptable.[2]

The EIA establishes the assessment process and details its main stakeholders: the owner (applicant), the competent authority, and the service for the protection of the environment. The *Manuel EIE, Directive de la Confédération sur l'étude de l'impact sur l'environnement* (FOEN, 2009) can be consulted for furthner details. The *Manuel EIE* outlines the applicable legal basis, gives information on the degree to which installations are subject to the EIA, and gives advice on the content of impact assessment reports. It also details and clarifies the steps involved in completing an EIA.

12.2.1 Process for completing an EIA (Hertig, 2005)

Initially, the owner begins a preliminary study to determine the impacts that the completed project may have on the environment and concurrently establishes the terms of reference for the study going forward (Figure 12.2). The preliminary study is designed to highlight the key issues, conditions, hypotheses, and requirements that will be answered in the EIA. Firstly it ensures that no major aspects are left out of the EIA, and secondly that not too much importance is attached to minor issues. In the preliminary study, a rather broad perimeter of study is set. Depending on the various elements of a water storage or hydropower scheme, the areas of study may be restricted to this set perimeter. If, based on the preliminary study, considerable impacts are expected to affect the environment, the terms of reference will facilitate the actual assessment itself. Should the project have no effect on the environment, the preliminary study is recognized as the impact assessment.

The terms of reference established during the preliminary study outline the environmental aspects that must be assessed. They then enable an EIA report to be written up for the first stage of project development (concession process stage). In this report the project is described and justified. It also takes stock of the initial status of the project and the infringements that may or may not be attributed to the project. It is important to be clear about this initial status, which is accepted as being the status at the beginning of the works. However, it should be emphasized that the impact assessment is often carried out many years before the construction of the hydropower scheme and dam begins (which can also be the case for other works). Anticipated additional disturbances must therefore be assessed before the beginning of the works.

This report describes all the planned protection measures that are designed to reduce harm to the environment as much as is possible. These must be achievable and meet suitability criteria (achieving the objective in the interests of the public), necessity (an intervention limited to what is strictly required), and proportionality (reasonable relationship between the target objective and the restriction in freedom required for achievement of the project). There are several levels of protection measures. First of all, to avoid harm to the environment, the most appropriate options and sites should be favored. An option study enables the choice of location for a dam to be justified, through its intended use, local conditions, and

---

[2] This section refers in part to the publication "EIE des aménagements hydroélectriques. Mesures de protection de l'environnement" (Information concernant l'étude d'impact sur l'environnement n° 8 du SAEFL (FOEN), Bern, 1998).

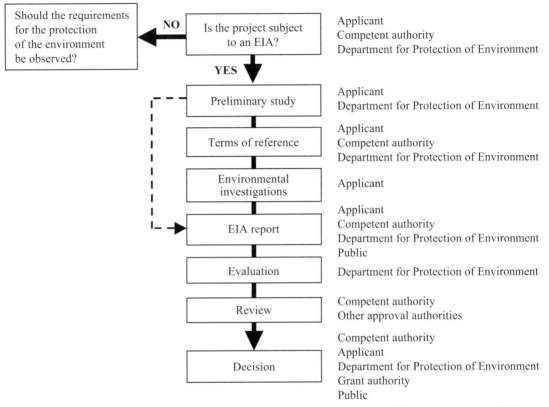

Figure 12.2 Simplified diagram of the process for completing an environmental impact assessment (EIA) (based on module 4 of the *Manuel EIE*)

technical specifications that must be respected. Avoiding, or at least reducing the harmful effects that could result from various aspects of the dam is recommended. If possible, the aim is to reinforce a work whose intended measures are not sufficient to prevent harm or destruction. Appropriate replacements should be planned in the case of inevitable and not entirely reparable harm. Finally, performance review and the environmental monitoring of the construction project helps to appropriately establish mitigation measures and monitor their execution.

The report from this first stage includes the terms of reference for the second stage of the EIA, which is the detailed design stage. The detailed design stage occurs at the project execution phase (authorization process for constructing and approval of plans and drawings).

12.2.2 Environmental aspects

The following impacts are covered in the EIA:
- Nature and natural landscapes
- National heritage sites (historical communication routes, constructed sites to be protected)
- Surface water and groundwater

- Forests
- Hunting and fishing
- Air
- Noise and vibrations
- Non-ionizing radiation (NIR)
- Ground
- Waste and contaminated sites
- Protection against catastrophes

In the case of dams, these impacts can be dealt with during project development, management of the construction site, and operation.

Protecting nature and natural landscapes means safeguarding biological diversity and conserving natural sites and landscapes. The main aspects include conserving natural resources and leisure spaces, and identifying and strengthening the relationship between local communities and natural landscapes.

The survival of plant and animal species can be ensured by protecting biotopes, creating new biotopes, and establishing buffer zones. Protecting biotopes helps stabilize natural populations and protect biodiversity. For a biotope replacement to be considered as acceptable, it must carry out the same functions and be located in the same topographical region. Finally, it is important to be aware of the existence of locally and regionally important biotopes indexed by the cantons. Swamps, marshy ground, wetlands, and alluvial zones of national importance are protected sites that must be conserved. Reservoirs can cover large surface areas and lead to a loss in ground cover. They can submerge biotopes and mountain agricultural land; these areas may, however, be recreated or replaced. However, it must also be admitted that bodies of water are a key element of the alpine landscape. Dams integrate harmoniously into the natural landscape and are an expression of human creativity. When reservoirs are drawndown in winter, the banks may appear barren and without vegetation but are often covered by snow (particularly in the Swiss Alps).

Protecting national heritage sites means safeguarding historical landscapes and places, and cultural and historical sites and monuments. Protecting these sites involves both the internal organization of urban areas and their architectural integration into the landscape.

Protecting water means safeguarding surface water and groundwater from all harmful effects. It also aims to guarantee a supply of water and maintain the effective evacuation and treatment of sewerage. Effects that are harmful to water must be analyzed, and water quality must be protected, as must the migration routes and reproduction cycles of aquatic flora and fauna. Groundwater must retain its capacity for regeneration.

In the case of a dam, the flow regime of a watercourse can be severely impacted by the capturing and storing of water in a reservoir. There is a modification in flow in downstream sectors, or even a flow transfer from summer to winter. This is why it is important to maintain residual, i.e., environmental flow in the watercourse. During operation, preventive measures must be planned for flushing or the emptying of the reservoir, or for the removal of debris (material should not be discarded directly downstream).

As for the ***conservation of forests***, all land area covered in trees or forest shrubs that directly fulfils a forest function (wood production, protection against natural dangers, leisure area, and ecological role) are considered as forests. Any clearing of forests must generally be compensated for in the same region, both quantitatively and qualitatively.

For ***hunting*** and ***fishing***, the aim is to guarantee the unimpeded passage of wild animals, install structures and maintain fundamental conditions so that biotopes are not cut off and isolated, and ensure that water containing fish stocks is protected.

Measures for *air protection* aim to diminish atmospheric pollution. They require limit values to be set during both construction and operation.

Protecting against noise is designed to prevent damaging and bothersome noise. Noise has a primarily local effect. The limit values for noise emission correspond to levels below which the public is not disturbed. In principle, noise emissions must be reduced at both a technical and operational level, while remaining financially viable. This means assessing zones in which measures must be taken during the construction period and during operation.

Protecting against radiation is related to the possible harmful or unpleasant effects of radiation. Industrial plant for the production and distribution of electricity generates non-ionizing electromagnetic fields in the direct vicinity. To avoid harm, especially to health, certain intensities may not be exceeded.

Protecting the ground aims to prevent irreversible damage caused by human activity, preventively protect the ground against damage, and maintain surveillance of open land. The construction of hydropower schemes and dams generally leads to a loss of land and long-term or temporary harmful effects. Ground that is used temporarily must be returned to its original condition.

With regard to *waste* and *special waste*, requirements concerning the reduction and treatment of water, as well as the organization and use of waste disposal facilities must be respected.

Protecting against major accidents and catastrophe aims to strengthen protective measures for people, property, and the environment against exterior influences such as natural hazards. To achieve this, risk must be reduced to an acceptable level. Checks must be carried out to ensure that the quantities of substances, products, and special waste are within accepted limits. If this is the case, the process is complete.[3]

For water-retaining facilities, section 5.4 and chapter 33 of this book deal with emergency planning to ensure the population is protected from the consequences of an uncontrolled release of water.

12.2.3 Reviewing the EIA

The competent authority responsible for approving the project must assess the compatibility of an installation with its environment, following a request from a specialized department for environmental protection. The request is based on the review of an impact report submitted by the applicant, as well as other documents submitted as part of the EIA.

This specialized department provides advice to the competent authority and the owner. It examines the terms of reference and then determines if the project meets the environmental regulations in effect. The cantonal department for the protection of the environment reviews the reports for installations that fall under the jurisdiction of the cantonal authorities. In the case of installation requiring federal sign-off, the cantons are heard by the competent authority.

The Federal Office for the Environment (FOEN) reviews reports for installations that require a decision from federal authorities.

---

[3] The Major Accidents Ordinance (OPAM) does not apply to dams, but does concern special substances, preparations, or waste, and rail installations, major transit routes, and river transport that is used to carry or transfer dangerous material.

IV. Concrete Dams

Depending on their shape, concrete dams can be grouped into several categories. Gravity dams, arch dams, and buttress dams can be defined by their specific features, including the amount of excavation and volume of placed concrete, the intensity of pressure, the amount of uplift acting on the dam, and their vulnerability to earthquakes or fluctuations in temperature. The choice of dam type also depends to a great degree on the site conditions (topography, geology).

Over time, technology, production, and placement of concrete has developed significantly and has in turn led to the construction of ever more audacious structures. Many dams in Switzerland are higher than 200 m, such as, for example, the Grande Dixence dam, which at 285 m is the highest concrete dam constructed to date, as well as the Mauvoisin (250 m), Luzzone (225 m), and Contra (220 m) dams.

Chapters 13 to 17 deal with the various types of concrete dams. They describe the dams' general characteristics and give a broad overview of verification approaches depending on their design (stability, loads and stresses, actions during earthquakes). Various construction aspects are also noted.

Chapter 13 is devoted to gravity dams, generally triangular-shaped massive structures whose weight is sufficient to resist water pressure. Chapter 14 describes buttress dams, structures made of relatively narrow, spaced elements that enable savings in the use of concrete. Chapter 15 deals with arch dams, whose curved, elegant shape enables water pressure to be redirected horizontally and vertically onto the abutments at the valley flanks. Chapter 16 looks at multiple-arch dams, which are related to buttress dams. They comprise cylindrical, inclined arches that are supported by triangular buttresses. The shape and spread of the valley, but also questions related to general stability are used as a basis for choosing the openings between the buttresses while achieving the correct arch distribution. Chapter 17 presents roller-compacted concrete (RCC) dams. Since the 1960s this innovative technique has brought new momentum to the construction of gravity dams. The initial idea was to employ methods used for embankment dams in the construction of concrete dams. The advantages of this solution include rapid execution, a reduction in cement content, and a decrease in cost per m^3. Although roller-compacted concrete is in some ways a new technology, it has now become the standard approach for large concrete gravity dams.

Chapter 18 addresses the most important aspects of concrete technology. The construction of concrete dams requires concrete that has a homogeneous structure and that performs consistently. High-quality concrete is vital, especially in terms of its workability, a factor that determines appropriate placing as well as its mechanical resistance characteristics. As such, a program of preliminary concrete trials is necessary, as is monitoring throughout the entire construction period. Test samples collected during operation allow the development of concrete characteristics to be assessed.

Chapter 19 examines various phenomena (fissuring, swelling, frost resistance) that, along with time, can directly affect mass concrete and facings. Some of these phenomena may produce minor deformations, but others result in damage to the body of the dam, jeopardizing its structural integrity and characteristics (for example, a loss of mechanical strength). Lastly, some have an effect on the visual appearance of the dam, causing superficial damage without impacting safety. This chapter is primarily drawn from the findings of the "Concrete" working group of the Comité suisse des barrages, published in the text *Le béton des barrages suisses: expériences et synthèse. Concrete of Swiss Dams: Experiences and Synthesis* (SwissCoD, 2000b).

13. Gravity dams

Gries gravity dam in Switzerland, height 60 m, year commissioned 1966 (Courtesy Ofima SA)

13.1 General Shape

The transversal vertical cross-section of a gravity dam is shaped very similarly to a triangle, often right-angled. The upstream face is vertical or slightly inclined in the upstream direction.

The downstream face is often rectilinear and has a 75 to 80% slope but may also be in the shape of a broken line (Figure 13.1). The peak of the triangle must in all cases reach the highest water level mark. The crest is formed by a thickening in the shape of a simplified triangle, which allows an access route to

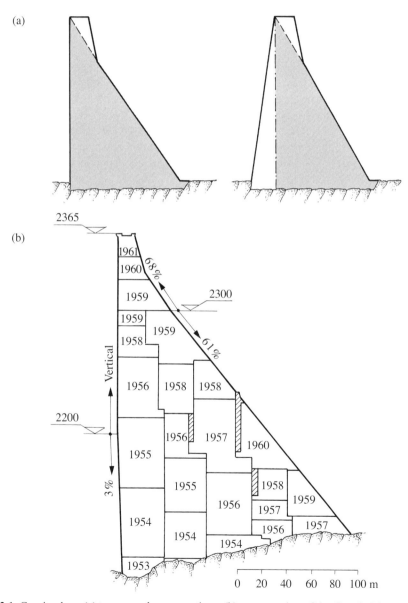

Figure 13.1 Gravity dam: (a) transversal cross-sections; (b) cross-section of the Grande Dixence dam

be integrated and provides additional freeboard (for waves) and resistance to ice load and floating debris. This upper part also contributes to the stability of the dam structure. Finally, to improve sliding strength, the foundation is sometimes inclined in the upstream direction.

Muttsee gravity dam

The new Muttsee dam is part of the recent pumped storage scheme (1000 MW) constructed between the Muttsee reservoir and the existing reservoir at the Limmern dam (Otto et al., 2016; Tognola and Balissat, 2011). The Limmern dam is in turn part of the Linth-Limmern hydropower scheme in operation since the 1960s.

As part of the pumped storage project, the capacity of the natural reservoir at Muttsee (9 mio m^3) was increased by 24 mio m^3 through the creation of a 1,954-meter-long gravity dam. Thanks to this dam, the natural level of the reservoir was raised by 28 m. The reservoir's current normal water level is at an altitude of 2,474 m.a.s.l.

The dam, whose crest level is at 2,476 m.a.s.l., is the highest water-retaining facility in the Alps. At this altitude, climatic conditions are harsh (a long winter period and relatively short summer period). This had a major impact on the supply of materials and the construction program, which lasted for a period of 6 years (2010 to 2015).

The new Muttsee dam is a conventional gravity dam comprising 68 independent blocks each with a length of 35 m. Its maximum height is 35 m, for a maximum foundation width of 27 m. The total volume of placed concrete is 225,000 m^3.

Muttsee gravity dam under construction (Courtesy A. Schleiss, December 2011)

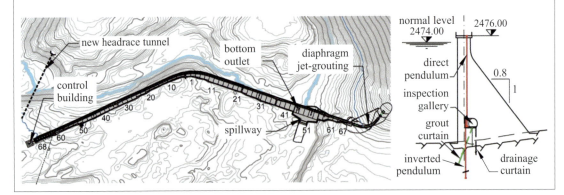

13.2 Loads and Key Verifications

Three fundamental principles govern static stresses in gravity dams:

- Self weight counteracts water pressure through resultant friction on the foundation
- Each block is itself stable (if the dam is totally rectilinear and the transversal joints are not grouted, no tridimensional arch effect contributes to stability)
- In principle, unreinforced dam concrete tolerates no tensile stress, in any case of loading

Furthermore, this very simple static system implies specific foundation conditions: the foundation must be made of rock, its deformation modulus high (low deformability), and its permeability low (watertightness).

Referring to Sections 11.1 and 11.2, as well as Figure 13.2, the main forces that come into play both for stability and the calculation of internal stress are self weight, water and sediment pressure, uplift, and the resultant dynamic forces from an earthquake. Uplift counteracts the stabilizing role of self weight, and it is therefore very important to take particular care when deciding on its distribution. Thermal effects have an impact on the state of the dam's internal stresses and deformation.

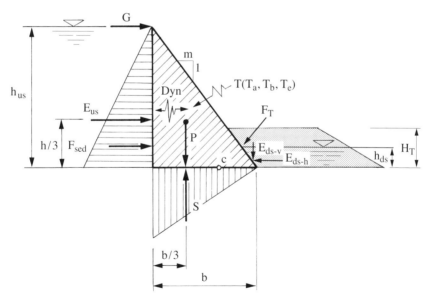

Figure 13.2 Schematic illustration of the main forces acting on a gravity dam (triangular cross-section). Key: P = self weight; E_{us} = upstream horizontal water pressure; E_{ds-h} = downstream horizontal water pressure; E_{ds-v} = downstream vertical water pressure; F_T = earth pressure; F_{sed} = sediment pressure; S = uplift; T = thermal effects; T_a = dam body temperature; T_b = air temperature; T_e = water temperature; Dyn = dynamic loading; G = ice load; h_{us} = upstream hydrostatic head; h_{ds} = downstream hydrostatic head; h_T = height of a downstream embankment.

With regard to the combination of loads, normal loading (type 1) takes into account the effects of self weight P, upstream and possibly downstream water pressure E, uplift S, possible sediment pressure F_{sed} (upstream) and earth pressure F_T (downstream), to which may also be added thermal effects T (due to seasonal variations and thermal gradients in the dam structure), as well as ice load in cold regions. The cases of both the empty reservoir (only P) and full reservoir (P + E + S + possibly F_{sed} and F_T) should be considered. In these cases of normal loading, no tensile stress in the concrete is accepted, and compressive stress remains relatively moderate (2 to 8 MN/m^2).

Cases of exceptional loading (type 2) correspond to that of normal loading, to which is added either the effect of a design flood or an avalanche.

Cases of exceptional loading (type 3) correspond to that of normal loading, to which is added either the effect of a safety flood or an earthquake. The emergence of slight tensile stresses in the concrete is accepted in these exceptional cases (the position of the resultant forces should, as a rule, be defined relative to the downstream extremity at a distance greater than 1/6 of the base length (Figure 13.4). In cases of an earthquake, slight tensile stresses are acceptable, given the dynamic and temporary nature of the load caused by the earthquake. Tensile stresses in the concrete may well cause cracks to open in the concrete, but these will soon close up again through dynamic cycles, and hydrostatic pressure will therefore not have time to build up. During post-event verification, attention must be paid to any modification in uplift.

The key aspects to be monitored are:

- Safety with regard overturning
- Safety with regard to sliding
- Safety with regard to failure

The two first elements to verify concern dam stability. Conditions for stability must be met both for plane sections and sections at the base of the structure, as well as along layers of rock in the foundation. The third element verifies that stresses in the dam structure and foundation are consistent with acceptable values.

13.3 Safety with Regard to Overturning

Generally speaking, a structure overturns when the resultant of all forces leaves the base of the structure's foundation (Figure 13.3). This definition does not provide enough detail for the analysis of the overturning of a gravity dam, as, at the limit of overturning according to this criteria, compressive stress becomes extremely high under the most loaded part of the foundation.

For gravity dams, there is another criterion guiding the verification of overturning: tensile stresses in the concrete are not admissible under any circumstances. Nor are they acceptable at the concrete/foundation rock interface. To meet this condition, the resultant of all forces taken into account must be contained within or up to the outer limits of the middle third (Figure 13.4). If this is the case, safety with regard to overturning is guaranteed. As for the slope of the resultant, this enables sliding stability to be assessed. This slope β, measured vertically, must lie in an order of magnitude between 27 and 42°.

To judge whether the slope m of the downstream face meets the above criteria, let's simplify the profile of the gravity dam and use a right-angle triangle with height h. The upstream face against which water pressure is exerted is assumed to be vertical, uplift distribution is triangular, and the uplift coefficient is k. This coefficient takes into account the actual distribution of uplift, influenced by the effect of the grout curtain and drainage under the foundation.

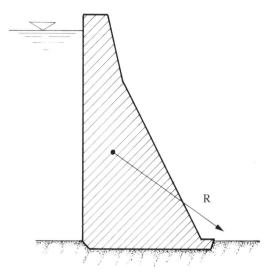

Figure 13.3 Non-stability with regard to overturning

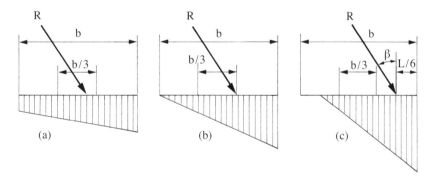

Figure 13.4 Relationship between the compressed base surface and the position of the resultant: (a) in the middle third; (b) at the downstream limit of the middle third; (c) at the downstream tolerable limit, taking into account the case of a maximum design earthquake (verification earthquake – see also Figure 13.25)

The application points of these three forces are illustrated in Figure 13.5. For this demonstration, other external forces exerted against the dam are assumed to be negligible (sediment pressure, downstream counterpressure of water).

The forces acting against a section of the dam with a width L are:

- Water pressure $\quad E = 1/2 \cdot \rho_E \cdot g \cdot h^2 \cdot L$
- Uplift $\quad S = 1/2 \cdot k \cdot \rho_E \cdot g \cdot b \cdot h \cdot L$

against which counteracts
- Self weight $\quad P = 1/2 \cdot \rho_B \cdot g \cdot b \cdot h \cdot L$

where
- h = dam height (and hydrostatic head in the case of a triangular profile)
- b = width of base
- ρ_E = unit density of water
- ρ_B = unit density of concrete
- k = uplift coefficient

For stresses to be zero at the toe of the upstream face, the resultant of forces must pass through point c indicated on Figure 13.5.

By establishing the moment equilibrium equation at point c, we obtain:

$$\sum M_C = E \cdot \frac{h}{3} + S \cdot \frac{b}{3} - P \cdot \frac{b}{3} = 0$$

however, $b = m \cdot h$.

The development of this equation enables m to be expressed:

$$m = \sqrt{\frac{\rho_E}{\rho_B - k\rho_E}}$$

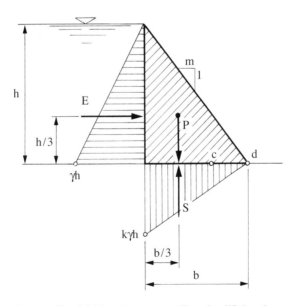

Figure 13.5 Forces due to self weight P, water pressure E, and uplift S acting against a gravity dam

The slope of the downstream face is thus independent of the dam height. This very simple equation is often called *Lévy's rule*.

If we assume $\rho_B/\rho_E = 2.5$ and $k = 0.85$ (1.0), we obtain a minimum value for the slope of the downstream face $m \geq 78\%$ (82%).

It thus becomes a simple matter to calculate the extreme vertical stresses on the foundation for each of the two load cases: the empty and full reservoir. In both cases the diagram is triangular, consistent with a foundation assumed to be rigid (Figure 13.6).

The factor of safety for global failure due to overturning of the profile around point d downstream of the foundation (Figure 13.5) can be calculated:

with
$$S_R = \frac{\sum M_{stab.}}{\sum M_{mobil.}} = \frac{M_P}{M_E + M_S} = \frac{P \cdot \frac{2}{3}b}{E \cdot \frac{1}{3}h + S \cdot \frac{2}{3}b}$$

$$b = m \cdot h \quad \text{and} \quad m = \sqrt{\frac{\rho_E}{\rho_B - k\rho_E}}$$

we obtain

$$S_R = \frac{\rho_B/\rho_E}{\frac{1}{2m^2} + k}$$

With the assumed values for $k = 0.78$ and $m = 0.78$, we obtain the factor of safety $S_R = 1.49$.

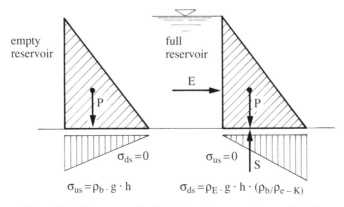

Figure 13.6 Stresses on the foundation in cases of normal loading

Effect of uplift on the slope of the downstream face for a triangular gravity dam

Let $\quad m = \text{tg } \phi = b/h$, from where we calculate $b = m \cdot h$, $\quad b' = \alpha \cdot b = \alpha \cdot m \cdot h$.

The forces are given by

$$P = 1/2 \cdot \gamma_b \cdot m \cdot h^2, \quad E = 1/2 \cdot \gamma_e \cdot h^2, \quad S = 1/2 \cdot k \cdot \alpha \cdot m \cdot \gamma_e \cdot h^2,$$

where
- h = dam height (triangular profile)
- γ_b = unit weight of concrete
- γ_e = unit weight of water
- k = uplift coefficient

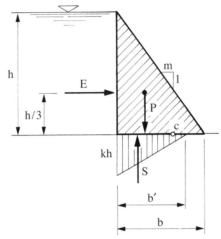

According to the criterion specifying that vertical upstream stresses must be zero for a full reservoir, the moment relative to point c is equal to zero. From this relation, we can calculate:

$$m = 1 / \{\gamma - \alpha \cdot k \cdot (2 - \alpha)\}\, 1/2.$$

For $\gamma_b = 2.4$ t/m³, we obtain:

k	$b' = b$, $\alpha = 1$	$b' = 0.5b$, $\alpha = 0.5$	$b' = 0.25b$, $\alpha = 0.25$
0.4	0.707	0.690	0.670
0.8	0.791	0.745	0.698
1.0	0.845	0.778	0.714

For $\gamma_b = 2.5$ t/m³, we obtain:

k	$b' = b$, $\alpha = 1$	$b' = 0.5b$, $\alpha = 0.5$	$b' = 0.25b$, $\alpha = 0.25$
0.4	0.690	0.674	0.656
0.8	0.767	0.725	0.682
1.0	0.816	0.756	0.696

Impact of the crest on the design

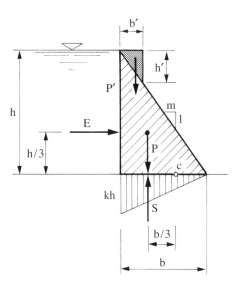

Let
$$b'/b = h'/h = \eta$$
$$B = m \cdot h$$
$$P' = 1/2 \cdot \gamma_b \cdot b' \cdot h' = \tfrac{1}{2} \cdot \gamma_b \cdot b \cdot h \cdot \eta^2 = \tfrac{1}{2} \cdot \gamma_b \cdot m \cdot h^2 \cdot \eta^2.$$

The sum of moments relative to c must be equal to zero,
$$\Sigma M_c = M_E - M_{(P-S)} - M_{P'} = 0$$
$$= h^3/6 - (m^2 \cdot h^3)/6 \cdot (\gamma_b - k) - 1/2 \cdot \gamma_b \cdot m \cdot h^2 \cdot \eta^2 \cdot 2/3 \cdot m \cdot h \cdot (1 - \eta) = 0.$$

From where we can calculate
$$m = 1/[(\gamma_b \cdot (1 + 2\eta^2 - \eta^3)) - k]^{1/2}.$$

For a simple triangular profile, we have
$$m = 1/[\gamma - k]^{1/2},$$

for $k = 1.0$:

η	0.0	0.1	0.2	0.3
m	0.831	0.819	0.788	0.750

13.4 Safety with Regard to Sliding

13.4.1 Definitions and principles of calculation

Let's take the simplified triangular profile of the gravity dam once again and examine the risk of sliding across the foundation under the effect of horizontal water pressure (Figure 13.7).

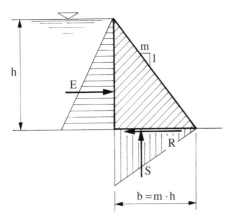

R: sliding resistant shear force,
A: unit sliding surface [m^2/m],
$\quad$ A = b = m · h

The sliding safety is defined as the ratio between the resisting forces and the (horizontal) pressure forces:

$$S_G = \frac{\text{resisting forces}}{\text{pressure forces}} = \frac{R}{E}$$

Figure 13.7 Calculating sliding stability

Resistance to sliding along a surface is expressed with Coulomb's equation from soil mechanics:

$$R = \Sigma V \tan \phi' + c' A_c,$$

where

$\quad \Sigma V \quad$ is the resultant of all forces perpendicular to the foundation
$\quad \phi' \quad$ is the effective angle of internal friction of the considered sliding surface (saturated)
$\quad c' \quad$ is the effective cohesion on the considered compressed sliding surface A_c (saturated).

The result is the equation:

$$S_G = \frac{(P - S) \tan \phi' + c'b}{E}$$

and with A = m·h and the relations describing forces P, E, and S described in the preceding section, we can deduce the following:

$$S_G = m \left(\frac{\rho_B - k\rho_E}{\rho_E} \tan \phi' + \frac{2c'}{\rho_E g h} \right)$$

In practice, cohesion is usually overlooked, either because it is very low or because the cracking of the rock or concrete makes it nonexistent.

The condition that compressive stress is only admitted under the foundation led in the previous section to Lévy's rule,

$$m = \sqrt{\frac{\rho_E}{\rho_B - k\rho_E}}$$

which gives

$$S_G = m \; \frac{\rho_B - k\rho_E}{\rho_E} \; \tan\phi' = \frac{1}{m} \tan\phi'$$

As a result, sliding safety is guaranteed if $S_G \geq 1$, which implies $\tan\phi' \geq m$.
This particularly simple equation is conditioned by the following simplifying hypotheses:

- The dam profile is drawn as a diagram in the form of a right-angled triangle, with a vertical upstream face and a horizontal foundation
- The uplift diagram is assumed to be triangular; uplift is zero at the downstream toe of the foundation
- The foundation is assumed to be rigid, and so the vertical stress diagrams on the foundation are linear
- No tensile stresses are accepted on the foundation (Lévy's rule)

In the early 1980s, the US Army Corps of Engineers (USACE) introduced the *Limit Equilibrium Method* to assess the stability of gravity dams against sliding. Developed from soil mechanics, this method defines a factor of safety as being equal to the relationship between the shear strength and shear stress applied along the slip surface:

$$FS = \tau_a/\tau,$$

where

τ = shear stress required to ensure equilibrium
τ_a = available shear strength determined by Coulomb's equation
FS = factor of safety

This method is more adapted to constructions built on soft ground, especially in the presence of non-cohesive wedge effects developing active and/or passive pressure (Figure 13.11) (USACE, 1983, 1995; Jansen, 1988).

13.4.2 Slip surfaces and their strength

Before determining the effective angle of internal friction of the slip surface, the potential slip surfaces must first be determined (Figure 13.8).

The calculation of mean friction angles in the rock and interface between the concrete and rock is not simple. The rock is subject to an entire network of joints and other discontinuities, which means that laboratory tests do not reflect the overall behavior of the rock mass. The quality of the concrete/rock interface depends to a great extent on the quality of contact grouting. In concrete structures, lift joints and possible shrinkage cracks create preferential sliding planes to which careful attention should be paid.

The following ranges of coefficient values for internal angles of friction can nevertheless be given.

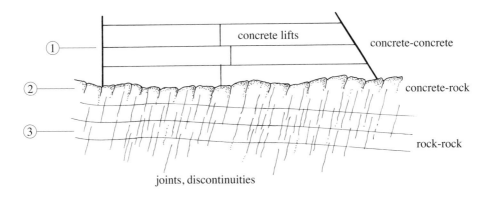

Figure 13.8 Potential slip surfaces and their characteristics

Table 13.9 provides various values for internal angles of friction ($\tan \phi'$) resulting from tests (Link, 1967).

Table 13.9 Indicative values for the angle of friction

Type of contact	Type of rock	$\tan \phi'$
Concrete-Rock	Limestone	0.8-1.3
	Sound gneiss (decomposed)	1.7 (0.5)
	Sound granite (decomposed)	1.5 (1.0)
	Sandstone	1.0-1.3
	Schist	1.0-1.6
Rock-Rock	Limestone	0.4-0.8
	Cracked gneiss	1.0
	Granite	0.8-1.9
	Sandstone	1.7

It should further be noted that tests and studies have been carried out to characterize peak shear strength for rock joints without filling. This strength depends in particular on normal effective stress acting on the contact surface (which can be associated with the sliding plane), as well as the roughness and morphology of the joint. The roughness of a joint's surface indicates the irregularities and undulations of the joint surface relative to its mean plane.

On the basis of test results on natural joints, Barton formulated an empirical criterion that enabled the angle of friction ϕ' to be calculated on a discontinuity, based on the traditional shear strength law $\tau = \sigma \, \text{tg} \, \phi'$ and by taking surface roughness into account:

$$\phi' = \phi_r + i = \phi_r + \{JRC \cdot \log 10\, (JCS/\sigma'_n)\}.$$

ϕ_r is the angle of friction that is residual or on a smooth joint, and i is the roughness component. JRC is the joint roughness coefficient; it was deduced from shear tests and from the comparison of 10 standard roughness profiles. A scale of value from 0 to 20, in steps of 2, was established, where value 0 corresponds to a straight, smooth surface and 20 to a very rough surface (Figure 13.10). JCS stands for joint wall compressive strength (which can be calculated using Schmidt's rebound number). Finally, σ'_n is the normal effective stress. If σ'_n nears JCS, the equation becomes the base equation for friction.

The expression given below reflects the cohesion of joint surface matching and corresponds to peak strength. When the rock mass moves (for example, following an earthquake), ϕ' can be modified and tends toward residual ϕ_r. A new value of ϕ_r must be introduced when performing post-event calculations.

Weathered rock near the surface is generally removed, sometimes to a considerable depth, so that the dam foundation is on the best possible quality of rock. The most critical slip surface is therefore often found just below the concrete-rock interface.

Excavating the foundation and replacing low quality rock with concrete allows a certain embedding of the dam and also creates an abutment block at the downstream toe. In preliminary stability calculations this block is generally not considered for dam sliding, given that a displacement in the dam structure is necessary for the block to become effective. It should nevertheless be noted that if the dam base is deeply embedded into the foundation or if a potential failure plane is located under the base, the passive strength of the foundation may sometimes be considered for sliding strength (Figure 13.11). Compressive and buckling strength of the layers must be sufficient to create a resistant wedge. It is also possible to create a heel perpendicular to the downstream toe. This passive pressure is not to be taken into account when considering

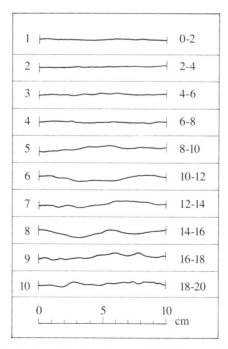

Figure 13.10 Standard profiles for estimating JRC (Barton and Choubey, 1997)

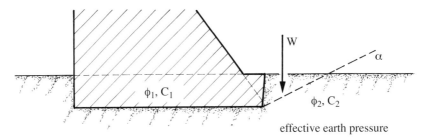

Figure 13.11 Foundation conditions when considering a downstream block (passive strength)

the risk of downstream scouring following overtopping. However, care should be taken when considering the effect of a toe block.

On the basis of these values, the factor of safety for sliding can be expressed as:

$$S_G = \frac{1}{m} \tan \phi'.$$

With a slope of the downstream face m = 0.80 and a horizontal foundation on good quality rock (ϕ' = 6°–62°), we obtain the factor of safety S_G = 1.8-2.3.

The generally accepted values for factors of safety, without taking cohesion into account, are the following (Table 13.12).

Table 13.12 Factor of safety for stability without considering cohesion (from SFOE (SFWG), 2002c)

| | LOADING | |
Normal	Exceptional	Extreme
1.5	1.3	1.1

13.4.3 Improvement in sliding safety

With poor quality rock, a low angle of friction value may mean that special provisions need to be taken to improve sliding safety.

By studying the criterion that describes the factor of safety for sliding

$$S_G = \frac{(P - S) \tan \phi'}{E},$$

three possibilities for increasing this factor become apparent:
- Increase vertical forces
- Reduce the hydrostatic force of horizontal pressure
- Improve the angle of friction value ϕ'

Several solutions can be implemented to reach these objectives.

Increasing vertical forces

Modifying the cross-section aims to increase self weight by increasing the volume of concrete (Figure 13.13 (a)). This measure is effective but expensive.

By sloping the upstream face of the dam, the hydrostatic pressure applied to a sloped plane is converted into a horizontal and vertical force corresponding to the weight of the water contained in the grey section (Figure 13.13(b)). This is a commonly implemented measure as it only requires a low additional volume of concrete.

As previously mentioned, uplift force may be reduced through grouting and drainage holes under the foundation (Figure 13.13 (c)). The creation of cavities in the dam joints (hollow gravity dam) guarantees perfect drainage in the foundation and considerably and reliably reduces uplift. However, cavities imply a reduction in self weight.

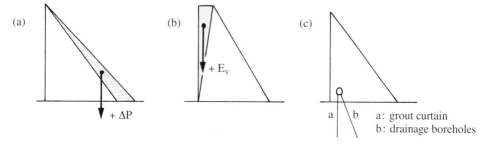

Figure 13.13 Measures for modifying the intensity of vertical forces: (a) increase the slope of the downstream face; (b) the slope of the upstream face; (c) reduce uplift

Reducing the hydrostatic force of horizontal pressure

It is not, of course, possible to reduce water pressure against the dam without reducing the water level (which is not the aim!). However, by sloping the dam foundation slightly upstream, a comparable result can be achieved (Figure 13.14). Self weight is converted into a force perpendicular to the foundation and a parallel upstream force that counteracts water pressure. In the same way, water pressure is converted into a force parallel to the foundation and a force perpendicular to the foundation.

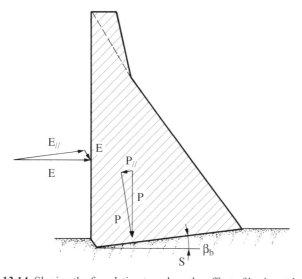

Figure 13.14 Sloping the foundation to reduce the effect of horizontal pressure

Therefore, for sliding that occurs along the foundation plane, the expression of the factor of security becomes:

$$S_G = \frac{(P\cos\beta_b + E\sin\beta_b - S)\tan\phi'}{E\cos\beta_b - P\sin\beta_b}.$$

As a rule, the foundation slope should not be greater than 10%, so that $\sin\beta \cong \tan\beta$ and $\cos\beta \cong 1$. The equation then becomes

$$S_G = \frac{(P + E\tan\beta_b - S)\tan\phi'}{E - P\tan\beta_b}.$$

It should be noted that if the rock is cracked or fissured, sliding may develop along a horizontal plane in the rock below the foundation (Figure 13.15 (a)). Stratigraphy may help to define potential slip surfaces (Figure 13.15 (b)). The engineer must ensure that accepted failure planes are kinematically possible.

The angle of friction on the rock-rock surface is generally higher than at the concrete-rock interface. It has also been observed that uplift develops fully in the network of cracks or joints, as well as the rules governing the implementation of an abutment block downstream of a sliding plane.

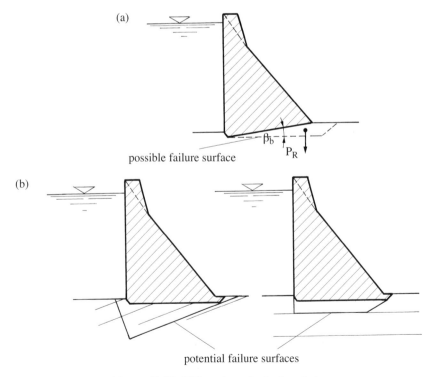

Figure 13.15 Sliding planes in the foundation

The weight P_R of the rock mass comprised between the foundation and the sliding plane may be taken into account so that the factor of safety becomes

$$S_G = \frac{(P + P_R - S) \tan \phi'}{E}.$$

Improving the angle of friction value ϕ'
This is carried out by way of the three following operations:
- The treatment of concrete joints (concrete)
- Stepped formwork (concrete)
- Grouting (rock)

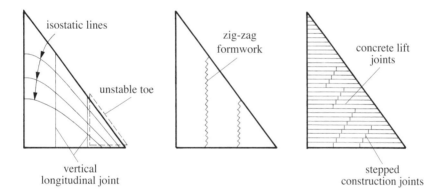

Figure 13.16 Example of transversal concrete stops at the Grande Dixence dam

Inside the dam structure, preferential sliding planes are created by the horizontal concrete lift joints, generally spaced every 2 to 3.5 m. To improve the shear strength of these surfaces, specific measures are taken to ensure high quality concrete lift joints:

- The surface is cleaned with water or pressurized air a few hours after setting to eliminate excess cement skin and thus make the surface rougher
- After concreting, the surface is meticulously cleaned to ensure that the new layer binds properly

For very high gravity dams, the transversal joints must be placed perpendicularly to the axis of the valley to limit the volume of concreting stages. As a rule, the concrete width of one single stage should not be greater than 30 to 40 m. These joints are laid out so that preferential sliding planes do not form. They are either formed by steps with sloped formwork or laid out directly in steps, depending on which layout is compatible with the main stress directions in the concrete (Figure 13.16). The orientation of steps is selected so that their surfaces are perpendicular to the isostatic lines.

For the foundation zone near to the dam, the rock should be treated with contact grouting and consolidation. Contact grouting ensures that the concrete-rock interface is solidly bound. Contact and consolidation grouting is designed to seal the cracks at the surface of the rock and is carried out with short boreholes spread throughout the foundation zone (see Part VII, Chap. 29).

13.4.4 Safety with regard to floating

There is a risk of instability due to floating in cases where uplift forces are greater than the sum of forces resulting from self weight, loads from vertical water pressure, and other loadings with a downward vertical component. A factor of safety is required and is given by the formula:

$$FS = \frac{(P_S + E_w)}{S},$$

where

P_S = weight of the structure and ancillary equipment (gates, bridge, footbridge, bridge crane, etc.)
E_w = weight of the water acting on the dam structure
S = uplift acting under the base of the dam structure

13.5 Safety with Regard to Failure

Compressive stresses in a gravity dam are generally low (directly related to height). In order to save costs, mass concrete therefore has a low cement content and includes large-sized aggregate. Compressive strength is also therefore low and is widely dispersed.

Tensile strength, for the same reasons and because there is a likelihood of cracks, is accepted as zero. When tensile stresses occur, openings in cracks can be observed, and if these cracks are in contact with the reservoir, hydrostatic pressure can become fully established. As a result, tensile stresses are not accepted in cases of normal loads.

Calculating stresses in a dam is today systematically carried out using computational models of two- or three-dimensional finite elements. The internal stresses caused by temperature variations and gradients must also be taken into account.

For a preliminary analysis, a simplified analytical approach, such as is described below, is often sufficient.

13.5.1 Loads and stresses in the dam

Let's assume a simplified triangular profile for a gravity dam and study the distribution of stress on an A-A horizontal cross-section located at depth z, as shown in Figure 13.17.

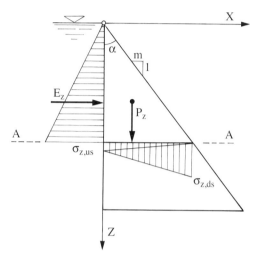

Figure 13.17 Distribution of vertical stress in a horizontal cross-section

Navier's hypothesis assumes that vertical stress varies linearly along a horizontal section. The distribution of stress thus takes the form of a trapezoid or a triangle.

The upstream face of the dam is assumed to be vertical, and the downstream face has a slope $m \geq m_{min}$, m_{min} as defined by Lévy's rule.

Let's first only consider self weight P.

Conditions of equilibrium in the section lead to the following stresses:

$$P \qquad \sigma_{z,us} = \rho_B\, g\, z \qquad \sigma_{z,ds} = 0.$$

Let's now add the effect of water pressure E:

$$P+E \qquad \sigma_{z,us} = g\, z\, (\rho_B - \rho_E\, m^{-2}) \qquad \sigma_{z,ds} = \rho_E\, g\, z\, m^{-2}$$

$$\sigma_{x,us} = \sigma_{x,av} = \rho_E\, g\, z$$

$$\tau_{xz,us} = 0 \qquad \tau_{xz,ds} = \rho_E\, g\, z\, m^{-1}.$$

And finally, by adding the effect of uplift S, according to the simplified triangular diagram outlined earlier:

$$S \qquad \sigma_{us} = k\, \rho_E\, g\, z \qquad \sigma_{ds} = 0.$$

The condition of having an equilibrium of forces on the cross-section leads to the following stresses:

$$P + E + S \qquad \sigma_{z,us} = g\,z\,[\rho_B - \rho_E\,(m^{-2} + k)] \qquad \sigma_{z,ds} = \rho_E\,g\,z\,m^{-2}.$$

If $m = m_{min}$ according to Lévy's rule, then $\sigma_{z,us} = 0$, in cases where load $P + E + S$.

13.5.2 Principal stresses

The principal directions are those in which stresses are maximal or minimal and for which tangential stresses are zero. The upstream and downstream faces provide the direction of principal stresses and are given by the equation

$$\sigma_{I,II} = \frac{1}{2}(\sigma_x + \sigma_z) \pm \sqrt{\frac{1}{4}(\sigma_x - \sigma_z)^2 + \tau_{xz}^2}\,.$$

When the principal stresses on the faces are known, it is then possible to calculate those between the faces; the state of stresses on any section is thus defined.

Upstream face

The upstream face is initially assumed to be vertical.

With an empty reservoir (load P), the stresses are:

$$\sigma_{z,us} = \rho_B\,g\,z,$$
$$\sigma_{x,us} = 0 \text{ et } \tau_{xz,us} = 0,$$

since hydrostatic pressure on the face is zero.

The resulting principal stresses are:

$$\sigma_I = \rho_B\,g\,z \text{ et } \sigma_{II} = 0.$$

With a full reservoir (load $P + E + S$), the stresses become:

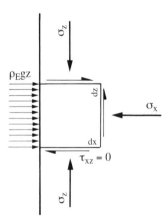

$$\sigma_{z,us} = g\,z\,[\rho_B - \rho_E\,(m^{-2} + k)],$$

$$\sigma_{x,us} = \rho_E\,g\,z,$$

$$\tau_{xz,us} = 0.$$

The principal stresses that result are:

$$\sigma_I = g\,z\,[\rho_B - \rho_E\,(m^{-2} + k)] \text{ and }$$

$$\sigma_{II} = \rho_E\,g\,z.$$

If $m = m_{min}$ according to Lévy's rule,

$$\sigma_I = 0 \text{ et } \sigma_{II} = \rho_E\,g\,z.$$

Downstream face

The downstream face is assumed to bear no exterior pressure.

In cases where load is P, the stresses are:

$$\sigma_{z,ds} = 0,$$

$$\sigma_{x,ds} = 0 \text{ et } \tau_{xz,ds} = 0,$$

since hydrostatic pressure on the face is zero.

The result is that principal stresses are zero.

In cases where load is P + E + S, the stresses become:

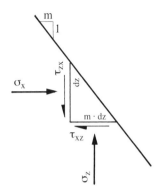

$$\sigma_{z,ds} = \rho_E \, g \, z \, m^{-2},$$

the result is

$$\tau_{xz,ds} = m \cdot \sigma_{z,ds} = \rho_E \, g \, z \, m^{-1}$$

and

$$\sigma_{x,ds} = m \cdot \tau_{xz,ds} = \rho_E \, g \, z.$$

The deduced principal stresses are:

$\sigma_I = 0$, parallel to the face, and

$$\sigma_{II} = (1 + m^2) \cdot \sigma_{z,ds} = (1 + m^{-2}) \, \rho_E \, g \, z$$

Principal stresses inside the dam

When at any point along a horizontal layer the stresses σ_z, σ_v, and τ_{xz} are known, it is possible to deduce the principal stresses at this point by using Mohr's circle (Figure 13.18). By considering forces in the directions of the principal stresses, it is then possible to draw two families of orthogonal curves: isostatic lines. The bisector directions are those where τ is maximum; they define the lines of maximum shear.

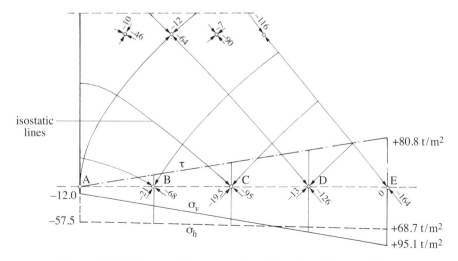

Figure 13.18 Diagrams of stresses and isostatics along a horizontal layer

The left diagram in Figure 13.19 shows the direction of principal stresses in a simplified triangular profile, based on the above calculation and assuming Navier's hypothesis. The right diagram shows the lines of equal minimum and maximum stresses determined by the finite elements on a real geometry, taking into account the distribution of the foundation reaction.

Finally, it can be useful to draw the lines of equal compressive stresses to determine the required strengths in the different zones of the dam (Figure 13.20).

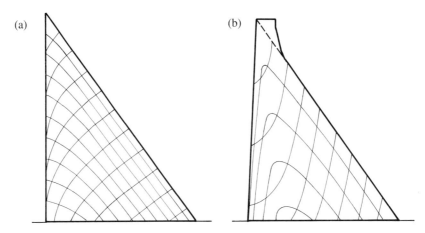

Figure 13.19 Principal stresses in the body of a gravity dam with a full reservoir: (a) triangular profile; (b) isostatic lines (lines of equal maximum and minimum compressive stresses) in a real cross-section

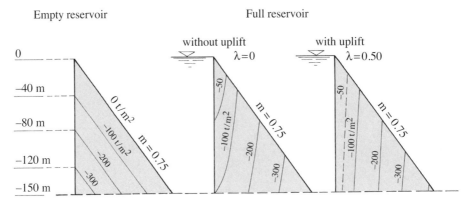

Figure 13.20 Lines of equal compressive stresses

13.5.3 Load-bearing capacity

Acceptable load-bearing capacity is defined as the maximum pressure that can be tolerated by a rock foundation while taking into account all the relevant factors of safety against failure of the rock mass or against such settlement that operation and safety of the dam structure are threatened. In fact, it is a question of ensuring that foundation displacement is compatible with structural deformations and the stability of the various elements of the foundation block (see Figure 13.15). The acceptable stresses must include a factor of security.

13.6 Loading During an Earthquake

13.6.1 True behavior of a dam during an earthquake

The true behavior of a dam during an earthquake is one of the most complex issues with which an engineer is confronted. There are several reasons for this:

- The dynamic behavior of such a massive structure is highly nonlinear and nonelastic
- The interaction between the mass of water and the dam structure must consider the compressibility of water, which may be considerable in some cases
- The interaction between the ground and the dam structure is vital from the point of view of energy dissipation

The formation of cracks in the concrete and the presence of joints between the blocks makes dam behavior highly nonlinear.

In gravity dams, cracks most often appear in the upper part of the cross-section, not far from the crest, at the point where the amplitudes are the largest.

Today some computational FEM models enable crack formation to be simulated, by applying mechanical failure theories. They require further development before an engineer who is not an expert in the dynamics of structures can use them. These models are the subject of ongoing research.

However, study on the issue of reservoir-structure interaction is noticeably more advanced.

Structure-fluid interaction is known to be greater when the dam's natural frequency, in conjunction with its foundation when the reservoir is empty, is close to the reservoir's natural frequency.

f_{dam}: dam's natural frequency when the reservoir is empty, depending on its geometry and materials.
$f_{reservoir}$: reservoir's natural frequency, depending on the shape of the reservoir, its depth, and the velocity of the pressure wave in the water.

The natural frequencies usually observed in concrete dams range from 1.4 to 5.1 s^{-1} [Hz] (Table 13.21). The reservoir's natural frequencies have been calculated to be between 1.3 and 4.4 s^{-1} [Hz].

Table 13.21 Water-structure interactions: examples of natural frequencies of dams and reservoirs: (1) arch dam; (2) gravity dam

Dam	h_{dam} [m]	ℓ_{crest} [m]	f_{dam} [Hz]	$f_{reservoir}$ [Hz]	$f_{dam}/f_{reservoir}$ [–]
Mauvoisin (CH) [1]	250	560	2.0	2.1	1.0
Kölnbrein (A) [1]	197	626	1.7	2.3	0.7
Emosson (CH) [1]	180	424	2.2	2.4	0.9
Morrow Point (USA) [1]	142	219	3.7	3.0	1.2
Pacoima (USA) [1]	113	180	5.1	4.4	1.2
Grand Dixence (CH) [2]	285	695	1.4	1.3	1.1
Pine Flat (USA) [2]	122	562	2.9	3.1	0.9

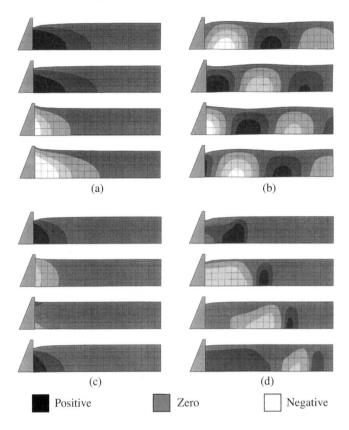

Figure 13.22 Dynamic water pressure acting on a gravity dam, depending on the compressibility of water and the natural frequency of the dam and reservoir (Bachmann, 1994): (a) water is compressible, harmonic movement: $f_{dam}/f_{reservoir} = 0.8$, water pressure is only affected close to the dam; pressure against the dam is seen to become negative (cavitation); (b) water is compressible, harmonic movement $f_{dam}/f_{reservoir} = 1.5$, in this case the dam becomes a wave generator and dynamic pressure occurs throughout the entire reservoir; (c) water is incompressible, harmonic movement $f_{dam}/f_{reservoir} = 1.5$, if the water is incompressible, dynamic pressure in the water will only occur close to the dam; (d) water is incompressible, dam movement like a shock, this is probably the most realistic case for an actual earthquake

With the help of several accelerographs installed in the Mauvoisin dam, and by continually analyzing its response to ambient vibrations, as well as recordings of two earthquakes, the resonance frequencies (natural frequencies) of the dam have been calculated according to the reservoir water level (Darbre et al., 2000; Darbre and Proulx, 2002; Chopra, 2020). Resonance frequencies were seen to increase as the water level of the reservoir rose up to almost two-thirds of its height and to decrease for water levels above this height.

Experience has shown that if $f_{dam}/f_{reservoir} < 0.7$, the effect of the compressibility of water can be disregarded, and the dynamic effect of the reservoir on the dam can be replaced by an inertia force of hydroseismic overpressure.

Figure 13.22 shows dynamic water pressure acting on a gravity dam.

13.6.2 Pseudo-static analysis

As noted previously, the pseudo-static approach has been used to carry out a simplified verification of dams in cases of earthquakes. If the earthquake acceleration is low, a < 0.15 g, and if water compressibility can be disregarded, $f_{dam}/f_{reservoir} > 0.7$, this method provides satisfactory results for the preliminary design phase. Remember that pseudo-static analysis can only evaluate the effect of horizontal acceleration oriented in the least favorable direction.

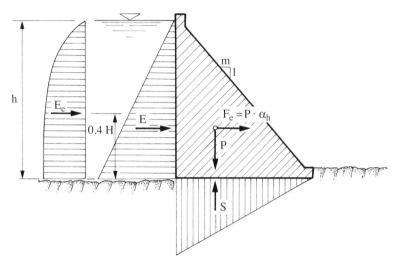

Figure 13.23 Seismic overloading based on a pseudo-static analysis

Forces involved

Figure 13.22 shows the principal forces involved in the load analysis for a full reservoir and an earthquake. To the forces P, E, and S already described are added the inertia forces of the dam and reservoir.

The dam's *force of inertia* F_e applied in the least favorable direction (that is, downstream) is therefore added to water pressure.

It is obtained with $F_e = \alpha_h\, gM = \alpha\, P$ (dam mass multiplied by horizontal acceleration) and the corresponding momentum becomes $M_e = F_e\, 1/3\, h = 1/3\, \alpha\, P\, h$ (triangular cross-section).

The dam induces movement in part of the reservoir water, whose acceleration produces pressure that must be taken into account. The water's *force of inertia* E_e is expressed by hydrodynamic overpressure on the upstream face of the dam, for which Westergaard formulated a parabolic equation (see § 11.5.2).

Cases of uplift

With regard to uplift, the opening of cracks at the upstream concrete/rock interface can be expected during an earthquake, and as a result, also a possible failure in the continuity of the grout curtain under the foundation, due to the appearance of tensile stresses in the zone of the upstream heel of the dam (Figure 13.24). In cases of earthquake loading, this possibility leads us to reject the decrease in uplift by a coefficient k and accept that uplift develops according to a triangular diagram whose upstream value equals hydrostatic

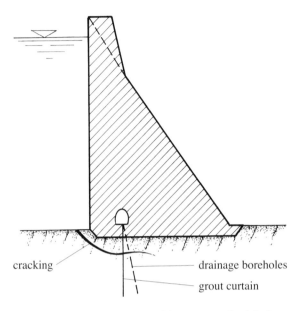

Figure 13.24 Possible cracking in the zone of the upstream heel during an earthquake

pressure. However, it is accepted that hydrodynamic overpressure cannot develop in cracks, due to the oscillating nature of the loading. It should be noted that drainage boreholes can, depending on the movements, remain effective during and after a seismic event.

13.6.3 Safety during an earthquake

The design earthquake is by its very nature an exceptional event. As such, a reduction in factors of safety is accepted compared to cases of normal loading. The duration of loading is also short:

- The factor of safety accepted on compressive stresses is reduced, even though this criterion is not generally a determining factor (concrete's compressive safety at 3, for example, instead of 4)
- Low tensile stresses ($\sigma_{\text{tensile stress}}$ < 1-2 N/mm^2) are tolerated along a small section of the upstream heel of the dam; these tensile stresses are lower than the resistance to tensile stress of uncracked concrete

Tensile stresses at the upstream heel of the dam will favor the development of open cracks during earthquake loading. The presence of these cracks will decrease the effective horizontal section and therefore overload the downstream toe of the dam. When the resultant of exterior forces leaves the middle third and reaches the lower sixth of the base (generally accepted limit), half of the contact surface is ineffective, and the compressive stress increases by 33%. The diagram in Figure 13.25 illustrates the increase in compressive stresses on the face through the opening of cracks on the opposite face, depending on the position of the normal resultant force. As long as the resultant remains in the middle third, there is no increase. The sliding factor of safety is reduced to 1.3 for preliminary design.

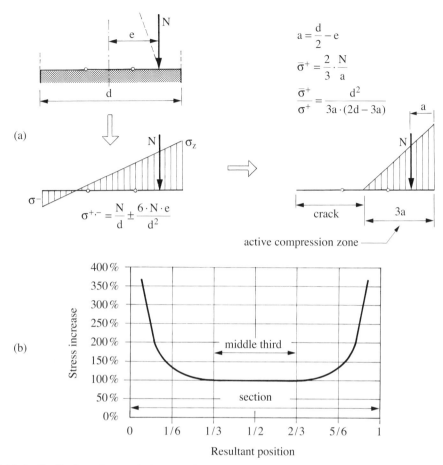

Figure 13.25 Redistribution of stress during cracking: (a) stress increase in the compressive edge when the opposite edge is under tensile stress; (b) increase in compressive stress depending on the position of the resultant

13.6.4 Dynamic analysis

When there is considerable seismicity at the site of the dam, that is, acceleration greater than 0.15 g can be expected, the pseudo-static analysis can no longer be considered sufficient, and a dynamic analysis must be undertaken.

A dynamic analysis considers a system of forces variable over time and takes into account the effects of inertia and damping. This analysis is based on a modeling of the system using the finite elements method (Darbre, 1993b).

These aspects are considered in more detail:

- The dam's dynamic response (in terms of displacement, velocity, acceleration, stress, and deformation)
- The actual ground-dam interaction
- The actual reservoir-dam interaction (depending on the dam's response, the reaction of the water may be greater than hydrodynamic pressure calculated according to Westergaard)

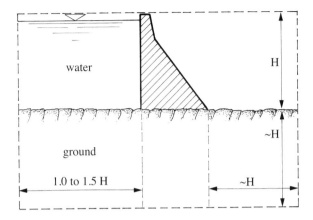

Figure 13.26 Recommended extension when establishing a model

To reach these objectives, the modeling system must include the dam, its foundation, and the reservoir. A restricted model is generally accepted, the extension to which is illustrated in Figure 13.26. Recent developments have enabled border finite elements to be integrated, which allows the behavior of the semi-infinite environment of the foundation to be simulated.

Eventually, the nonlinear and nonelastic behavior of the concrete in the dam will be taken into account by integrating the mechanisms of cracking and damage to the concrete into the model. Much research is still being devoted to this field of study.

Another major difficulty in dynamic modeling is calculating the reference loading to be considered. As a rule, there is never an accelerogram of an extreme earthquake that occurred at the site of the dam. If seismic activity is known to occur in the region, accelerograms of earthquakes with a lower magnitude recorded reasonably close to the site may be available. In Switzerland, many analyses have been carried out based on measurements taken during the Frioul earthquake that occurred in Italy in 1976.

Calculating design acceleration for the analysis of a dam at a given site is carried out by a statistical analysis of accelerograms of events, followed by the extrapolation of lower occurrence probabilities ($1/100$ to $1/1000$ years^{-1}). This approach is, of course, verified and completed by regional analysis and geological study.

The chosen reference accelerogram is deformed through homothety to achieve design acceleration.

When the dam's statics are spatial, as is the case for arch dams, the three dimensions of the accelerogram must be considered.

For gravity dams, a 2D analysis in the transversal vertical plane is generally sufficient. It is, however, important to analyze not only the highest section of the dam, but also the lateral sections (which sometimes lie on sloped foundations), whose various natural frequencies could be less favorable.

Today dynamic analysis poses no particular difficulties as long as linear structural behavior is accepted. The development of nonlinear models integrating the mechanics of the formation and propagation of cracks (failure mechanisms) have faced two obstacles, resulting in two major lines of research: improved knowledge of failure mechanisms in dam concrete and, from a computational perspective, the correction of grids of finite elements over time to take into account the position and development of cracks (Figure 13.27).

13.6.5 Brief description of methods suited to dynamic analysis

Finite element method (FEM)

The finite element method (FEM) enables both static and dynamic linear and elastic analyses to be carried out, as well as nonlinear and nonelastic analyses. The advantage is that complex geometrical structures can be modeled, as well as the foundation, by considering geomechanical particularities. The major parameters are the deformation moduli for rock and concrete. It is also possible to introduce different material characteristics.

For gravity dams, a two-dimensional analysis is an appropriate approach. Depending on the project specificities (joints with shear keys, major variations in the characteristics of the rock throughout the valley), a three-dimensional analysis may be required.

No matter how the stresses are distributed, verifying that the equilibrium of forces and moments are satisfied is recommended. This method provides the opportunity for redoing the calculations by eliminating the elements under tension.

Dynamic method

The solution to the problem consists in analyzing loads over a limited time while considering the structure's deformation mode and the characterization of the extent of damage caused by an earthquake. The functionality of the dam structure must then be ensured by considering the new situation and the change to loading that the earthquake may cause (uplift, sediment pressure). The most commonly used methods include the pseudo-dynamic method, developed by Professor A. Chopra, and the response spectrum (*modal dynamic response*).

Pseudo dynamic method

This is a simplified method for analyzing the response spectrum, in the most basic deformation mode due only to the horizontal component of ground movement. This method allows compressive stresses and tensile stresses to be calculated at the level of the foundation. Depending on the results, the level of damage can perhaps be estimated and introduced into a post-event stability analysis.

Modal dynamic method

This method is a modal analysis using finite element calculations. The principal deformation modes are calculated, and the structure's response is found by combining the response from several of these modes. There are two approaches: response spectrum analysis and time history analysis.

Response spectrum analysis

Response spectrum is the maximum dynamic response of an oscillating excited mass around its attachment point by an earthquake. The spectrum is represented by its natural frequency and the damping of the oscillating mass.

The dynamic response of a dam can be determined using the response spectrum method by overlaying several deformation modes. With this analysis, it is possible to evaluate the dam's maximum response on the basis of the earthquake response spectrum.

A modal analysis carried out using a model representing the studied dam enables the oscillation modes and corresponding natural frequencies to be calculated. The forces resulting from the earthquake are replaced by substitution loads that initially lead to identical deformation and the same loading. These substitution loads, which are calculated by combining the maximum response of several deformation modes, are subtracted from the amplification values corresponding to the natural frequency. Given the fact that the peaks of various modes will not occur simultaneously, the modal responses cannot be added algebraically.

The overlaying occurs according to the combination of the square root of the responses of individual modes (SRSS method).

The response spectrum method is applied to small and medium height dams.

Calculation method for transient conditions
This method takes different accelerograms as input data and calculates the evolution over time of the dam's response for the total duration of ground movement caused by an earthquake. This analysis provides the maximum stress values, among others. The response of each vibration mode caused by a specific acceleration is calculated at each point. All the modal responses are added algebraically for each time step over the course of the seismic event. The type of accelerogram influences the results, and this is why the use of several accelerograms is recommended.

Block rocking analysis
If the dynamic analysis calculations indicate that the concrete may crack, a block rocking analysis must be carried out. This analysis is useful for estimating the stability of the dam structure or a part of the structure, if it is believed that the crack is going to continue to grow and that a block may become free. Instead of trying to calculate the depth of the crack, it can be accepted that the base of the block is totally cracked, so much so that the block could pivot upstream or downstream. The dynamic behavior of the block in question can be determined by adding the moments relative to the pivot point situated at the upstream or downstream edge. Calculating the moment takes into account the resultant forces from horizontal and vertical acceleration, the self weight of the block, the effect of the reservoir, and the horizontal and vertical components of hydrodynamic pressure. To solve the problem, the sum of the moments is accepted as being equal to zero.

The aim of the RS DAM program (Renversement et glissement seismique des barrages en béton – Seismic Rocking and Sliding of Concrete Dams), established by the Ecole Polytechnique de Montréal, was to create a tool that could analyze the transient response for a totally cracked section of a concrete dam subjected to seismic loading. RS DAM is based on the rigid body dynamic equilibrium.

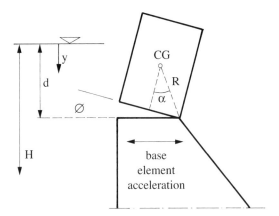

(Sources: Ghrib et al., 1997; FERC, 2002; SFOE (SFWG), 2003)

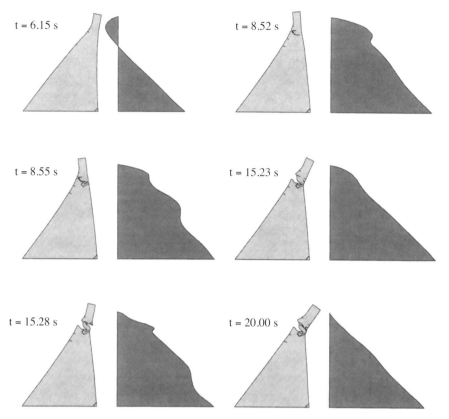

Figure 13.27 Dynamic nonlinear analysis of the Pine Flat gravity dam in California for the Taft earthquake with $a_{max} = 1.8$ m/s^2 with a full reservoir (Bachmann, 1994)

13.7 Earthquake Verification of Gravity and Masonry Dams according to current Swiss directives[1]

13.7.1 Methods of calculation

Depending on the dam category defined in Section 11.5.2, the Swiss guidelines relating to the safety of water-retaining facilities (SFOE (SFWG), 2002b) establish minimum requirements for the method of calculation for concrete dams:

- For category III dams, empirical models and approaches may be applied. The characteristic values of materials may be estimated by referring to similar dam structures and examples in the relevant literature.
- For category II dams, dynamic properties must be deduced from a specific modeling of the studied dam. The dam's dynamic forces can be calculated with the help of simplified methods. The dynamic

---

[1] The notes from the guidelines have been retained.

characteristic values of materials must be calculated using static tests specific to the studied dam and completed by knowledge acquired on similar dams and examples presented in the relevant literature.
- For category I dams, dynamic properties and dynamic forces must be assessed using a detailed model of the dam and its foundations. The dynamic characteristic values of materials must be calculated using static tests specific to the studied dam and completed by characteristics drawn from the relevant literature and similar dams.

Gravity dams and buttress dams can generally be modeled in two dimensions. The critical transversal cross-section must be defined and its choice justified. If the geometric and/or construction conditions of a gravity dam or buttress dam are such that its behavior during an earthquake is apparently three-dimensional, the corresponding model must be three-dimensional. Arch dams are generally calculated using a three-dimensional model.

13.7.2 Process for the verification of category II water-retaining facility

As an example, this section describes the various earthquake verification stages for concrete dams put forward in the directives and detailed in the main documentation for the earthquake verification of water-retaining facilities (SFOE (SFWG), 2003). The verification flow chart is presented in Figure 13.28.

With regard to Figure 13.28, the elements necessary for calculating seismic loads are outlined in Section 11.5.2. If the verification of safety is not satisfied at any point during the process, certain actions may need to be taken (for example, a more detailed calculation, construction measures, operational measures). By "Other verifications" we mean water release facilities necessary for safety, the abutments, and the foundations.

13.7.2.1 Calculating seismic loading
As mentioned in Section 13.6.4, a statistical analysis enables the design accelerogram to be determined. For dams in Switzerland, existing isoseismal maps established for various return periods may be used (see Figures 11.24 and 16.25) (Sägesser and Mayer-Rosa, 1978). The known determining MSK intensity is transformed into peak horizontal acceleration (see § 11.5.2).

13.7.2.2 Assessment of dam geometry and condition
Geometry
The effective geometry of the dam, including the exact line of the contact zone between the dam and the foundations must be surveyed and documented in a comprehensive manner. If there is any uncertainty regarding the line along the contact zone, in situ surveys must be carried out to remove these uncertainties.

Condition
While assessing the dam condition, annual safety reports and important information on the safety of the dam facility must also be assessed and any weak points concerning the earthquake safety verification must be considered.

Attention should be paid to any extraordinary events that occurred during construction as well as other aspects of the construction phase necessary for the verification of earthquake safety.

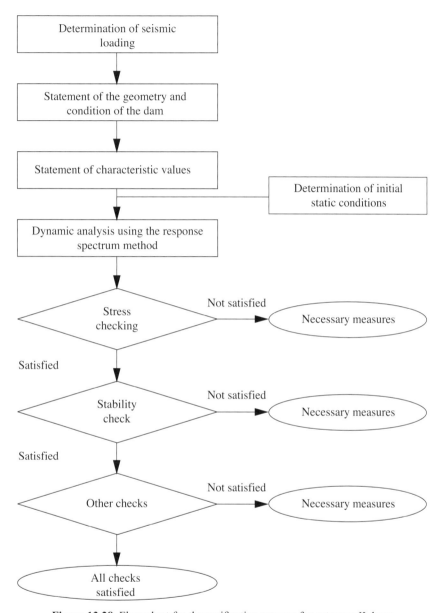

Figure 13.28 Flow chart for the verification process for category II dams

13.7.2.3 Assessment of characteristic values of materials
Characteristic deformation values and dam dimensions
In order to ensure the correct representation of dam behavior, the most probable characteristic values (median values) of the parameters are used.

For category II dams, assuming a linear-elastic isotrope material with viscous damping is sufficient. The following values are therefore required:

- Dynamic elastic modulus E_d
- Poisson's ratio ν
- Density ρ
- Damping of the material ζ

The dynamic characteristic values obtained through laboratory tests or vibration measures are introduced into the calculations. If the results of such tests are not available, the dynamic elastic modulus can be determined by increasing the static elastic modulus E_s in line with the following equation:

$$E_d = 1.25 \cdot E_s \text{ [MPa]}.$$

The characteristic value of the static elastic modulus E_s must be specific to the dam in question. The value may be drawn from the results of tests carried out during construction, if they are appropriate for the age of the dam. In the absence of test results, the static or dynamic elastic modulus must be determined through tests on the dam itself. Poisson's ratio values ν and density values ρ can be assessed through experience.

The accepted value for the critical damping of the material must not be greater than 5%. This damping parameter influences the dynamic amplification of the response spectrum.

Deformation properties of the foundation subsurface

In order to ensure the correct representation of dam behavior, the most probable characteristic values (median values) of the deformation properties for the foundation subsurface are used. For foundation subsurfaces in category II dams, assuming a linear-elastic isotrope material without mass and with viscous damping is sufficient. The following values are therefore required:

- Dynamic elastic modulus, resp. dynamic deformation modulus E_d [MPa]
- Poisson's ratio ν
- Damping of the material ζ

Dam concrete strength

The dynamic, uniaxial compressive strength f_{cd} and tensile strength f_{td} of concrete are applicable during analyses of stress conditions in category II dams, determined on the basis of a linear-elastic calculation with viscous damping.

In order to avoid underestimating the dam's concrete strength reserves, the strength values taken into account must be conservative. Using mean values is only allowable if a sufficiently large series of tests that has been statistically analyzed is available. If there are only a small number of test samples, a value lower than the mean must be applied.

The values employed must be compatible with the age of the dam (in line with the verification).

Dynamic concrete strength can be calculated empirically from static strength:

- Dynamic compressive strength f_{cd} as a function of static compressive strength f_{cs}:

$$f_{cd} = 1.5 \cdot f_{cs} \text{ [MPa]}$$

with respect to dynamic tensile strength f_{td} as a function of static tensile strength f_{ts}:

$$f_{td} = 1.5 \cdot f_{ts} \leq 4\text{MPa}.$$

- Dynamic tensile strength f_{td} as a function of dynamic compressive strength f_{cd}:

$$f_{td} = 0.1 \cdot f_{cd} \leq 4\text{MPa}.$$

Dynamic tensile strength calculated with an empirical formula must not be greater than 4 MPa.

Characteristic strength values must be calculated specifically for the dam. Tests carried out during the construction phase are accepted as base data for this calculation.

Surface strength of the dam-foundation subsurface interface

The dam-foundation subsurface interface can be accepted as plane. If the body of the dam is embedded into the foundations, the plausibility of this effect must be demonstrated. If this demonstration is missing or if the execution documents are incomplete, the effect of this embedding must be disregarded.

For foundations on rock, the following parameters (median values) of the interface between the dam and foundations can be estimated on the basis of the relevant literature:

- Angle of friction φ (angle of friction for concrete material/rock)
- Dilation angle (dilation angle of the mechanical imbrication of dam foundations/rock)
- Cohesion c (cohesion of concrete materials/rock due to micro-imbrication)

13.7.2.4 Initial static conditions

The verification earthquake should be considered as an extraordinary load. The corresponding loads must therefore be added to those due to usual operating static loads, which are:

- Self weight
- Water pressure
- Temperature (corresponding to the maximum reservoir level)
- Earth pressure due to downstream fill or sediment in the reservoir
- Ice load (where applicable) (corresponding to the maximum reservoir level)

The case with a full reservoir is sufficient for safety verification. The accepted reservoir level corresponds to the maximum operating level (reservoir water level).

Uplift acting on the interface between the dam and the foundation subsurface is only considered for stability verification. Uplift distribution can be based on effective uplift measurements with a full reservoir or assumed to be linear between upstream and downstream in the absence of measurements. Uplift acts perpendicularly to the dam abutment surface.

An estimation for the distribution of temperature in the body of the dam is sufficient. Temperature distribution can be assumed as linear along the transversal cross-section.

Assessed deformation and loading correspond to the initial static conditions for earthquake verification. Loading due to normal operating loads are added to those due to the earthquake.

13.7.2.5 Specific modeling for the response spectrum method

Process

The minimum requirement for verifying the earthquake safety of category II dams is a specific analysis of the dam in question using the response spectrum method. This analysis is carried out as follows:

- Geometric modeling
- Model calibration under normal static loads
- Calculation of natural frequency, modal damping, and oscillating masses
- Calculation of maximum modal deformation and corresponding loading

Geometric modeling of a gravity dam
Category II gravity dams can be analyzed using a two-dimensional calculation based on the transversal cross-section critical for earthquake safety. This critical transversal cross-section must be evaluated and its choice justified. As a minimum, the foundation subsurface must be modeled as a system with dampers and springs. The basis for calculating the corresponding parameters (spring rigidity and damping values) is the dynamic rigidity of the foundations (for example, that corresponds to the initial natural frequency of the dam).

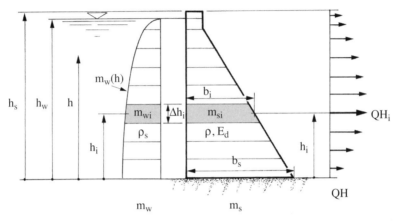

Figure 13.29 Two-dimensional model of a gravity dam; transversal cross-section and essential basic concepts. Key: m_s = dam mass; m_{si} = dam mass within band i; m_w = water mass; m_{wi} = water mass within band i; h_s = dam height; h_w = water height; h = height (variable); h_i = band height i; Δh_i = depth of band i; b_s = dam width; b_i = width of band i; ρ = dam density; ρ_w = water density; E_d = dynamic elasticity modulus; QH = horizontal substitution for seismic forces; QH_i = substituted seismic forces acting against band i.

For category III gravity dams, simple modeling is possible. This can involve a model with bars (cantilever with a variable cross-section). Figure 13.29 illustrates this transversal cross-section, as well as the most important basic concepts for the calculation.

Calculating the hydrodynamic effect of water
The oscillating water mass, with the dam representing the hydrodynamic effect of water on the dam with an approximately vertical upstream face, is calculated based on height h in accordance with Westergard's following equation:

$$m_w(h) = \frac{7}{8} \cdot \rho_w \cdot h_w \cdot \sqrt{1 - \frac{h}{h_w}} \quad [kg/m].$$

For practical reasons, the dam is divided into several horizontal bands. The choice of this division is made depending on the shape of the dam, the water level, and the level of precision desired for the calculation. The bands can be of varying thicknesses. Band i is described as follows:

$$m_{wi} = \frac{7}{8} \cdot \rho_w \cdot h_w \cdot \sqrt{1 - \frac{h_i}{h_w}} \cdot \Delta h_i \quad [kg/m].$$

If the study is being carried out for several transversal cross-sections of the dam, the corresponding height h_w must be used for each band.

Empirical calculation of the first natural frequency of a gravity dam
For gravity dams with a triangular transversal cross-section, the first natural frequency (base frequency) f_s can be initially calculated using the below formula. A dam with a slightly different transversal cross-section can be approximated by using a triangle of the same height and same surface as the effective transversal cross-section.

$$f_s = \alpha \cdot \frac{b_s}{h_s^2} \sqrt{\frac{E_d}{\rho}}, \text{ but to a maximum of 10 Hz.}$$

The limit of 10 Hz is based on observations during which the flexibility of the foundations becomes critical.

The fundamental period T_s in seconds is expressed as:

$$T_s = \frac{1}{f_s} \quad [s],$$

where α is a form factor that depends on the shape of the dam. It is represented in Table 13.30 as a function of ratio b_s/h_s and for cases of both a full and empty reservoir. For intermediary cases, the coefficient can be interpolated.

Table 13.30 Form factor for calculating the first natural frequency (base frequency)

b_s/h_s	α (empty reservoir)	α (full reservoir)
0.6	0.19	0.13
0.8	0.17	0.12
1.0	0.15	0.11

Spectral acceleration
The determining spectral acceleration can be set by using a response spectrum from the first natural period (fundamental period, resonance period) (Figure 13.31).

If it is not possible to empirically calculate the fundamental period, it must be assumed that it is located in the maximum amplification domain of the response spectrum.

A substantial amount of seismic loading is taken into account by the calculation of the first natural frequency and corresponding spectral acceleration. A correction factor is taken into account for higher modes. Figure 13.31 shows that the highest natural frequencies are not amplified if the period is smaller than a

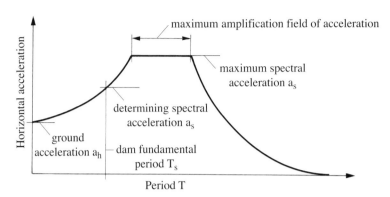

Figure 13.31 Calculating horizontal acceleration using the response spectrum

limit value. Spectral acceleration is then equal to ground acceleration. As a result, the influence of the first natural frequency increases with the difference between spectral acceleration and ground acceleration, that is, with the relation between a_s and a_h. If the ratio between a_s and a_h decreases, higher natural frequencies gain in importance. The corresponding correction factor is illustrated in Figure 13.32.

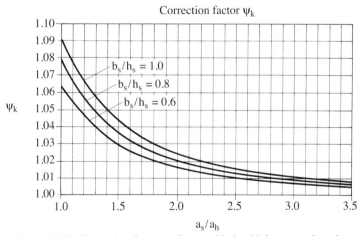

Figure 13.32 Correction factor ψ_k for considering higher natural modes

Empirical calculation for the first deformation mode for gravity dams
The first deformation mode for a gravity dam with typical geometry is the shaking of the dam, with the maximum amplitude occurring at the level of the crest. The form of the deformation is similar for all ratios b_s/h_s and may be described by the following formula:

$$\psi_i = 0.69 \cdot \left(\frac{h_i}{h_s}\right)^3 + 0.14 \cdot \left(\frac{h_i}{h_s}\right)^2 + 0.17 \cdot \left(\frac{h_i}{h_s}\right)$$

The resulting calculation of the shape factor ψ_i represents deformation at height h_i relative to the maximum deformation at the level of the crest (height h_s). It is used for the distribution of seismic loading across the height of the dam and is represented graphically in Figure 13.33.

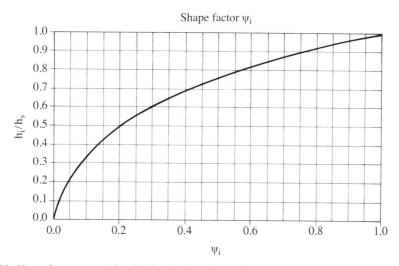

Figure 13.33 Shape factor ψ_i used for the distribution of substituted seismic load over the height of the dam

Calculating seismic load with the simplified response spectrum method (one mode)
In a two-dimensional model, attention should be paid to the horizontal and vertical component of the earthquake. The calculation of substituted horizontal seismic load is described below and refers to Figure 13.29. For practical reasons, the dam is divided into several horizontal bands. As already mentioned, the choice of this division is made depending on the shape of the dam, the water level, and the level of precision desired for the calculation. The bands can be of varying thicknesses.

The dam mass in band i can be calculated as follows:

$$m_{si} = \rho_s \cdot b_i \cdot \Delta h_i \ [kg/m].$$

With the water mass oscillating with the dam m_{wi}, we obtain the total mass of band i by:

$$m_i = m_{wi} + m_{si} \ [kg/m].$$

This mass is reduced by a mass coefficient indicating the proportion of the total mass oscillating with the first natural frequency. The mass coefficient ψ_m is outlined in Table 13.34.

Table 13.34 Mass coefficient for the first natural frequency (base frequency)

b_s/h_s	ψ_m (empty reservoir)	ψ_m (full reservoir)
0.6	0.39	0.41
0.8	0.39	0.43
1.0	0.40	0.44

The substituted total seismic load acting on the dam is calculated from the spectral acceleration, the mass coefficient, the correction factor, and the total mass of the dam using the following equation:

$$QH_{tot} = a_s \cdot \psi_k \cdot \Psi_m \cdot \sum m_i \quad [kN/m].$$

This load is distributed over the height of the dam structure by using the shape factor of the deformation:

$$QH_i = QH_{tot} \cdot \frac{m_i \cdot \psi_i}{\sum m_i \cdot \psi_i} \quad [kN/m].$$

The representative single force on each band calculated using this method is to be introduced as a static load in the bar model, with respect to the finite elements model. In this way, the dam loading due to the horizontal component of the earthquake can be calculated.

The substituted vertical seismic load is calculated by taking into account the dam mass only. Water does not affect dam oscillations in the vertical direction. Frequencies for oscillations in the vertical direction are generally so high that the entire dam is loaded by the vertical ground acceleration. There is no amplification. Vertical acceleration a_v is calculated according to Section 11.5.2.

The substituted vertical seismic load can be calculated as follows from the oscillating dam mass:

$$QV_{tot} = a_v \cdot \sum m_{si} \quad [kN/m].$$

The value for a single band is:

$$QV_i = a_v \cdot m_{si} \quad [kN/m].$$

The substituted vertical seismic load can result – depending on its direction – in a decrease or an increase in dam self weight. The eccentricity of the resultant force of vertical load on each band relative to the transversal cross-section of the calculation must be taken into account, as well as the two active directions (upward and downward).

Calculating seismic load using the pseudo-static method with a uniform deformation mode
If it is not possible to calculate base frequency and the first deformation mode using an empirical or more detailed method (for example, using the formula given above in "Empirical calculation of the first deformation mode for gravity dams") or if the geometry of the dam cannot be approximated by a triangle, the load due to an earthquake can be calculated using the pseudo-static method with a uniform deformation mode. Attention must be paid to both the horizontal and vertical components of the earthquake. The calculation of substituted horizontal seismic load is described below and refers to Figure 13.29.

The dam mass in band i can be calculated as follows:

$$m_{si} = \rho_s \cdot b_i \cdot \Delta h_i \quad [kg/m].$$

With the water mass oscillating with the dam m_{wi}, we obtain the total mass of band i by:

$$m_i = m_{wi} + m_{si} \quad [kg/m].$$

Lastly, the substituted total horizontal load can be calculated from the total oscillating mass and spectral acceleration a_s using the following equation:

$$QH_{tot} = a_s \cdot \sum m_i \ [kN/m].$$

The value for a single band is:

$$QH_i = a_s \cdot m_{si} \ [kN/m].$$

The substituted vertical seismic load is calculated in the same way as that described above in "Calculating seismic load with the simplified response spectrum method (one mode)."

13.7.2.6 Stress verification

Stress verification (strength verification) involves demonstrating that the maximum load due to a combination of initial static loads (self weight, hydrostatic pressure, uplift) and loads due to an earthquake are not greater than the dynamic strength of the materials and joints (both tensile and compressive strength). The principal stresses are calculated for each case from stress components. The minimum and maximum principal stress values obtained from the combination of seismic loads according to Table 13.35 (that is, cases 1 to 4 for the two-dimensional models and cases 1 to 8 for the three-dimensional models) are compared to the dynamic strength of the material. It should be noted that, in principle, the dam is loaded by the earthquake along the two horizontal orthogonal axes and the vertical axis. Only horizontal loading in the direction of the watercourse is to be considered for category III dams.

If this verification is not satisfied, it must be additionally demonstrated that:
- A redistribution of stress in the neighboring zones is possible, and
- Damage to the dam (cracks) does not lead to an uncontrolled release of a large volume of water

If the required verification cannot be satisfied, appropriate corrective construction or operational measures must be taken.

Table 13.35 Directions of seismic loading

Number of combinations of seismic loads	Two-dimensional model		Three-dimensional model		
	horizontal	vertical	upstream-downstream	left-right	vertical
1	+	+	+	+	+
2	+	−	+	+	−
3	−	+	+	−	+
4	−	−	+	−	−
5			−	+	+
6			−	+	−
7			−	−	+
8			−	−	−

13.7.2.7 Stability verification
Verification of stability involves ensuring that no sliding or overturning of the dam or part of the dam will occur during an earthquake. Both verifications are made using a rigid body model with the effective geometry of the dam and foundations.

Sliding
To verify sliding safety, the maximum strength between the dam and the foundations must be greater than the total shear, according to the following equation:

$$c + \sigma_m \cdot \tan(\varphi + i) \geq \tau_m,$$

where
- φ = concrete/rock angle of friction along the slip surface
- i = dam/rock dilation angle along the slip surface
- c = concrete/rock cohesion along the contact interface and possibly the foundation excavations
- σ_m = normal mean effective stress at the contact interface
- τ_m = mean shear stress at the contact interface

The embedding of the dam into the foundation subsurface (foundation excavations) must only be taken into account for resistance through the intermediary of a cohesion, if a composite action (imbrication) between the dam and rock can be demonstrated. If this effect cannot be demonstrated, only the strength on the horizontal contact interface (abutment surface against the dam) must be considered.

The factor of sliding safety can also be assessed using the equation:

$$S_G = [(\Sigma V + QV_{tot}) \cdot \tan(\varphi + i) + c \cdot b]/(\Sigma H + QH_{tot}),$$

where
- ΣV = sum of vertical static forces (self weight, uplift)
- QV_{tot} = substituted vertical seismic load
- ΣH = sum of horizontal static forces
- QH_{tot} = substituted horizontal seismic load (dam mass and water mass oscillating against the dam)
- b = surface of the base

The factor of safety S_G must be equal to or greater than 1.1 (Table 13.11, extreme case), or even up to 1.3 for preliminary design (see § 13.6.3).

Overturning
Verification must indicate that by combining the initial static loads (self weight, hydrostatic pressure, uplift) and seismic loading, the stresses at the interface between the dam and the foundations are smaller than the extreme concrete strength values. The values to take into account are outlined in Section 13.7.2.3 "Dam concrete strength." Dynamic tensile strength calculated using an empirical formula must not be greater than 2 MPa (4 MPa is the upper limit). Should the level of acceptable dynamic tensile stresses at the upstream heel be exceeded, the opening of the upstream foundation joint should not lead to compressive stresses greater than the compressive strength at the downstream toe.

Calculating the position of the resultant (determined by the relation between the sum of the moments from the center of gravity of the section on the sum of vertical forces, including uplift) is an efficient means of rapidly assessing conditions of stability with regard to overturning. If this force is located in the middle third, there are no tensile stresses in the section.

Another indicator of stability against overturning is given by the ratio between the sum of the stabilizing moments compared with downstream (ΣM_{stab}) and the sum of the overturning moments (ΣM_{mobil}) (see Section 13.3). This ratio must be equal to or greater than 1.1.

$$S_R = \Sigma M_{stab} / \Sigma M_{mobil}.$$

Foundation stability
Verification of foundation stability involves ensuring that during an earthquake no local failure of the foundation occurs that might endanger dam stability.

Verification deemed to be provided
If stability is not satisfied, the following checks must be made:

- Ensure that the overall stability of the dam is not threatened (taking into account partial instabilities)
- Ensure that damage to the dam (cracks, instability of individual blocks, etc.) does not lead to an uncontrolled release of a large volume of water

If the required verifications are not met, appropriate construction or operating measures must be taken.

13.7.2.8 Other verifications
Verification of bank stability
If there may be potentially unstable slopes or other similar zones along the reservoir banks, verification must ensure that no landslides or rockslides will lead to unacceptable overtopping of the dam crest.

Verification of the correct functioning of appurtenant structures vital for safety
After an earthquake, all appurtenant structures that are important for dam safety must be checked, particularly water release structures. They must continue to operate or must be immediately reinstated.

13.7.2.9 Examples of computer-aided calculations and analyses
In the SFOE (SFWG) publication entitled *Guideline for Earthquake Verification: Examples for Small Water Retaining Facilities (March 2003)*, there is the complete seismic verification of a category II gravity dam, based on the same model described in this chapter.

The Ecole Polytechnique de Montréal developed the CADAM and RS DAM programs on the basis of seismic loading for calculations for gravity dams (Leclerc et al., 2003). The aim of CADAM is to provide technical support to the study of the principles of structural stability analysis for gravity dams. CADAM is based on rigid body equilibrium and Euler-Bernoulli beam theory. It enables stability to be calculated under static and dynamic loads. It contains various options concerning geometry, uplift, drainage, and the development of fissuring. The aim of the RS DAM program (seismic rocking and sliding of concrete dams) is to provide a tool that analyzes the transient response for a totally cracked section of a concrete dam subject to seismic loading. RS DAM is based on rigid body dynamic equilibrium.

13.8 The Effect of Temperature

13.8.1 Solar radiation

Solar radiation onto a concrete surface produces considerable heat. Depending on its orientation and climatic conditions, the downstream face of a gravity dam can be subject to substantial solar radiation. However, when the reservoir is full, the upstream face is in contact with noticeably colder water. The downstream face experiences thermal expansion while the upstream face contracts. This results in deformation of the dam and an upstream displacement of the crest (Figure 13.36).

The thermal inertia of the mass of the gravity dam is such that the temperature inside the mass starts to build up after construction to reach a mean value that remains practically constant throughout the whole year.

The faces, however, undergo a cyclical variation that mainly depends on the seasons and the operating cycle of the reservoir. This means that there is a considerable state of thermal stress in the dam, particularly around the faces, which cannot be neglected and to which is added the stresses previously calculated for other loads.

Longitudinal deformation at the crest due to the effects of temperature can be significant. The monitoring of dams during operation is based to a significant extent on the measurement of deformation. These measurements are compared to calculated deformations and in order to check that dam behavior is appropriate. It is therefore important to assess with a sufficient degree of certainty the effect of these temperature variations. This aspect is especially important for dam structures prone to alkali-aggregate reactions, as the effects of temperature are superimposed upon those of chemical swelling.

The finite elements method enables the range of temperature in the dam to be estimated and therefore also the range of thermal expansion.

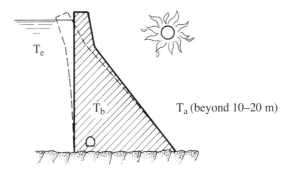

Figure 13.36 Considered temperatures. Key: T_e = water temperature (we can assume T_e = 4°C from a certain depth, which varies depending on local conditions); T_b = concrete temperature in the dam body; T_a = air temperature near the downstream face, may reach 60 to 70°C.

Simplified calculation of crest displacement due to water pressure and temperature variation

In isostatic gravity dams, load inside the dam structure is not influenced by deformation. However, it can be useful to know these values for the monitoring of dam behavior.

Effect of water pressure

H. Juillard (1961) devised theoretical formulae giving the horizontal and vertical displacements of a point anywhere in a triangular gravity dam. By choosing as coordinates crest S with horizontal axes x (positive in a downstream direction) and vertical z (positive downward), we obtain:

$$\delta_h = \frac{\gamma z^2}{E_b}\left(\frac{1}{m^3} - \frac{x}{z}\cdot\left(1+\frac{v}{m^2}\right)+\left(\frac{x}{z}\right)^2\cdot\frac{v}{m^3}\right)$$

$$\delta_v = \frac{\gamma z^2}{E_b}\left(\frac{v}{2}+\frac{1}{2m^2} - \frac{x}{z}\cdot\frac{2}{m^3}+\left(\frac{x}{z}\right)^2\cdot\left(\frac{1}{2}-\frac{1}{m^2}-\frac{v}{2m^2}\right)\right)$$

where
E_b = concrete elasticity modulus
v = Poisson's ratio
m = tg ϕ.

Horizontal displacement Δx at the crest: Let's assume a rectangular dam profile embedded into the rock and that only the lengths h and s of the faces vary by Δh and Δs. In addition, we can assume that the faces remain rectilinear under the effect of concrete deformation, and we can disregard variation in the base length.

From the formulae put forward by H. Juillard, we can deduce (with all simplifications made) deformation of the upstream face (for x = 0 and z = h):

$$\Delta h = \delta_v = +\frac{\gamma h^2}{2E}\left(\frac{1}{m^2}+v\right)$$

and deformation of the downstream face (for x = b and z = h):

$$\Delta s = \delta_v \cdot \cos\varphi + \delta_h \sin\varphi = -\frac{\gamma h^2}{2E}\cdot\frac{1}{\sin^2\gamma\cos\gamma}.$$

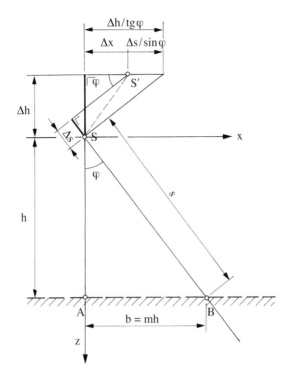

We can therefore estimate horizontal displacement at the crest, which is given with the equation:

$$\Delta x = \Delta h / \mathrm{tg}\,\phi + \Delta s / \sin\phi.$$

By replacing Δh and Δs, we end up with

$$\Delta x = \frac{\gamma h^2}{2E} \cdot \left(\frac{1 + (1+m^2)^2}{m^3} + \frac{v}{m} \right).$$

Rock deformation: Rock deformation due to the effects of water pressure can be estimated from Vogt's formulae. The rotation of the base to calculate horizontal displacement of crest S is given by

$$\delta w = \frac{1}{E_r} \left(k_\tau \cdot \frac{T}{b} + k_\mu \cdot \frac{M}{b^2} \right),$$

where

- E_r = deformation modulus of the rock
- T = unit shear force due to water pressure
- M = unit bending moment due to water pressure
- k_τ = coefficient characterizing shear force
- k_μ = coefficient characterizing deformation due to bending moment

given that

$$T = \gamma h^2/2; \quad M = \gamma h^3/6; \quad b/h = \operatorname{tg} \phi = m$$

Displacement Δx_r equals:

$$\Delta x_r = \frac{\gamma h^2}{2 E_r} \left(\frac{k_\tau}{m} + \frac{k_\mu}{3 m^2} \right)$$

(as a first approximation: $k\tau = 0.57$; $k_\mu = 5.18$).

Effect of a variation in temperature

By postulating that most dam displacement due to a change in temperature is determined by the thermal condition of zones located near to the faces, we can accept that the dam reacts to variations in temperature broadly as an assemblage of two rigid bars free to pivot around their base.

The deformations of the faces under the effect of temperature variation are:

- downstream: $\Delta s = s \cdot \Delta T_s \cdot \beta = (h/\cos \phi) \cdot \Delta T_s \cdot \beta$
- upstream: $\Delta h = h \cdot \Delta T_h \cdot \beta$

where

- ΔT_s = temperature variation of the upstream face
- ΔT_h = temperature variation of the downstream face
- β = thermal dilation factor for concrete

By combining the deformation of the faces, horizontal displacement equals:

$$\Delta x_\Theta = h \cdot \beta \left[\frac{\Delta T_s - \Delta T_h}{\cos \varphi \cdot \sin \varphi} + \Delta T_h \cdot \operatorname{tg} \varphi \right].$$

References

JUILLARD, H., 1961. Le développement de la construction des barrages-poids en Suisse. *Cours d'eau et énergie*. Special issue 6/7.

JOOS, B. and KOLLY, J.-C., 1995. D'un simple modèle de détermination des déformations d'un barrage-poids sous l'influence de la température. "Research and Development in the Field of Dams" Conference, Crans-Montana.

13.8.2 Concrete heating during hardening

13.8.2.1 Introduction

During the construction of a structure as massive as a gravity dam, concrete heating through cement hydration can be considerable and cause major cracking if specific precautions are not taken.

Estimating temperature rise

It can be assumed that a volume of concrete located in the middle of a gravity dam is surrounded by concrete in a thermal state that is practically identical and that, as a result, heat exchange is very slow. The rise in temperature over the few days required for the cement to harden may be considered as adiabatic and be estimated using the following equation:

$$\Delta T = \frac{W \cdot D}{\rho_B \cdot c_B},$$

where
- W = heat from cement hydration [kJ/m³]
- D = cement content [kg/m³]
- ρ_B = concrete density [kg/m³]
- c_B = heat specific to the concrete [kJ/°C·kg]

Assuming the use of standard Swiss Portland cement and a cement content of 250 kg/m³, we have
- W = 335 [kJ/m³],
- D = 250 [kg/m³],
- ρ_B = 2450 [kg/m³],
- c_B = 0.84 [kJ/°C·kg],

which leads to a rise in temperature of

$$\Delta T = 40.7°C.$$

This rise in temperature is reached after 5 to 7 days and is added to the initial temperature of the fresh concrete. Temperature in the body of a block in the dam becomes

$$T_{max} = T_o + \Delta T,$$

where T_o is the temperature of fresh concrete and is calculated using the equation:

$$T_o = \frac{A\,T_A c_A + C\,T_C c_C + E\,T_E c_E}{A\,c_A + C\,c_C + E\,c_E},$$

where
- A, C, E = respective volume of aggregate, cement, and water, in [kg]
- T_A, T_C, T_E = respective temperature of aggregate, cement, and water, in [°C]
- c_A ≅ c_C = 0.2 [kJ/°Ckg]
- c_E = 1.0 [kJ/°Ckg]: heat specific to aggregate, cement, and water

In an alpine environment, the result is fresh concrete temperature varying between 8 and 15°C, and therefore a maximum temperature in the body T_{max} in the order of 50 to 65°C.

In arid countries, the temperature of aggregates can exceed 40°C. If no measures were taken, the temperature of fresh concrete would reach 80°C!

Consequences of concrete cooling

Hardened concrete whose temperature ranges from 50 to 65°C will gradually cool until reaching its equilibrium temperature, generally between 5 and 10°C in a temperate climate.

Due to this cooling of more than 40°C, the dam contracts. As concrete is not highly deformable, and because this cooling occurs more quickly near the faces than in the center of the mass, concrete cracking is practically inevitable.

This is why the dam is divided into blocks separated by vertical transversal joints. The opening of these vertical joints between two blocks can be estimated:

$$\delta = \beta \cdot \Delta T \cdot l \cong 10^{-5} \, [°C^{-1}] \cdot 40 \, [°C] \cdot 16 \, [m] = 6.5 \, [mm] \, .$$

Measures to accelerate concrete cooling

To limit cracking of the concrete as it cools, two measures can be taken:

- Lowering the maximum temperature reached after hardening
- Accelerating cooling so that the contraction occurs while the concrete is still young and deformable

Figure 13.27 illustrates the difference in temperature evolution between concrete that cools naturally and concrete artificially refrigerated.

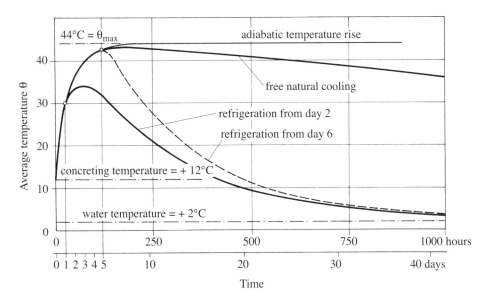

Figure 13.37 Evolution in concrete temperature

13.8.2.2 Measures for the cooling of concrete

Natural cooling

When the dam is slender, that is, up to 10 or 12 meters thick, natural cooling through the two faces is sufficiently quick – providing that climatic conditions are favorable. However, apart from the crest zone, gravity dams are substantially thicker. In this case, artificial cooling is recommended.

Artificial cooling during hardening

An appropriate solution for evacuating the heat from concrete hydration involves installing a cold-water circulation system through a network of pipes laid into the fresh concrete as the works progress (Figure 13.38). This artificial cooling measure is implemented during the construction of nearly every massive dam (except for RCC dams), in conjunction with other measures.

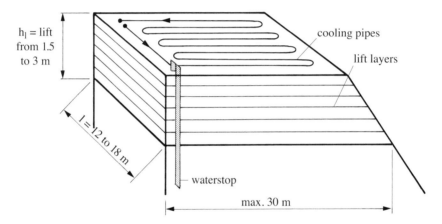

Figure 13.38 Layout of cooling pipes

On high-altitude dam constructions sites, where the average air temperature is between 4 and 20°C, natural water with a summer temperature range between 5 and 6°C and winter range of 1 to 2°C can be used. Refrigeration must occur, if not continuously, at least near-continuously. Summer refrigeration means that thermal peaks can be controlled and an initial lowering of the temperature can be achieved. Further cooling then takes place in winter.

The circulation of cold water takes place in metal pipes laid out on the surface of each lift. The choice of tube diameter generally takes into account thermal aspects and loss of load. Tubes with a diameter of between ϕ ¾" (or 19/22 mm) and ϕ 1" (or 25/29 mm) are frequently used. With regard to supply, the diameter of the main tube in a horizontal duct is in the order of 4" to 6", while the distribution pipes are between 2" and 3".

Ideally, the cooling pipes should be spread out at equal heights and widths to provide a uniform arrangement of cooling sources. However, in practice, the cooling pipes must be laid on a hard surface and the concrete lifts are thus used, which determines their spacing. The cooling pipes are laid on the surface of the concrete lift after it has been stripped and cleaned. The spacing of cooling pipes depends on when the concreting was carried out. The cooling pipes will be closer together for concrete placed near the end of construction so that the lifts reach a similar temperature at the same time.

The cold water first circulates in the middle and then, having warmed slightly, the edges. Refrigeration is completed by the contact the faces have with the air. As the water temperature is higher at the end of the cooling pipes than at the beginning, the direction of water circulation must be reversed every day in order to equalize the temperatures. In cold regions with a risk of frost, the outer coils should not be too close to the faces.

The water supply for the cooling pipes comes through a horizontal gallery and feeds the cooling pipes on all the lifts above it, until the next gallery. The cooling pipes on a concrete lift are fed as a group, but the exit point must be individual and visible. This enables visual checks to ensure that water is circulating in the cooling pipe layers. Using galleries within the dam for the conveyance and evacuation of water is the most convenient solution.

Monitoring concrete temperature is an important operation. Water circulation is interrupted for 24 hours while the water in the cooling pipes takes on the temperature of the concrete. The water is then collected and its temperature measured.

The initial cooling of fresh concrete

In hot countries, in addition to artificial cooling during hardening, the concrete is also cooled prior to its placement.

The various components of the concrete are treated prior to mixing:

- the aggregates – are protected from solar radiation
  – are cooled with cold water or with a jet of compressed and cooled air
- water – is refrigerated
  – is replaced by ice flakes
- the cement – refrigerated air is blown into storage silos

Furthermore, fresh concrete is transported in refrigerated containers. The travel time between the concrete silo and the dam should be as short as possible. This constraint can be very important when assessing options for organizing work-site installations.

These preliminary cooling measures will allow the initial temperature of the fresh concrete to be lowered by ten or so degrees in the best-case scenario.

Given the specific heat of each component, refrigerating the mixing water is the most effective solution.

Use of low heat cement

To limit heating during concrete hardening, low heat cement such as pozzolanic cement or blast furnace slag is often used. In some cases, a significant proportion of fly ash is also added.
One of the characteristics of these cements is that hardening occurs more slowly, thus delaying formwork removal. In addition, concrete compressive strength at 90 days is noticeably lower. However, hardening of the concrete continues to develop and final strength requirements will be met.

Slower heating makes artificial cooling measures more effective, and maximum observed temperatures are lower than with standard cement.

Another advantage is that concrete hardening takes place more slowly. Shrinkage occurs on more deformable concrete and there is less cracking.

In order to establish the appropriate mix or mixes for a specific dam, construction always begins with a major round of preliminary tests carried out on possible aggregates and cements selected for the dam.

13.9 3D Stability Analyses

Where possible, gravity dams are erected in U-shaped or open valleys. In these situations, the height of the blocks are roughly equal along most of the valley. A traditional 2D stability calculation of the various independent elements of uniform width located across several points is generally carried out. For a narrow valley with relatively steep slopes or with varied geological conditions, it can be judicious to use 3D calculations (Lombardi, 2007).

Stability along the longitudinal direction (bank-center) of the blocks on the banks must be verified, while taking into account the slope of the foundation (Figure 13.39). Increased uplift force and reduced self weight can be noted along the surface of the foundation. It is possible that the factor of safety may be lower than that obtained when calculating along the transversal direction. If the blocks are not freestanding, the neighboring block must provide lateral support. To increase the factor of safety by enabling the transfer of force from the abutments at the valley flanks toward the center, continuity should be ensured between the blocks by grouting the joints and /or installing shear keys (Figure 13.44). Block stability can be further increased by including, for example, steps or a by giving a specific shape to the foundation surface that increases shear strength. These conditions may be modified following an earthquake.

Analyzing the transfer of forces can be a complex process. The three components X, Y, and Z must be taken into account at each stage of the calculation (Figure 13.40). A complete 3D analysis using finite elements for the entire structure may be carried out.

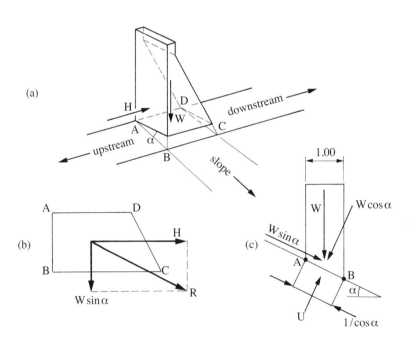

Figure 13.39 Equilibrium of an independent block: (a) block on an inclined foundation; (b) surface at the foundation level; (c) cross-section with inclined base and principal forces (from Lombardi, 2007).
Key: α = slope; W = self weight; U = uplift; β = uplift coefficient (equal to 1 for a triangular distribution without drains)

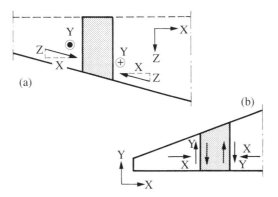

Figure 13.40 Forces that are usually transferred through the joints from one block to another: (a) longitudinal cross-section; transferal of compressive forces from one block to another; (b) plan view with forces transferred from one block to another through the joints (from Lombardi, 2007)

Lombardi (2007) proposed a simplified calculation for the factor of safety. Firstly, based on the load factor method, actions due to hydrostatic pressure, uplift, sediment pressure, and earthquakes (for example, in a pseudo-static form) are multiplied by the required factor of safety n, while the effective values of resistive forces (self weight, friction, and cohesion at the foundation level) are taken into account. Another approach involves taking the effective forces into account and dividing the geotechnical characteristics by the required factor of safety. The partial factor of safety method can also be considered, which combines the two previous methods.

The value of the forces of shear strength can easily be calculated on the basis of the angle of internal friction ϕ and cohesion on the surface; only its orientation is not known in advance. Two extreme directions and intermediary directions can be considered, as illustrated in Figure 13.41.

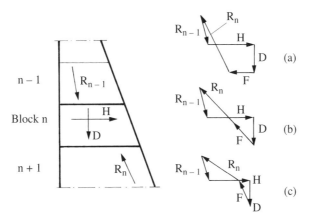

Figure 13.41 Three typical cases for the equilibrium of a block (from Lombardi, 2007).
Key: R_{n-1} = action of the lower block; H = horizontal pressure; D = inclined component of self weight; F = resistance through friction and foundation cohesion; R_n = reaction of the upper block.

1st case

It is accepted that for each block the inclined component is directly carried over to the next block through compressive force. Foundation shear strength must take up horizontal pressure H. If this strength is insufficient, the difference is assumed to be transferred to the next block through the shear force acting in the joint.

2nd case

It is accepted that for any block the foundation shear strength is oriented in the opposite direction to the resultant forces acting directly on the block in question. In this case the equilibrium must also be reestablished by transferring the missing forces to the following block through the joint.

Other hypotheses

It may be assumed, depending on the nature and configuration of joints, that foundation shear strength will be the first to take the inclined component of the forces. If necessary, shear strength and the joint's compressive strength must intervene to ensure the downstream stability of the block with the required factor of safety.

Naturally, the calculation is conducted from the upper block toward the valley bottom. Each block is analyzed by including the inclined forces that come from the upper part. Additional virtual joints can be introduced to take into account the modification of geotechnical conditions or the slope of the foundation.

13.10 Specific Construction Aspects

13.10.1 Shape and dimension of blocks

As previously mentioned, a dam is made up of successive blocks, whose dimensions are restricted by the concrete installation at the construction site and the placement of the concrete. The standard main dimensions are given in Figure 13.42.

If the dam is wider than 30 m, the block should be split into an upstream and downstream block in order to limit the volume for the concreting stages.

Sometimes lift heights are set at 1.5 m to benefit from natural cooling of the horizontal surfaces. This height is also particularly favorable for use near to the concrete-rock interface, as the rock restricts the thermal deformation of the concrete. However, this approach affects the works program, as a delay of 4 to 5 days

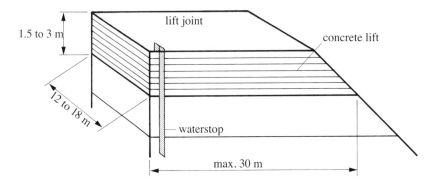

Figure 13.42 Standard dimensions for concrete lifts

is required before the most recent lift can be covered. The advantage of a 3-meter lift is that the horizontal galleries can be covered in one go.

13.10.2 Precautions for major earthquakes

When a gravity dam is constructed in a high-risk seismic area, the design based on a dynamic analysis must be accompanied by certain precautionary construction details.

Shape of the crest
To limit the concentration of stress and the opening of cracks, the transition area from the crest to the downstream face should be rounded (Figure 13.43).

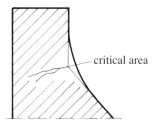

Figure 13.43 Transition zone of downstream face near the crest

Joints between the blocks
To improve the stability of blocks under a transversal dynamic load, the joints are filled with cement grout. Transversal forces can thus be transferred from one block to the neighboring block and absorb lateral acceleration. The drawback to this filling is that annual variations in temperature can create a state of additional transversal stress.

Block stability can be improved by placing shear keys in the joints between the blocks, as illustrated in Figure 13.44. The joints are then filled with grout to improve load transfer. The stability of the overall dam is not affected, as each block has its own geometry (the height changes) and, as a result, has different natural frequencies.

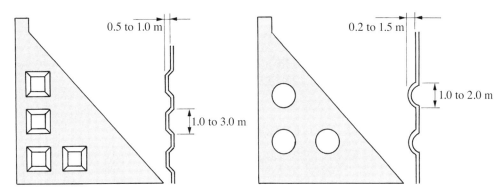

Figure 13.44 Shear keys in the joints

13.10.3 Concrete quality, cement content

Stresses within a gravity dam are low. Depending on the loads considered, only the faces bear higher levels of load (upstream for an empty reservoir, downstream for a full reservoir). Thermal stresses and the risk of cracking are also limited in zones near the faces.

The only criterion limiting cement content is the workability of fresh concrete during its placement. Cement contents from 130 to 180 kg/m^3 are frequently observed.

On the faces, however, the cement content is higher, often between 250 and 280 kg/m^3 for a thickness varying from 1.5 to 3.0 m. In addition to static considerations, this higher quality concrete is necessary to ensure the watertightness of the upstream face and to protect against frost and weathering.

The low compressive strength necessary at the center of the dam and the thermal stability of the concrete mass means that a lower minimum cement content can be chosen for the dam core. This low cement content favors the dissipation of hydration heat and saves costs (Figure 13.45).

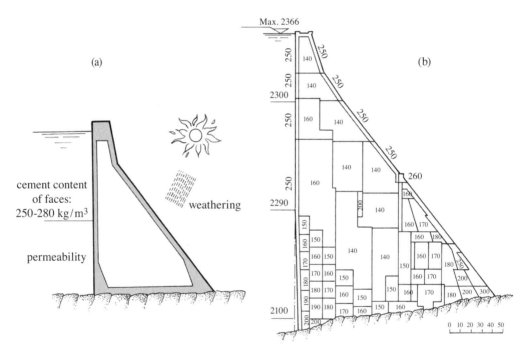

Figure 13.45 Cement content: (a) principles for cement content distribution; (b) distribution of cement content in the Grande Dixence dam

13.10.4 Construction joints and waterproofing system

Generally speaking, the surfaces of transversal joints are plane, as the blocks should be individually freestanding (Figure 13.46 (a)). However, as mentioned in Section 13.9.2, shear keys can be employed to improve stability in case of an earthquake (Figure 13.44). The transversal joints in the Grande Dixence dam, for which the surfaces were not plane but indented, had an injection system. After several

 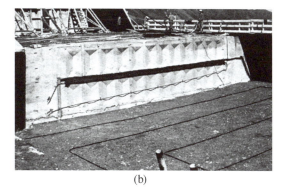

Figure 13.46 Example of transversal joints: (a): plane surface of the Panix dam; (b): indented surface of the Grande Dixence dam

successive grout injections a monolithic block was created (Figure 13.46 (b)), whose greater stability and improved strength was nevertheless not considered in the calculations.

In addition, a specific sealing system must be implemented so that the vertical transversal joint separating two blocks does not create a preferential path for seepage. This joint, located near the upstream face and installed from the foundation right up to the crest, must take up the entire load from the reservoir's hydrostatic pressure. Given the importance of this element, and the impossibility of repairing it without drawing down the reservoir to the level of the defect, the joint is made up of two waterstop strips (in synthetic rubber or copper) between which a draining tube is placed. This tube collects water from leaks that have made it past the first barrier and thus gives good information about the state of the joint. Should the drain become blocked or leak profusely, the second sealing strip is capable of taking up the total amount of pressure (Figure 13.47 (a)).

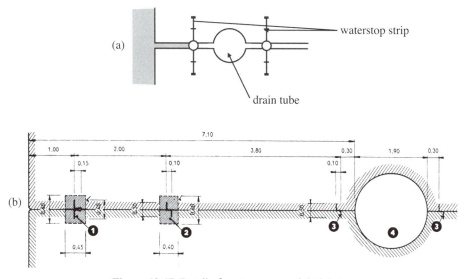

Figure 13.47 Detail of upstream watertight joints

At the Grande Dixence dam (Figure 13.47 (b)), the grout curtain encased in the concrete is made up of a ❶ 1.5 mm-thick, V-shaped copper sheet located 1 meter away from the upstream face and followed up 2 meters beyond by a second ❷ Z-shaped copper sheet of the same thickness; lastly, 3.80 m from the last copper sheet, ❸ a steel sheet, also Z-shaped, ensures the joint's perfect watertightness. To identify possible leaks, these sheets are connected horizontally every 16 meters by a Z-shaped sheet. One end of this sheet leads to a ❹ control shaft situated 7.10 m from the upstream face. The Z- and V-shapes enable the blocks to move without breaking the sealing system.

13.10.5 Freeboard

By definition, freeboard is the vertical distance between the normal maximum operating level and the crest (Figure 13.48). The following criteria are given as an example and are taken from Minor (1998). They are essentially based on hydraulic considerations.

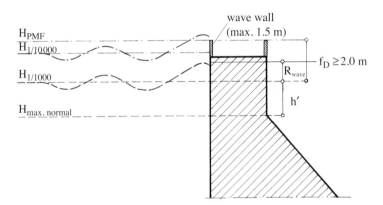

Figure 13.48 Freeboard required for concrete dams (from Minor, 1988). Key: h' = headwater height; R_{wave} = wave runup height; f_D = freeboard.

- Design flood HQ_{1000}
  - rule n – 1 for water release structures
  - waves and wave run-up are not considered
  - freeboard must be $f_D > 2.0$ m
- 10,000-year flood $HQ_{10\,000}$
  - all water release systems are operating
  - waves and wave run-up are not considered
  - water level (without waves) must be lower than the crest
  - a wave wall of maximum 1.5 m may be erected for protection against waves
  - safety flood (deluge) (PMF)
  - water level without waves must be lower than the crest; if the foundation is not threatened, an overtopping of the crest of 1 m max may be accepted

13.11 Gravity dam heightening

13.11.1 Reasons and prerequisites

Until recently, most dams that have undergone heightening were gravity dams. Several reasons may lead to the decision to heighten a dam:

- Undersized reservoir:
  - the hydrology that served as the basis for design was not well known or was not representative
  - the hydrology has changed (glacier melt, urbanization)
- Reduction in reservoir storage capacity:
  - either due to reservoir sedimentation or a modification in operating constraints
- Aim to improve use of water inflow:
  - an increase in the reservoir volume would provide an economic advantage, for example, a better use of the storage by transferring water from the summer to the winter
- Enabling multi-year water transfers:
  - for example, an increased population in some arid countries requires larger volumes to be transferred from a wet year to a dry year
- Increase in inflow to the catchment area:
  - by creating new intakes and water transfer structures
  - through artificial inflow by pumping
- Creation of a retention volume for flood control:
  - aimed at improving protection against floods in the dam's downstream area

For dam heightening to be worth considering, a certain number of conditions must be met:

- There must be a complete record of dam behavior since impoundment, and it must indicate no anomalies
- There must also be a complete record of foundation behavior with no anomalies
  The behavior analysis of the foundation during operation is an important and additional source of information; it completes the information acquired during dam construction (boreholes, exploratory galleries, observation of the foundation prior to concrete placement, absorption tests for grout)
- The uplift regime must be fully understood; its increase can be controlled through new technical measures (supplementary grouting, drains)
- The concrete quality must tolerate an increase in stresses

And lastly, when a dam requires rehabilitation works because of its overall condition and stability conditions, this can be the opportunity to heighten it and increase the reservoir storage capacity.

13.11.2 Heightening methods

13.11.2.1 Slight increase in height

When the height increase is low compared to the dam height, the increase in load will also be low.
If H is the dam height and F the forced exerted by water pressure, we obtain:

$$\text{for } \Delta H = 10\% \cdot H, \; E = (1.1)^2 \cdot H, \text{ hence } \Delta E \cong 20\% \cdot E.$$

Based on today's criteria, some older gravity dams are overdesigned and can tolerate supplementary load while remaining within acceptable levels of safety. Often the most important case to verify is sliding safety on the foundation.

Heightening may be carried out by simply strengthening the crest. The new concrete must be secured to the whole structure by appropriate surface treatment and by installing short post-tensioning anchors (Figure 13.49). Generally, passive anchors are not used. When designing these anchors, it is important to consider dynamic forces during an earthquake.

In simple cases where the required increase in water level is low (< 2 m), the construction of a simple parapet wall that resists water pressure may be sufficient.

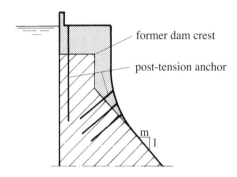

Figure 13.49 Heightening of the crest only

13.11.2.2 Major increase in height

Heightening is considered major if it reaches or exceeds 10% of the dam's original height. In this case, complex interventions are necessary to guarantee safety for the dam and reservoir. Several measures may be implemented.

Strengthening of the upstream face

The cross-section of the gravity dam is adapted to the new height by adding concrete upstream (Figure 13.50). The thickness of the concrete is uniform across the entire surface of the face and is equal to $E = m \cdot \Delta H$, where m is the slope of the downstream face.

The new concrete is connected to the old concrete with shear dowels. The combined action taking up the internal load from the former structure and its strengthening will only be effective for load applied after the strengthening has been carried out. The interface between the old and new concrete (which is a preferential path for seepage) must be carefully drained. The joints between the blocks must extend into the new concrete, and waterstops must be inserted near the new upstream face.

Due to its position, strengthening of the upstream face can only occur when the reservoir is empty. As a rule, the reservoir can only be emptied to the level of the bottom outlet, which itself must be sufficiently high to avoid siltation. It is therefore rare for a reservoir to be emptied right down to the foundation level. Furthermore, when the reservoir is empty, the greatest stresses on the foundation occur in the upstream face. During strengthening, it is not possible to load the greater part of the foundation and therefore to ensure the continuity of stresses in the foundation. Because of these reasons and also due to economic considerations, this method of heightening is only rarely employed.

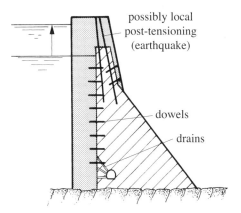

Figure 13.50 Heightening by strengthening the upstream face

Strengthening the downstream face
The cross-section of a gravity dam is in this case adapted to the new height by adding concrete downstream (Figure 13.51). The thickness of the concrete is uniform across the entire downstream surface and is equal to $E = m \times \Delta H$, where m is the slope of the downstream face. The new concrete is connected to the old concrete with shear dowels. The joints between the blocks are extended into the new concrete and drained.

As the downstream face is generally accessible in all situations, carrying out strengthening right down to the foundation creates no major problems. The bottom part of the strengthening will be carried out while the reservoir is empty and when the stresses on the downstream toe of the foundation are at their lowest. Particular attention must be paid to the quality of the placed concrete; the lower part of the downstream face is the part of the dam structure that bears the most load.

To ensure stability of the new crest in cases of an earthquake, it is often connected to the concrete of the former dam with post-tensioning anchors.

This solution does, however, have the major drawback of requiring appurtenant structures such as spillways to be lifted if they are integrated into the body of the dam.

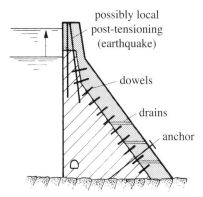

Figure 13.51 Heightening by strengthening the downstream face

Strengthening by covering the upstream and downstream faces
This solution, where new concrete covers both the upstream and downstream faces, enables the strengthening of an old dam that no longer completely meets the required safety criteria. Covering the upstream and downstream faces not only allows the dam to be heightened, but it also further strengthens the zones that bear the most load and improves watertightness, should this be required (Figure 13.52).

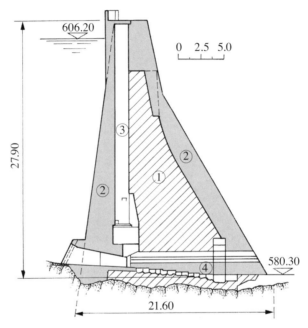

Figure 13.52 Strengthening by covering the upstream and downstream faces (Muslen dam). Key: ① existing dam body; ② envelope of new concrete; ③ access shaft; ④ bottom outlet gallery.

Crest strengthening and post-tensioning
An additional vertical load is applied to the heightened crest of the dam through the installation of post-tensioning anchors. The crest also acts as a distribution beam, allowing a homogeneous introduction of post-tensioning load (Figure 13.53).

The great advantage of using post-tensioning anchors for heightening is that they can be installed independently of reservoir operation. The time required to complete the installation (including works on the crest, boreholes, installing anchors, and tensioning) is a complex parameter to manage.

Sealing of the post-tensioning anchors occurs below the dam foundation. The holding force of the fixings, which depends on local geology, is therefore one of the determining criteria for the feasibility of this solution. Considerable prestressing force often needs to be applied, and the use of anchors with more than 10,000 kN is often essential. As the anchor heads are situated on the dam crests, the cable length can be up to 1.5 to 2 times the total height of the dam, which makes them difficult to handle.

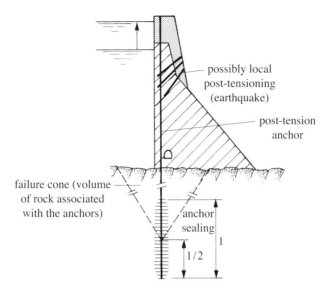

Figure 13.53 Strengthening the crest with post-tensioning

13.11.3 Use of post-tensioning

13.11.3.1 Required post-tensioning force

The aims of post-tensioning are the following:

- No tensile stresses on the upstream face with a full reservoir
- Limited tensile stresses on the downstream face with an empty reservoir
- Maximum compressive stresses are lower than acceptable limits on the foundation
- Sliding safety and shear safety are sufficient
- Stability of the entire dam and its foundation is achieved
- Durability of anchors (permanent protection against corrosion)

The two first criteria determine the required post-tensioning force as well as the position of the heads and the slope of the anchors.

To avoid tensile stresses on the upstream face when the reservoir is full, the most cost-effective approach is to position the anchors as far upstream as possible. However, to limit tensile stresses on the downstream face when the reservoir is empty, the axis of the post-tensioning force must be located in the middle third of the section, that is, much further in toward the center of the dam structure. The post-tensioning force required to avoid the appearance of tensile stresses for both an empty and full reservoir would be excessive and would soon be unrealistic. One solution is to add anchors on the downstream side. Again, maximum compressive stresses on the foundation would soon become considerable. For these reasons, and providing the geology is suitable, low tensile stresses at the downstream toe are tolerated when the reservoir is empty. These tensile stresses can lead to a detachment or decompression of the downstream toe of the dam when the reservoir is empty.

Let's assume a simplified triangular profile for the gravity dam and apply an increase in hydrostatic pressure without increasing self weight. Uplift will also increase in the same way as water pressure. Figure 13.54 presents diagrams with vertical stress on the foundation for a non-heightened and heightened dam, to which a post-tensioning force has been added.

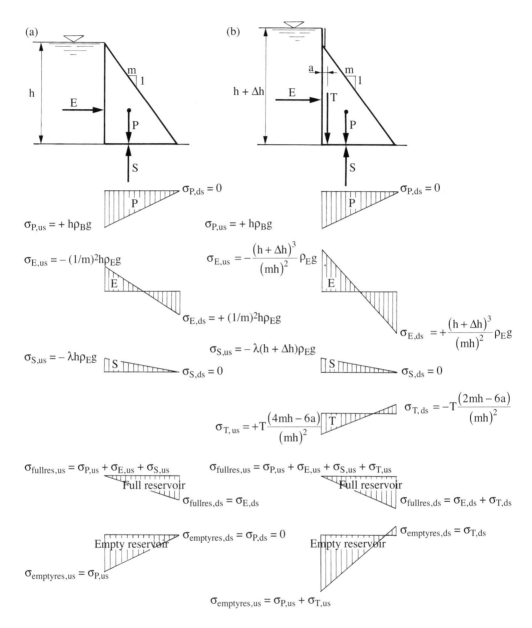

Figure 13.54 Diagram of the vertical stresses on the foundation: (a) before heightening; (b) after heightening and post-tensioning

Let's also assume that the dam prior to heightening was designed by applying Levy's rule, so that the stresses with a full reservoir against the upstream face are zero:

$$m = \sqrt{\frac{\rho_E}{\rho_B - k\rho_E}}.$$

The load due to self weight, water pressure, and uplift becomes:

$$P = \frac{1}{2} \cdot \rho_B \cdot g \cdot h \cdot mh$$

$$E = \frac{1}{2} \cdot \rho_E \cdot g \cdot (h + \Delta h)^2$$

$$S = \frac{1}{2} \cdot k \cdot \rho_E \cdot g \cdot mh \cdot (h + \Delta h).$$

Additionally, the added post-tensioning force has the same objective, with a heightened profile. Calculating the stresses gives us a minimal post-tensioning force per linear meter T_{min}:

$$T_{min} = \frac{\rho_E g \left[k (h + \Delta h)(mh)^2 + (h + \Delta h)^3 - \left(\frac{\rho_B}{\rho_E}\right) h (mh)^2 \right]}{4mh - 6a},$$

where a is the distance between the post-tensioning axis and the upstream face (Figure 13.54). As a rule, $a = 2$ to 3 m.

Post-tensioning also has a direct impact on sliding safety:

$$S_G = \frac{(P + T - S) \tan \phi'}{E}.$$

P and ϕ' are not affected by heightening, however, E and S increase.

This simple approach allows us to determine the post-tensioning required for stability at a given raised height (Figure 13.55). If the dam profile is not modified, raising the dam height by 20% requires a minimum post-tensioning in the order of 25% of self weight. Depending on the accepted angle of friction for the sliding safety calculation, post-tensioning can even reach 35% (Laffite, 1985).

Post-tensioning is often associated with crest strengthening through the placement of a large mass of new concrete. This increase in self weight must also be included in the calculation, which in some cases enables post-tensioning to be reduced to approximately 15% of self weight.

As a rule, the ratio between post-tensioning force per meter in one section and self weight is approximately the ratio between the heightening and the initial height of the dam:

$$\frac{T}{P} \approx \frac{\Delta h}{h}.$$

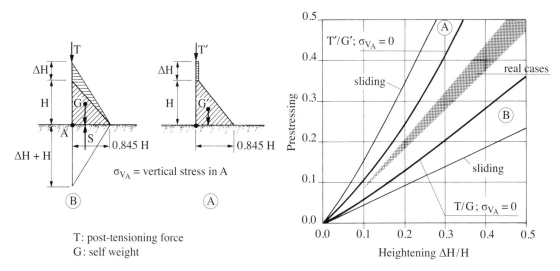

Figure 13.55 Necessary post-tensioning for the stability of a heightened gravity dam compared to the self weight of the dam (Laffite, 1985)

In addition to the post-tensioning cables installed at the crest, inclined cables may be required from the front of the downstream face in order to increase strength along the sliding plane (Figure 13.51). In this latter case, the question is to find out whether the horizontal component of the post-tensioning force must be considered as an active or passive force for the calculation of sliding safety. As an active force, it is added to the denominator of the other exterior horizontal forces. As a passive force, it is added to the sliding strength resulting from vertical forces.

13.11.3.2 Technical limitations to prestressing

The maximum working force per anchor is between 10,000 and 13,000 kN, which represents 50 to 60% of bearing capacity (approximately 20,000 kN). The minimum space between the cables is in the order of 2 to 3 m to avoid adverse interactions. The lengths and slope of the neighboring anchors can also be varied, as illustrated in Figure 13.56 with the example of the Maigrauge dam (Ammann et al., 2003; Viret et al., 2003).

Post-tensioning involves the handling of thick cables with a maximum length of between 120 and 150 m, which in many cases and for both practical and construction reasons limits the feasible height for raising the dam. Currently, the maximum feasible height for the raising of a dam is 30 to 40% of the initial height.

Lastly, it is a good idea to verify the possibility of cable failure under tension-shear conditions for the hypothesis that a violent earthquake may occur.

GRAVITY DAMS 309

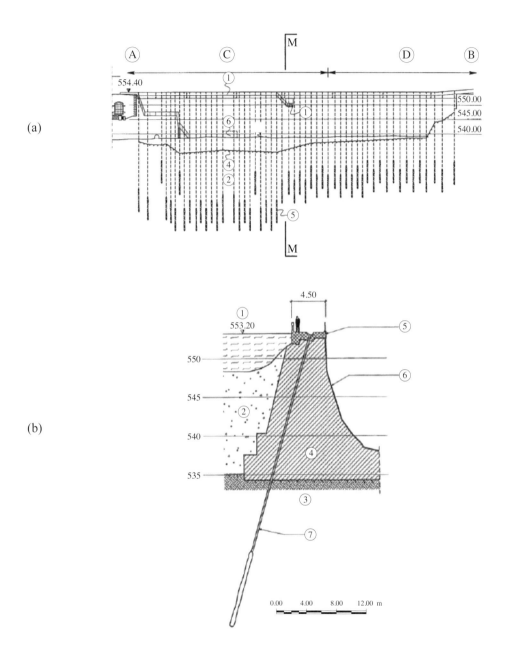

Figure 13.56 Maigrauge dam: (a) Elevation layout of prestressed rock anchors: A *Right bank, B Left bank, C Curved section, D Rectilinear section, 1. New crest, 2. Foundation rock, 3. Alluvium, 4. Concrete-rock interface, 5. Prestressed rock anchors, 6. Old bottom outlet, 7. Inverted pendulum chamber*; (b) M-M cross-section of the rehabilitated dam: *1. Normal reservoir water level, 2. Sediment, 3. Molasse, 4. Dam, 5. New crest, 6. Repair of the existing gunite, 7. Post-tension rock anchor $P_0 = 1600$ kN (168 mm diameter)*

13.11.3.3 Length and depth of anchor sealing

The length of the anchor sealing depends on the anchoring force and the quality of rock in the fixing zone. The whole length of the anchor tends to be placed in layers with approximately the same strength and deformability characteristics. This length can be estimated through calculations, but the hypotheses demonstrate that the holding force of the sealing must be checked through in situ tests. Failure modes for the sealing depends on shear stresses in the rock along the seal. These must remain low to avoid major creep (1 to 2 N/mm^2).

The depth of the sealing is determined by the stability of the whole dam-foundation system (Figure 13.57). Furthermore, there must be a sufficient volume of rock associated to the anchors (cone of influence) so that the reaction is achieved once the cables are tensioned. Analysis of the stability of a non-heightened gravity dam presupposes a plane of low tensile strength or sliding strength located either in a horizontal concrete lift joint (concrete-concrete), at the concrete-rock interface, or in a system of discontinuities not far below the foundation level (rock-rock). With post-tensioning heightening, these planes are strengthened by the force of the post-tensioning.

However, a fracture surface may appear at a deeper level, where the prestressing force disappears, that is, at the level of the sealing. Overturning safety and sliding safety for the entire dam and for the rocky wedge compressed by post-tensioning must be verified. The length of the anchor sealing must be selected so that the level of safety is sufficient.

When the rock is assumed to be isotropic and homogeneous, the failure mode illustrated in Figure 13.55 can be used. Computational and physical models have demonstrated that failure occurs by overturning of the dam and the dihedral ABC around the downstream toe of the dam (point C, Figure 13.58).

If, however, the rock has cracks or a system of fissures, these planes with little strength will determine the failure mode. In this case, several failure modes are defined depending on the analysis of overturning and sliding safety.

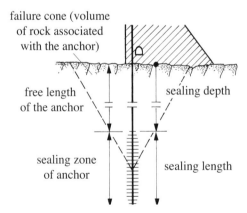

Figure 13.57 Diagram of the anchor installation

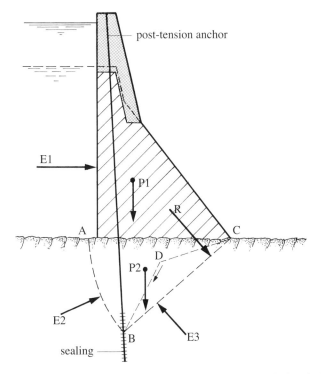

Figure 13.58 Failure model for a heightened dam on a homogeneous isotropic foundation (Laffite, 1985)

13.11.3.4 Some technical aspects of post-tension rock anchors

Figure 13.59 shows an ideal post-tension rock anchor, such as those used for heightening dams. It comprises a double protection against corrosion.

The placement of anchors is carried out in the following way:

- From the dam crest, a hole is drilled into the concrete and then into the foundation rock until the deepest part of the anchor sealing
- A PE (polyethelene) tube is inserted into the hole down the whole of its free depth, followed by a notched metal tube fitted with grouting sleeves in the sealing zone
- A fixed packer is installed at the top end of the sealing zone
- The sealing zone is grouted in several stages from a mobile double packer
- The post-tension rock anchor is installed; the bare strands are set into the sealing zone; however, they are sheathed in the free zone
- The tube is grouted at low pressure along the entire length
- Lastly, the anchor head is installed, and the entire system is tensioned.

The free zone of the anchor (Figure 13.57) must be secured by pre-sheathed or greased strands. The space between the strands and the sheath is grouted in order to fill in empty spaces. For some anchors, the filling of the free zone in the tube is carried out after tensioning with a viscous resin protecting against corrosion instead of cement grout. This solution is used for controllable anchors. These anchors can be re-tensioned.

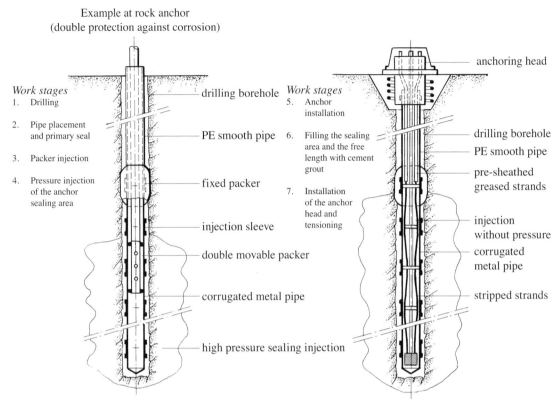

Figure 13.59 Details of the post-tension rock anchors used to heighten dams (from Lafitte, 1985)

Ensuring permanent and sound protection of anchors against corrosion is major concern. Specific points of the anchor are particularly vulnerable, such as the anchor head, the contact between the free zone and the sealing, as well as the seal itself. Watertightness of the sealing zone is often achieved through a double system. The sheaths surrounding the anchors and cables must be carefully protected against water, as well as the rock in the sealing zone, so that no water circulates against the prestressed steel (Lafitte, 1985).

To confirm the calculation hypotheses and validate the process for installing the anchors, it is strongly recommended to carry out test anchors in similar conditions to those used for the dam anchors. Tests enable the appropriate length of the sealing zone to be established and also mean that any installation issues can be solved, particularly those related to water circulation inside the rock mass of the foundation. Test anchors can be fitted with measuring cells that track anchor behavior across time and quantify various effects, such as steel relaxation and creep in the body of the anchor. The number of planned test anchors for a given dam will depend on the size of the project and the risks involved should the anchor fail (von Matt, 1997).

The anchor tensioning process must be carefully prepared. The behavior of the anchors and the entire dam structure must be measured and inspected in great detail. Post-tensioning force generally implies a considerable density of anchors. The order in which the anchors are tensioned must aim to gradually and uniformly load each block in the dam. It is advisable to have a certain number of anchors that can be controlled during operation. To this end, hydraulic, mechanical, or electrical devices measuring tension have been developed and have proven to be reliable. The number of measuring and control anchors depends on the specific conditions of the dam. The combined number of measuring and control anchors can be in the order of at least 5% of the total number of anchors (this percentage is suggested by the SIA 267, 2013a, b. Géotechnique).

Checking the bearing capacity of anchors over time
It is important to guarantee the durability of rock anchors in use by carrying out regular checks using several methods. A regular monitoring program includes:

- Visually inspecting the heads (by removing the protective cap) and the supporting zone
- Checking the tension of anchors performing this function
- Verifying the electrical insulation of cables in a non-destructive manner by measuring electrical resistance; electrical insulation protects the prestressed steel from the action of stray electric currents and monitors the watertightness of the protective envelope

In addition, the usual monitoring of dam behavior will give indications as to the condition of post-tensioning; as cables relax, this can result in dam deformation, water seepage, or a modification in uplift. Monitoring the position of anchor heads can also be included in the program for geodetic measurements.

The main causes of observed damage are due to insufficient or faulty protection against corrosion, the inflow of aggressive water, or cracking corrosion under tension (hydrogen embrittlement). Under the action of stray electric currents, defects in the sheath may create entrance or exit points for current, depending on the direction of the current. The exit of stray currents may lead to anodic corrosion (metal dissolution); the entrance of currents leads to a cathodic reaction (the production of hydrogen). Appropriate protective measures guarantee bearing capacity over time. A loss in steel strength through fatigue should not occur as the anchors undergo practically no modification in tension while the dam is operating. In the case of an operating basis earthquake, care should be taken that cracking does not reach the cables, as corrosion protection could be compromised.

With regard to a decrease in strength in the sealing zone, practice seems to show that no negative effects of ageing appear. If the sealing zone is in an environment with aggressive water, precautions must be taken regarding the grouting material.

Should specific observations be made during operation, the risk should be assessed and any necessary measures taken (restricting operation). The following stage involves making calculations and examining the anchors, after which point decisions can be made around increasing monitoring, replacing anchors, or installing new ones.

13.12 Strengthening Gravity Dams

There are many reasons (insufficient stability, calculation hypotheses that do not comply with the applicable regulations, etc.) why it may be necessary to implement strengthening measures on existing dams. There are various solutions available to address observed defects.

These include strengthening the dam profile by adding concrete downstream or upstream, in the same way as outlined above. In this case, particular care must be taken when binding the old and new concrete. It is also possible to improve stability conditions by employing post-tensioning methods or by constructing a concrete block or large embankment at the downstream toe. Installing anchors during the rehabilitation of old concrete dams may require special provisions to distribute the occasional forces occurring on poor-quality concrete or masonry. Furthermore, the installation of post-tensioned cables on a dam structure susceptible to an alkali-aggregate reaction must take into account the fact that the tensile stresses in the cable will increase as the concrete swells.

Dam operating conditions will often dictate the choice of one or other solution.

14. Hollow Gravity Dams and Buttress Dams

Lucendro buttress dam in Switzerland, height 69 m, year commissioned 1947 (Courtesy D. Quinche)

14.1 From the Traditional Gravity Dam to the Buttress Dam

14.1.1 Traditional gravity dam vs. hollow gravity dam

Gravity dams have several advantages, but also a few drawbacks. Some of their advantages include:
- A simple shape that adapts readily to the topography
- Low stresses
- Concrete that is under relatively little stress and has a low cement content
- Thanks to their massive shape, they resist extraordinary events better than thin dams

However, the disadvantages must also be listed:
- The placement of a large volume of concrete
- Substantial uplift, which affects stability as the foundation is completely covered
- Reservoir filling and emptying cycles imply a complete switching of states of stress (as illustrated in Figure 14.1); this stress on the foundation can lead to plastic deformation
- Considerable heat generation during concrete curing due to the massive nature of the structure; as a rule, artificial concrete cooling is required

These considerations also demonstrate that overall, the material is not used in the best possible manner. The compressive strength of most of the structure is excessive compared to the stress it undergoes.

As the cost of a gravity dam is directly related to the volume of placed concrete, engineers have sought to eliminate concrete from areas where it is not well utilized. Hollow gravity dams were developed out of a desire to address these concerns, while retaining a triangular shape. A 3- to 5-meter-deep cavity with a maximum width of 20 m is formed across the widest possible surface between two blocks. The cavity, flanked by the upstream and downstream faces, may be closed off at its base by an apron fitted with drains to limit uplift (Figure 14.2 (b1)). One variation consists in leaving the cavity open so that the foundation remains entirely uncovered. This option enables uplift to be drastically reduced, while simultaneously permitting observation of the foundation's bearing capacity, as well as of any drains drilled into the rock (Figure 14.2 (b2)). Due to the decrease in concrete volume, sliding and overturning stability are reduced compared to a traditional gravity dam. To improve this situation, and in particular to ensure sliding safety, an inclined upstream face is employed, which widens the base of the foundation and benefits from the vertical component of water pressure (Figure 14.2 (c1) and Figure 14.2 (c2)). Figure 14.3 illustrates a dam structure of this kind.

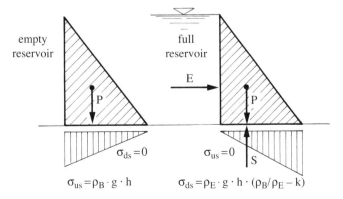

Figure 14.1 Gravity dam: stresses under the foundation during normal loading

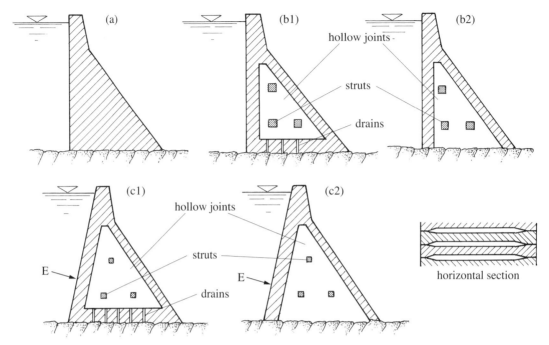

Figure 14.2 From the traditional gravity dam to the hollow gravity dam – general layout: (a) traditional gravity dam; (b) hollow gravity dam with a vertical upstream face with (1) or without (2) an apron at the base of the cavity; (c) hollow gravity dam with an inclined upstream face with (1) or without (2) an apron at the base of the cavity

Figure 14.3 The Albigna hollow gravity dam: (a) during construction; (b) view of a cavity

Compared to a full wall, hollow gravity dams have the following advantages:
- Concrete savings: 10 to 15%
- Easier cooling
- Drainage simple to install and monitor
- Decrease in uplift (Figure 14.4)

However, they do require:
- A larger formwork surface
- Strengthening through buttress struts (to ensure stability in the transversal direction in case of an earthquake)

14.1.2 Buttress dam

To further reduce the volume of concrete and increase drainage in the foundation, the cavities are open at the downstream end and continue between the blocks down to the level of the foundation. These improvements have led to the buttress dam, which is illustrated in its traditional layout in Figure 14.4.

The advantages and disadvantages of buttress dams can be summarized as follows:

- The vertical component of water pressure created by the slope of the upstream face substantially increases stability
- Uplift is weak (Figure 14.6) and easily controlled, through the surface of the foundation left open between the blocks
- The volume of concrete is reduced by 25 to 30% compared to a gravity dam
- The dissipation of hydration heat is facilitated; artificial cooling is generally not required
- It is also possible to integrate a spillway supported by the buttresses and a bottom outlet between the buttresses

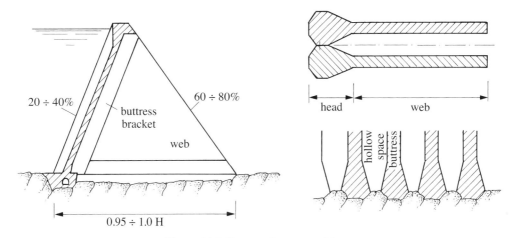

Figure 14.4 Buttress dam: general layout

Figure 14.5 Lucendro (CH) (a) and Al Massira (Morocco) (b) buttress dams

However,
- There is a large formwork surface, and the forms are complicated, which diminishes the advantage of the reduction in volume by approximately a third
- The interface surface between the two buttress heads is small
- Transversal stability during an earthquake may be critical, due to the absence in continuity of the downstream face
- Integrating the foundation of the buttresses at the abutments is challenging
- The dam is relatively vulnerable to malicious acts and rock falls

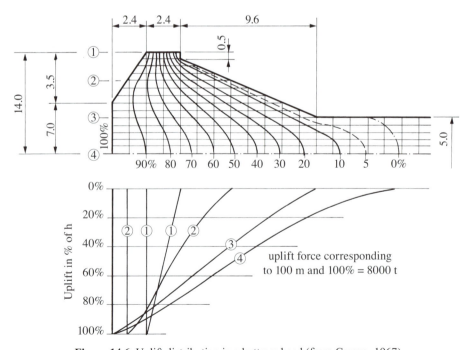

Figure 14.6 Uplift distribution in a buttress head (from Crespo, 1967)

This relative vulnerability and the complex aspects inherent to buttress dams explains their low numbers in Switzerland.

After the Second World War, buttress dams designed at the time included a downstream slab, created by widening the buttresses. This arrangement was not satisfactory, however, because if one cell (the empty space between two buttresses) came into contact with the reservoir, the hydrostatic pressure that was created would create transversal load that could threaten the entire dam structure.

Based on this knowledge, the Swiss authorities decided to modify two buttress dam projects. The Cleuson dam, which had been started in 1947, was transformed into a hollow gravity dam, even though construction was already well under way. Every second cavity was filled in and the unfilled blocks were strengthened with a 12-meter concrete mass in the upstream part of the cavity (Figure 14.7). At the Lucendro dam (1947), longitudinal buttress struts were added many years after dam operation had begun, in order to improve its lateral stability.

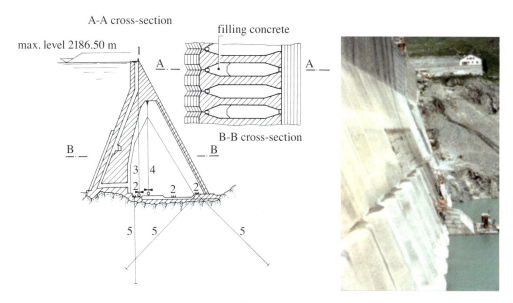

Figure 14.7 Strengthening the Cleuson buttress dam with fill concrete: (1) polygonal pillar; (2) clinometers; (3) pendulum with telerecorder; (4) pendulum with optic measurement system; (5) telerockmeters (cross-section from SwissCoD, 1985). In the photo on the right, the change in profile of the upstream face is visible. This was introduced during the works.

14.1.3 Buttress dam with closed downstream facing

The buttress dam with closed downstream facing is a particular form of the buttress dam; it has a continuous downstream slab (Figure 14.8).

From a strictly static perspective, the downstream slab is not indispensable. However, it does offer the following advantages:

- The effect of temperature in hot regions is reduced
- The buttress web is protected against frost in cold regions
- Lateral stability in case of an earthquake is increased
- Vulnerability of dam structure to malicious acts is reduced

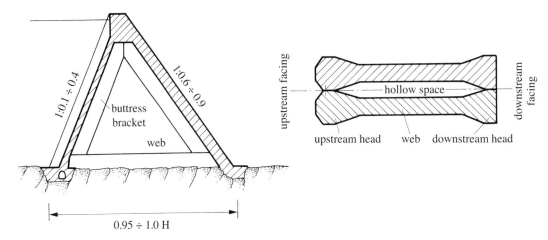

Figure 14.8 Buttress dam with closed downstream facing: general layout

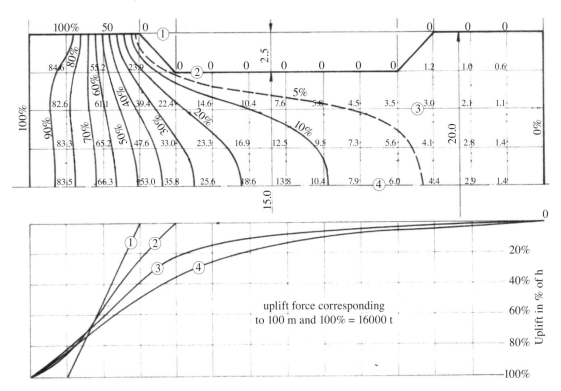

Figure 14.9 Uplift distribution for a hollow gravity dam (from Crespo, 1967)

14.1.4 Comparing different profiles through figures

As mentioned previously, compared to a traditional gravity dam, buttress dams with a closed downstream facing save concrete, but require the installation of a larger formwork surface. In addition, for a buttress dam, the slopes of the upstream and downstream faces need to be adapted to meet sliding stability parameters, implying an increased footprint. Uplift distribution is also different. To calculate the most economical profile, as well as the space between the buttresses, several options must be studied.

Table 14.10 Some points of comparison between a hollow gravity dam and a buttress dam (Crespo, 1967)

	Height (in m)	Traditional gravity dam	Hollow gravity dam	Buttress dam
Self weight P (in t)	50 m	48,970	41,000	23,510
	75 m	108,870	86,110	45,200
	100 m	192,530	147,160	73,690
Uplift S (in t)	50 m	16,580	6,800	3,400
	75 m	37,250	10,200	5,100
	100 m	66,300	13,600	6,800
Stability H/V	50 m	0.772	0.731	0.661
	75 m	0.785	0.741	0.712
	100 m	0.792	0.755	0.743
Volume of concrete	50 m	100%	83.7%	68.5%
	75 m	100%	79.1%	59.3%
	100 m	100%	76.4%	54.7%
Formwork surface	50 m	100%	112.9%	146.8%
	75 m	100%	122.2%	169.0%
	100 m	100%	129.0%	186.6%

Legend: H = sum of horizontal forces (self weight P + uplift S); V = water pressure.

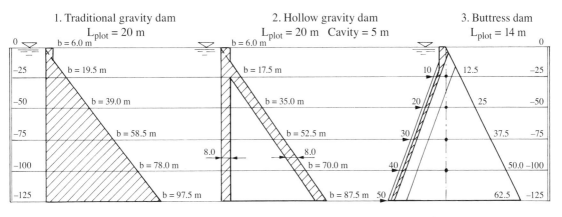

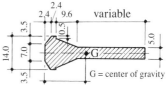

G = center of gravity

324 CONCRETE DAMS

Table 14.10 (and the accompanying figures) illustrate calculations for a gravity dam (center distance between the blocks: 20 m triangular uplift distribution), a buttress dam with closed downstream facing (center distance between the blocks: 20 m; cavity width: 5 m; uplift distribution as per Figure 14.9), and a buttress dam (center distance between the buttress heads: 14 m; uplift distribution as per Figure 14.6).

14.1.5 Other types of buttress dams

Some buttress dams also comprise an upstream face made of a succession of flat slabs, generally in reinforced concrete. These slabs comprise independent elements resting on corbels integrated into the buttresses on the upstream side, which results in a flexible structure. The slab is subject to water pressure and so works like a beam with two supporting ends. The corbels and upstream faces of the buttresses are

Figure 14.11 Cross-sections of buttress dams with upstream slab: (a) independent slabs; (b) with continuous upstream slab

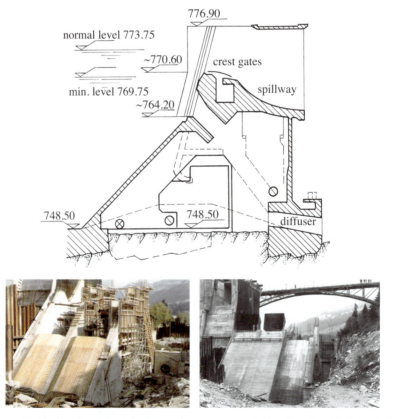

Figure 14.12 Grout curtain at the Lessoc barrage power station (Courtesy H. Pougatsch)

strengthened with steel reinforcement. Ensuring watertightness (waterstop) between the slab and the pillar is vital to avoid leaks (Figure 14.11 (a)). The upstream face can also be continuous, which makes the structure more rigid (Figure 11.11 (b) and 14.12). The slab works in bending and is reinforced on both faces. The usual rules should be applied to all reinforced concrete elements.

Generally speaking, the buttresses are embedded into the rock foundation, which creates a favorable effect with regard to sliding resistance.

The multiple arch dam is a buttress dam whose upstream face is comprised of a series of cylindrical arches. This type of dam is covered in chapter 16.

14.2 Stresses in Buttress Dams

14.2.1 Stresses on the faces

As with gravity dams, when Navier's hypothesis is applied, extreme stresses occurs on the faces.

In cases of load P + E + S, self weight + water pressure + uplift, the greatest compressive stresses are located at the toe of the downstream face, while the lowest stresses in the same cross-section are located on the upstream face.

The minimum stress on the upstream face must in all cases remain compressive:

$$\sigma_{min,us} = f(P, E, S) \geq 0.$$

The main stresses are normal and parallel to the faces (Figure 14.13). By referring once again to the equations:

with a full reservoir, on the downstream face:

$$\sigma_I = \sigma_{max,ds} = \sigma_{z,ds}(1 + \tan^2\alpha) \qquad \sigma_{II} = 0$$

and on the upstream face:

$$\sigma_I = \sigma_{min,us} = \sigma_{z,us}(1+ \tan^2\delta) - \sigma_{Ez}\tan^2\delta \qquad \sigma_{II} = \sigma_{Ez},$$

as a result, to satisfy the stress

$$\sigma_{min,us} \geq 0,$$

the following condition must be met:

$$\sigma_{z,us}(1 + \tan^2\delta) \geq \sigma_{Ez}\tan^2\delta,$$

which can be expressed by

$$\sigma_{z,us} \geq \sigma_{Ez}\sin^2\delta.$$

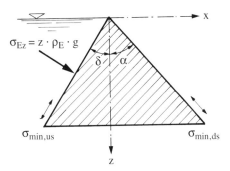

Figure 14.13 Stresses on the faces of a buttress

14.2.2 Vertical stresses

As with gravity dams, given the rigidity of the structure, we can assume that Navier's hypothesis is realistic. This hypothesis leads us to accept that vertical stresses vary linearly along a horizontal section. For each section, the center of gravity, the surface, and the downstream and upstream resistance moments must be calculated (Figure 14.14).

In this case, we obtain the vertical stresses with the equation,

$$\sigma_z(x) = \frac{N}{A} \pm \frac{M_y}{J_y} \cdot x,$$

and the stresses on the upstream and downstream edges are given by the following equations:

$$\sigma_{z,us} = \frac{N}{A} + \frac{M_y}{W_{us}} \qquad \sigma_{z,ds} = \frac{N}{A} - \frac{M}{W_{ds}}.$$

where
- N = normal load resulting from weight, water pressure, and uplift
- A = section of the buttress
- M_y = moment relative to the y axis
- J_y = moment of section inertia relative to the y axis
- W = resistance moment in upstream (us) and downstream (ds) direction
- x = distance to the neutral axis of the section

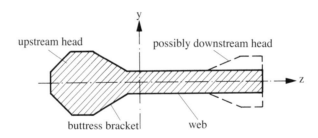

Figure 14.14 Geometry of a buttress section with an upstream head

14.2.3 Stresses inside the buttress

Once the exterior loads have been calculated, the same interior equilibrium conditions apply as for a gravity dam. However, the calculation is more complicated than for gravity dams, given the complex geometry of the buttress (whose shape varies depending on the height) and the various thicknesses of one horizontal section.

Today the finite elements method is mostly used to calculate stresses insides the buttress.

For structures of a comparable height, the principal stresses of a buttress dam are greater than those of a gravity dam. Tensile stresses are often observed in the upstream part of a buttress even though vertical stresses are compressive in this zone. These tensile stresses, added to those from changes in temperature, are partially responsible for the diagonal cracks that can form in the buttresses.

14.2.4 Shape of the upstream head

Let's now examine the shape of the upstream head of the buttress in more detail and analyze the stresses that develop within it, while still assuming that Navier's hypothesis is valid.

Rectangle head buttress

Simple geometry, with a sealing system near the upstream face (Figure 14.15 (a))

Under hydrostatic pressure σ_{Ez}, the rectangular shape of the head leads to tensile stresses at point A of the A-B section. These tensile stresses are even greater if the cantilever beam d is large. There are two solutions for avoiding these tensile stresses:

- Shifting the sealing system downstream
- Modifying the shape of the head

The aim of these two options is to create lateral water load that compresses the head and makes the tensile stresses disappear.

Shifting the sealing system downstream

Shifting the sealing system downstream (Figure 14.15 (b) and (c)) prevents tensile stresses from occurring at point A.

The joint between the two blocks needs to be loaded by water pressure. As this joint is only a few millimeters thick, there is a risk of it becoming blocked by concrete efflorescence or organic material. Widening the joint or creating a groove guarantees the loading by water pressure. However, the concrete forms are more complicated.

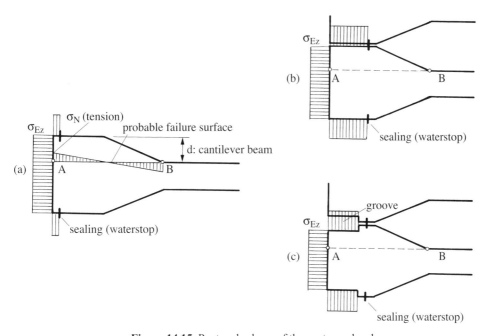

Figure 14.15 Rectangle shape of the upstream head

Rounded head and diamond head buttresses

Rounded heads better orient forces in the head and limit the concentration of stresses. However, the disadvantage is that their complicated concrete geometry requires special formwork.

The geometry of polygonal (or diamond-shaped) heads is simpler to create than rounded heads. Most buttress dams built since 1950 employ the polygonal shape head (Figure 14.16).

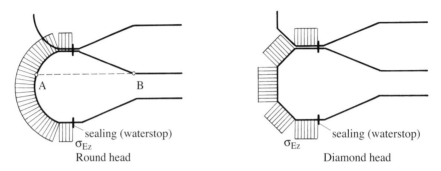

Figure 14.16 Round head and diamond head buttresses

14.2.5 Optimizing the diamond head

Due to the slope of the upstream face, self weight includes a component that is normal to the face. This component applied on the cantilever of the heads leads to the appearance of tensile stresses, under the simple effect of self weight, as illustrated in Figure 14.17. These stresses are low; they are tolerated with an empty reservoir. However, with a full reservoir, it is essential that these tensile stresses are compensated for by the compression created by water pressure.

Let us now analyze the stresses due to this effect and due to hydrostatic pressure on the upstream face of the head.

The fundamental requirement is $\sigma_{min,am} \geq 0$ in the direction of the principal stresses, that is, parallel to the face.

Stress σ_N is a function of self weight, water pressure, uplift, and the geometry of the head. This is an effective stress.

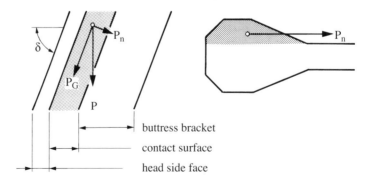

Figure 14.17 Effect of self weight on the buttress head

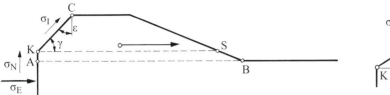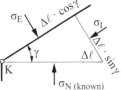

Figure 14.18 Load on the buttress head

From stress σ_N, water pressure σ_E, and the geometry, it becomes possible to calculate the principal minimum stress at point K, along the K-C inclined plane. This relation can be expressed as:

$$\sigma_I = \sigma_{\min} = \sigma_N \cdot \left(1 + \tan^2 \varepsilon\right) - \sigma_E \cdot \tan^2 \varepsilon.$$

However

$$\tan^2 \varepsilon = \frac{1}{\tan^2 \gamma},$$

hence

$$\sigma_I = \sigma_{\min} = \sigma_N \cdot \left(1 + \frac{1}{\tan^2 \gamma}\right) - \sigma_E \cdot \frac{1}{\tan^2 \gamma}$$

and

$$\sigma_I = \sigma_{\min} = \sigma_N - \left(\sigma_E - \sigma_N\right) \cdot \frac{1}{\tan^2 \gamma}.$$

The fundamental requirement (compression) can only be satisfied if σ_N is large and if the term to be subtracted is small (therefore $\tan^2 \gamma$ is large).

As previously mentioned, the simplest geometry of the head, comprising an upstream sealing joint, is subject to tensile stresses under the effect of water pressure on the upstream face. These tensile stresses are a function of the load and cantilever d (see Figure 14.15).

To cancel out these tensile stresses, the most effective geometric propositions place the waterstop further downstream. In other cases, water pressure also acts by compressing the A-B section, which becomes the most likely location for failure. The polygonal head is the simplest option to implement.

To respect the condition, the angle γ must be as large as possible. This condition reduces the compression effect of the A-B section due to the effect of water pressure, which is not advisable. To resolve this apparent contradiction, it is a question of selecting the shape of the head so that the front part is wider than the thickness of the webs. The typical geometrical characteristics of a diamond head are illustrated in Figure 14.19. The stresses on the A-B and K-S sections (Figure 14.18) must be verified, as must the principal stress σ_I at point K, parallel to the inclined plane of the face (principal minimal stress). Figure 14.20 illustrates other shapes.

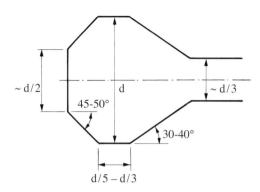

Figure 14.19 Typical proportions for an upstream polygonal head

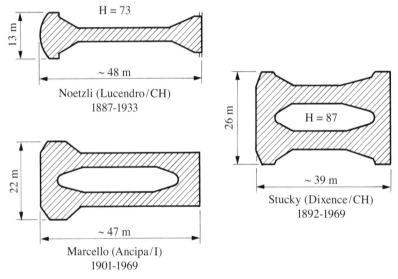

Figure 14.20 Special buttress heads named after their designers

14.3 Overturning and Sliding Safety

14.3.1 Methods of calculation

Verifying the overturning and sliding safety for hollow gravity dams and buttress dams is carried out in the same way as for a gravity dam. They must fulfil the same criteria. If the slopes of the faces have been carefully selected, overturning stability is rarely critical. Sliding stability must be systematically verified. It can be improved by creating an active downstream abutment block with jacks (Figure 14.21) or by installing sloped rock anchors, both of which are expensive operations. A counter-sloped foundation for the buttress also improves the situation.

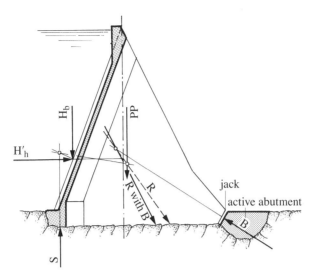

Figure 14.21 Improving the sliding stability of a buttress using a jacked block (from Crespo, 1967)

As for geometry, it is of course the center of gravity for the horizontal sections and the core that must be determined, while taking into account the exact form of the buttress, which, unlike the gravity dam, is no longer a rectangle.

14.3.2 Hypothesis for considering uplift

One of the principal characteristics of a buttress dam is the reduction in uplift, due to the free space left between the buttresses downstream of their heads.

The two vertical sections along the axis of the buttress and between two buttresses can be clearly distinguished. DIN standards have proposed uplift distributions, which are illustrated in Figure 14.22.

14.3.3 Buckling of the buttress web

Verification should check that there is no risk of buckling in the compressed part of the buttress web. Should there be a joint or longitudinal fissure in the buttress web, the latter could be considered as a plate or column.

14.4 Earthquake Behavior

Buttress dams are particularly vulnerable to earthquakes, which is why this type of dam structure should be avoided in zones with strong or moderate seismicity. However, even in zones with a low seismic risk, earthquake behavior must be analyzed.

As a rule, the forces due to an earthquake are comparable to those considered for a gravity dam. The calculations must be performed in the transversal (upstream-downstream) and longitudinal direction (left bank-right bank). An approach using the pseudo-static method can be applied during the preliminary design

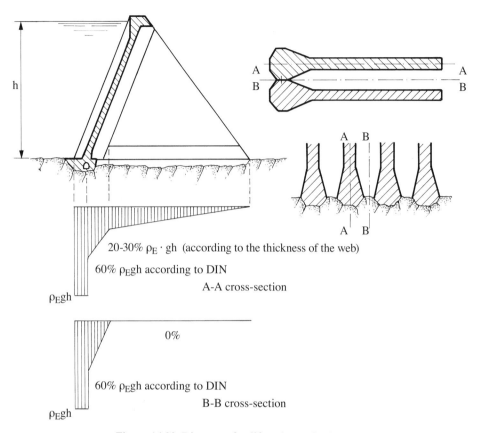

Figure 14.22 Diagram of uplift underneath a buttress

phase. However, owing to the slope of the upstream face, the increase in hydrostatic pressure must be taken into account, for which Zangar (1952) proposed an equation that enables this overpressure to be calculated at each level, depending on the slope of the face.[1] To carry out a dynamic analysis, the finite elements method is used. It is also useful for calculating the dam's dynamic response. Using the time history calculation method as well as the response spectrum method, is also possible.

To simplify the description of the various phenomena, we will outline two cases: along the longitudinal and transversal directions of loading relative to the dam axis.

14.4.1 Transversal loading

The upper part of the buttress head is most threatened by vibrations, due to the massive nature of this zone of the crest. Depending on the selected shape, large earthquakes can cause horizontal cracks in the concrete (Figure 14.23).

---

[1] This equation and the curves are presented in the book *Design of Small Dams* (1987), published by the Bureau of Reclamation.

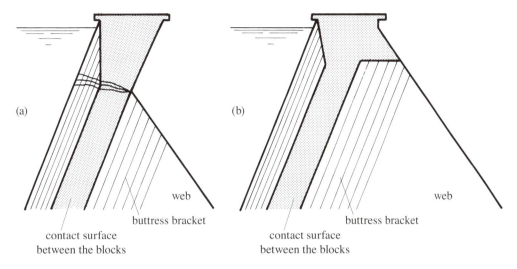

Figure 14.23 Crest: (a) traditional solution; (b) preferable solution

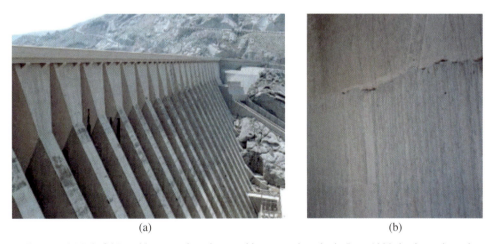

Figure 14.24 Sefid Ruud buttress dam damaged by an earthquake in June 1990: horizontal crack at the level of a concreting lift joint

14.4.2 Longitudinal loading

Acceleration in the longitudinal direction is most critical as buttress stability in this direction is very low.

Several measures help stabilize the buttress, by preventing or limiting the ability of a block to shift toward its neighbor:

- Implementation of a downstream facing by widening the downstream part of the buttresses
- Implementation of a joining foundation slab between the buttresses (which implies specific measures to guarantee drainage of the foundation) or continuous braces
- Implementation of buttress struts between the blocks, similar to those in Figure 14.2

14.5 The Effects of Temperature

During construction, buttress dams have large formwork surfaces which allow simple natural cooling of the concrete. Artificial cooling measures are not normally necessary.

However, variations in temperature play a major role in the state of internal stresses in the dam under normal loading. Buttresses are relatively thin structures, and so they react immediately to variations in temperature. In cases of extreme climatic conditions, temperature gradients between the head and the buttress web can be considerable. Substantial tensile stresses can therefore arise.

Figure 14.25 (a) illustrates the temperature distribution in concrete on a horizontal cross-section situated along the axis of a dam buttress in an arid climate. The temperature is 15°C on the upstream side, near the face in contact with the reservoir. Downstream, where the web is thin, in contact with the atmosphere, and subject to the sun's rays, the temperature reaches nearly 40°C.

The state of stress and the deformation resulting from this temperature range were calculated using the finite elements method, while assuming a linear elastic material. Tensile stresses develop in the head and on the upstream face. These tensile stresses can reach 1 MPa in the upper part and cause horizontal lift joints to open. Figure 14.25 (b) illustrates a buttress that has warped due to the temperature gradient.

The simplest way to avoid these temperature gradients is to close the space between the buttresses with a downstream facing. This can be done by widening the web (in which case it becomes load-bearing and integrated within the structure) or by installing an insulating, non load-bearing material.

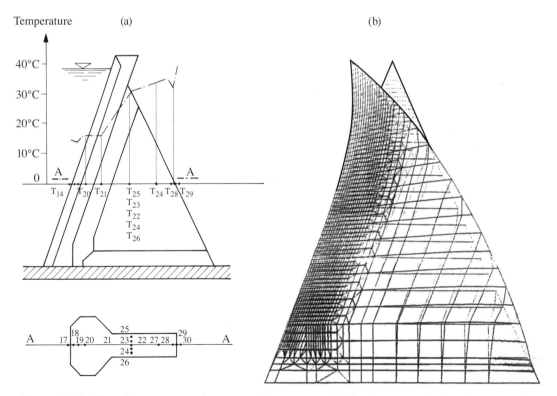

Figure 14.25 Effect of temperature on buttresses: (a) temperature distribution across a horizontal cross-section; (b) the deformation of a buttress under the effect of temperature

14.6 Specific Issues

14.6.1 Buttress foundations

As with any dam structure, the foundation of a buttress dam must also be of a very high quality. To decrease ground compression, it can sometimes be necessary to widen the base of the buttress web. When the valley flanks are steep, the transversal sliding of an individual buttress may occur. This sliding is more likely to occur when layers dip in an unfavorable direction. To avoid this sliding, excavation should always take place in steps (Figure 14.26 (a)).

If the stability of an individual block is not guaranteed along a transversal axis, several actions may be taken:

- Installation of beams or braces between the buttresses (Figure 14.26 (b))
- Widening of buttress web toes until they join up (Figure 14.26 (c))

In this latter case, specific measures must be taken to prevent uplift developing underneath the web foundations. These measures may include drainage boreholes (which require regular monitoring) or longitudinal galleries leading to the downstream face.

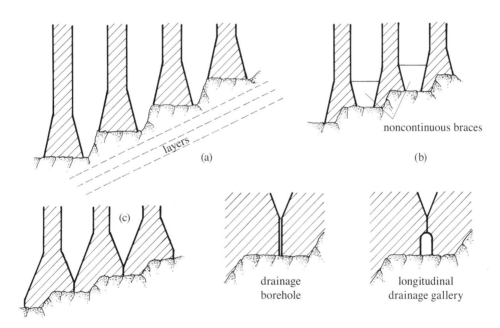

Figure 14.26 Arrangement for the buttress foundation

14.6.2 Construction stages

As with gravity dams, concreting stops must be defined so as not to create preferential sliding planes. Concreting lift joints should therefore be parallel to the foundation.

Furthermore, concrete thermal effects and shrinkage are accentuated in buttress dams, due to the large surface area of the blocks. Hydration heat escapes more quickly, and also the concrete drying process is faster. To avoid the risk of cracking, the buttress should be constructed in stages (staggered across several months) to allow the concrete to deform while it is still young (Figures 14.27 and 14.28).

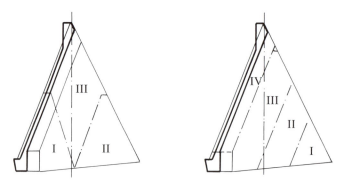

Figure 14.27 Concreting stages for a buttress (from Crespo, 1967)

Figure 14.28 Construction of Al Massira buttress dam (Morocco)

14.6.3 Sealing system

The sealing system between the buttress heads is similar to that of a gravity dam. Two waterstop bands are generally installed, with a drainage well between the two. The upstream waterstop can be monitored by checking the drain (Figure 14.29). It is also possible to seal a possible leak in the upstream band by filling the drain with clay.

At the upstream heel of the dam, the continuity of this system with the grout curtain is ensured by contact grouting and pressure relief boreholes installed from one of the lower galleries (Figure 14.30).

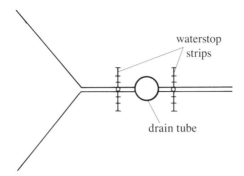

Figure 14.29 Detail of the sealing system between the buttress heads

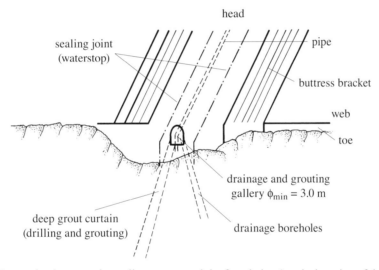

Figure 14.30 Connection between the sealing system and the foundation (vertical section of the upstream toe of buttress head)

15. Arch Dams

Valle di Lei arch dam in Switzerland, height 143 m, year commissioned 1961 (Courtesy A. Siegfried).

15.1 General Shape and Advantages

An arch dam is a dam that is curved horizontally and vertically and that transmits water pressure to the abutments at the valley flanks through arch action. The installation and construction of these types of dams must meet, as we have previously outlined, very strict criteria, both topographical (valley shape) and especially geological (rock quality, stability of abutments), due in particular to the considerable load transferred to the lateral abutments.

When the principal conditions are met, arch dams have several clear advantages compared to other types of dams:

- The volume of concrete is reduced (base thickness is in the order of 15–20% of the height, compared to 75–80% for a gravity dam)
- The price per m³ of concrete, with more cement content, is only slightly higher, as is the price per m² for formwork
- The three-dimensional hyperstatic behavior of the static system results in a high reserve of bearing capacity
- The effect of uplift is reduced, given the foundation's reduced thickness; however, seepage gradient under the foundation is considerable

15.2 Main Types of Arch Dams

Arch dams can be either single or double curvature. The first arch dams built prior to the Second World War were mostly single curvature dams. Since the 1950s (except for a few exceptions), all arch dams have been built with a double curvature.

15.2.1 Single curvature arch dams

A single curvature arch dam is a dam whose profile develops along a curvilinear axis that is identical from the toe to the crest. The dam forms part of a cylinder whose thickness is constant across any given level (Figure 15.1).

This type of dam is suitable for very narrow valleys, where the width is practically the same over the entire height (canyon). For V-shaped valleys, the arches gradually become smaller from the top to the

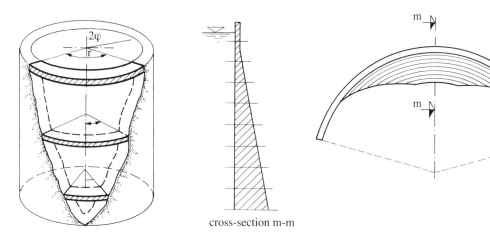

Figure 15.1 Principle of the single curvature arch

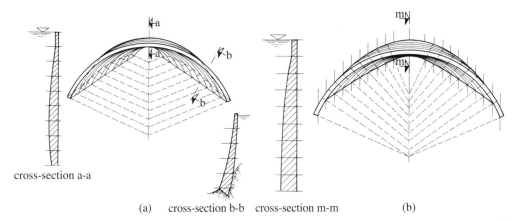

Figure 15.2 Geometric definition of a single curvature arch dam: (a) constant opening angle dam; (b) variable radius and opening angle dam

bottom. As the radius is constant, the bearing capacity of the lower, much shorter arches is low. To address this, the thickness of the arches varies across the height, as illustrated in Figure 15.1.

To avoid lower arches from having a narrow opening, variable radius and constant opening angle arch dams were developed. These types of dams have a more complex geometry and provide improved load transfer to the lateral rock abutments (Figure 15.2). In this case, the lower arches bear much of the load and take up almost all hydrostatic pressure. A compromise was found with variable radius dams and variable opening angle dams, as illustrated in Figure 15.2 (b). By optimizing the geometry of the latter, the double curvature arch dam was finally developed.

15.2.2 Double curvature arch dams

The double curvature arch dam's name comes from the curved section with variable thicknesses of the vertical cross-sections of the dam, as well as the curved section and generally variable thickness of the

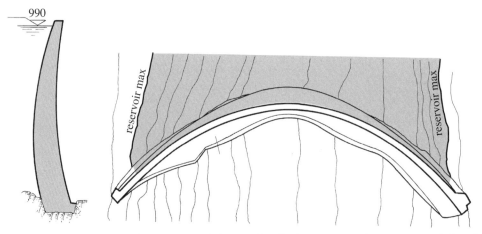

Figure 15.3 Example of a double curvature arch dam: Malvaglia dam, cross-section at crown and plan view (project Prof. A. Stucky)

ARCH DAMS

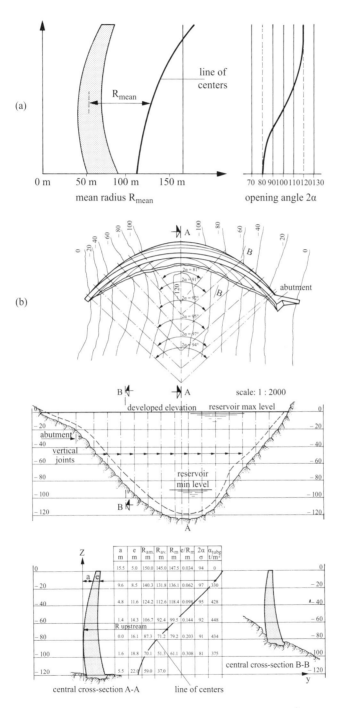

Figure 15.4 Principal geometric characteristics of an arch dam: (a) graph showing the center lines of the mean radius and opening angles of the arches; (b) example of data required to define the cross-section at crown and the arches (from Crespo, 1967)

horizontal sections. The dam's geometry is determined by the distribution of load and stress in the dam and its foundation, under various loading situations (Figure 15.3).

As a rule, the radii and their opening angles decrease from the top to the bottom, as shown in Figure 15.4. The geometry of the arches comprises

- Circular arches
- Parabolic segments
- Elliptical arches
- Logarithmic spiral segments

15.3 Choosing the Initial Shape

15.3.1 Dam height

The crest level of a dam, no matter the type of structure, is determined firstly by the topography and geology and secondly by economic considerations.

The optimum height is obtained by analyzing the costs and revenue associated with the water-retaining facility (that is, the reservoir), while also considering environmental aspects.

The costs of the dam include:

- Construction costs, interim interest accrued during construction, and corresponding financial depreciation
- Interest on the debt
- Operating and rehabilitation costs
- Tax, charges, and duties

To assess revenue, the various aims of the water-retaining facility need to be considered:

- Energy production (the main objective in Switzerland, but often secondary in dry countries)
- Improving agricultural productivity (irrigation)
- Risk reduction (for dams used as flood protection)
- Water supply for local communities, etc.

Quantifying revenue is often complex. Hydrological uncertainty must be taken into account, for example, by simulating production across a number of representative years. Recent experience has shown that the medium-term projection of expected returns is difficult. Several scenarios must be envisaged, and a sensitivity analysis of the principal parameters is necessary.

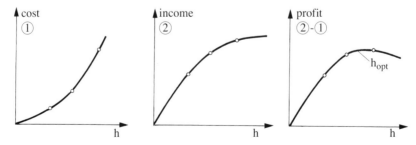

Figure 15.5 Optimizing dam height

Profit is the difference between income and expenses. The relationship between the reservoir level and profit can therefore be established. The optimum height is that which corresponds to the most profit, while, of course, respecting the limits determined by the topography and geology.

15.3.2 Shape of horizontal sections

15.3.2.1 Circular arches

Identifying the most appropriate choice of geometry for a double curvature arch dam is an iterative process. As a first step, we assume that the horizontal sections (the arches) are formed by circular arches.

Figure 15.6 illustrates the main descriptive parameters for arches. The preliminary geometry must satisfy the following conditions:

- The curvature radius decreases continuously from the top to the bottom
- At the crest level, the center opening angle 2α is in the order of 120 to 130°
- At mid-height, the center opening angle 2α is greater than 85°
- At the base, the center opening angle 2α is approximately equal to 80°; this condition is relatively easy to satisfy in V-shaped valleys, but much more difficult in U-shaped valleys
- The arch/rock angle of incidence β_a must be at least or greater than 30°; this essential criterion is sometimes difficult to meet and must be assessed while taking the geology into account (cracks, strike and dip).

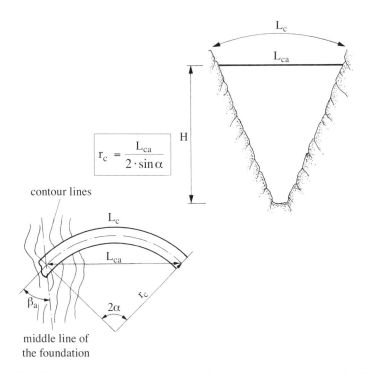

Figure 15.6 Geometric definitions for the arches: 2α = arch opening angle, β_a = arch/rock angle of incidence, r_c = arch radius, L_c = developed arch length, L_{ca} = chord length

Several solutions are available to compensate for this shortfall in the angle of incidence, by correcting the initial choice of the constant radius on the arch:

- While retaining the circular radii, modify the radius by maintaining a continuous relationship between the radius and the height; uptake of the lateral load can often be improved by choosing larger radii, to the detriment of the opening angle
- By waiving the definition of circular arches, several other solutions are possible (described in § 15.3.2.2)

15.3.2.2 Parabolic and elliptical arches or arches with logarithmic spiral segments

Introduction to the arch geometry toward the abutment

With the aim of improving the arch/rock angle of incidence, arches toward the abutment normally have a radius greater than the arch's main radius (three-centered arch) (Figure 15.7(a)). By choosing a radius smaller than the main radius, an arch thickness that gradually increases toward the arch springing is obtained (Figure 15.7(b)). This solution introduces a discontinuity in the arch curvature, which is less favorable from a stress perspective. Often this discontinuity can be seen with the naked eye on the downstream face of the arch.

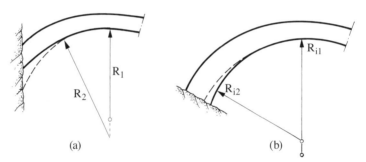

Figure 15.7 Example of arch geometry toward the abutment

Arches comprising parabolic segments

Many arch dams in Switzerland have been designed with parabolic arches. This solution has three major advantages:

- *Strength*
  For highly open profiles, the calculations show that water pressure against the arch is lower near the springing than it is at the arch crown (an effect due to the proximity of the foundation). The shape of the parabola can take this effect into account.
- *Topography*
  The upstream and downstream faces are often defined by two parabolas whose osculating circles have the same center at the summit. This layout causes the thickness to gradually vary toward the arch springing and therefore adapts well to dissymmetrical valleys.
- *Geology*
  Sometimes during excavation, the foundation has to be deepened locally (if the geological expectation has not been verified). The parabolic shape of the arches enables the arch to be extended into the foundation without overly modifying the direction of the resultant force. With circular arches, this deepening almost always indicates a new geometric definition for the overall arch, as the condition $\beta_a \geq 30°$ is no longer respected.

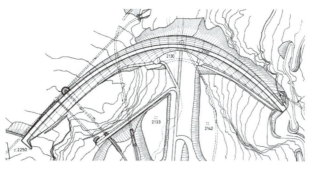

Figure 15.8 Moiry dam: parabolic arches with a wing gravity wall on the right bank extending the arch continuously (Project Prof. A. Stucky) (drawing from SwissCoD, 1964)

Degree 2 parabolic arches are entirely defined when the position of the crown (at the parabola summit) and the springing is fixed, leaving no freedom for the arch section to be adapted to the stress fields. Parabolas toward the abutment are often used to add a degree of freedom to the geometric definition of the arches (Figure 15.8).

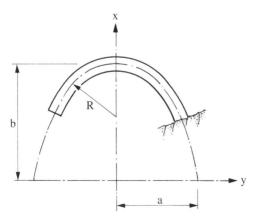

geometrical definition

$$\frac{x^2}{a^2} + \frac{y^2}{b^2} = 1$$

local radius of curvature

$$R = \frac{1}{a \cdot b}\left[\frac{x^2 \cdot b^2}{a^2} + \frac{y^2 \cdot a^2}{b^2}\right]^{3/2}$$

Figure 15.9 Definition of elliptical arches

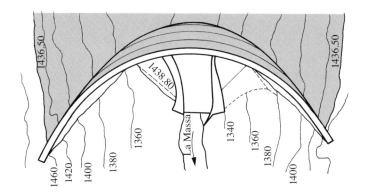

Figure 15.10 Gebidem dam: example of a dam with elliptical arches (Project H. Gicot)

Arches comprising elliptical arches
Elliptical arches have an additional degree of freedom compared to parabolas. The radius of the osculating circle can be selected at the crown or the angle of incidence to the springing, without having to employ arches made up of several segments of curves. In practice, the radius of curvature at the springing is 2 to 3 times greater than at the crown (Figure 15.9; Figure 15.10).

Arches with segments of logarithmic spirals
Spiral arches have the same advantages as parabolic and elliptical arches. However, their geometric definition is more complex (Figure 15.11). The curvature radius varies proportionally with the curvilinear distance to the arch crown. As a rule, the curvature radius at the springing is 1.5 to 3 times greater than at the crown.

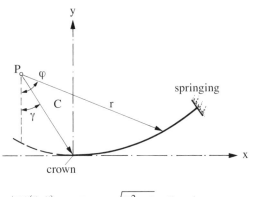

$r = C \cdot e^{\tan \gamma (\varphi - \gamma)}$ $R = r \cdot \sqrt{C^2 + 1}$ (local curvature radius)

Figure 15.11 Definition of arches with segments of logarithmic spirals

Figure 15.12 McKays Point dam in California: example of an arch dam with segments of logarithmic spirals

15.3.3 Shape of the vertical sections

The vertical section of an arch dam is called a cantilever (sometimes also the wall). The crown cantilever is located at the dam axis. The process of geometrically defining the crown cantilever must occur simultaneously with the establishment of the horizontal sections, the arches. As illustrated in Figure 15.13, the horizontal and vertical sections are not independent. The process is iterative, and its conclusion must respect the conditions established for the arches and cantilevers.

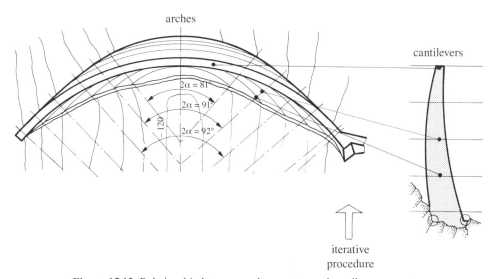

Figure 15.13 Relationship between arch geometry and cantilever geometry

Cantilever geometry seeks to satisfy the two following conditions:
- With a full reservoir, no vertical tensile stresses must develop on the upstream face
- With an empty reservoir, the tensile stresses on the downstream face are limited to $|\sigma_{av}| \leq 1/4\sigma_{am}$

The first criterion is satisfied if self weight counteracts the tensile stresses created by hydrostatic pressure. At the upstream heel of the dam, this condition cannot be satisfied if self weight creates an upstream bending moment, reinforcing compressive stresses.

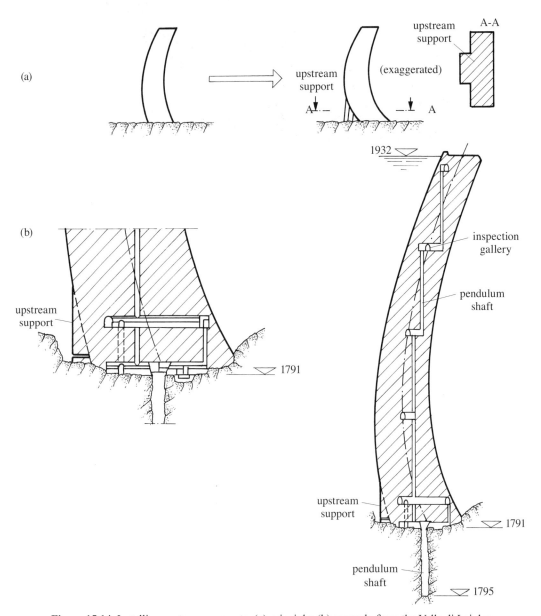

Figure 15.14 Installing upstream supports: (a) principle; (b) example from the Valle di Lei dam

This condition contradicts the second criterion, because if the cantilever slopes too much upstream, tensile stresses appear at the downstream toe of the dam under the sole effect of self weight (empty reservoir). A commonly employed method for limiting these tensile stresses is to install an upstream heel support (or crutch) at the heel of the upstream face (Figure 15.14).

Bear in mind that arch dams are built as a series of juxtaposed and independent vertical blocks, whose width is limited to 16 or 18 m in order to avoid shrinkage cracks. The length must also be limited and can reach 30 to 35 m. At the Mauvoisin dam, whose base is 54 m, a stepped longitudinal annular joint was installed. The height of the lifts is generally 3.2 or 1.5 m. The form of the joints is helical so that the various elements of the dam interlock together. A ½ grade per meter can be reached without causing any difficulties in the installation of formwork.

The cantilevers are only secured together at the end of construction by cement grout injected into the joints separating the blocks. This operation, which makes the arch hyperstatic, is known as joint closure or joint grouting.

However, during the construction phase, each block that will be part of the dam after grouting must be stable on its own and independent of adjacent blocks.

The stability of blocks near the crown, as well as at the dam abutment sides is generally easily guaranteed. However, the two blocks on either side of the crown are often more problematic (Figure 15.15).

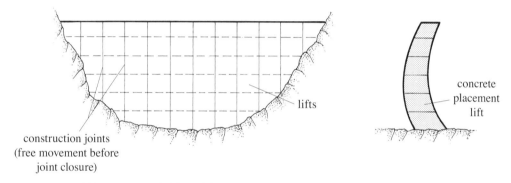

Figure 15.15 Division of the arch dam into blocks and construction stages

Most often, the solution is to correct the arch and cantilever geometry, which implies a further iteration in the process of defining the geometry.

The intermediary phases of construction are often very important, especially if the cantilevers are highly curved (Figure 15.16).

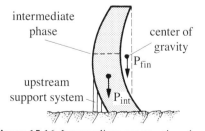

Figure 15.16 Intermediary construction phase

15.3.4 Thickness of the cantilever at the crown

There are several approaches for the preliminary design of the thickness of the cantilever at the crown.
In Switzerland, the following simple rules are commonly used, based on the sizes defined in Figure 15.17:

	Wide valley (U-shaped)		Narrow valley (V-shaped)
Crest thickness d_c	$d_c = H / 15$		$d_c = H / 20$
Base thickness d_b	$d_b = L_c / 20$	(in exceptional cases $L_c / 30$)	$d_b = L_c / 15$

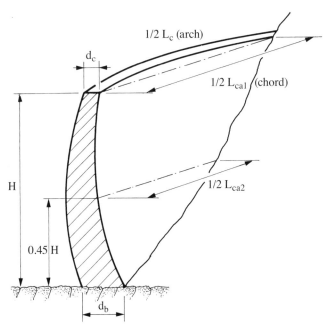

Figure 15.17 Cross-section of a cantilever at the crown. Key: H = static height of the dam; L_c = developed length of the crest; L_{ca} = chord length of an arch; d_c = crest thickness; d_b = thickness at the base.

Empirical equations established on the basis of a statistical analysis carried out by the US Bureau of Reclamations define the thickness across 3 levels of the crown section, that is, from the crest d_c to the base d_b and at a level equal to 0.45 of the height (Figure 15.17). The suggested equations result in a very thick base and a relatively narrow crest:

$$d_c = 0.01 \cdot (H + 1.2 \cdot L_{ca1})$$

$$d_{0.45} = 0.95 \cdot d_b$$

$$d_b = \left[0.0012 \cdot H \cdot L_{ca1} \cdot L_{ca2} \cdot \left(\frac{H}{122} \right)^{(H/122)} \right]^{1/3}.$$

ARCH DAMS

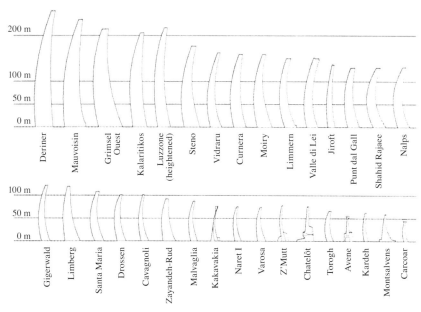

Figure 15.18 (a) Cross-sections at the crown of various arch dams designed by Swiss engineering firms (from Stucky SA)

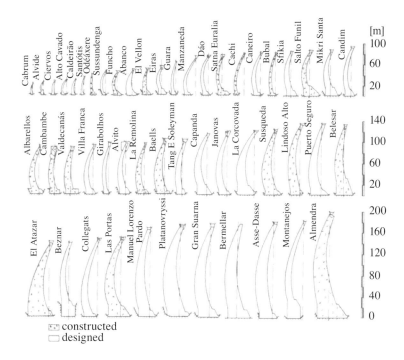

Figure 15.18 (b) Cross-sections at the crown of various arch dams designed by Portuguese and Spanish engineering firms (from Serafim, 1989)

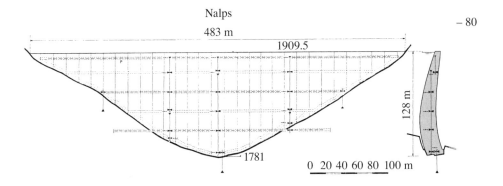
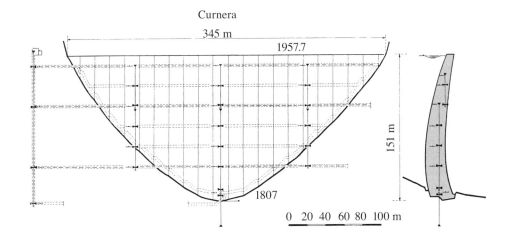
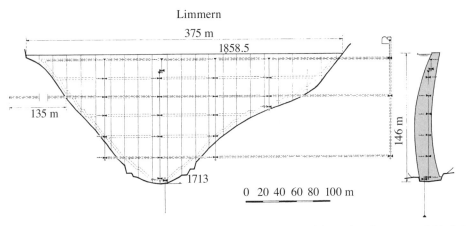

Figure 15.19 Cross-sections at the crown and longitudinal cross-sections of the Nalps, Curnera, and Linth-Limmern dams (from SwissCoD, 1964)

If the crest must include a road designed for the passage of road traffic – as is the case in Switzerland, for example, on the Rossens and Schiffenen dams (FR), its dimensions will need to take the standardized road width into account (Gicot, 1985). It should also be noted that the criteria corresponding to protection requirements in cases of armed conflict may dictate minimum dimensions and drawdown modes for the reservoir, which will need to be respected.

Figures 15.18 (a) and (b) illustrate various dam profiles, organized by height, mostly designed by Swiss (a) and Portuguese and Spanish (b) engineering firms (Serafim, 1989; Serafim et al., 1990). Figure 15.19 compares three dams built in Switzerland, showing the cross-section at the crown and the longitudinal cross-section of the valley.

15.3.5 Slenderness coefficient

The definition of the slenderness coefficient for an arch dam was proposed by Dr. Giovanni Lombardi (world-renowned Swiss expert on dams and Doctor honoris causa of the École polytechnique fédérale de Lausanne, EPFL).

The slenderness coefficient is defined by the equation:

$$C = \frac{S^2}{V \cdot H},$$

where
- C = slenderness coefficient
- S = average surface of dam
- V = concrete volume
- H = height on the foundation

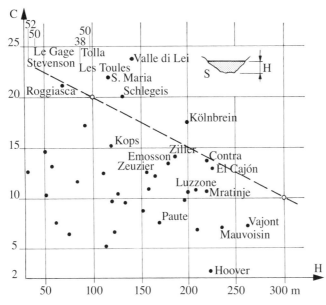

Figure 15.20 Relationship between slenderness coefficient C and height H for some existing arch dams (Lombardi, 1988)

Lombardi demonstrated that for a very slender arch dam, shear forces become very high and can lead to cracking in the concrete.

Given current technological advances, a slenderness coefficient equal to 20 is adequate and a value of 15 should not be an issue.

15.3.6 Stability of the rock abutments

Given the considerable loads that can be transferred to them and the presence of water, the stability of the rock abutments is vital for the overall safety of an arch dam. It is worth paying particular attention to this aspect by carrying out a specific verification. First of all, a survey of the structural discontinuity systems of the rock mass (joints, diaclases) allows identification of the large plane areas that may contain potential weaknesses. A plane, or even the intersection of two or several planes may define a zone (or a wedge) with a risk of sliding should forces be applied to it. The failure mode is given by the orientation of cracks or stratification. Once the boundary of the zone has been established, the engineer can calculate the volume and self weight of the mass that may be prone to movement.

The forces that come into play in the stability analysis are the self weight of the zone in question, the reactions of the dam (from static or dynamic calculations), uplift due to hydrostatic forces acting on the outer limits of the zone, and dynamic forces created by the accepted design earthquake. Calculations should be made for cases of normal, exceptional, and extreme loading.

Various failure modes taken from rock mechanics also allow useful verifications to be made. A method developed by Pierre Londe is one solution if the discontinuity system is well defined (Figure 15.21).

As a possible measure, a drainage system (gallery, drainage holes) or anchors may be installed (see Section 30.5, Strengthening Abutments).

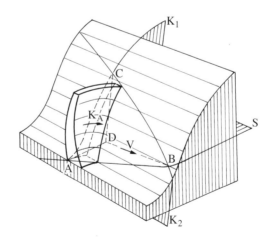

Figure 15.21 Londe method (calculating the wedge). Key: K_1, K_2 = joints, diaclases; S = schistosity, stratification; K_A = force transmitted by the dam to the abutments; A-B-C-D = possible failure model (wedge); V = direction of sliding.

15.4 Methods of Calculation

15.4.1 General information

Although engineers have been building arch dams for almost 90 years, the static calculation of such a complex structure is still continually evolving. Today, the finite element method is used systematically. Current developments aim to integrate into the calculations the strongly nonlinear behavior of the material (the concrete), the structure (including joint closure, concreting lift surfaces, cracks, etc.), and the nonhomogeneous rock foundation, whose geomechanical characteristics may vary significantly.

Lastly, the finite element method also enables study of the dynamic behavior of the structure under seismic loading.

However, before reaching the stage of static dam verification using finite elements, more simple approaches can be applied in the preliminary design of the structure, resulting in a satisfactory geometry with a minimum of effort. At this stage in the design, it is therefore possible to gain the best possible understanding of the load-bearing system and have all the initial elements necessary for validating the calculations through finite elements. Furthermore, it can also be useful to draw on preliminary design from existing dam structures.

In this respect, we will develop an approach based on an analogy with the membrane, followed by a simplified version of the trial-load method (see § 15.4.3.7). All arch dams constructed up until the 1980s, and in particular the large arch dams built in Switzerland, were designed using this method. Today, with the computer technology available, its use can only be justified for the preliminary design phase.

15.4.2 Tube formula, membrane

The tube formula, which treats the arch dam as a membrane, can be used for the very first estimation of loads and stresses in the arches.

The membrane theory disregards arch dam bending resistance. Stress distribution in the arches is therefore uniform.

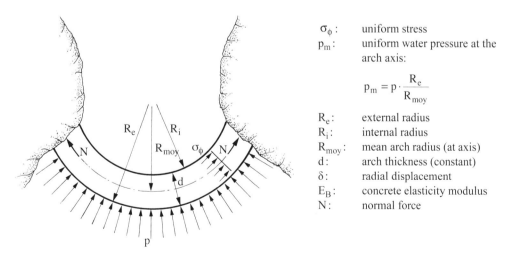

σ_ϕ : uniform stress
p_m : uniform water pressure at the arch axis:

$$p_m = p \cdot \frac{R_e}{R_{moy}}$$

R_e : external radius
R_i : internal radius
R_{moy} : mean arch radius (at axis)
d : arch thickness (constant)
δ : radial displacement
E_B : concrete elasticity modulus
N : normal force

Figure 15.22 Arch dam with circular arches

The dam is subdivided into independent horizontal segments (Figure 15.22), and these arches are deemed as being free to deform along a horizontal plane. As a further simplification, it is accepted that these arches have a constant thickness and that their radii are constant (circular arch).

This approach is clearly not exact, as it overlooks all transfer of load through shearing, but it does have the advantage of giving an initial sense of the required dimensions.

For the first evaluation, it is accepted that the arches are not embedded.

The uniform stress in the arch is therefore obtained using the tube formula:

$$\sigma_\phi = N/d = p_m \cdot R_{moy}/d.$$

Radial displacement becomes

$$\delta = 1/E_B \cdot R_{moy} \cdot \sigma_\phi.$$

If this simple method is used for the preliminary design of an arch dam, the maximum acceptable stresses in the middle section of the dam (Figure 15.23) should not exceed $\sigma_\phi \leq 3.5 \div 4.0$ MPa.

As a rule, by carrying out more detailed calculations, the maximum stresses on the downstream face of an arch dam must be lower than $7 \div 8$ MPa (compression), and no tensile stresses must appear on the upstream face. These requirements are most often met if the acceptable values indicated above using the tube formula are considered.

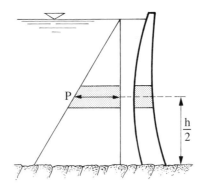

Figure 15.23 Definition of the middle crown section

15.4.3 Trial-load method

15.4.3.1 General Information

The trial-load method enables a more precise preliminary design of the arch dam, as its bearing capacity is analyzed in two directions. Before calculations using finite elements was developed, this method was systematically used for detailed calculations. It was developed in the 1930s in the United States. Until the appearance of the first computers, considerable human resources were required to carry out the calculations.

15.4.3.2 Beam system

The basic principle that enables the method to be implemented is the replacement of the three-dimensional shell structure that comprises the arch dam with an orthogonal system of beams, made up of horizontal arches and vertical cantilevers. The arches and cantilevers are considered to be partially embedded into the foundation (depending on the deformability of the foundation rock, characterized by E_R). Figure 15.24 provides an illustration of this model.

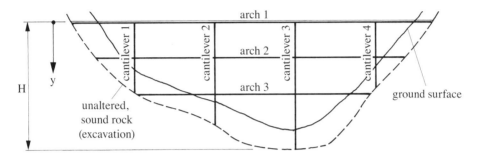

Figure 15.24 Model of a dam with three arches and four cantilevers

The trial-load method presupposes a distribution of water pressure onto the horizontal arches and vertical cantilevers. The result is elastic deformation of the arches and cantilevers across the three dimensions (6 degrees of freedom). This deformation must be equal at each intersecting point of the arches and cantilevers. If the deformation of the cantilevers does not coincide with that of the arches, then the distribution of hydrostatic pressure between arches and cantilevers must be adapted. This gave rise to the name of the method Trial-Load.

In many cases of preliminary design, a symmetrical model comprising three arches and one cantilever provides satisfactory results through simply adjusting radial deformation only.

To illustrate this method and enable its rapid application, a few choices and hypotheses have been made in the following sections.

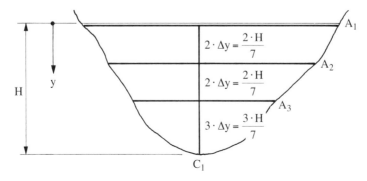

Figure 15.25 Preliminary design with three arches and one cantilever

Shape of the arches and layout of the model

The arches are assumed to be circular and of a constant thickness. This hypothesis simplifies in the following sections the calculation of the stresses and deformation of the arch but is not necessary for the trial-load method.

In most cases, the arches are of a variable thickness and their shape is parabolic, elliptical, or even composed of spiral segments. The choice of three arches and their dam layout from top to bottom in Figure 15.25 issues from the following reasonable compromise:

Arch	Depth y
A1	0
A2	$2\Delta y = 2H/7$
A3	$4\Delta y = 4H/7$

The first arch is at the level of the crest, the second at a depth of $2H/7$ and the third at $4H/7$.

Position of the cantilever

If the valley is almost symmetrical, the cantilever (or wall) is located where the valley is at its highest. If the valley is markedly asymmetrical, the position of the cantilever will be chosen by compromising between the position of maximum height and the axis of symmetry of the crest arch.

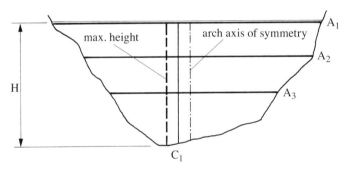

Figure 15.26 Location of the cantilever for an asymmetrical valley

15.4.3.3 Distribution of water pressure between the arches and cantilevers

Water pressure acting on the arch dam is distributed between the three arches and the cantilever. It can be broken down into two parts, one that loads the arches and the other that loads the cantilever:

$$p_E = p_A + p_C = \gamma_E \cdot y,$$

where

p_E = water pressure in position y
p_A = proportion of p_E taken up by the arch
p_C = proportion of p_E taken up by the cantilever

and $\gamma_E = \rho_E \cdot g$.

Experience has shown that the distribution of water pressure on the cantilever may almost be treated as a parabolic function of y (Figure 15.27). However, this distribution can also take on other forms depending on the valley shape, dam characteristics (thin or thick arch), and the deformability of the rock (Figure 15.28).

At the three intersecting points between the arches and the cantilever, this function of the cantilever load takes the following values: $\overline{p_1}, \overline{p_2}, \overline{p_3}$

The remaining water pressure is taken up by the arches, which are accepted as being uniformly loaded by the following values: p_{A1}, p_{A2}, p_{A3}.

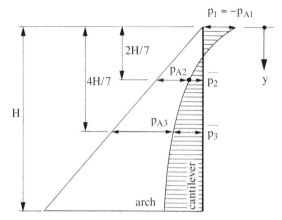

Figure 15.27 Parabolic function of the division of water pressure between the arches and the cantilever (typical distribution shape for the wide V-shaped valley)

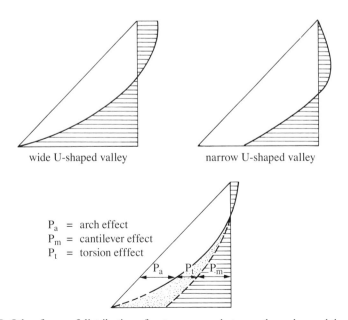

Figure 15.28 Other forms of distribution of water pressure between the arches and the cantilever

As illustrated in Figure 15.27, the repartition of water pressure occurs as follows:

$$\underbrace{\text{Load on cantilever}}_{\text{parabolic function}} + \underbrace{\text{Load on arch}}_{\text{remainder}} = \text{water pressure}$$

$$p_E(1) = p_{A1} + \overline{p_1} = 0 \implies p_{A1} = -\overline{p_1}$$

$$p_E(2) = p_{A2} + \overline{p_2}$$

$$p_E(3) = p_{A3} + \overline{p_3}.$$

The values $\overline{p_1}$, $\overline{p_2}$, $\overline{p_3}$ are located on a parabolic curve in line with the hypothesis outlined above (Figure 15.27).

As the upper arch A1 most certainly takes up a part of positive water pressure, the cantilever load is negative at the intersecting point between this arch and cantilever A_1–C_1, that is, at the point at which the cantilever lies against the arch.

15.4.3.4 Deformation compatibility

The repartition of water pressure between the arches and the cantilever is calculated by applying the condition of radial deformation compatibility to each intersecting point between the two elements (Figure 15.29). The following conditions must be met:

$$\delta_A = \delta_C$$

$$\delta_{A1} = \delta_{C1}$$

$$\delta_{A2} = \delta_{C2}$$

$$\delta_{A3} = \delta_{C3}.$$

For the preliminary design of an arch dam, an adjustment in radial deformation at the insertion points of the arches and cantilever may be sufficient. Apart from an increase in the number of arches and cantilevers, a further improvement consists in adjusting all the deformations at each insertion point of the arches and cantilevers (there are six in total). In practice, however, only the three most important deformations are usually adjusted: radial, tangential in the horizontal direction, and angular to the vertical axis. In the central section of a symmetrical dam, the two latter are null and void (through symmetry). For a simplified system with three arches and one central cantilever, only the compatibility of the radial deformation needs to be checked.

The radial deformation of the cantilever and arches are calculated at three intersecting points by applying the theory of virtual work. The radial deformation is then adjusted by obtaining three equations for the three unknown values:

$$\overline{p_1}, \overline{p_2}, \overline{p_3}.$$

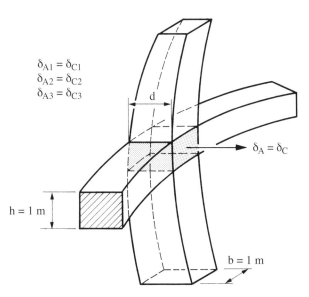

Figure 15.29 Compatibility between radial deformation of the cantilever and the arch at their intersecting point

Deformation of the arch at the center δ_A

For an arch loaded with a uniform radial force, radial deformation at the center (at the crown) is calculated by the equation

$$\delta_A = \beta_i \cdot \frac{p_A \cdot R_{moy}^2}{d \cdot E_B}$$

and

$$\beta_i = f\left(2\alpha, \frac{R_{moy}}{d_A}, \frac{E_B}{E_R}\right),$$

where
- R_{moy} = mean arch radius (at the axis)
- d = arch thickness
- E_B = concrete elasticity modulus
- E_R = deformation modulus of the foundation rock
- 2α = arch opening

As a reasonable approximation for preliminary design, it is accepted that the value of the deformation modulus of the foundation rock is equal to half the concrete elasticity modulus:

$$E_R \approx \frac{1}{2} \cdot E_B.$$

The value β is given in Table 15.30 for partial embedding ($E_R \approx 1/2 \cdot E_B$) and for perfect rigid embedding ($E_R \to \infty$).

364   CONCRETE DAMS

Table 15.30 β values: for $E_R \approx 1/2 \cdot E_B$ and $E_R \to \infty$ (perfect embedding)

E_R	α R_{moy}/d	30°	35°	40°	45°	50°	55°	60°	65°
$E_R = \frac{1}{2} E_B$	3	1.631	2.056	2.432	2.728	2.936	3.067	3.137	3.08
	5	1.781	2.175	2.454	2.623	2.712	2.747	2.749	2.75
	7	1.953	2.267	2.446	2.533	2.564	2.563	2.545	2.55
	10	2.106	2.303	2.390	2.416	2.413	2.396	2.373	2.35
	20	2.182	2.217	2.215	2.200	2.182	2.164	2.146	2.13
	40	2.094	2.084	2.070	2.055	2.043	2.032	2.023	2.01
$E_R \to \infty$	3	0.5037	0.6980	0.897	1.085	1.249	1.388	1.500	
	5	0.7671	1.0227	1.243	1.416	1.545	1.640	1.708	
	7	1.0152	1.273	1.464	1.596	1.686	1.748	1.791	
	10	1.293	1.506	1.641	1.726	1.780	1.816	1.841	
	20	1.677	1.766	1.814	1.842	1.860	1.872	1.881	
	40	1.824	1.851	1.866	1.875	1.881	1.886	1.891	

Deformation of the cantilever δ_C

The deformation of the cantilever at any point y under the effect of horizontal forces transmitted by the arches may be calculated by using the principle of virtual work:

$$\delta_C = \sum \frac{MM'}{EJ} \cdot \Delta y + \sum \frac{QQ'}{GF'} \cdot \Delta y + \delta^*,$$

where

$$E = E_B$$

$$J = \frac{b \cdot d^3}{12} \; ; \text{ for } b = 1 \text{ m} \Rightarrow J = \frac{d^3}{12} \text{ (bending)}$$

$$F' = \frac{5}{6} \cdot b \cdot d \; ; \text{ for } b = 1 \text{ m} \Rightarrow F' = \frac{5}{6} \cdot d \text{ (shear)}$$

$$\frac{E_B}{G_B} = 2 \cdot (1 + \upsilon_B) \; ; \text{ for } \upsilon_B = \frac{1}{6} \Rightarrow \frac{E_B}{G_B} = 2.333$$

δ_C = radial deformation of the cantilever (at any point y)
M', Q' = moment, shear force caused by unit load $P' = 1$ at a given point
M, Q = moment, shear force caused by real loading
F' = reduced section of the rectangular section accepted for the calculation (cf. Figure 15.37)

δ* is the deformation that results from the deformability of the foundation rock. It issues from an upstream-downstream transversal deformation and an angular deformation of the foundation at the toe of the cantilever under the effect of a bending moment and shear force:[1]

$$\delta^* = \delta_R + \alpha_R \cdot (H - y),$$

where
- δ_R = transversal deformation of the rock at the toe of the cantilever
- $\alpha_R \cdot (H - y)$ = deformation at point y through the rotation of the foundation
- H = dam height
- y = calculation point height of the searched deformation

δ_R and α_R are calculated with the help of equations established by Vogt derived from the theory of three-dimensional elasticity:

$$\delta_R = \frac{C_2 \cdot M_R}{E_R \cdot d_b} + \frac{C_3 \cdot Q_R}{E_R} \quad \text{and} \quad \alpha_R = \frac{C_1 \cdot M_R}{E_R \cdot d_b^2} + \frac{C_2 \cdot Q_R}{E_R \cdot d_b},$$

where
- M_R, Q_R = moment, shear force at the toe of the cantilever
- d_b = dam thickness at the toe of the cantilever
- C_1, C_2, C_3 = coefficients without dimensions that depend on Poisson's ratio for the rock µ and the shape of the foundation[2]

As previously mentioned, for a preliminary calculation, $n = E_R / E_B = 1/2$ can be adopted, and for an initial approximation, the following Vogt coefficients can be accepted:

$$C_1 = 5.2, \quad C_2 = 0.8 \quad \text{and} \quad C_3 = 2.0.$$

The deformation of the cantilever at point y is therefore obtained by applying the principle of virtual work using the following equation:

$$\delta_C(y) \cdot E_B = \sum \frac{MM'}{J} \cdot \Delta y + \sum \frac{QQ'}{F'} \cdot \frac{E_B}{G_B} \cdot \Delta y + \delta_R \cdot E_B + \alpha_R \cdot (H - y) \cdot E_B$$

$$\delta_C(y) \cdot E_B = 12 \cdot \sum \frac{MM'}{d^3} \cdot \Delta y + 2.8 \cdot \sum \frac{QQ'}{d} \cdot \Delta y$$

$$+ (H - y) \cdot \left[\frac{10.4 \cdot M_R}{d_b^2} + \frac{1.6 \cdot Q_R}{d_b} \right]$$

$$+ \left[\frac{1.6 \cdot M_R}{d_b} + 4.0 \cdot Q_R \right]$$

---

[1] Furthermore, normal loading causes vertical deformation, and a moment of torsion causes an angular movement in the foundation plane.

[2] Diagrams to determine these coefficients have been provided in *Design of Arch Dams* published by the US Bureau of Reclamation (1977). Some values are given in this chapter in a text box (§ 15.4.3.6).

The moments and shear forces loading the cantilever can be expressed based on the parabolic load distribution = f(y).

Parabolic load distribution can be estimated using the following formula:

$$\overline{p}(y) = \overline{p}_1 + \frac{-3 \cdot \overline{p}_1 + 4 \cdot \overline{p}_2 - \overline{p}_3}{4} \cdot \frac{y}{\Delta y} + \frac{\overline{p}_1 - 2 \cdot \overline{p}_2 + \overline{p}_3}{8} \cdot \frac{y^2}{\Delta y^2}.$$

This formula enables the cantilever loading to be easily obtained in seven horizontal sections divided up the height of the dam, by using the values in Table 15.31:

$$\overline{p}(y) = \mathbf{a_p} \cdot \overline{p}_1 + \mathbf{b_p} \cdot \overline{p}_2 + \mathbf{c_p} \cdot \overline{p}_3.$$

Table 15.31 Parabolic load distribution of a cantilever split into seven horizontal sections

y	a_p	b_p	c_p
$0\Delta y$	1.0000	0.0000	0.0000
$1\Delta y$	0.3750	0.7500	−0.1250
$2\Delta y$	0.0000	1.0000	0.0000
$3\Delta y$	−0.1250	0.7500	0.3750
$4\Delta y$	0.0000	0.0000	1.0000
$5\Delta y$	0.3750	−1.2500	1.8750
$6\Delta y$	1.0000	−3.0000	3.0000
$7\Delta y$	1.8750	−5.2500	4.3750

With $\overline{p}(y)$, shear force Q(y) becomes:

$$Q(y) = \overline{p}(y) \cdot y$$

$$Q(y) = \overline{p}_1 \cdot y + \frac{-3 \cdot \overline{p}_1 + 4 \cdot \overline{p}_2 - \overline{p}_3}{8} \cdot \frac{y^2}{\Delta y} + \frac{\overline{p}_1 - 2 \cdot \overline{p}_2 + \overline{p}_3}{24} \cdot \frac{y^3}{\Delta y^2}.$$

This equation describes the distribution of shear force along the height of the cantilever. Table 15.32 provides the values of reduced shear force $Q(y)/\Delta y$ for a cantilever split into seven horizontal sections:

$$\frac{Q(y)}{\Delta y} = \mathbf{a_Q} \cdot \overline{p}_1 + \mathbf{b_Q} \cdot \overline{p}_2 + \mathbf{c_Q} \cdot \overline{p}_3.$$

The moment M(y) is also calculated based on the shear force Q(y) and the load ( ):

$$M(y) = \int Q(y) \cdot dy$$

$$M(y) = \frac{\overline{p}_1 \cdot y^2}{2} + \frac{-3 \cdot \overline{p}_1 + 4 \cdot \overline{p}_2 - \overline{p}_3}{24} \cdot \frac{y^3}{\Delta y} + \frac{\overline{p}_1 - 2 \cdot \overline{p}_2 + \overline{p}_3}{96} \cdot \frac{y^4}{\Delta y^2}.$$

Table 15.32 Distribution of reduced shear force in a cantilever split into seven horizontal sections

y	a_Q	b_Q	c_Q
$0\Delta y$	0.0000	0.0000	0.0000
$1\Delta y$	0.6667	0.4167	–0.0833
$2\Delta y$	0.8333	1.3333	–0.1667
$3\Delta y$	0.7500	2.2500	0.0000
$4\Delta y$	0.6667	2.6667	0.6667
$5\Delta y$	0.8333	2.0833	2.0833
$6\Delta y$	1.5000	0.0000	4.5000
$7\Delta y$	2.9167	–4.0833	8.1667

Table 15.33 provides the values of the reduced moment $M(y)/\Delta y^2$ for a cantilever split into seven horizontal sections:

$$\frac{M(y)}{\Delta y^2} = \mathbf{a_M} \cdot \overline{p_1} + \mathbf{b_M} \cdot \overline{p_2} + \mathbf{c_M} \cdot \overline{p_3}.$$

Table 15.33 Distribution of reduced moment in a cantilever split into seven horizontal sections

Y	a_M	b_M	c_M
$0\Delta y$	0.0000	0.0000	0.0000
$1\Delta y$	0.3854	0.1458	–0.0312
$2\Delta y$	1.1667	1.0000	–0.1667
$3\Delta y$	1.9688	2.8125	–0.2813
$4\Delta y$	2.6667	5.3333	0.0000
$5\Delta y$	3.3854	7.8125	1.3021
$6\Delta y$	4.5000	9.0000	4.5000
$7\Delta y$	6.6354	7.1458	10.7188

The deformation searched for at each intersecting point is obtained by calculating the loading of the cantilever under a unit load for each point. Table 15.34 gives the loading of the cantilever, that is, the moment and shear force at each intersecting point under a unit load ($\overline{p} = 1$).

By multiplying the values in Tables 15.32 and 15.33 with those in Table 15.34, the expressions $M \cdot M'$ and $Q \cdot Q'$ are obtained. The result of this multiplication can be deduced from Table 15.35:

$$\frac{QQ'(y)}{\Delta y} = \mathbf{a_{QQ'}} \cdot \overline{p_1} + \mathbf{b_{QQ'}} \cdot \overline{p_2} + \mathbf{c_{QQ'}} \cdot \overline{p_3}.$$

368   CONCRETE DAMS

Table 15.34 Distribution of the moment and shear force under a unit load applied to the mean sheet at the arch-cantilever intersection point 1 (y = 0), 2 (y = 2Δy) et 3 (y = 4Δy)

	\multicolumn{7}{c	}{Level of unit load}	
	\multicolumn{2}{c	}{At point 1: y = 0}	\multicolumn{2}{c	}{At point 2: y = 2Δy}	\multicolumn{2}{c	}{At point 3: y = 4Δy}	
y	$\frac{M'}{\Delta y}$	Q'	$\frac{M'}{\Delta y}$	Q'	$\frac{M'}{\Delta y}$	Q'	y
0Δy	0	1	0	0	0	0	0Δy
1Δy	1	1	0	0	0	0	1Δy
2Δy	2	1	0	1	0	0	2Δy
3Δy	3	1	1	1	0	0	3Δy
4Δy	4	1	2	1	0	1	4Δy
5Δy	5	1	3	1	1	1	5Δy
6Δy	6	1	4	1	2	1	6Δy
7Δy	7	1	5	1	3	1	7Δy

Table 15.35 Factor for $Q \cdot Q'(y)/\Delta y$ at the intersection point 1 (y = 0, 2 (y = 2Δy) and 3 (y = 4Δy)

	\multicolumn{9}{c	}{Level of load}
	\multicolumn{3}{c	}{At point 1: y = 0}	\multicolumn{3}{c	}{At point 2: y = 2Δy}	\multicolumn{3}{c	}{At point 3: y = 4Δy}
y	$a_{QQ'}$	$b_{QQ'}$	$c_{QQ'}$	$a_{QQ'}$	$b_{QQ'}$	$c_{QQ'}$	$a_{QQ'}$	$b_{QQ'}$	$c_{QQ'}$
0Δy	0	0	0	0	0	0	0	0	0
1Δy	0.6667	0.4167	−0.0833	0	0	0	0	0	0
2Δy	0.8333	1.3333	−0.1667	0.8333	1.3333	−0.1667	0	0	0
3Δy	0.7500	2.2500	0.0000	0.7500	2.2500	0.0000	0	0	0
4Δy	0.6667	2.6667	0.6667	0.6667	2.6667	0.6667	0.6667	2.6667	0.6667
5Δy	0.8333	2.0833	2.0833	0.8333	2.0833	2.0833	0.8333	2.0833	2.0833
6Δy	1.5000	0.0000	4.5000	1.5000	0.0000	4.5000	1.5000	0.0000	4.5000
7Δy	2.9167	−4.0833	8.1667	2.9167	−4.0833	8.1667	2.9167	−4.0833	8.1667

Before integration, each factor in Table 15.35 must be multiplied with $\Delta y^2/F(y)$, which varies across the height of the cantilever. The sum of the form

… ARCH DAMS …

$$\sum_{y=0}^{7\Delta y} \frac{QQ'}{F} \Delta y$$

is then introduced into the equation of virtual work.

Similarly, the result of the multiplication $M \times M'(y)/\Delta y^3$ can also be deduced from the following tables (Table 15.36):

$$\frac{MM'(y)}{\Delta y^3} = \mathbf{a}_{MM'} \cdot \overline{p_1} + \mathbf{b}_{MM'} \cdot \overline{p_2} + \mathbf{c}_{MM'} \cdot \overline{p_3}.$$

Table 15.36 Factor for $M \cdot M'(y)/\Delta y3$ at the intersection point 1 ($y = 0$), 2 ($y = 2\Delta y$) et 3 ($y = 4\Delta y$)

	At point 1: y = 0			At point 2: y = 2Δy			At point 3: y = 4Δy		
y	$a_{MM'}$	$b_{MM'}$	$c_{MM'}$	$a_{MM'}$	$b_{MM'}$	$c_{MM'}$	$a_{MM'}$	$b_{MM'}$	$c_{MM'}$
0Δy	0	0	0	0	0	0	0	0	0
1Δy	0.3854	0.1458	−0.0312	0	0	0	0	0	0
2Δy	2.3333	2.0000	−0.3333	0	0	0	0	0	0
3Δy	5.9062	8.4375	−0.8437	1.9687	2.8125	−0.2812	0	0	0
4Δy	10.6667	21.3333	0.0000	5.3333	10.6667	0.0000	0	0	0
5Δy	16.9271	39.0625	6.5104	10.1563	23.4375	3.9062	3.3854	7.8125	1.3021
6Δy	27.0000	54.0000	27.0000	18.0000	36.0000	18.0000	9.0000	18.0000	9.0000
7Δy	46.4479	50.0208	75.0312	33.1771	35.7292	53.5937	19.9063	21.4375	32.1562

Before the addition, each factor in Tables 15.34, 15.35, and 15.36 must be multiplied by $\Delta y^4 / J(y)$, which varies across the height of the cantilever. The sum of the form

$$\sum_{y=0}^{7\Delta y} \frac{MM'}{J} \Delta y$$

is then introduced into the equation of virtual work.

NOTES:

- When doing the integration using Tables 15.31 to 15.36, the factors at the point $y = 7\Delta y$ must be multiplied by $\Delta y / 2$ and not Δy, because the size of the interval of integration is $\Delta y/2$ at the intersection with the foundation.
- The width of the cantilever and arches is unitary (1 m), and so loads p, Q, and M must be calculated for a 1 m width.

Lastly, the radial deformation of the cantilever at each point that intersects with the three arches can be expressed as a function of the load $\overline{p}$.

$$\delta_{C1} = f_1(\overline{p_1}, \overline{p_2}, \overline{p_3})$$
$$\delta_{C2} = f_2(\overline{p_1}, \overline{p_2}, \overline{p_3})$$
$$\delta_{C3} = f_3(\overline{p_1}, \overline{p_2}, \overline{p_3})$$

where

$$E_B \cdot \delta_{C1} = a_{11} \cdot \overline{p_1} + a_{12} \cdot \overline{p_2} + a_{13} \cdot \overline{p_3}$$
$$E_B \cdot \delta_{C2} = a_{21} \cdot \overline{p_1} + a_{22} \cdot \overline{p_2} + a_{23} \cdot \overline{p_3}$$
$$E_B \cdot \delta_{C3} = a_{31} \cdot \overline{p_1} + a_{32} \cdot \overline{p_2} + a_{33} \cdot \overline{p_3}.$$

Calculating repartition of water pressure between the arches and the cantilever

Water pressure is split into one part that is supported by the arches and one part that is supported by the cantilever. This repartition is given by the following equations:

$$(\gamma_E = \rho_E \cdot g)$$

$$\overline{p_1} + p_{A1} = p_E(1) = 0$$
$$\overline{p_2} + p_{A2} = p_E(2) = \gamma_E \cdot 2\Delta y \cdot \frac{R_2 + d_2/2}{R_2}$$
$$\overline{p_3} + p_{A3} = p_E(3) = \gamma_E \cdot 4\Delta y \cdot \frac{R_3 + d_3/2}{R_3}.$$

The compatibility of radial deformation at the three intersecting points results in three equations with three unknowns $\overline{p_1}, \overline{p_2}, \overline{p_3}$:

- Intersection point 1 (y = 0):

$$(E_B \cdot \delta_{A1}) = \beta_1 \cdot \frac{p_{A1} \cdot R_1^2}{d_1} = (E_B \cdot \delta_{C1}) = a_{11} \cdot \overline{p_1} + a_{12} \cdot \overline{p_2} + a_{13} \cdot \overline{p_3}$$

$$p_{A1} = -\overline{p_1}$$

$$\left(a_{11} + \frac{\beta_1 \cdot R_1^2}{d_1}\right) \cdot \overline{p_1} + a_{12} \cdot \overline{p_2} + a_{13} \cdot \overline{p_3} = 0$$

- Intersection point 2 (y = 2Δy):

$$(E_B \cdot \delta_{A2}) = \beta_2 \cdot \frac{p_{A2} \cdot R_2^2}{d_2} = (E_B \cdot \delta_{C2}) = a_{21} \cdot \overline{p_1} + a_{22} \cdot \overline{p_2} + a_{23} \cdot \overline{p_3}$$

$$p_{A2} = p_E(2) - \overline{p_2}$$

$$a_{21} \cdot \overline{p_1} + \left(a_{22} + \frac{\beta_2 \cdot R_2^2}{d_2}\right) \cdot \overline{p_2} + a_{23} \cdot \overline{p_3} = \frac{\beta_2 \cdot R_2^2}{d_2} \cdot p_E(2)$$

- Intersection point 3 ($y = 4\Delta y$):

$$(E_B \cdot \delta_{A3}) = \beta_3 \cdot \frac{p_{A3} \cdot R_3^2}{d_3} = (E_B \cdot \beta_{C3}) = a_{31} \cdot \overline{p_1} + a_{32} \cdot \overline{p_2} + a_{33} \cdot \overline{p_3}$$

$$p_{A3} = p_E(3) - \overline{p_3} \qquad (9)$$

$$a_{31} \cdot \overline{p_1} + a_{32} \cdot \overline{p_2} + \left(a_{33} + \frac{\beta_3 \cdot R_3^2}{d_3}\right) \cdot \overline{p_3} = \frac{\beta_3 \cdot R_3^2}{d_3} \cdot p_E(3)$$

The solution to this system of three equations provides the three unknown values ($\overline{p_1}$, $\overline{p_2}$, $\overline{p_3}$).

Calculating shear force and moments in the cantilever
Once the proportion of water pressure supported by the cantilever has been calculated, the distribution of shear force and moment can be calculated using the following equations (parabolic distribution of load):

$$Q(y) = \overline{p_1} \cdot y + \frac{-3 \cdot \overline{p_1} + 4 \cdot \overline{p_2} - \overline{p_3}}{8} \cdot \frac{y^2}{\Delta y} + \frac{\overline{p_1} - 2 \cdot \overline{p_2} + \overline{p_3}}{24} \cdot \frac{y^3}{\Delta y^2}$$

$$M(y) = \frac{\overline{p_1} \cdot y^2}{2} + \frac{-3 \cdot \overline{p_1} + 4 \cdot \overline{p_2} - \overline{p_3}}{24} \cdot \frac{y^3}{\Delta y} + \frac{\overline{p_1} - 2 \cdot \overline{p_2} + \overline{p_3}}{96} \cdot \frac{y^4}{\Delta y^2}.$$

15.4.3.5 Load and stress at the toe of the cantilever

Stresses with an empty reservoir
When the reservoir is empty, the self weight of the arch dam is transferred to the ground through the cantilevers (walls) alone, as the dam is built in independent vertical sections. These sections are only secured together at the end of construction by cement grout injected into the joints. This operation, called joint closure or grouting, makes the structure hyperstatic. During construction, self weight increases gradually, but it always acts on the same horizontal sections. It is therefore logical to calculate the effect of self weight in the same way as for a gravity dam.

The vertical stress at the toe of the cantilever (per meter of width for a section assumed as rectangular) under the effect of self weight with the notations given in Figure 15.37 is calculated by:

$$\sigma_{v,am} = \frac{P}{d} + \frac{6 \cdot M_P}{d^2}$$

$$\sigma_{v,av} = \frac{P}{d} - \frac{6 \cdot M_P}{d^2}.$$

372   CONCRETE DAMS

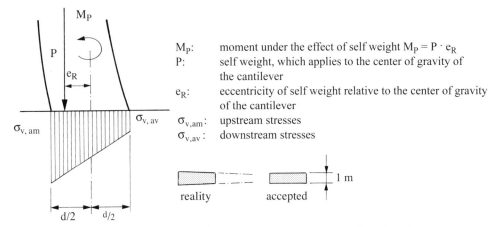

Figure 15.37 Distribution of stresses at the toe of the cantilever under the effect of self weight

Stresses with a full reservoir
As per the notations in Figure 15.38, with a full reservoir the stresses at the toe of the cantilever become:

$$e_R = \frac{M_P - M_E}{P}$$

$$\sigma_{v,am,av} = \frac{P}{d} \pm \frac{6 \cdot P \cdot e_R}{d^2}.$$

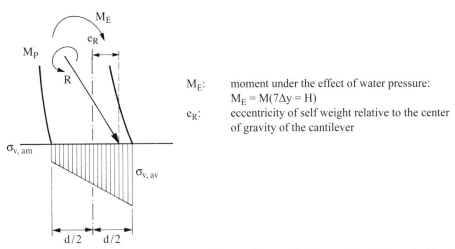

Figure 15.38 Distribution of stresses at the toe of the cantilever under the combined effect of self weight and water pressure

15.4.3.6 Load and stress in the arches
Given the hypothesis that the proportion of water pressure supported by an arch is constant along its length, the moments and normal load can be calculated at the crown and springing of an arch using the following formulas.

Arch crown
Normal load is equal to:

$$N_C(y) = \beta_n \cdot R_{moy} \cdot p_A.$$

Table 15.39 Coefficient β_n for calculating normal load at the crown

α	30°	35°	40°	45°	50°	55°	60°
R_{moy}/d							
3	0.2675	0.3803	0.4957	0.6024	0.6935	0.7672	0.8245
6	0.4912	0.6336	0.7438	0.8212	0.8759	0.9125	0.9374
10	0.7006	0.8109	0.8794	0.9213	0.9473	0.9637	0.9744
20	0.8961	0.9410	0.9647	0.9777	0.9854	0.9901	0.9930
40	0.9709	0.9841	0.9906	0.9942	0.9962	0.9974	0.9982

The moment is equal to:

$$M_C(y) = \beta_m \cdot R_{moy} \cdot d \cdot p_A.$$

Table 15.40 Coefficient β_m for calculating the moment at the crown

α	30°	35°	40°	45°	50°	55°	60°
R_{moy}/d							
3	0.1885	0.2025	0.2030	0.1925	0.1751	0.1544	0.1333
6	0.2047	0.1907	0.1670	0.1414	0.1179	0.0977	0.0809
10	0.1760	0.1457	0.1175	0.0944	0.0762	0.0620	0.0509
20	0.1084	0.0816	0.0626	0.0490	0.0390	0.0316	0.0259
40	0.0567	0.0415	0.0314	0.0245	0.0195	0.0158	0.0130

Arch springing
Normal load is equal to:

$$N_N(y) = \overline{\beta_n} \cdot R_{moy} \cdot p_A.$$

Table 15.41 Coefficient $\overline{\beta_n}$ for calculating normal load at the springing

α	30°	35°	40°	45°	50°	55°	60°
R_{moy}/d							
3	0.3656	0.4922	0.6131	0.7189	0.8030	0.8665	0.9122
6	0.5594	0.6999	0.8037	0.8736	0.9202	0.9498	0.9687
10	0.7407	0.8451	0.9076	0.9444	0.9661	0.9792	0.9872
20	0.9100	0.9517	0.9730	0.9842	0.9906	0.9943	0.9965
40	0.9748	0.9870	0.9928	0.9959	0.9976	0.9985	0.9991

The moment is equal to:

$$M_N(y) = \overline{\beta_m} \cdot \overline{R_{moy}} \cdot d \cdot p_A.$$

Table 15.42 Coefficient β_m for calculating the moment at the springing

α	30°	35°	40°	45°	50°	55°	60°
R_{moy}/d							
3	−0.1059	−0.1337	−0.1510	−0.1569	−0.1534	−0.1434	−0.1300
6	−0.2043	−0.2069	−0.1926	−0.1728	−0.1481	−0.1262	−0.1069
10	−0.2251	−0.1963	−0.1647	−0.1361	−0.1121	−0.0928	−0.0771
20	−0.1700	−0.1318	−0.1026	−0.0816	−0.0653	−0.0528	−0.0441
40	−0.0992	−0.0735	−0.0566	−0.0435	−0.0348	−0.0285	−0.0230

Lastly, the horizontal stresses at the crown and springing of the arches are calculated by:

$$\sigma_{h,am,av} = \frac{N}{d} \pm \frac{6 \cdot M}{d^2},$$

where

$\sigma_{h,am}$ = upstream horizontal stresses
$\sigma_{h,av}$ = downstream horizontal stresses
N = normal load at the crown, with respect to the springing
M = moment at the crown, with respect to the springing

Displacement due to foundation deformation (according to Vogt)

Normal load, shear force, and the bending moment transferred by the dam at the level of the foundation results in

- Vertical settlement
- Upstream-downstream displacement at the toe of the dam
- Rotation in the vertical plane of the embedded section

Vogt established formulas for calculating these deformations due to the unit load acting on the infinite half-space of a homogeneous media.

- Rotation under the effect of bending moment $\alpha' = C_1 \cdot M_R/E_R \cdot d_b^2$
- Vertical settlement under the effect of normal load $\beta' = C_4 \cdot N_R/E_R$
- Upstream-downstream displacement under the effect of shear force $\gamma' = C_3 \cdot Q_R/E_R$
- Rotation under the effect of shear force $\alpha'' = C_2 \cdot Q_R/E_R \cdot d_R$
- Upstream-downstream displacement under the effect of bending moment $\gamma'' = C_2 \cdot M_R/E_R \cdot d_b$

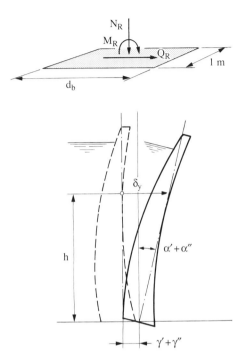

where

- N_R = normal load at the toe of the cantilever
- M_R, M_R^*, Q_R = bending and torsion moments, shear force at the toe of the cantilever
- d_b = dam thickness at the toe of the cantilever
- C_1, C_2, C_3 = coefficients without dimensions that depend on Poisson's ratio μ and the shape of the foundation (relation b/a where b = width of the dam structure and a = thickness of the foundation) (see figure below).

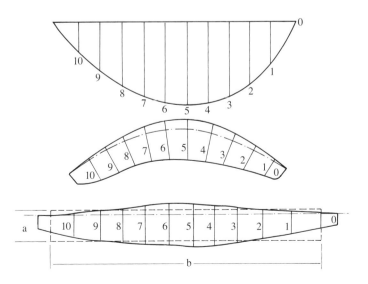

Vertical settlement at the toe is equal to:

$$\delta_V = \beta' = C_4 \cdot N_R/E_R.$$

Transversal displacement at the toe is equal to:

$$\delta_R = \gamma' + \gamma'' = \frac{C_2 \cdot M_R}{E_R \cdot d_R} + \frac{C_3 \cdot Q_R}{E_R}.$$

Rotation in a vertical plane at the toe is equal to:

$$\alpha_R = \alpha' + \alpha'' = \frac{C_1 \cdot M_R}{E_R \cdot d_b^2} + \frac{C_2 \cdot Q_R}{E_R \cdot d_b}.$$

Rotation causes displacements that are linearly proportional to the height on the foundation of the point in question.

If we accept Poisson's ratio $\mu = 0.20$, the coefficients C1, C2, C3 and C4 have the following values:

b/a	C1	C2	C3	C4
2	4,86	0,46	1,38	1,28
4	5,15	0,55	1,77	1,62
6	5,26	0,59	2,00	1,86
8	5,32	0,62	2,17	2,02
10	5,36	0,63	2,30	2,16
15	5,40	0,65	2,54	2,40
20	5,43	0,67	2,72	2,57

NOTE:
- The weight of the reservoir water and the presence of a grout curtain can also have an impact on deformation at the level of the foundation.
- However, the various measurements carried out for the first impoundment and during operation give a clear view of foundation deformation.

By using the finite element method, the foundation characteristics can be defined more precisely than by using Vogt's coefficients.

15.4.3.7 Example of a calculation for an arch dam

To illustrate the calculation for central adjustment, the Punt dal Gall arch dam, in Grisons, is given below as an example.

ARCH DAMS

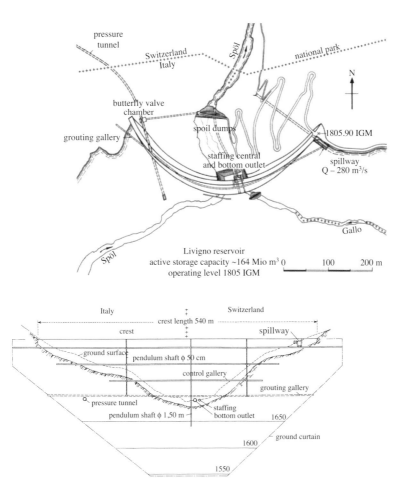

Figure 15.43 Punt dal Gall dam (GR), H = 133 m, L_c = 540 m, L_ca = 450 m; plan view; front elevation of dam and grout curtain; typical cross-section (from SwissCoD, 1985)

Model using an arch-cantilever system

Figure 15.44 brings together the geometric data useful for calculating central adjustment.

Place (position) y [m]	Distance to the reference axis e [m]	Thickness d [m]	Mean radius R_{moy} [m]	Exterior radius R_e [m]	Half-opening angle α [°]	Arch chord L_{ca} [m]	Arch length L_c [m]
0	1.97	10	260	265	60	450	544
19	−3.96	14	245	252	58	415	495
38	−7.99	17	227	235.5	55	372	436
57	−10.86	20	199	209.0	52	313	360
76	−11.86	23	160	171.5	49	242	274
95	−9.71	25	127	139.5	45	180	200
114	−5.62	27	97	110.5	40	125	136
133	0.00	29	27	–	34	30	32

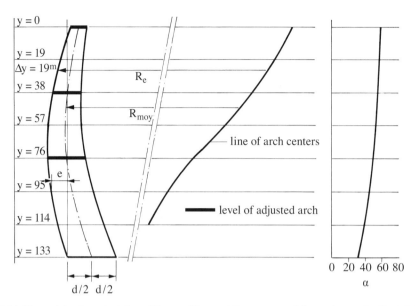

Figure 15.44 Geometric characteristics of the cantilever at the crown and the arches at the Punt dal Gall dam

Radial deformation of the arch at the crown δ_A

With the values β_i in Table 15.30, the radial deformation of the 3 arches at the crown can be calculated using equation (1).

Table 15.45 Radial deformation of the arch at the crown

Arch no.	Position	α	R_{moy}/d	β	$E_A \cdot \delta A$
1	y = 0	60°	26.0	2.10	$14\,196 \times p_{A1}$
2	y = 2Δy	55°	13.4	2.31	$7002 \times p_{A2}$
3	y = 4Δy	49°	7.0	2.57	$2860.5 \times p_{A3}$

Deformation of the cantilever at the level of arch 1 ($y_1 = 0$): δ_{C1}

Initially, the values $J(y)$ and $\Delta y^4 / J(y)$ are calculated for each position y, then multiplied by the factors in Table 15.36.

Table 15.46 Factor for $MM'(y)/\Delta y$ at the intersecting point 1 (y = 0)

y [m]	$J(y)$ [m^4]	$\Delta y^4/J(y)$	$a_{MM'}$	$b_{MM'}$	$c_{MM'}$
0	83.3	1563.9	0	0	0
19	228.7	569.9	219.6	83.1	−17.8
38	409.4	318.3	742.8	636.6	−106.2
57	666.7	195.5	1154.5	1649.3	−164.9
76	1013.9	128.5	1370.9	2741.8	0
95	1302.1	100.1	1694.3	3909.9	651.6
114	1640.4	79.5	2145.5	4291.1	2145.5
133	2032.4	64.1	2978.9	3208.1	4812.1

The values from Table 15.47 are then calculated with the factors in Table 15.35.

Table 15.47 Factor for QQ′(y)/Δy at the intersecting point 1 (y = 0)

y [m]	d(y) [m]	Δy²/d(y)	a_{QQ'}	b_{QQ'}	c_{QQ'}
0	10	36.1	0	0	0
19	14	25.8	17.2	10.7	−2.1
38	17	21.2	17.7	28.3	−3.5
57	20	18.1	13.5	40.6	0
76	23	15.7	10.5	41.9	10.5
95	25	14.4	12.0	30.1	30.1
114	27	13.4	20.1	0	60.2
133	29	12.4	36.3	−50.8	101.7

Table 15.32 in y = 7Δy gives:

$$Q_R = Q(7\Delta y) = 55.4 \cdot \overline{p}_1 - 77.6 \cdot \overline{p}_2 + 155.2 \cdot \overline{p}_3$$

and Table 15.33 in y = 7Δy:

$$M_R = M(7\Delta y) = 2395.4 \cdot \overline{p}_1 + 2579.6 \cdot \overline{p}_2 + 3869.5 \cdot \overline{p}_3.$$

The deformation of the cantilever at point y = 0 is obtained using the principle of virtual work, according to equation (4):

$$\sum \frac{MM'}{J} \cdot \Delta y = 8817.0 \cdot \overline{p}_1 + 14\,915.8 \cdot \overline{p}_2 + 4914.2 \cdot \overline{p}_3$$

$$2.8 \cdot \sum \frac{QQ'}{d} \cdot \Delta y = 305.6 \cdot \overline{p}_1 + 353.3 \cdot \overline{p}_2 + 408.4 \cdot \overline{p}_3$$

$$133 \cdot \left[\frac{10.4 \cdot M_R}{d_R^2} + \frac{1.6 \cdot Q_R}{d_R} \right] = 4346.1 \cdot \overline{p}_1 + 3673.3 \cdot \overline{p}_2 + 7503.0 \cdot \overline{p}_3$$

$$\left[\frac{1.6 \cdot M_R}{d_R} + 4.0 \cdot Q_R \right] = 353.8 \cdot \overline{p}_1 - 168.1 \cdot \overline{p}_2 + 834.3 \cdot \overline{p}_3$$

$$\Rightarrow E_B \cdot \delta_{C1} = 13\,822 \cdot \overline{p}_1 + 18\,774 \cdot \overline{p}_2 + 13\,660 \cdot \overline{p}_3.$$

The same calculation is carried out for the levels of intersection with the other arches.

Deformation of the cantilever at the level of arch 2 (y₂ = 2Δy): δ_{C2}

Using Table 15.36, we obtain the values for Table 15.48.

Table 15.48 Factor for MM′(y)/Δy at the intersecting point 2 (y = 2Δy)

y [m]	J(y) [m⁴]	Δy⁴/J(y)	a_{MM'}	b_{MM'}	c_{MM'}
0	83.3	1563.9	0	0	0
19	228.7	569.9	0	0	0
38	409.4	318.3	0	0	0
57	666.7	195.5	384.8	549.8	−55.0
76	1013.9	128.5	685.4	1370.9	0
95	1302.1	100.1	1016.6	2345.9	391.0
114	1640.4	79.5	1430.3	2860.8	1430.3
133	2032.4	64.1	2127.8	2291.5	3437.2

Using Table 15.35, we obtain the values for Table 15.49.

Table 15.49 Factor for $QQ'(y)/\Delta y$ at the intersecting point 2 ($y = 2\Delta y$)

y [m]	d(y)[m]	$\Delta y^2/d(y)$	$a_{QQ'}$	$b_{QQ'}$	$c_{QQ'}$
0	10	36.1	0	0	0
19	14	25.8	0	0	0
38	17	21.2	17.7	28.3	−3.5
57	20	18.1	13.5	40.6	0
76	23	15.7	10.5	41.9	10.5
95	25	14.4	12.0	30.1	30.1
114	27	13.4	20.1	0	60.2
133	29	12.4	36.3	−50.8	101.7

Table 15.32 in $y = 7\Delta y$ gives:

$$Q_R = Q(7\Delta y) = 55.4 \cdot \bar{p}_1 - 77.6 \cdot \bar{p}_2 + 155.2 \cdot \bar{p}_3$$

and Table 15.33 in $y = 7\Delta y$:

$$M_R = M(7\Delta y) = 2395.4 \cdot \bar{p}_1 + 2579.6 \cdot \bar{p}_2 + 3869.5 \cdot \bar{p}_3.$$

Deformation of the cantilever at point $y = 2\Delta y$ is obtained using the principle of virtual work, according to equation (4):

$$\sum \frac{MM'}{J} \cdot \Delta y = 4581.0 \cdot \bar{p}_1 + 8273.2 \cdot \bar{p}_2 + 3484.9 \cdot \bar{p}_3$$

$$2.8 \cdot \sum \frac{QQ'}{d} \cdot \Delta y = 232.7 \cdot \bar{p}_1 + 283.6 \cdot \bar{p}_2 + 419.4 \cdot \bar{p}_3$$

$$95 \cdot \left[\frac{10.4 \cdot M_R}{d_R^2} + \frac{1.6 \cdot Q_R}{d_R} \right] = 3104.4 \cdot \bar{p}_1 + 2623.8 \cdot \bar{p}_2 + 5359.3 \cdot \bar{p}_3$$

$$\left[\frac{1.6 \cdot M_R}{d_R} + 4.0 \cdot Q_R \right] = 353.8 \cdot \bar{p}_1 - 168.1 \cdot \bar{p}_2 + 834.3 \cdot \bar{p}_1$$

$$\Rightarrow E_B \cdot \delta_{C2} = 8272 \cdot \bar{p}_1 + 11\,012 \cdot \bar{p}_2 + 10\,098 \cdot \bar{p}_3$$

Deformation of the cantilever at the level of arch 3 ($y_3 = 4\Delta y$): δ_{C3}

Using Table 15.36, we obtain the values for Table 15.50.

ARCH DAMS

Table 15.50 Factor for MM'(y)/Δy at the intersecting point 3 (y = 4Δy)

Y [m]	J(y) [m⁴]	Δy⁴/J(y)	$a_{MM'}$	$b_{MM'}$	$c_{MM'}$
0	83.3	1563.9	0	0	0
19	228.7	569.9	0	0	0
38	409.4	318.3	0	0	0
57	666.7	195.5	0	0	0
76	1013.9	128.5	0	0	0
95	1302.1	100.1	338.9	782.0	130.3
114	1640.4	79.5	715.1	1430.4	715.1
133	2032.4	64.1	1276.1	1374.9	2062.3

Using Table 15.35, we obtain the values for Table 15.51.

Table 15.51 Factor for QQ'(y)/Δy at the intersecting point 3 (y = 4Δy)

y [m]	d(y) [m²]	Δy²/d(y)	$a_{QQ'}$	$b_{QQ'}$	$c_{QQ'}$
0	10	36.1	0	0	0
19	14	25.8	0	0	0
38	17	21.2	0	0	0
57	20	18.1	0	0	0
76	23	15.7	10.5	41.9	10.5
95	25	14.4	12.0	30.1	30.1
114	27	13.4	20.1	0	60.2
133	29	12.4	36.3	−50.8	101.7

Table 15.32 in y = 7Δy gives:

$$Q_R = Q(7\Delta y) = 55.4 \cdot \overline{p}_1 - 77.6 \cdot \overline{p}_2 + 155.2 \cdot \overline{p}_3$$

and Table 15.33 in y = 7Δy:

$$M_R = M(7\Delta y) = 2395.4 \cdot \overline{p}_1 + 2579.6 \cdot \overline{p}_2 + 3869.5 \cdot \overline{p}_3.$$

Deformation of the cantilever at point y = 4Δy is obtained using the principle of virtual work, according to equation (4):

$$\sum \frac{MM'}{J} \cdot \Delta y = 1692.3 \cdot \overline{p}_1 + 2899.8 \cdot \overline{p}_2 + 1876.5 \cdot \overline{p}_3$$

$$2.8 \cdot \sum \frac{QQ'}{d} \cdot \Delta y = 155.4 \cdot \overline{p}_1 + 71.6 \cdot \overline{p}_2 + 409.8 \cdot \overline{p}_3$$

$$57 \cdot \left[\frac{10.4 \cdot M_R}{d_R^2} + \frac{1.6 \cdot Q_R}{d_R} \right] = 1862.6 \cdot \overline{p}_1 + 1574.3 \cdot \overline{p}_2 + 3215.6 \cdot \overline{p}_3$$

$$\left[\frac{1.6 \cdot M_R}{d_R} + 4.0 \cdot Q_R \right] = 353.8 \cdot \overline{p}_1 - 168.1 \cdot \overline{p}_2 + 834.3 \cdot \overline{p}_1$$

$$\Rightarrow E_B \cdot \delta_{C3} = 4064 \cdot \overline{p}_1 + 4378 \cdot \overline{p}_2 + 6336 \cdot \overline{p}_3$$

Calculating water pressure
Water pressure at the level of each of the three arches is calculated according to page 370:

$$\overline{p_1} + p_{A1} = p_E(1) = 0$$

$$\overline{p_2} + p_{A2} = p_E(2) = 0.3942 \text{ N/mm}^2$$

$$\overline{p_3} + p_{A3} = p_E(3) = 0.8146 \text{ N/mm}^2.$$

Solving the system of equations
The linear system of three equations to three unknowns $\overline{p_1}$, $\overline{p_2}$, and $\overline{p_3}$ can be established through the equations on pages 370 and 371. Loading on the cantilevers can be calculated at the level of the three arches of calculation:

1. $(13\,822 + 14\,196) \cdot \overline{p_1} + 18\,744 \cdot \overline{p_2} + 13\,660 \cdot \overline{p_3} = 0$
2. $8272 \cdot \overline{p_1} + (11\,012 + 7002) \cdot \overline{p_2} + 10\,098 \cdot \overline{p_3} = 2760$
3. $4064 \cdot \overline{p_1} + 4378 \cdot \overline{p_2} + (6336 + 2860) \cdot \overline{p_3} = 2330,$

which gives as a result:

$$\overline{p_1} = -0.1978 \text{ N/mm}^2$$

$$\overline{p_2} = +0.0723 \text{ N/mm}^2$$

$$\overline{p_3} = +0.3064 \text{ N/mm}^2.$$

Calculating stresses in the arches using the tube formula
Mean loads and stresses in the arches can be quickly calculated using the tube formula.

Table 15.52 Calculating stress in the arches using the tube formula, while taking into account the distribution of arch-cantilever pressure resulting from the adjustment calculation

y [m]	d [m]	$p_E = p \cdot R_a/R_m$ [N/mm²]	$\overline{p}$ [N/mm²]	p_A [N/mm²]	N_A [kN/m]	$\overline{\sigma_{\phi A}}$ [N/mm²]
0	10.0	0	−0.1978	0.1978	51428	5.14
19	14.0	0.1917	−0.0583	0.2500	61242	4.37
38	17.0	0.3867	0.0723	0.3144	71378	4.20
57	20.0	0.5873	0.1939	0.3934	78290	3.91
76	23.0	0.7991	0.3064	0.4927	78840	3.43
95	25.0	1.0237	0.4100	0.6137	77943	3.12
114	27.0	1.2740	0.5045	0.7695	74640	2.76
133	29.0	2.0054	0.5901	1.4154	38215	1.32

Table 15.53 Using the tube formula to calculate stresses in arches deemed independent

Arch no.	p [N/mm²]	d [m]	R_{moy} [m]	R_e [m]	p_m [N/mm²]	N [kN/m]	$\sigma\Phi$ [N/mm²]
0	0.00	10.0	260.0	265.0	0.0000	0	0
1	0.19	14.0	245.0	252.0	0.1954	4788	3.42
2	0.38	17.0	227.0	235.5	0.3942	8949	5.26
3	0.57	20.0	199.0	209.0	0.5986	11913	5.96
4	0.76	23.0	160.0	171.5	0.8146	13034	5.67
5	0.95	25.0	127.0	139.5	1.0435	13253	5.30
6	1.14	27.0	97.0	110.5	1.2987	12597	4.67
7	1.33	29.0	27.0				

Calculating stresses in the arches while taking embedding into account

Using Tables 15.39 to 15.42, the normal loads and moments in the arches, at the crown, and at the springing can be calculated, for the load case from water pressure.

Table 15.54 Stresses in the arches at the crown and the springing

y [m]	dA [m]	$N_{clé}$ [kN/m]	$M_{clé}$ [kNm/m]	$\sigma_{clé,am}$ [N/mm²]	$\sigma_{clé,av}$ [N/mm²]	N_{naiss} [kN/m]	M_{naiss} [kNm/m]	$\sigma_{naiss,am}$ [N/mm²]	$\sigma_{naiss,av}$ [N/mm²]
0	10.0	51148	11330	5.79	4.44	51288	−19424	3.96	6.29
19	14.0	60409	29983	5.23	3.40	60795	−48468	2.86	5.83
38	17.0	69418	62864	5.39	2.78	70254	−96331	2.13	6.13
57	20.0	74616	111190	5.40	2.06	76008	−164123	1.34	6.26
76	23.0	69647	203770	5.34	0.72	72773	−261838	0.19	6.13
95	25.0	58777	306065	5.29	−0.59	64394	−327214	−0.57	5.72
114	27.0	40657	394771	4.75	−1.74	48572	−320868	−0.84	4.44
133	29.0	7151	226053	1.86	−1.37	12441	−82210	−0.16	1.02

Accepted stresses 7 to 8 MPa in accordance with 14.4.2.

Calculating stresses in the cantilever

Stresses in the cantilever due to self weight are recapitulated in Table 15.55(a). Furthermore, stresses in the cantilever due only to water pressure, and also combined with self weight, are presented in Table 15.55(a).

Table 15.55 (a) Stresses in the cantilever due only to self weight

y [m]	d [m]	e [m]	N_{PP} [kN/m]	e_R [m]	$\sigma_{PP,am}$ [N/mm²]	$\sigma_{PP,av}$ [N/mm²]
0	10.0	1.97				
19	14.0	−3.96	5700	2.80	−0.08	0.90
38	17.0	−7.99	13063	4.08	−0.34	1.88
57	20.0	−10.86	21850	4.71	−0.45	2.64
76	23.0	−11.86	32062	4.05	−0.08	2.87
95	25.0	−9.71	43462	1.13	1.27	2.21
114	27.0	−5.62	55813	−2.76	3.33	0.80
133	29.0	0.00	69112	−7.30	5.98	−1.22

Table 15.55(b) Stresses in the cantilever under the effect of water pressure alone and by combining loading from water pressure and self weight

y [m]	d [m]	M_E [kNm/m]	Q_E [kN/m]	$\sigma_{E,am}$ [N/mm^2]	$\sigma_{E,av}$ [N/mm^2]	$\sigma_{E+PP,am}$ [N/mm^2]	$\sigma_{E+PP,av}$ [N/mm^2]
0	10.0						
19	14.0	−27 165	−2418	0.83	−0.83	0.75	0.06
38	17.0	−75 648	−2271	1.57	−1.57	1.23	0.31
57	20.0	−98 291	272	1.47	−1.47	1.02	1.16
76	23.0	−51 217	5039	0.58	−0.58	0.50	2.29
95	25.0	106 197	11 858	−1.02	1.02	0.25	3.23
114	27.0	411 323	20 560	−3.39	3.39	−0.05	4.19
133	29.0	898 312	30 972	−6.41	6.41	−0.43	5.19

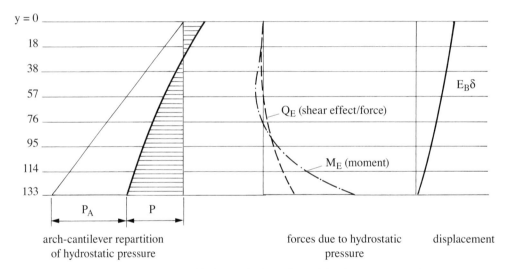

Figure 15.56 Graph with the results from the radial adjustment example

15.4.4 Finite element method (FEM)

The trial-load method enables the preliminary design of an arch dam. However, to verify the operational ability of the dam in the context of final design, the finite element method is most commonly used today. This method allows the body of the dam structure and foundation to be modeled, while also including their specificities. It is also indispensable for analyzing the behavior of the dam under cases of dynamic loading from an earthquake.

As previously mentioned, a more detailed analysis must take into account the following phenomena and characteristics:

- The mass concrete forming the arch has a highly nonlinear and nonelastic behavior (Gunn, 2001)
- The behavior of the structure is also nonlinear, due to the cracking and possible opening of joints

- Construction phases and joint closure must be considered to obtain appropriate distribution of internal loading
- Progression of the temperature field within the concrete is important
- Interaction between the ground and the dam structure must be considered
- Interaction between the reservoir water and the structure affects dynamic behavior, due to the compressibility of water (Hohberg et al., 1992)

The calculation model must therefore be based on a system comprising the dam, its foundation, and the reservoir. For a purely static analysis, only the dam and a band of the foundation rock are necessary. However, care should nevertheless be taken to avoid introducing into the calculation a deformation of the rock mass under the effect of its self weight, which would lead to unfavorable stresses in the arch. If initial stresses are to be allowed for, preliminary loading is necessary.

Figure 15.57 illustrates the model established for calculating the Salam Farsi dam (Iran), a 125-meter-high thick arch.

Initially, the natural frequencies of the first four modes of the structure were determined for an empty reservoir (linear model, Figure 15.58).

On this basis, geometric and structural provisions were taken to optimize the shape and:

- Increase natural frequencies
- Remove asymmetric modes

Recent developments have enabled the dynamic analysis of dams to be carried out while taking certain nonlinear effects into account. Figure 15.59 illustrates a nonlinear model of the same dam, to which four vertical shear joints have been added.

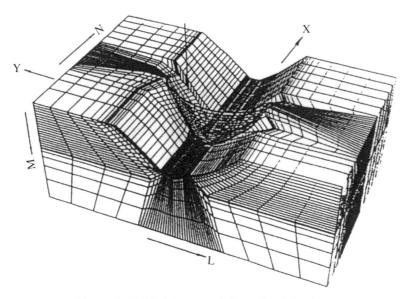

Figure 15.57 Modeling an arch dam using finite elements

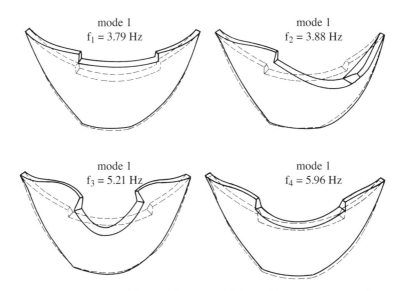

Figure 15.58 Natural frequencies of an arch dam with an empty reservoir

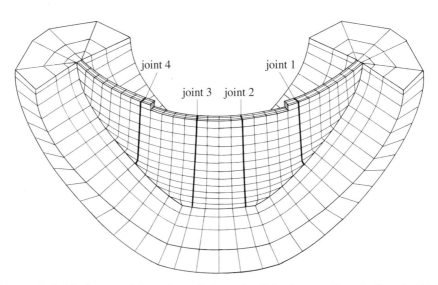

Figure 15.59 Nonlinear modeling of an arch dam using finite elements (from Lofti et al., 1995)

The nonlinear effects issue firstly from the creation of cracks and secondly from the presence of vertical construction joints. Although these joints have been filled with cement grout through closure injections, they undeniably constitute a weak point should an earthquake occur (loss of arch action).

Figure 15.60 illustrates the deformations of the dam from Figure 15.59, while considering the nonlinear effects for different coefficient values of joint shear strength. All the cases in Figure 15.61 illustrate the first oscillation mode, under a dynamic horizontal load whose extreme acceleration reaches $2.0 \cdot g$ in the upstream-downstream direction and with a full reservoir.

Recent models also enable the interaction between the dam and the reservoir to be taken into consideration (Figure 15.61).

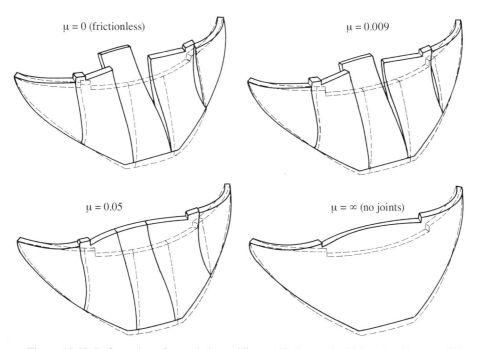

Figure 15.60 Deformation of an arch dam while considering vertical joints (nonlinear model)

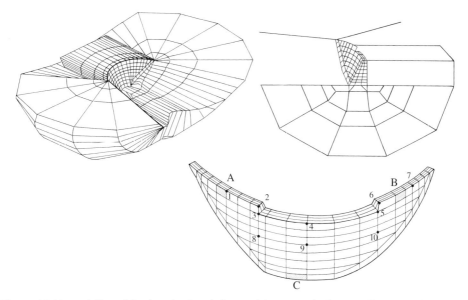

Figure 15.61 Modeling of the dam, its foundation, and the reservoir (from Malla and Wieland, 1995)

15.5 The Effects of Temperature

First of all, it should be noted that there are different periods within which the internal temperature of an arch dam plays a prominent role. The first significant thermal effects appear as a release of heat during the setting or hardening of concrete in the construction phase (see § 13.8.2). Thereafter, natural cooling occurs, which can be artificially accelerated through the installation of a system that circulates cold water (see § 15.7.3). Then, at a time to be determined that takes into account the appropriate thermal conditions and future implications of temperatures on stresses, the joints are closed to create a monolithic structure (see § 15.7.4). Lastly, during operation, ambient temperatures (air, water) have a cyclical effect on dam deformation (see § 32.5.2).

15.5.1 Consequences of meteorological conditions

As arch dams are relatively slender structures, they are subject to exterior meteorological conditions, which have an effect on the temperature conditions inside the dam.

Temperature inside the concrete is affected by:
- Air temperature and wind on the downstream face
- Water or air temperature on the upstream face
- Solar radiation on the downstream face

These three parameters enable the thermal state of the dam (which varies throughout the year) to be determined. Annual variation in air temperature can be determined through daily in situ measurements, which are used to establish mean temperature and amplitudes, as well as to produce a sinusoidal curve equivalent to the variation from this seasonal effect. In the reservoir, water temperature at a certain depth remains almost constant. At the surface, the water temperature varies depending on atmospheric conditions. For an alpine reservoir the temperature of the water in winter is around 2 to 3°C for the upper few meters and 4°C at a depth lower than 10 m. Water reaches its maximum density at a temperature of 4°C and therefore tends to sink to the bottom of the reservoir. Lastly, to account for the effect of solar radiation, air temperatures are increased by a few degrees in summer and decreased by a few in winter.

15.5.2 Hypotheses relative to the internal distribution of temperature[3]

In concrete, these temperature variations are noticeably lower: they are generally out of phase and reduced due to the thermal inertia of mass concrete.

The temperature of the concrete on the downstream face can vary from 5°C to 25°C, due to heat exchange with the atmosphere and to solar radiation. The upstream face is either in contact with the reservoir water or with the air when the reservoir is empty or the water level has been lowered.

Let's analyze the temperature field in a horizontal section of the dam. As a rule, thermal inertia at the heart of the dam is considerable. It undergoes a very slow temperature cycle, generally across one year. On the faces, the variations in temperature are noticeably greater. As a result, the temperature diagram is quite complicated (Figure 11.18).

To simplify calculations, the temperature diagram from any section can be represented by a linear relationship between the two faces (trapezoidal shape), as illustrated in Figure 15.62. A linear distribution of temperature can be accepted as long as the linear distribution diagram and the real distribution

---
[3] See also § 11.4.4.

have the same surface area and their centers of gravity are at the same distance from the axis of the section. Once the sinusoidal variations on the faces are known, the engineer can determine the fictive extreme temperatures on the face sufficient for project calculations (Stucky and Derron, 1957; Dungar, Zakerzadeh, 1992).

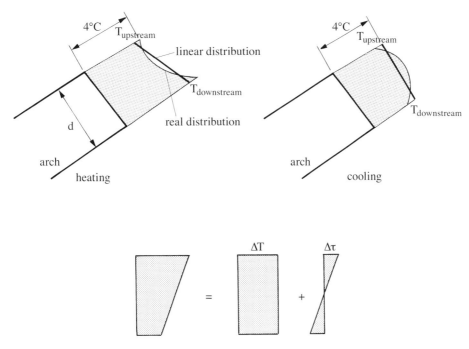

Figure 15.62 Temperatures in an arch dam, linear approximation

This trapezium can be broken down into
- Uniform variation in temperature relative to a reference temperature
- Temperature gradient between the two faces

Uniform variation in temperature across the whole section corresponds to the variation in temperature between a reference date and the date in question. The reference must correspond to the period when arch closure by joint grouting is carried out. As the closure temperature is not uniform, the mean annual temperature is sometimes used during the reassessment of existing dams. In fact, it is during this process that the arch becomes hyperstatic. Under the initial temperature conditions during closure, no hyperstatic load is added due to the effect of temperature.

Choosing when to carry out joint closure is therefore an important decision. Joint closure is carried out when the concrete is as cold as possible. The positive variation in temperature that is generally observed relative to the reference introduces horizontal compression in the hyperstatic arches, which plays a major prestressing role by limiting the appearance of tensile stresses.

15.5.3 The effects of two modes of temperature load

15.5.3.1 Uniform variation in temperature

Uniform variation in temperature occurs on both the horizontal and vertical planes.

The vertical cantilevers are free to deform under the effect of temperature variation. A vertical displacement of the crown is thus observed without the appearance of additional stresses:

$$\Delta H = \int \beta_t \cdot \Delta T(y) \cdot dy \quad \text{and} \quad \sigma_{V,\Delta T} = 0,$$

where β_t is the thermal dilatation coefficient of the concrete: $\beta_t \cong 10^{-5}\ [°C^{-1}]$.

If the horizontal arches were free to deform, a variation in the mean arch radius would occur, without the appearance of additional stresses.

In reality, the arches are at least partially embedded in their foundation and can therefore not be lengthened. A hyperstatic force applied to the elastic center of the arch is the result, which has the following value (Figure 15.63):

$$N_H(\Delta T) = \pm \Delta T \cdot \beta_t \cdot E_B \cdot d \cdot k,$$

where

- ΔT: uniform variation in temperature
- E_B: concrete elasticity modulus
- d: arch thickness
- k: arch form factor, which also depends on the extent to which the arch is embedded

$$k = 1/(C_1 \cdot \lambda^2 + C_2)$$

$$\lambda = R_{moy}/d,$$

where

2α	C_1	C_2
90°	0.10	1.51
120°	0.33	1.92
180°	1.78	3.1416

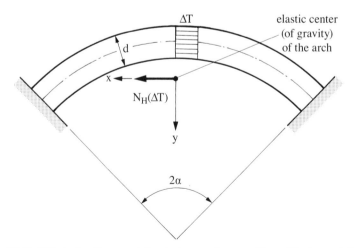

Figure 15.63 Effect of uniform variation in temperature: position of the hyperstatic force

This load is accompanied by bending moments in the arches, due to the hyperstatic behavior. The stresses are:

$$\sigma_t = N/F \pm M/W = N_H(\Delta T) \cdot \cos \alpha / d \pm (N_H(\Delta T) \cdot y)/W$$

and

$$W = 1 \cdot d^2 / 6.$$

An upstream deformation of the dam crown is observed on the arches if they are slender and highly curved.

15.5.3.2 Temperature gradient

In an arch subject to temperature gradient, the middle sheet of the arch undergoes no variation in its thermal state. The result is that this mean sheet does not deform and as a result, the arches experience no deformation.

However, the outer fiber lengthens and the inner fiber shortens. The sections, which are unable to turn, bear the load from the uniform bending moment, the effect of which is to cancel the relative rotations. If the sections were able to shift freely in relation to each other, the rotation of the two neighboring sections, at a unit length distance, would have the following value:

$$\Delta \omega = 2 \cdot \Delta T \cdot \beta_t / d,$$

where

ΔT: linear temperature difference between the faces
β_t: thermal dilatation coefficient of the concrete

The bending moment that cancels this rotation is expressed as:

$$M = \Delta \omega \cdot I \cdot E_B,$$

where

$I = 1 \cdot d^3 / 12$ and E_B is the concrete elasticity modulus

$$M_{\Delta T} = (2 \cdot \Delta T \cdot \beta_t / d) \cdot (1 \cdot d^3 / 12) \cdot E_B$$

$$M_{\Delta T} = 1/6 \cdot \Delta T \cdot \beta_t \cdot d^2 \cdot E_B.$$

The normal stresses on the face have the following values:

$$\sigma_{us,ds} = \pm \beta_t \cdot \Delta T \cdot E_B.$$

15.5.3.3 Analysis

As the arch dam is a relatively thin structure, temperature variations between the upstream and downstream faces cause considerable stress and deformation. These can reach several mm at the crest of the crown in the upstream-downstream direction.

As part of dam monitoring, the measured deformation is compared with the calculated deformation to verify the dam's behavior. This analysis must, of course, take into account the effect of temperature in order to carry out the comparison and so that the conclusions are reasonable (Bossoney, 1994).

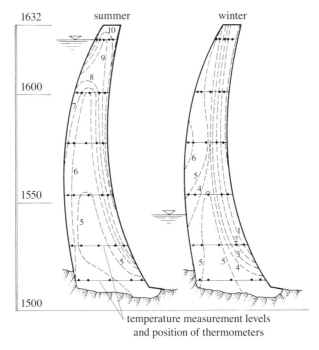

Figure 15.64 Distribution of summer and winter concrete temperature in a vertical section

The effect of uniform variation and temperature gradient can be calculated on the hyperstatic structure that represents the arch dam. Until the 1980s, the same trial-load method described above was used to carry out this calculation. Today, a calculation using the finite element method allows the internal distribution of concrete temperature to be determined and therefore gives results that are closer to reality. The study can focus on the vertical section at the crown and, depending on the size of the dam project, on 1 to 2 additional lateral sections (Figure 15.64), based on a 2D finite element or finite difference model for category I and II dams (SFOE, 2017). Stationary or transient states can also be determined.

The main difficulty encountered during calculations is defining the cases of thermal load that must be considered. These depend on the materials, climatic conditions at the site, the orientation of the faces and the shade cast by the sides of the valley, the reservoir operating mode (water level on the upstream face), thermal adsorption parameters, and the thermal conditions of the dam during joint closure.

15.6 Loads and Stresses

Usual loading (normal, exceptional, and extreme), such as self weight, hydrostatic and hydrodynamic pressure, and temperature effects lead to the introduction of internal loading in both the cantilevers and the arches: normal shear loading, bending and torsion moments, as well as shear force acting transversally on the mean surface (Figure 15.65).

Calculating the stresses in the cantilevers and arches presupposes a linear distribution between the upstream and downstream faces. The surfaces remain plane after bending. Figure 15.66 provides an example of the orientation of principle stresses on the upstream and downstream faces under the effect of self weight and water pressure.

ARCH DAMS

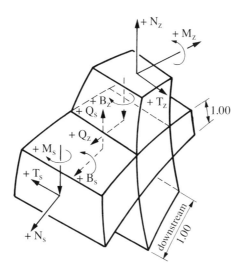

Vertical radial face		Horizontal face	
N_s	normal force	N_z	normal effort
T_s	radial shear force	T_z	radial shear force
Q_s	vertical tangential force	Q_z	vertical tangential force
M_s	bending moment of the arch	M_z	bending moment of the cantilever
B_s	torsion moment of the arch	B_z	torsion moment of the cantilever

Figure 15.65 Interior loading on the cantilevers and arches

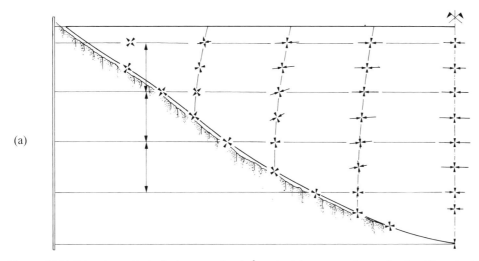

Figure 15.66 Directions of principal stresses in t/m^2 on the (a) upstream face under the effect of self weight and water pressure of a 127-meter-high arch dam (source: EPFL course notes, Prof. J. P. Stucky)

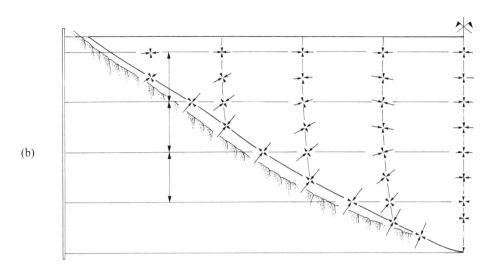

Figure 15.66 Directions of principal stresses in t/m² on the (b) downstream face under the effect of self weight and water pressure of a 127-meter-high arch dam (source: EPFL course notes, Prof. J. P. Stucky)

15.6.1 Localizing compressed zones and zones under tension

Some fundamental differences distinguish arch dams from gravity dams in terms of loading and stress analysis.

Gravity dams

The choice of a triangular shape for the gravity dam section means that the stress distribution cannot be influenced. The greatest compression is usually located:
- At the upstream heel when the reservoir is empty
- At the downstream toe when the reservoir is full

The lowest stresses always affect the opposite face.

The most important condition is that tensile stresses under normal loading must be avoided as these stresses favor the creation of cracks, which tend to worsen due to the isostaticity of the section (instability). The principal method for avoiding tensile stresses is to select the appropriate slope for the downstream face.

In cases of exceptional load, low tensile stresses are acceptable, but they must be limited to the concrete's tensile strength ($1 \div 4$ [N/mm²]).

Arch dam

For arch dams, the aim is to obtain compressive stresses wherever possible. To attain this objective, all the geometric parameters of the shell can be freely selected.

The dam's hyperstaticity enables tensile stresses to be accepted in some zones and under some loading cases, even if the concrete's tensile strength cannot be counted on. Even though the construction joints are injected with cement grout to ensure the transfer of horizontal loading, their tensile strength is practically zero.

In an arch dam, cracks are not due to the direct action of exterior forces but to the hyperstatic loading of the embedding in the abutments. They result from the rigidity of the arches embedded at its abutments. When tensile stresses appear, a crack opens, which locally reduces the rigidity of the arch. The effect of this decrease in rigidity is to reduce the embedding, which thus prevents the crack from further developing. In addition, the local diminution in rigidity accentuates the deformation of the section in question. The continuity of deformations in the hyperstatic structure mean that a proportion of load is taken up by the other sections in the dam, which tends to reduce the load in the zones under the highest stress (transfer of forces from the cantilever mode to the arch mode).

Concrete tensile strength does not need to equal exterior forces.

Figure 15.67 shows the typical position of compressive and tensile zones in an arch dam, as well as the position of the resultant, taking into account the effect of torsion (R*) or not (R). The hatched area indicates the zones under tension. As for zones that are basically compressive, they define the active zones of the arch and cantilever.

Under the effect of water pressure, typical zones of compressive and tensile stresses in the arches become apparent. These tensile stresses appear:

- On the upstream face, near the springings
- On the downstream face, at the crown

A compressive zone forming an active arch can be identified.

Depending on the extent of the zone under tension, the intensity of compressive stresses in the active arch can increase greatly, as shown in Figure 15.68.

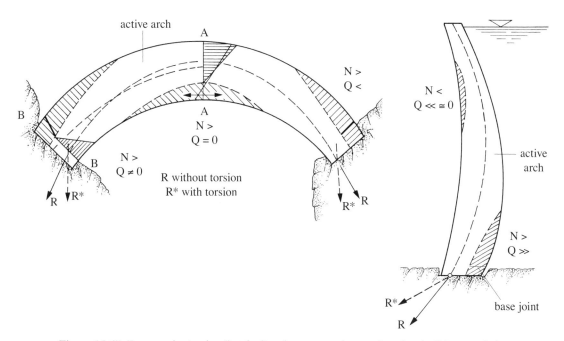

Figure 15.67 Zones under tension (hatched) and compressed zones (non-hatched) in an arch dam (from Lombardi, 1988)

CONCRETE DAMS

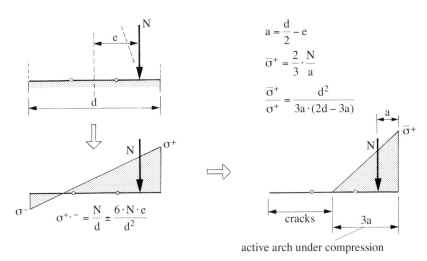

Figure 15.68 Creation of a crack and increased compressive stresses

It is thus possible to calculate the increase in compressive stresses based on the position of the normal resultant force:

$$a = \frac{d}{3} \qquad \frac{\bar{\sigma}^+}{\sigma^+} = 1,$$

$$a = \frac{d}{6} \qquad \frac{\bar{\sigma}^+}{\sigma^+} = 1\frac{1}{3},$$

$$a = \frac{d}{12} \qquad \frac{\bar{\sigma}^+}{\sigma^+} = 2\frac{2}{7},$$

$$a \to 0 \qquad \frac{\bar{\sigma}^+}{\sigma^+} \to \infty.$$

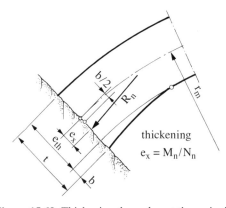

Figure 15.69 Thickening the arches at the springing

In practice, it is accepted that the resultant can leave the middle third of the section and reach a = d / 6, which leads to a 33% worsening in the maximum compressive stress on the edge and the development of a cracked zone reaching half-way through the total thickness (see also Figure 13.25).

At the arch springing, the sections are heavily compressed at the intrados and under tension at the extrados. Locally and gradually thickening the arches downstream enables this stress state to be improved (Figure 15.69).

15.6.2 Effects of load

15.6.2.1 Normal load

Normal loads are those issuing from the combination of self weight, water pressure, and the effects of temperature.

Under this loading, maximum compressive stress is located:
- On the downstream face at the arch springing and the toe of the dam
- On the upstream face, at the crown

As a rule, it is accepted that compressive stresses must not exceed 7 to 8 MPa, and that tensile stresses are not permitted.

15.6.2.2 Exceptional load

When considering the action of an earthquake in addition to normal loading, the appearance of tensile stresses is tolerated. These tensile stresses lead to the opening of cracks and the worsening of compressive stresses. Nevertheless, the resultant must remain in the middle two-thirds of the section ($a \geq d/6$).

15.6.3 Effect of shear force

Until now, we have analyzed stresses due to the combination of normal loading with a bending moment. Through the hyperstaticity of the shell and the elasticity of the material, the concentration of load leads to the redistribution of loads throughout the entire arch.

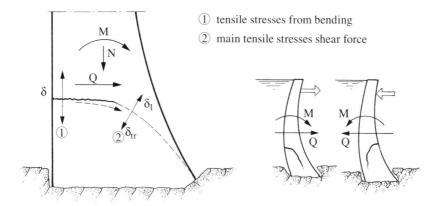

Figure 15.70 Development of a combined crack and examples of combinations of forces with damage to the dam (from Lombardi, 1988)

However, the appearance of cracks due to high shear force rather than normal loading is much more worrying. This high level of shear force develops primarily in zones near the foundation. A crack due to tensile stresses caused by a bending moment can grow because of the occurrence of shear force on the principal stresses (Figure 15.70).

Figure 15.71 shows the different types of cracks that can be found on operating arch dams and illustrates the stability of these cracks depending on the normal load or shear force that caused them.

Figure 15.71 calls for a few comments. At the crown of an arch, the normal loading is high, while shear force is zero. In case A, the zone under tension located on the downstream face and any cracks that might appear pose no danger to the dam structure. Case H illustrates a crack perpendicular to the foundation. This crack, which affects the downstream face, is not rare and is not an issue as it should not take up the shear force. However, as mentioned above, should shear force predominate with regard to normal loading, the situation becomes more complex and the risk of continuous cracks forming increases. Cases E, G, and K are in these respects representative. Lastly, the limit between stable and unstable cracks is given by the principal tensile stress values σ_1 due to tensile forces.

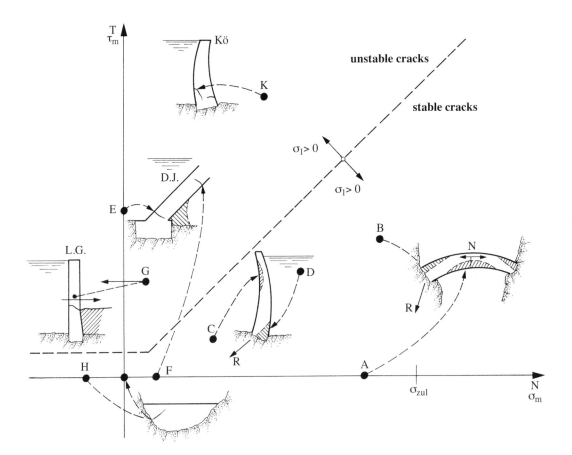

Figure 15.71 Stable and unstable cracks in an arch dam (from Lombardi, 1988). Key: N = normal loading; T = shear force; σ_m, τ_m = respective stresses; Kö = Kölnbrein, D.J. = Daniel-Johnson, L.G. = Le Gage.

In documentation concerning dams, three typical cases of fissuring due to shear force can be found:
- The Kölnbrein arch dam in Austria
- The Daniel-Johnson multiple arch dam in Canada
- The Le Gage cylindrical arch dam in France

In Switzerland, the case of the Zeuzier arch dam has been very well documented. This 154-meter-high dam is located in Central Valais, near to the border with the canton of Bern. During construction of an exploratory gallery 1.4 km away from the dam and 400 m below the lower level of the riverbed (with the aim of later building a road access below the Rawyl Pass), major cracks started to appear on the dam, as shown in Figure 15.72 (see also § 2.2.3).

This fissuring resulted from an exceptional stress state caused by the differential settlement of the dam abutments and by the two valley sides of the dam coming together at the level of the crest. These foundation movements were due to drainage occurring in the rock mass when an exploratory gallery for a road tunnel was excavated and to the consolidation of fill material in some rock fissures, which enabled this drainage.

This situation required the immediate and precautionary drawdown of the reservoir. Major rehabilitation works were carried out, mainly through the grouting of cracks with synthetic resins. An increase in the dam monitoring system enabled the gradual re-impoundment of water into the reservoir. Since then, the behavior of the dam has been satisfactory. In the end, the road access tunnel project was abandoned.

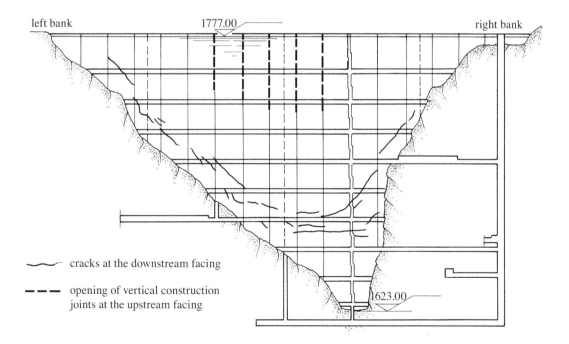

Figure 15.72 Zeuzier dam, fissuring due to foundation movement

15.7 Structural Details

15.7.1 Configuration of the toe of the dam

When the valley is relatively wide and the slenderness coefficient is high (> 15 to 20), the influence of shear forces becomes essential. Furthermore, tensile stresses at the toe of the upstream face are often difficult to prevent and become excessive. To avoid high tensile stresses, the bearing capacity of the cantilevers is reduced, thereby transferring more stresses to the arches.

There are two possibilities for achieving this: the peripheral joint and the upstream joint.

15.7.1.1 The peripheral joint

The most commonly employed solution involves installing a peripheral joint that follows the dam foundation. The dam is then set on a structure known as a Pulvino.

Figure 15.73 provides the example of the Osiglietta dam in Italy (Dolcetta et al., 1991). The dam is clearly separated from its foundation by a joint, which prevents the foundation from taking up the shear force. This solution was first developed in Italy, hence its name.

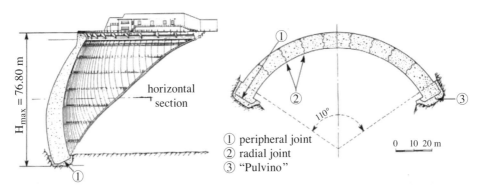

Figure 15.73 Dam foundation with Pulvino, overview of the system (from Dolcetta et al., 1991)

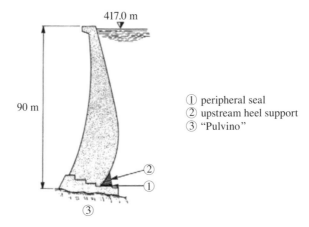

Figure 15.74 Dam foundations with Pulvino, installation of a support at the upstream heel (from Dolcetta et al., 1991)

ARCH DAMS

During the construction phase, stability of the single blocks is often guaranteed by the instalment of supports at the upstream heel, as illustrated in Figure 15.74 (Dolcetta et al., 1991).

Figure 15.75 (Dolcetta et al., 1991) shows a typical Pulvino system. The peripheral joint is generally combined with a base gallery in the Pulvino, designed to enable monitoring of joint behavior and the foundation. Furthermore, near the joint, significant surface reinforcement is often installed both in the Pulvino and the dam (Figure 15.75).

One of the main difficulties with the Pulvino system is guaranteeing watertightness of the peripheral joint. Figure 15.76 illustrates a commonly employed watertightness system.

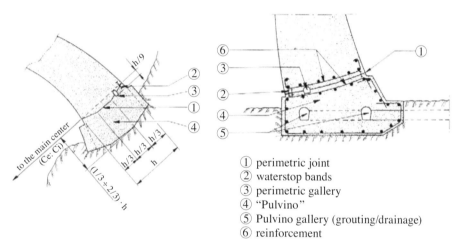

① perimetric joint
② waterstop bands
③ perimetric gallery
④ "Pulvino"
⑤ Pulvino gallery (grouting/drainage)
⑥ reinforcement

Figure 15.75 Dam foundation with Pulvino, typical layout (from Dolcetta et al., 1991)

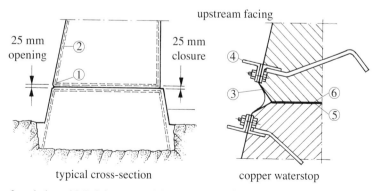

Figure 15.76 Dam foundation with Pulvino, watertightness system: ① waterstop; ② grout; ③ copper foil; ④ 63×63 mm steel angle profiles; ⑤ graphite paste; ⑥ asbestos sheet impregnated with graphite (from Dolcetta et al., 1991)

15.7.1.2 The upstream joint

The second solution involves a joint parallel to the foundation installed between the upstream face and a drainage gallery in the dam, close to the middle plane. This joint artificially reduces the rigidity of the section near the foundation, just like a crack would (Rescher, 1993). The installation of such a base joint can be limited to the part of the upstream heel that is under tension and in the lower part of the valley. The presence of this joint does not affect the dam's monolithic state. A watertightness system installed upstream prevents the reservoir from placing pressure on the joint. In Switzerland, Les Toules dam was the first to include an upstream joint (Figure 15.77). An upstream joint is also found at the Schiffenen, Gebidem, and Hongrin dams.

Depending on how the works progress, grouting the foundations can sometimes result in slight upheaval, which may complicate design estimation of possible joint movement.

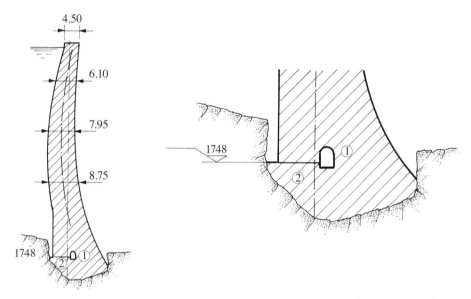

Figure 15.77 Les Toules dam: example of a dam with an upstream base joint ① base gallery; ② base joint

15.7.2 Galleries and shafts

All concrete dams, providing the sections are large enough, have a network of galleries and shafts. Figure 15.78 presents the system installed in the Emosson arch dam.

In particular, note:

- Vertical shafts containing monitoring pendulums
- A vertical elevator shaft
- Horizontal inspection galleries in the dam structure and in the foundation
- A base gallery close to the foundation

During construction, this system of shafts and galleries is used for:

- The refrigeration of concrete during construction (post-cooling)
- Joint grouting
- The installation of measuring instruments

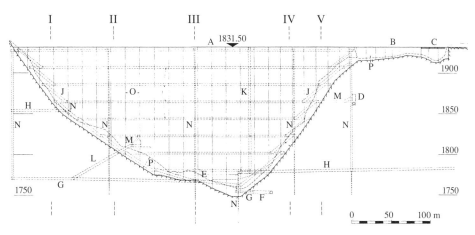

Figure 15.78 Network of galleries and monitoring system in the Emosson dam. Key: G = drainage gallery; H = exploratory gallery; I = monitoring gallery; J = base gallery; K = elevator shaft; L = access gallery; M = connecting gallery; N = pendulum shaft.

Once operation has begun, these galleries allow:

- Access to instruments that measure deformation (pendulums, extensometers, ultra-precise leveling bench marks) and pressure (piezometers, manometers)
- Measurement of uplift in the rock (from the base gallery)
- Collection and measurement of seepage flow; evacuation of this water
- Visual inspection of the condition of the dam (and especially variations in this condition)
- Interventions in the foundation rock after commissioning (grouting, drainage, installation of measuring systems)
- Access to specific safety and operational equipment (bottom outlet gates, headrace intake gates, etc.)

The distance between the various levels of horizontal galleries varies from one dam to another, depending on requirements. It is usually between 20 and 30 m. The shafts depend on the position of the pendulums. They are most often spaced by 50 to 80 m. Some pendulum shafts are extended under the dam foundation to a depth of 20 to 50 m, depending on dam height and geology. Inverted pendulums are placed in these types of shafts (see Figure 32.12).

The shape of inspection galleries depends on certain criteria:

- Static: to avoid stress concentrations
- Structural: to enable the complete execution of the gallery in a single concrete lift
- Functional: to enable all necessary equipment to pass through the galleries, especially a small drilling machine for grouting or additional drainage

Lastly, passage through the galleries must be as comfortable as possible, to encourage regular inspections by the dam operating staff.

Horizontal inspection galleries are generally around 1.50 m wide and 2 to 2.30 m high. Galleries always have an evacuation channel for drainage water, either on the upstream side or in the center.

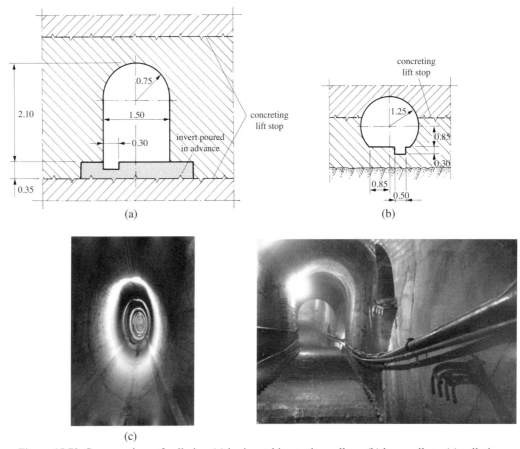

Figure 15.79 Cross-sections of galleries: (a) horizontal inspection gallery; (b) base gallery; (c) galleries in the Grande Dixence dam (Courtesy H. Pougatsch)

Galleries are generally positioned as closely as possible to the upstream face:
e = 3.5 m for H = 50 m
e = 4.0 m for H = 100 m
e = 5.0 m for H = 150 m
e = 6.0 m for H = 200 m

The base gallery is often located 2 to 3 meters from the contact surface between the concrete and the rock. It follows along this interface point at a distance of 2 to 6 m from the upstream face. The slope of this gallery depends, of course, on the slope of the abutments. It is almost horizontal at the base of the excavation and can reach a slope of 400% if the excavation is almost vertical.

The base gallery has stairs or metal ladders if the slope is too steep. Its section is larger than that of horizontal galleries, as a small drilling machine must be able to be installed for drilling down toward the foundation.

The diameter of vertical pendulum shafts is ϕ_{min} = 80 cm for accessible shafts (ladder) and ϕ_{min} = 40 cm for inaccessible shafts.

15.7.3 Artificial cooling of hard concrete (post cooling)

With regard to arch dams, it is important to limit the rise in concrete temperature due to cement heat of hydration and to reach thermal equilibrium for the dam before joint closure by grouting, not only to prevent cracks originating from excessive thermal gradients, but also and especially to ensure closure of the dam in the best possible conditions. It is important to carry out joint closure before the effects of exterior heating appear inside the concrete mass.

As with gravity dams, evacuating cement heat of hydration occurs by the circulation of cold water through a network of metal cooling pipes laid out on the surface of each lift (see § 13.8.2.2 and Figure 13.37). The cold water initially circulates in the center, then, slightly warmer, passes along the edges where cooling is completed by the contact of the faces with the air (it is therefore not necessary to regularly reverse the direction of water circulation). The water supply for the cooling pipe system comes through a horizontal gallery and feeds the pipes on all the lifts above it, until the next gallery (Figure 15.80). Monitoring the concrete temperature is also an important activity.

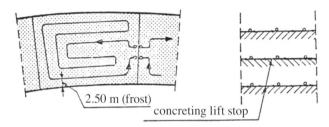

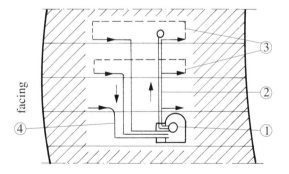

Coil supply system

① Main cold water pipe 4" to 6" laid in the accessible horizontal gallery
② Single supply line, 2 to 3", embedded in concrete, initially equipped with a tap (in the horizontal gallery)
③ Coils, 3/4 or 1", placed on the horizontal lifts surface (folded down vertically in the figure)
④ Insulated return pipes, 3/4 or 1", embedded in the concrete leading into the channel of the horizontal gallery, equipped with a tap at the outlet (the tap is used to adjust the flow rate of each coil.)

Figure 15.80 Artificial cooling of concrete: installation and supply of the cooling pipe system

15.7.4 Treatment and grouting of joints

Reasons behind joint closure

In an arch dam, joint closure between the blocks is an important operation that is an integral part of construction planning. Grouting the joints is designed to join the blocks and obtain a monolithic structure, which enables the arch effect. Planning for this operation must be based on considerations relative to the stress state and progression of the works. A numerical analysis establishes the closure temperature. Indeed, joint closure conditions the thermal state of the dam during operation and is calculated based on the closure temperature or sometimes, for the reassessment of existing dams, on the mean annual temperature. Furthermore, the engineer can seek to minimize cracks caused by the effects of temperature. Lastly, grouting is carried out when the opening of the joints is sufficient (1.5 to 2.5 mm) to ensure the correct application of a generally thick grout.

Method for joint grouting
In gravity dams, the joints between the blocks are open due to concrete shrinkage. A watertightness joint is placed upstream to avoid seepage and the loading of the joint. In an arch dam, the joint must be filled with cement grout to ensure the transfer of horizontal load from one block to its neighbor. The role of watertightness is not just to insulate the joint from the reservoir, but also and especially to demarcate the surface to be grouted. Previously, copper foil was used as a watertightness system, but it is not easy to install. Today, 25-, 32-, or 40-cm-wide synthetic rubber seals called waterstops are used.

Joint grouting takes place in several successive stages from the inspection galleries. To aid the grouting process, the joints are divided into panels 20 to 30 m high using waterstops. Before starting to grout the joints, the ramps and collectors need to be checked by running pressurized water through them. The grout is injected from the bottom toward the top from the lower gallery and from the center toward the sides of the dam. Vents ensure that air trapped in the joint can escape. The composition of the cement grout must be such that it can penetrate the joint and fill all empty spaces. Generally, the water:cement ratio is 2:1 or 1:1. While injecting, pressure must be limited to avoid excessive deformation of the blocks and the opening of the joint in the injected section below the joint. Grout take varies between 10 and 15 kg/m^2. The grouting of one joint takes several goes. It may even be necessary to carry out re-grouting after several years have passed. After each operation, the pipes must be carefully flushed.

Before concreting, the joint is equipped with pipes with sleeve or grouting valves according to the diagrams in Figure 15.80 and Figure 15.81. These elements are sufficiently supple to allow the pressurized grout to pass through during grouting and to close up again as soon as the pressure drops.

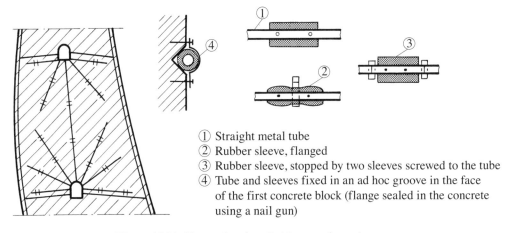

① Straight metal tube
② Rubber sleeve, flanged
③ Rubber sleeve, stopped by two sleeves screwed to the tube
④ Tube and sleeves fixed in an ad hoc groove in the face of the first concrete block (flange sealed in the concrete using a nail gun)

Figure 15.81 Sleeve pipes installed in a star formation

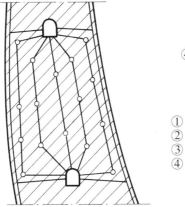

① Metal tube ∅ 1″, with bifurcation
② Rubber stopper
③ Flexible metal hoop/ring
④ Tab designed to produce a swirl when washing the hose

Figure 15.82 Pipes with valves, installed in series

15.8 Heightening Arch Dams

15.8.1 Reasons for heightening

Increasing reservoir storage capacity and shifting the use made of inflow volume from summer to winter is a major factor in guaranteeing Switzerland's energy supply. The heightening of existing dams with this aim is an option that was implemented for the Mauvoisin and Luzzone dams (Schleiss, 1999, 2012).

In addition, given current efforts at energy transition and the developments in new renewable volatile energy sources (solar and wind energy), hydropower production companies have begun studying pumped storage options, which would allow water to travel between an upper and lower reservoir depending on the demand and supply of electricity. The close proximity of the Emosson and Vieux-Emosson dam reservoirs (part of the Nant-de-Drance project) has been utilized to create such a system.

15.8.2 Heightening the Mauvoisin dam (VS)

The Mauvoisin arch dam (VS) was heightened by 13.5 m to reach a height of 250 m (Figure 15.83 and Figure 15.84). This enabled the total reservoir volume to be increased by approximately 30 mio. m³ and an additional winter production of 100 GWh. Studies have shown that direct water inflow is sufficient to guarantee the filling of the additional reservoir capacity, even in a dry year.

Heightening took place between 1989 and 1991 through the construction of an additional arch in a continuation of the dam faces so as not to modify its appearance. The volume of concrete required was 80,000 m³, which represents barely 4% of the previously existing concrete. The right bank abutment is free due to the presence of a corridor to evacuate any blocks of ice that fall from the Giétroz glacier. The joints between the blocks were fitted with shear keys to improve transmission of forces during earthquakes (for an example, see Figure 13.44).

The spillway was refurbished and fitted with automatic gates that gradually open once the water in the reservoir reaches a certain level. The existing water release and intake structures were able to tolerate the slight increase in hydraulic head without any issue.

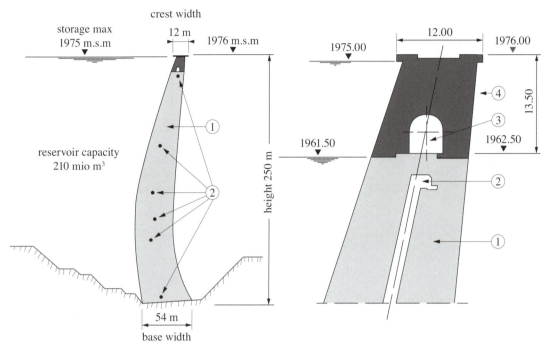

Figure 15.83 Diagram of the Mauvoisin dam heightening 1. Initial dam; 2. Inspection galleries; 3. Access gallery to the Chanrion power station (located in the right valley flank); 4. Heightened section

Figure 15.84 Heightening of the Mauvoisin dam from 236.5 m to 250 m (Courtesy W. Hauenstein).

The existing monitoring system was completed by the installation of instruments measuring rock deformation, by lengthening the pendulums into the new, upper part, and by installing measuring stations for clinometers and additional deformation. The behavior of the dam continues to be regular and complies with safety criteria.

15.8.3 Heightening of the Luzzone dam (TI)

The Luzzone arch dam (TI) was heightened by 17 m to reach a height of 225 m (Figure 15.85). This resulted in an increase in reservoir storage capacity of approximately 20 mio. m^3 and an additional winter production of around 60 GWh.

The heightening of the Luzzone dam took place through the addition of an arch whose face geometry continued from that of the existing profile. A gallery was included in this new section. On the left bank, the heightened section does not rest against the rock, and the horizontal forces of arch pressure are directed downward. The works were carried out from 1995 to 1999, and 80,000 m^3 of concrete was placed (Baumer, 2012).

With regard to monitoring, additional devices were installed (new pendulums, thermometers, joint behaviour measuring devices, exensometers, piezometric cells). The behavior of this heightened dam has met expectations.

Figure 15.85 Heightening of the Luzzone dam from 208 m to 225 m (Courtesy Ofima SA)

15.8.4 Heightening of the Vieux-Emosson dam (VS)

The heightening of the crest of the Vieux-Emosson dam by 21.5 m was carried out as part of the Nant-de-Drance pumped storage power station project. It results in double the storage capacity of the Vieux-Emosson lake and thus ensures sufficient flexibility to fully exploit the power station's 900 MW capacity.

The normal maximum reservoir level was raised by 20 m, going from the former altitude of 2,205 m.a.s.l. to the new altitude of 2,225 m.a.s.l. This increase corresponds to an economic and structural optimum that also takes the existing conditions into account (geometry of the first dam, foundations, site topography) (Valotton, 2012).

The freeboard was established based on the flood level, in particular the risk of a flood wave caused by the arrival of an avalanche into the reservoir or the rapid rise in water level due to an untimely pumping activity.

The heightened structure consists of a double curvature arch dam whose horizontal and vertical sections are made of parabolic segments. The heightened section rests against the first arch gravity dam at 2,195 m.a.s.l. upstream and 2,185 m.a.s.l. downstream. The specific geometry of this stepped interface was largely dictated by reasons of geometric matching (Figure 15.86). Two seasons (2013 and 2014) were necessary for the concreting of the heightened section. Closure of the vertical joints took place early 2015.

The installation of the dam's monitoring system has taken place as the works progress.

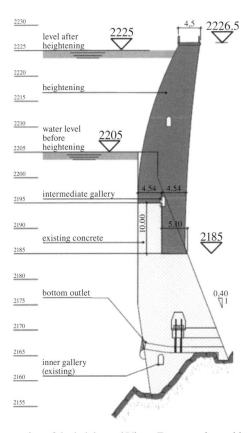

Figure 15.86 Vertical cross-section of the heightened Vieux-Emosson dam with the concrete from the first dam

Figure 15.87 Heightening of the Vieux-Emosson dam (Courtesy A. Schleiss, November 2013)

16. Multiple Arch Dams

Hongrin arch dam in Switzerland, height (north/south) 123/95 m, year commissioned 1967
(Courtesy © Swiss Air Force)

16.1 General Introduction

Multiple arch dams are a kind of buttress dam. They comprise cylindrical, inclined arches supported by triangular buttresses that transfer load from the arches to the foundation ground. In the past, dams were built with a reduced opening between the buttresses, which increased the number of narrow arches; the buttresses and arches were in reinforced concrete. Advances in dam construction techniques have led to an increase in space between the buttresses. Buttress and arch thicknesses have increased, formwork surfaces have decreased, and the placing of concrete has been facilitated.

The shape and spread of the valley, but also issues related to general stability, are used as a basis for choosing the size of the opening between the buttresses while achieving correct arch distribution. Through its design, this type of dam is particularly suitable for wide valleys. It requires a low volume of concrete, which results in considerable financial savings compared to a traditional gravity dam; however, formwork installation is more costly.

The first concrete multiple arch dam was built in 1908 in California. Ten Mile Creek dam is 19 m high and has 12 arches. In the following decade, a dozen multiple arch dams were constructed. In 1924, Fred A. Noetzli proposed the freestanding double wall buttress, for which no tensile stresses were tolerated (Figure 16.2). The extension of the arch intrados into the exterior face of the buttresses favors the transfer of load. The join between the two faces takes up water pressure and the moments transmitted by the arches.

Figure 16.1 Les Marécottes dam (H = 19 m/1925/VS) and Faux-la-Montagne dam (H = 19 m/1953/France), designed by Alexandre Sarrasin

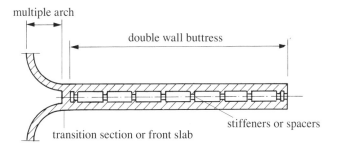

Figure 16.2 Noetzli system (from Jansen, 1988).

During the 1920s, two low multiple arch dams also began operating in Switzerland: Les Marécottes dam (VS/1925/H = 19 m) and the Oberems dam (VS/1927/H = 11 m), which were designed by Alexandre Sarrasin (Figure 3.10, Figure 16.1). The buttresses at Les Marécottes dam have a trapezoid shape and are 25 cm thick. The toes of the buttresses are embedded and are braced with struts. Every second strut has been replaced with an arch in order to absorb lateral expansion. The reinforced gunite arches are circular and their thickness ranges from 12 cm at the base to 8 cm at the top. They are embedded in the buttresses (Sarrasin, 1939; Brühwiler and Frey, 2002).

In 1928, a 76-meter-high dam with double curvature arches was built in Arizona. The Beni-Bahdel and Meffrouch dams in Algeria can also be mentioned (Figure 16.3), which were designed by the A. Stucky engineering firm (Stucky, 1937). After the Second World War, several multiple arch dams were commissioned in France, in particular the Granval dam in 1959, which is 88 m high and whose central arch has a 50-meter-high chord. The largest multiple arch dam in the world was built between 1961 and 1968 in Canada. With a height of 214 m, the Daniel-Johnson dam (Manic 5) has a central arch resting on buttresses 162 m apart and on lateral arches with a span of 76 m (Figure 16.4).

Figure 16.3 Meffrouch dam (Algeria)

Figure 16.4 Daniel-Johnson dam (Canada)

16.2 Principal Characteristics and Sizes

16.2.1 The arches

The arches are generally cylindrical and in principle are identical so that the resultant from pressure is directed into the plane of the buttresses.

Choosing the slope of the arches and the center opening angle, and establishing seasonal variations in temperature (if they are considerable), are the essential elements determining the design of the project.

The arches are usually at a 45 to 50° angle to the horizontal. This slope will depend on the accepted angle of friction following verification of sliding stability. For a low accepted angle of friction, the slope may tend toward 45°. The arch slope introduces an element of vertical pressure directed downward, which has a positive effect on the general stability of the structure.

The parallel buttresses allow for the possibility of center angles varying from 115 to 135°, or even up to 180°. Depending on their size and the presence of tensile stresses, the arches may need to be reinforced.

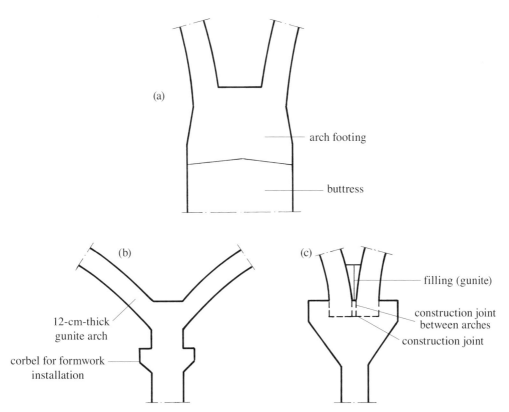

Figure 16.5 Examples of arch connections to a buttress (without indicating reinforcement): (a) Beni-Bahdel dam (thick buttress without connecting reinforcement with the arches) (from Stucky, 1937); (b) Les Marécottes dam (thin buttress) (from Brühwiler and Frey, 2002); (c) Florence Lake dam (From Jansen, 1988)

The arches can be connected to the buttresses, which makes the structure rigid and vulnerable to damage in cases of slight foundation movement (Figure 16.5). Depending on the size of the dam, arches with low bearing capacity (chord up to 10 m) and the buttress are concreted at the same time. This type of structure is suitable for valleys with more or less steep flanks.

For wide, U-shaped valleys, a dam with arches of high bearing capacity (chord from 20 to 30 m), unconnected to the generally thick and unreinforced buttresses (providing there are no tensile stresses) can be envisaged. In this case, the structure becomes more flexible. As there is no connection between the arch and the buttress, the latter must be stable on its own.

The arches grow gradually thicker from the top toward the bottom to take into account the increase in hydrostatic pressure. The inside face of the arch is cylindrical to simplify the formwork. Any variation in adapting the thickness occurs on the outside face, and a conical shape can be chosen (Figure 16.6). To ensure ample placement of traditional concrete, the thickness should not be smaller than 50 cm. To improve the stress state and facilitate the arch-buttress interface, an increase in arch thickness toward the embedding can be included in the design.

The base of the arches is horizontal and embedded into the cut-off. This is a complex zone, as the arches are not able to freely deform. It is a good idea to ensure a gradual transition between the downstream face and the foundation slab by inserting a gusset (Figure 16.7).

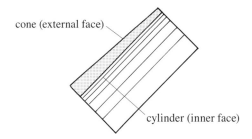

Figure 16.6 Shape of the arch

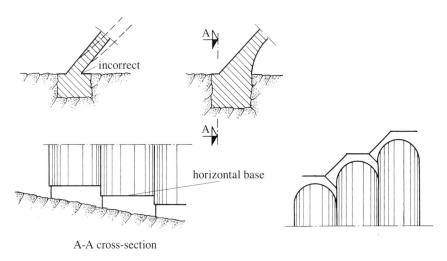

Figure 16.7 Embedding the arch at the base

16.2.2 The buttresses

The buttresses form the supporting element for the arches and transfer load to the foundation. They extend out from the arches. Their thickness is usually constant directly underneath the arch abutment and sometimes increases in the downstream direction. Buttresses with a constant thickness are also possible and are generally equal to 1/10 of the spacing.

General stability criteria determine the size of the buttresses. The slopes of the upstream and downstream buttress faces are selected to contribute to the overall stability of the dam.

Depending on the thickness of the buttresses, bracing may need to be installed to ensure lateral stability. These elements are sensitive to temperature variation and shrinkage, which can lead to substantial loading, especially if the bracing is continuous. In this case, the formula of every second brace can be used (Figure 16.8) or a rectilinear brace can be alternated with a curved brace (Figure 16.1). Another solution employed at the Beni-Bahdel dam consists of placing articulated braces on every second bay to alleviate shrinkage effects.

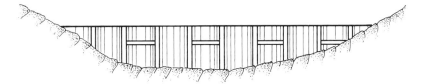

Figure 16.8 Noncontinuous bracing distribution

If strong lateral loading can be expected during an earthquake, a conventional abutment or thrust block can be placed at the base of the buttresses. Depending on the slope of the abutments at the valley flanks, braces or possibly joining slabs should be incorporated.

16.2.3 The foundations

As with any concrete dam, the foundation rock determining the layout of the dam design must be of good quality. The buttress foundations are similar to those of traditional buttresses. The base of the buttress may be anchored in a narrow excavation, which also provides the effect of an abutment or thrust block. To improve sliding stability, the excavation line can be stepped and slope upstream, depending on the nature of the rock layers.

Installing a slab allows acceptable stresses to be obtained. Ground stresses may be weak and well distributed. If the resultant of forces falls near to the middle of the base, with either a full or empty reservoir, the variation in stresses, depending on loading, is low. Settlement will be all the more uniform and constant.

16.3 General Stability Assessment

16.3.1 Sliding stability

Sliding stability for multiple arch dams must be systematically verified on the basis of the same safety criteria applied to gravity dams and buttress dams. Due to the relative slenderness and lightness of these structures, sliding stability is not guaranteed solely by concrete self weight. As previously mentioned, due to the slope of the arches, it also draws on the contribution of the vertical component directed toward the base of the hydrostatic pressure. The potential slip surfaces are naturally located at the level of the concrete lift surfaces and the base of the structure, as well as at the foundation, depending on the presence of preferential geological planes.

If the arches are closely connected to the buttresses, the buttress and the halves of the 2 adjacent arches are taken into account for the calculation. Sliding stability is determined by the relation between the sum of horizontal forces (water pressure) and vertical forces (buttress self weight, vertical component of self weight of the 2 half-arches, and water pressure, to which is added the effect of uplift). Cohesion is overlooked.

Uplift has very little effect due to the gap between the buttresses. Uplift can be considerable if there is an apron between the buttresses. Its distribution is then similar to that of a gravity dam (Figure 16.9). Depending on the design, the presence of water downstream between the buttresses located in the zone of the watercourse may need to be taken into account. Lastly, the geological condition of the foundation (nonhomogeneity) should be considered as it can affect the shape of the uplift diagram.

420 CONCRETE DAMS

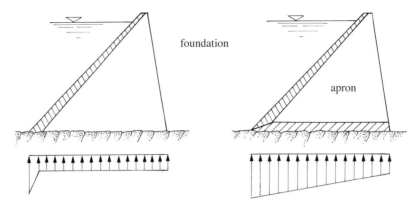

Figure 16.9 Uplift distribution (from FERC, 1997)

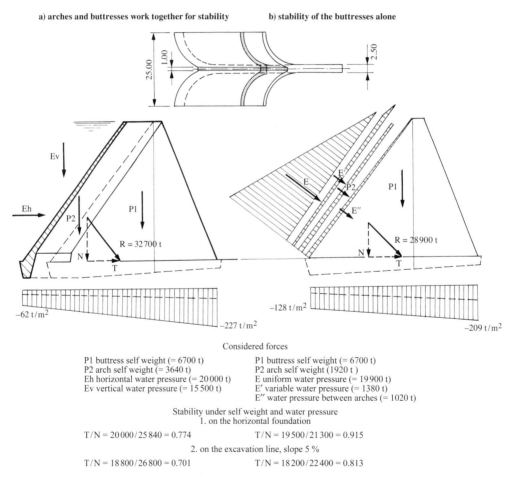

Considered forces

P1 buttress self weight (= 6700 t)   P1 buttress self weight (= 6700 t)
P2 arch self weight (= 3640 t)   P2 arch self weight (1920 t)
Eh horizontal water pressure (= 20 000 t)   E uniform water pressure (= 19 900 t)
Ev vertical water pressure (= 15 500 t)   E′ variable water pressure (= 1380 t)
  E″ water pressure between arches (= 1020 t)

Stability under self weight and water pressure
1. on the horizontal foundation

T/N = 20 000/25 840 = 0.774   T/N = 19 500/21 300 = 0.915

2. on the excavation line, slope 5 %

T/N = 18 800/26 800 = 0.701   T/N = 18 200/22 400 = 0.813

Figure 16.10 Example of stability calculations for a multiple arch dam (from Crespo, 1967)

Although the arches are supposed to be supported by the buttresses, the latter are subject to forces resulting from their self weight, normal pressure at the axis of the arch due to its self weight, hydrostatic pressure both on the upstream face and transferred by the arch (variable and uniform pressure), as well as uplift. It is therefore important to determine beforehand the load transferred by the arch. Sliding stability is also established by the relation of the sum of horizontal and vertical forces.

The acceptable value of the relation between the horizontal and vertical forces is selected depending on the nature of the interface between the two faces of the sliding plane in question (concrete-concrete, concrete-rock, rock-rock).

Limit values of this relation between 0.75 and 0.80 are suggested in the literature. Higher values may, however, be accepted, providing they do not go beyond 1.00.

Figure 16.10 presents a summary of the resultants from a sliding stability calculation taking into account (or not) the connection between the arch and the buttress (Crespo, 1967).

There are several solutions for improving sliding stability. First of all, as with gravity dams and buttress dams, the foundation can be sloped (see also § 13.4.3). In addition, abutment or thrust blocks can be placed downstream or anchor rods installed in the foundation (this latter solution is particularly appropriate for the strengthening of an existing dam). At the Beni-Bahdel dam (Figure 16.11), the decision to raise the water level was made during construction. In order to ensure transversal stability, this required, among other things, the installation of active abutment blocks at the downstream heel of the buttress (Stucky, 1970).

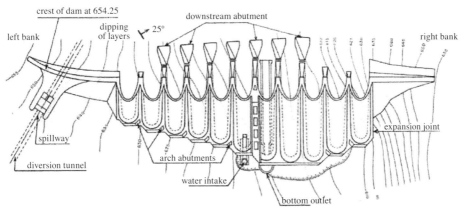

Figure 16.11 Beni-Bahdel multiple arch dam (Algeria) with downstream abutment blocks

16.3.2 Overturning stability

Overturning stability is less critical than sliding stability, however, appropriate verification must nevertheless be carried out. As for gravity dams and buttress dams, it is a question of calculating the resultant of the forces at play and determining its application point at the level of the base section. This takes into account the connection (or not) between the arch and the buttress: the section is limited to that of the buttress if the cut-off is independent or also includes 2 half-arches, in the case of a connection with the cut-off.

16.4 Loads and Stresses

Due to the existence of sections under considerable loads and stresses, the calculations must be detailed. Analysis covers all the specific components of the dam, such as the upstream parts, the buttresses, the connection between the upstream cut-off and the buttress, the supporting elements (support base, corbels), etc.

The actual calculations can be carried out based on traditional material strength formulae (linear distribution of normal stress) and the finite element method. The M, N, and T forces must be known in order to calculate the reinforced concrete sections. The acceptable level of reinforced concrete strength and the calculation of reinforcing bars follow the usual rules.

This type of dam structure is highly sensitive to variations in temperature, and there is therefore a risk of cracking in the buttresses and arches (see example E, Figure 15.71).

As a rule, the forces due to an earthquake are comparable to those considered for gravity dams and buttress dams. The calculations must be performed in the transversal (upstream-downstream) and longitudinal direction (left bank-right bank). For the preliminary design phase or for a zone with low seismicity, an approach using the pseudo-static method can be taken. However, owing to the slope of the upstream face, the increase in hydrostatic pressure must be taken into account. Zangar (1952) proposed an equation for this that enables the overpressure to be calculated at each level, depending on the slope of the face.[1] To carry out a dynamic analysis, the finite element method tends to be used, as it is particularly effective for calculating the dam's dynamic response. It is possible to carry out a 2D analysis by considering the buttress and arches separately. The 3D calculation of a multiple arch dam is, however, a complex operation.

And lastly, using the time history method of calculation and the response spectrum method is also possible.

16.4.1 Arches

Under the effect of different cases of loading, the arch works in bending. The calculation of loads and stresses in the arch due to self weight, hydrostatic pressure, thermal effects, and shrinkage also occurs by assuming these to be split into unit height bands perpendicular to the center lines. Load in the arches is determined by considering the level of embedding directly underneath their abutments. The formulae and monographs established to calculate load in the arches of arch dams can be used.

16.4.1.1 Self weight

The vertical self weight of one component of the arch is broken down into two forces: one parallel to the arch generatrix, compressing it down to the foundation, and the other perpendicular, taken up by the arch (Figure 16.12).

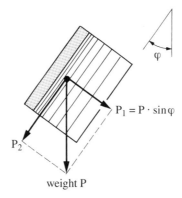

Figure 16.12 Breakdown of arch self weight

---

[1] This relation and the curves are presented in the book *Design of Small Dams* (1987), published by the Bureau of Reclamation (USBR).

16.4.1.2 The effect of water pressure

Due to the slope of the arch, each calculated band bears a hydrostatic pressure that varies from the crown (h_1) to the springing (h_2), as the springings are lower than the crown (Figure 16.13). For the calculation, it is broken down into a uniform pressure equal to hydrostatic pressure h_1 and a variable pressure of 0 at the crown to ($h_2 - h_1$) at the springing.

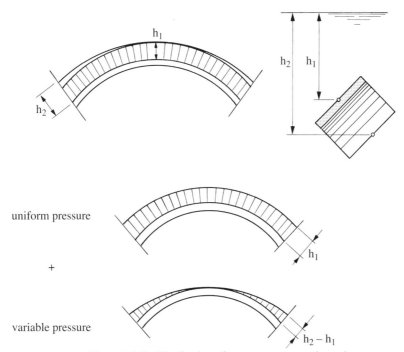

Figure 16.13 Distribution of water pressure on the arch

16.4.1.3 The effect of temperature variation and shrinkage

As the arch is relatively thin, the concrete follows annual variations in temperature, but the thermal inertia of the concrete also plays a role in daily variations.

The effect of shrinkage is treated as a reduction in temperature that is assessed while considering the climate, arch thickness, the type of cement, and other factors. For large arches, joints are included in the design to significantly reduce this effect.

Although self weight and hydrostatic pressure cause compressive stress, thermal effects and shrinkage (generally treated as a reduction in temperature for the purposes of calculation) lead to considerable tensile stresses. It is therefore very important to use strong reinforcement.

16.4.1.4 Lower and upper parts of the arch

In the lower part, the arch is embedded in the cut-off and, because of this, the embedding moments will develop in the arch generatrix due to the effects of water pressure and temperature variation. This part, under loading by the aforementioned bending moments and normal compressive loads due to self weight, must be reinforced.

Lastly, particular care should be taken with regard to the top of the arch, which is subject to variable water pressure that can cause a considerable bending moment. Once this part is carefully reinforced (Figure 16.14), the top of the arches generally stiffens.

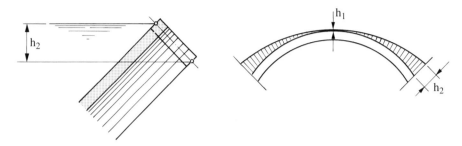

Figure 16.14 Distribution of water pressure at the top of the arch

16.4.2 Buttresses

Loading is calculated by considering self weight and the reactions of the arch. The calculation is carried out in the same way as a traditional buttress (see Section 14.2). Stability with regard to buckling of the buttress must also be verified.

Concrete reinforcement by steel bars must be planned for in case of tensile stresses along the upstream face of the buttress. In many cases, tensile stresses are produced near the upstream face due to the geometry (slope) and forces applied. Indeed, to avoid possible damage, the strength of the buttress head in relation to the load caused by arch reactions (water pressure, temperature) must be verified. Tensile stresses and shear stresses must be calculated. Depending on the structural layout of this part, a base must be created and as a result, reinforcing bars are necessary.

17. Roller-Compacted Concrete Dams (RCC)

Construction of the Rialp roller-compacted concrete gravity dam in Spain, height 101 m, year commissioned 1999 (Courtesy A. Schleiss)

17.1 General Description[1]

17.1.1 A brief history

After some early projects completed in the 1960s, particularly in Italy (Alpe Gera dam with a height of 172 m) and Canada, the technique of roller-compacted concrete (abbreviated as RCC[2]) developed strongly from the beginning of the 1980s: firstly in Japan (an 89-meter-high dam whose upstream and downstream faces were in conventional concrete) and then in the United States (Willow Creek dam, Oregon, 331,000 m³ placed in fewer than 5 months) and Great Britain. From 1974 to 1982, major rehabilitation works on the flood spillway channel were carried out at the Tarbela dam in Pakistan; a volume of 350,000 m³ of RCC was placed in 42 days. Since then, many dam projects have been constructed all over the world. As the years have passed and in view of the experience gained, the size of RCC dams has gradually increased: the maximum height of constructed dams went from 100 m in the 1980s to 160 m in the 1990s, to reach approximately 200 m from the year 2000 (Mason and Dunlop, 2003). As of late 2017, more than 700 RCC dams have been built in the world. The highest can be found in Colombia (Miel I, H = 188 m, in operation since 2002) and China (Longtan, H = 217 m, completed in 2009). Statistics indicate that dams with a height of between 30 and 50 m form the majority of RCC dams.

In 2018, RCC dams had been built or were being built in 70 countries. RCC dams have been constructed in regions with varying climates (arid, cold, or humid) (Durand et al., 1998). RCC dams have also proven to be the optimum solution in areas where rock embankment dams and double curvature arch dams in conventional concrete were already proposed.

Although in 2003 roller-compacted concrete was in some ways considered a new technology, it has now become the standard approach for large concrete gravity dams (ICOLD, 2018b, 2020).

17.1.2 Main characteristics and construction process

The initial idea behind RCC dams was to use the method of construction from embankment dams in the construction of concrete dams. This innovative process gave new momentum to the construction of gravity dams, particularly due to its quick execution (construction progresses by 2.5 to 3 m in height per week for large dams), a reduction in cement content, and therefore a cost per m³ that is considerably less than traditional concrete. Furthermore, it only requires a limited formwork surface. On a technical level, RCC dams must meet the same stability criteria as gravity dams, and they require similar analysis methods. Monitoring galleries and shafts can also be included in the design.

Compared to embankment dams, RCC dams are more impermeable and more resistant to erosion. Their volume is also smaller, and as a result the construction period is also greatly reduced. The placement rate for RCC dams can reach 10,000 m³/day. A further advantage of RCC dams is that appurtenant structures can be incorporated (see § 17.2.3). Their reduced footprint means that river diversion structures are shorter, as are any conduits crossing through the dam.

---

[1] To ensure that the most recent technological advances are included, this chapter 17 is based on the indications and information contained in the ICOLD Bulletin 177, *Roller-compacted Concrete.*
[2] In Japan, Roller-compacted Concrete dams (RCC) dams are known as Rolled-Compacted Dams (RCD).

It should be noted that RCC dams have proven to be sound with regard to internal erosion, overtopping, and earthquakes. During construction, the risk of damage from floods and cofferdam overtopping is lower. The Kerville dam in Texas (USA), which was overtopped in 1985 by a water depth of 4.4 m only 30 days after construction was completed, is a good example.

However, several negative aspects exist in opposition to these many advantages, such as the quality of the faces and the cost of cement transport.

In addition to the construction of new structures and cofferdams, RCC is also suitable for the rehabilitation of an existing dam, protection against overtopping of embankment dams, the strengthening of concrete dams (for example by building buttresses), and the heightening of conventional concrete dams.

The choice of site (see Section 17.2) is determined by the existence of a rock foundation and the local availability of a sufficient quantity of aggregate and water.

Generally speaking, RCC is a stiff material that does not settle. Its manufacturing process and properties once hardened are equivalent to those of traditional concrete. RCC is made up of aggregate (the maximum diameter of the particles generally varies from 50 to 76 mm), water, and water binders, to which are added mineral admixtures (fly ash, pozzolan). There is a proportion of 3 to 8% fines, which enable the cement content to be reduced; this improves the workability of the RCC and prevents segregation. RCC, which has a low cement content and reduced moisture content, is placed across large surface areas in consecutive thin layers using the usual earthworks machinery. Lift heights are between 25 and 40 cm; however, a depth of 30 cm is standard accepted practice. The placement of RCC is carried out by trucks, belt conveyors, bulldozers, and vibrating rollers that operate practically continuously (24 hours a day, 7 days a week). This is one of the reasons why issues relating to transportation and the preparation of aggregate are so important.

Furthermore, to ensure clean execution of the facings and also impermeability, an upstream slab is made out of traditional cast concrete or by using concrete elements that are precast or poured on site—the former serving as formwork.

Figure 17.1 presents a cross-section illustrating the various elements of an RCC dam.

In addition to the rectilinear gravity dam, whose traditional triangular profile is generally used, other shapes have been adopted for RCC dam projects. Initially, RCC arch gravity dams (suitable for U-shaped valleys) were built in South Africa. As for the geometry of these dams, their upstream facing is vertical and the downstream face or facing is sloped (H:V) by 0.6:1 and 0.5:1 (Figure 17.2). Methods for the grouting of transverse contraction joints have also been developed.

The first RCC dams were built and completed in South Africa and China, but it is in the latter country that this technology has rapidly grown. From 1990 until today, almost 220 RCC dams have been completed. Other RCC dams are also in operation today in Panama, Pakistan, Puerto Rico, Laos, and Turkey, among other countries.

In one case, the RCC was placed directly against the upstream and downstream formwork. As a precaution and to improve impermeability, the upstream face was covered with a geomembrane. Figure 17.3 shows the cross-section of the design of an RCC arch dam. Evaluating the thermal regime of an RCC arch dam is difficult, and three-dimensional calculations using the finite element method are necessary.

ROLLER COMPACTED CONCRETE DAMS (RCC)

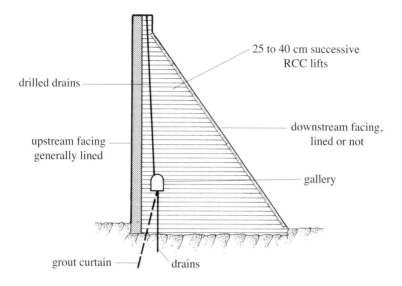

Figure 17.1 Main elements in an RCC dam

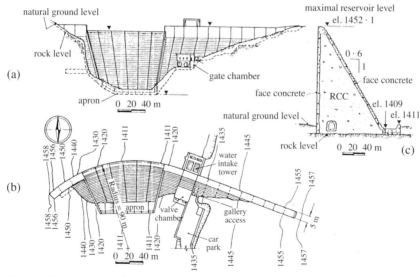

Figure 17.2 Example of arch gravity dam: Wolwedans dam (H = 70 m), South Africa (from Hollingworth et al., 1989)

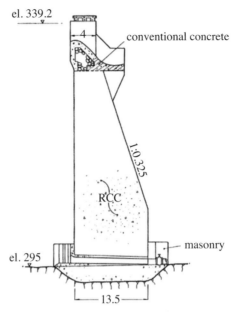

Figure 17.3 Cross-section of the design of an RCC arch dam in China (from Dunstan, 1989)

17.2 Design and Structural Layout

17.2.1 Site conditions

From a topographical point of view and as with gravity dams, U-shaped and wide valleys are well suited to the construction of RCC dams. For a site to be selected, the RCC dam needs to be set on rock foundations of good quality with sufficient bearing capacity and, if possible, limited settlement. As with other types of dams, the surface of the foundation must be regular, and should the need arise, any irregularities must be removed to avoid hard spots, which may be at the origin of cracks and the site of stress concentrations. Depending on its geological characteristics (presence of faults, joints, cracks), the foundation is treated in the same way as for a gravity dam.

It is very important that borrow pits capable of providing the entire volume of required aggregate can be found in close proximity.

17.2.2 Choice of profile for an RCC gravity dam

The transversal, vertical section of an RCC dam can have a geometric shape similar to a triangle, with a vertical upstream face, or a trapezoidal shape with sloped upstream and downstream faces (Figure 17.4). The hard embankment dam with a symmetrical profile proposed by Londe and Lino (1992) can be envisaged for situations where the foundations are of lesser quality, as loading on the foundation is low. This type of dam is also of interest from a seismic point of view, as the dynamic stresses, in particular tensile stresses, are substantially lower than those apparent in the traditional gravity dam profile (Londe and Lino, 1992; CFGB, 1997; ICOLD, 2003) (see also Section 17.6).

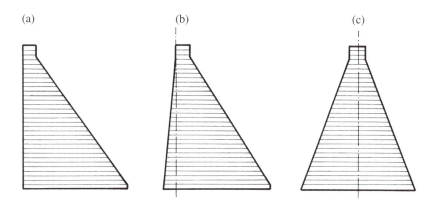

Figure 17.4 Various examples of cross-sections: (a) triangular shape with vertical upstream face; (b) trapezoidal shape; (c) symmetrical shape (cross-sections from examples in CFGB, 1997)

The choice in crest width must enable the mechanical placement of RCC. The dam profile can also include an upstream cofferdam.

Of course, the selected profile must meet the usual general stability criteria and acceptable stresses must also be respected (see Section 17.3).

17.2.3 Integrating appurtenant structures

RCC dams have facilitated the use of stepped chutes along the downstream face, which simplifies the integration of a spillway with a free weir at the crest (Figure 17.5). This type of weir allows highly efficient dissipation of energy and only requires a relatively shallow and short plunge pool to release water into the

Figure 17.5 Kholong Tha Dan RCC dam in Thailand with integrated spillway (dam height H = 93 m, crest length 2720 m, volume of RCC 5.47 mio m^3) (Courtesy M. Wieland)

watercourse. Aeration systems have recently been installed on weirs upstream of stepped chutes built in RCC for very high unit flow rates (50 m^3/s/m) and when the flow rate is greater than 15 m/s (Ditchey et al., 2000, Terrier et al., 2021). Conduits can also cross the dam from one side to the other. Avoiding elements with angles opening inward is recommended, as these can lead to cracks. Broadly speaking, care must be taken to design appurtenant structures that do not hinder the continuous placement of RCC. Appurtenant structures equipped with gates should ideally be located outside of the RCC dam body; they can be incorporated into a conventional concrete structure directly in contact with the RCC.

17.2.4 Facing and impermeability of the upstream face

Due to its method of construction, water can infiltrate into the body of RCC dams. To limit seepage, the upstream face can be covered with a suitable facing. Firstly, the durability of the structure must be guaranteed by protecting it against climatic effects (bad weather, freeze/thaw cycles) which can lead to its deterioration. By installing an upstream facing, the exterior appearance is improved, and regular surfaces are obtained.

Until the early 1990s, the most common method for protecting the upstream face of an RCC dam was to use conventional concrete, which had the disadvantage of requiring the installation of a conventional concrete manufacturing plant and specific transport. The greatest challenge lay in the separate placement of the RCC and conventional concrete.

Later, new methods were developed for producing the upstream faces of RCC dams. These involved adding a RCC grout to increase its workability and then pervibration. This was initially called GERCC (Grout-Enriched RCC) and later GEVR (Grout-Enriched Vibratable RCC). Table 17.24 presents examples of types of RCC applied against the formwork of arch dams.

GERCC, developed in China, involves pouring a liquid grout onto the surface of noncompacted RCC as soon as it has been spread. The grout permeates throughout the layer of RCC as it is pervibrated. GEVR, which was developed in Europe, involves spreading liquid grout onto the surface of the most recent RCC layer, before it is covered by the next layer (see § 17.4.1.4). GEVR (Grout-Enriched Vibratable RCC) is commonly used for the upstream facing of arch dams and RCC gravity-arch dams.

Of course, other solutions for covering the upstream face or installing an upstream slab also exist, some of which are briefly described below.

RCC facing with formwork

Should a completed upstream face require facing with RCC, this vertical surface can be lined with formwork. The upstream surface can also be covered with a geomembrane to improve impermeability (Figure 17.6).

Contiguous placement of conventional concrete and RCC

The different layers of RCC are applied directly onto the conventional concrete, which has already been placed (Figure 17.7). The placement of the two types of concrete occurs at practically the same time. However, it should be noted that RCC can be placed more quickly than conventional concrete and therefore its placement schedule needs to be adapted. To ensure good interpenetration between the two concretes, they are pervibrated. Care should also be taken to avoid segregation and gaps.

This process is also used in zones next to rock abutments, against which conventional concrete is used.

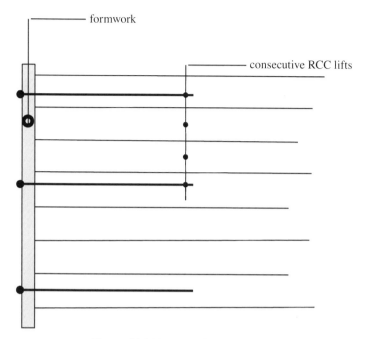

Figure 17.6 Upstream face with a rough RCC surface

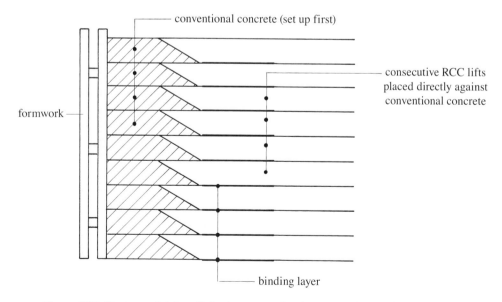

Figure 17.7 Upstream slab installed using conventional concrete placed jointly with RCC

Creation of a vertical face in conventional vibrated concrete

In this case, the upstream facing of vibrated concrete serves as the formwork for the RCC in the dam body. The structural layout for this conventional concrete wall is similar to that of a concrete dam (Figure 17.8). The designer should include dilatation joints equipped with sealing strips (waterstops) and drains. The horizontal concrete lift joints of the facing should also have a sealing system.

General impermeability can be improved by installing a PVC membrane attached using metal profiles or an exterior coating. The sealing system must be capable of absorbing joint movement.

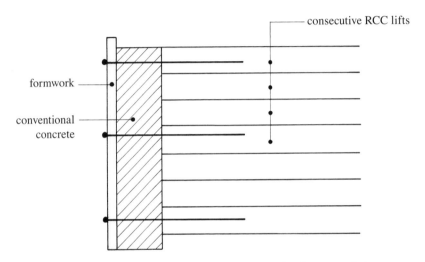

Figure 17.8 RCC formwork made from a conventional concrete facing

The use of precast concrete panels with or without a watertight barrier

Panels can be prepared on site or be transported in from elsewhere. They are designed for the construction of a vertical upstream face and provide a tidy final appearance for the upstream face. The RCC is placed directly against the panels which are supported and held by bars set into the lower layers of the RCC already placed (Figure 17.9).

Some panels are not sufficient to prevent seepage, while others are able to achieve this thanks to a layer of PCV membrane laid on this inner face of the panel.

Use of concrete elements poured on site or blocks of precast concrete

Precast concrete elements are overlaid and poured on site using a sliding formwork. They can also be precast and interlocked (Figure 17.10). A conventional concrete air-entraining admixture is used. It has advantages with regard to impermeability, resistance to erosion, and the freeze/thaw cycle.

RCC, which can only be used for limited lift heights, is placed directly against the precast elements. The correct bond with the upstream slab requires appropriate RCC composition and efficient compaction to avoid gaps at the interface.

This system can be applied to a vertical, stepped, or sloped surface (particularly for the downstream face). This method of application ensures high-quality construction of the faces.

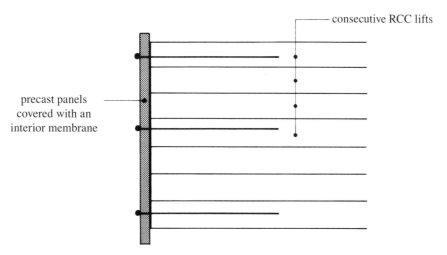

Figure 17.9 Precast concrete panels with or without a watertight barrier

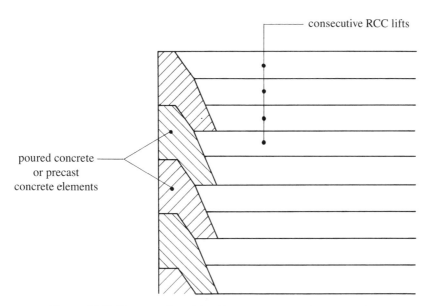

Figure 17.10 Upstream face slab with precast or poured concrete elements

Rough surface of the upstream face

Depending on the type of reservoir (no reservoir or permanent reservoir) and its use (protective structure, flood suppression), an upstream face without a slab could possibly function, though there are some disadvantages to having a rough surface. Due to the difficulty in carrying out compaction at slope edges, including an overprofile to allow for the inevitable deterioration of the surface (crumbling, degradation) is recommended. Ultimately, it results in a sloped, irregular, permeable face.

17.2.5 Finishing and facing of the downstream face

Some dams have a downstream face with a rough surface. In this case, the slope of the face is equal to the natural slope of the RCC, between 45 and 65° (which equals slopes of 1.0:1.0 and 0.8:1.0). It goes without saying the compaction at slope edges is not ideal, and resistance to bad weather and erosion over time is not high. In order to compensate for these disadvantages, an overprofile can also be included to account for possible deterioration. However, this no longer seems to be a preferred method.

Figure 17.11 illustrates mechanical means of compaction using a vibrating plate mounted on a tractor used at Les Olivettes dam in France to create the downstream face.

Applying a conventional concrete facing formed with steps or on a sloped plane (Figure 17.12) improves the situation and can, in some cases, be vital. This is the case, for example, for slopes that are steeper than the natural slope (especially if the slope is greater than 0.8H:1.0V).

As for the final look and durability, the placement system for elements in poured or precast and interlocked concrete is identical to that used for the upstream face.

Should the RCC dam have a weir with high flow speeds, the sill will need to be lined with conventional concrete. In principle, steps anchored in RCC of a height equal to a multiple (2 or 3) of the lift must be allowed for. Installing steps has the advantage of reducing the residual energy to be dissipated at the downstream toe of the dam (André, 2004, Stojnic et al., 2021).

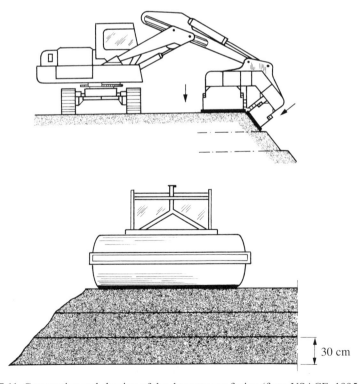

Figure 17.11 Compaction and shaping of the downstream facing (from USACE, 1995, 2000)

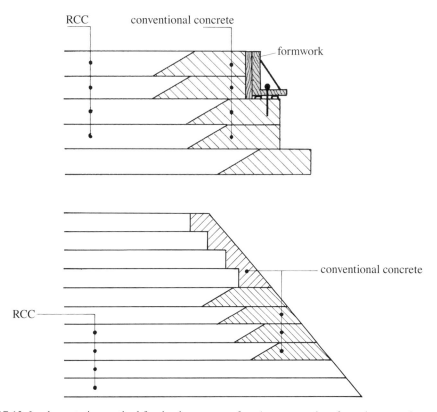

Figure 17.12 Implementation method for the downstream face (top: stepped surfaces; bottom: plane surfaces)

17.2.6 Treatment of horizontal work surfaces

As mentioned previously, an RCC dam is built by the consecutive placement of thin layers. Before continuing to pour RCC, the surface of the last layer must be treated carefully, as its quality depends on its stability and impermeability. In regard to the latter, it is important to limit the risk of seepage and water loss. To ensure the monolithic nature of the structure, the interface and cohesion across the entire horizontal surface play a major role. The lift surfaces must reach the levels of shear strength (sliding stability) and tensile strength (particularly important for bearing seismic load) determined in the project design.

During construction, the wait time between two layers must be as short as possible so that the following layer is placed on a surface that is still damp. There is a risk that the surface quality could be reduced if it is exposed to the air for too long. Placing a new layer while the air is cool and damp is therefore recommended.

To ensure the proper implementation of the connection between the layers and to guarantee impermeability, possible measures include cleaning the surface awaiting the new pour and applying a layer of mortar before placing the next layer of RCC.

First of all, the lift surfaces must be clean, that is, free of any debris and dirt, before placement can continue. This is why the concrete always undergoes a cleaning process. The various methods for cleaning depend on the amount of time that the surface is to remain in the open air: for a duration shorter than 24 hours, the debris and water are simply removed with aspiration; between 24 and 48 hours, cleaning is done with a metal brush; more than 72 hours, cleaning is carried out with a sand blaster and pressurized water (jet cleaning should be done with low pressure, approx. 7 kg/cm^2).

Then, a layer of mortar is spread across all or part of the surface to be covered. The mortar is made up of large aggregate. It has a high cement content and "high" slump. The extent of the mortar layer depends on the degree of use of the reservoir and on the results from a structural analysis. This layer of mortar also increases shear strength and tensile strength, as well as cohesion between the layers. In a zone with seismic activity, mortar is applied across the whole surface. Limited use of mortar can be envisaged for a dam located in a region with low seismicity. Without mortar, strength at the level of the horizontal joint is lower than that of the RCC itself.

The time between the application of mortar and the RCC must be as short as possible during hot periods and can be longer if the weather is cold. Use of a setting retarder is recommended in hot regions, especially if mortar has not been included in the design.

During bad weather, make sure that puddles of water do not form. Placement should be paused or even postponed in cases of heavy rain and low temperatures (< 0°C). Some RCC dams have been built with layers sloped from the right bank to the left bank or upstream-downstream to drain rainwater (see also § 17.4.4.4).

17.2.7 Vertical transverse contraction joints

As a rule, transverse contraction joints are an important structural element that are designed to help control cracking, especially due to the effects of temperature and confinement caused by the foundation, as well as irregularities in the foundation. They are of value across the whole section. However, it should be noted that some low-height dams have been constructed without joints. Some reservations have been expressed, however, due to the fact that the execution of transverse joints slows down the placement of RCC and increases costs.

There are several methods for installing these joints. They can be located near to the upstream face in a section made of conventional concrete and must be fitted with a sealing system and drainage. The most widespread solution consists in driving a frame with a steel plate covered in a plastic sheet into the RCC using vibration. After the RCC has been placed and compacted, the plate is removed and the plastic remains in place. The panel gradually shifts downstream. Another solution involves installing the contraction joint using formworks, in the same way as for traditional concrete dams. And finally, it is also possible to place the uncompacted RCC on both sides of a panel covered with a sheet of PVC. The RCC is compacted after the panel is removed, leaving behind the sheet of PVC.

The position and number of joints are determined based on a thermal analysis, any construction issues, and the foundation profiles. For the latter, it is preferable to include a transverse joint if different conditions or any irregularities exist in the foundations. The space between the joints varies greatly from one dam to another. According to the literature, it can range from 15 m (Japan) to 40–100 m (USA). This space depends on the type of RCC and its cement content (Table 17.13). A distance of 20 to 45 m is considered an appropriate interval.

Table 17.13 Spacing between contraction joints based on the type of RCC

Type of mix	Cement mix [kg/m^3]	Space between the joint [m]
Dry and lean mix	< 99	30 and over
Mix with medium-paste content	100–149	15–50
Mix with high-paste content	> 150	20–75

17.2.8 Galleries and shafts

The galleries and shafts in RCC dams fulfil the same roles as those in traditional concrete dams. The presence of a perimetric gallery enables grouting works to be carried out, drains to be drilled, seepage water to be collected, and concrete surfaces to be inspected. In RCC dam projects with a height lower than 30 m, engineers may decide not to include a gallery.

Galleries are often installed in the body of an RCC dam near to the upstream face at a distance sufficient to ensure that construction equipment can operate in the zone or at the upstream heel to guarantee the connection between the upstream facing and the foundation (Figure 17.14 (a)). There are several different ways of integrating a gallery into an RCC dam without disturbing the work in progress.[3] One solution involves the conventional use of ordinary formwork or precast elements; once removed these provide a way to look at the RCC dam. It is also possible to plan for the installation of a gallery in a section of uncemented RCC that will later be excavated; this approach has the advantage of not impacting the works program. Executing a gallery in conventional concrete does not allow the inspection of RCC while the dam is operating and therefore could potentially delay the detection of seepage water or cracks. Depending on the method chosen, care must be taken that construction machinery does not damage the initial RCC layers. A precast slab (Figure 17.14 (b)) or a near horizontal reinforcing beam can be placed above the gallery as a strengthening measure. To avoid right angles, the gallery ceiling can also be given the shape of a broken line (2 sloped sections with possibly one central horizontal section); the selected slopes must allow for adequate placement of concrete.

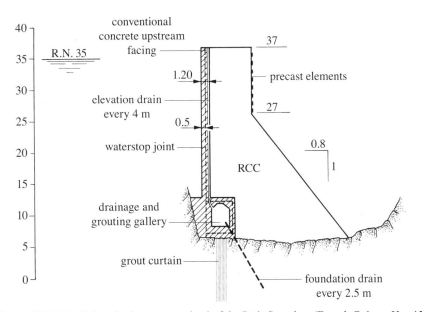

Figure 17.14 (a) Gallery in the upstream heel of the Petit-Saut dam (French Guiana, H = 45 m)
(from Dussart et al., 1992)

---

[3] A reduction in productivity of 10 to 15% for layers crossed by a gallery (ICOLD, 2003).

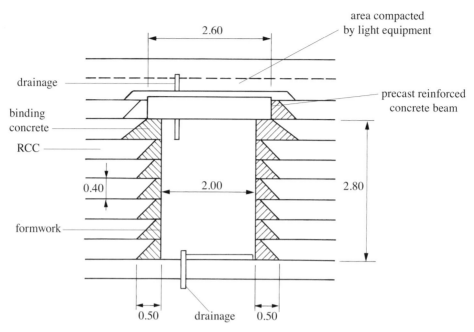

Figure 17.14 (b) Method for installing an inspection gallery (from Giovagnoli et al., 1992)

17.2.9 Drainage and seepage control

The aim of a drainage system, as for any concrete water-retaining facility, is to intercept downstream flowing seepage, restrict uplift to improve conditions of stability, and prevent excessive pressure against the downstream facing. Water can be collected by way of vertical drains integrated into the upstream facing and vertical drains drilled into the concrete mass from the crest or perimeter gallery (3 m spacing). Drilled, horizontal drains opening out at the downstream toe enable seepage water to be evacuated. Traditional drains made of porous pipes placed in the plane of a lift or in the foundation also enable water to be directed toward a collector.

17.2.10 Grouting

As is usually the case, grout can be injected from a perimetric gallery. However, to create a worksite that is independent from the dam worksite, this grouting can be carried out from an upstream platform during the construction of the dam. This means that more space is available than in a gallery and that the presence of the dam acts as an overload.

17.3 Essential Aspects in Design Analysis

17.3.1 Analysis of general stability

The principles applied for verifying the stability of gravity dams are also valid for RCC dams. However, there are some differences, particularly with regard to the hypotheses and specific conditions necessary for ensuring sliding safety at the interface of each consecutive layer. As a rule, calculations should be carried out based on a two-dimensional structure.

Stability must be verified at the level of the horizontal construction joints, at the foundation interface, and along discontinuity zones in the foundation itself. The usual factors of safety for gravity dams are to be used as a reference and must be respected.

Shear strength, especially at the interface of the layers, is an important element for the verification of stability for RCC dams. It is depicted using Coulomb's law:

$$R = c + \sigma \cdot \tan\phi = c \cdot A + \{(\Sigma V - \Sigma u)/A\} \cdot \tan\phi,$$

where
- R = shear strength [MPa]
- c = cohesion [MPa]
- σ = normal stress [MPa]
- $\tan\phi$ = angle of friction [°]
- A = surface [m^2]
- u = uplift [MPa]

hence the factor of safety $FS = R/\Sigma H$.

What emerges is that in the first approximation, cohesion can be accepted at 5% of the compressive strength as long as a layer of mortar is applied between the layers of RCC. Should this not occur, a zero-cohesion value must be introduced. The internal angle of friction can vary between 40 and 60°; 45° is generally accepted as a first approximation (USACE, 2000). A slight upstream slope in the layers can improve conditions for shear strength.

With regard to uplift in the body of an RCC dam, a linear variation similar to that of traditional gravity dams is accepted. The percentage of hydrostatic pressure to consider will depend on the permeability of the RCC, the treatment of horizontal joints (whether there is a layer of mortar), the means used for the placement of the RCC, and the construction method for the facing (whether there is a sealing system). If there is no layer of mortar between the RCC layers, uplift equal to 100% of upstream and possibly downstream hydrostatic pressure must be assumed. If the upstream face has a waterproof facing possibly combined with a drainage system, a reduction in uplift can be envisaged. Care should be taken over the type of downstream facing, which (if it is watertight) can have an impact on the uplift value in its near vicinity. Lastly, at the level of the foundations, drainage conditions must be considered in order to establish uplift distribution. As has been mentioned previously, for RCC dams higher than 30 m, a base gallery is recommended, as is the integration of drains drilled into the dam body to reduce uplift and control seepage.

17.3.2 Loads and stresses

Calculations are carried out as for two-dimensional structures, based on a traditional stress calculation in the plane field or a finite element calculation. The calculation of RCC dams under seismic loading is similar to that of traditional concrete dams.

RCC dams are unreinforced structures and therefore rely on their compressive strength, tensile strength, and especially shear strength to counteract the forced applied to them and to the considerable effects of temperature. RCC dams must be designed so that no tensile stresses appear in situations of normal loading. It goes without saying that tensile stresses are tolerated in situations of exceptional and extreme loading. As for acceptable stresses, these follow similar criteria as those applied to traditional vibrated concrete (CVC).

Tensile stress failure at the level of the lift joints is a critical characteristic of RCC. Several tests are possible to check tensile strength; the direct tension test is the most appropriate. For dynamic calculations, dynamic tensile strength is accepted to be equal to 1.5 times direct tensile strength.

17.3.3 Cracking

Cracks can appear for several different reasons, in particular due to a variation in volume (concrete contraction following cooling), excessive tensile stresses caused by an earthquake, stress concentration at the openings, an alkali-aggregate reaction, or differential or rapid deformation in the foundation. Cracks also depend on the amount of stress on the foundations, climatic conditions, and the geometry of the structure.

Cracks can begin to form at the concrete-foundation interface and then propagate throughout the body of the dam, both upward and toward the interior. Transversal cracking can be an issue across the whole of the section and thus affect the monolithic nature of the dam. Figure 17.15 presents different types of cracks that can be found in both an RCC dam and a conventional CVC dam. Cracks can also appear on the surface if it is subject to rapid and substantial daily fluctuations in ambient temperature.

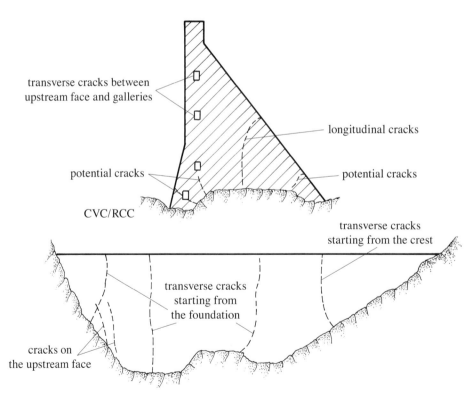

Figure 17.15 Typical cracks in a conventional concrete dam and an RCC dam (from Sarkaria and Andriolo, 1995)

17.3.4 The effect of RCC's internal temperature

How the dam behaves in response to the release of concrete heat during hardening is an important element because thermal stress develops, which ends up causing tensile stresses. During construction, the concrete temperature may be considerable in the center of the dam. The air temperature also has an effect near the upstream and downstream faces of the dam; indeed, concrete temperatures react strongly to ambient conditions.

It is therefore usual and essential to carry out a detailed thermal analysis of an RCC dam's stresses using the finite element method. It is recommended that calculations take into account the method of construction and works program, as well as external conditions (such as temperature during placement, climatic variations in temperature, the presence of water upstream of the dam, etc.). The results provide the basis for minimizing and controlling the development of tensile stresses (thermal cracking), and for establishing the distribution of contraction joints.

To limit and prevent cracking as much as possible, the difference in temperature between the peak value reached during concrete placement and the final stabilized value should be reduced. The possibility of lowering the temperature should be studied. Different means exist, as in the case of conventional concrete, to control peak temperatures: opting for a low cement content, using a cement with low heat generation, choosing appropriate ratios of mineral admixtures (pozzolan, fly ash), cooling the aggregates, replacing water introduced into the concrete mixer with cooled water or ice, choosing an appropriate time period for the placement of RCC (constructing complex elements in winter or at night), cooling surfaces of compacted RCC, or including contraction joints.

To better track the development over time of RCC temperatures, fiber-optic cables can be installed in the body of the dam.

17.4 Constructing RCC Dams

17.4.1 Composition of RCC

The aim of the following sections is not to present universal recipes for manufacturing RCC, but to highlight some key principles.

17.4.1.1 Types of RCC

In general practice, three types of RCC dams exist (Dunstan, 1994, 1999):

- An RCC dam with a dry, lean-mix and a cement content of less than 100 kg/m^3. This type of RCC (LCRCC) is relatively permeable, particularly at the level of the horizontal joints (which suggests that upstream watertightness protection will be required) and has moderate levels of strength, elasticity modulus, and thermal stresses. Cohesion between the layers is low, and the layers should only have a short exposure time during construction.
- RCD dams (roller-compacted dam)
  The RCD dams constructed in Japan are similar to traditional gravity dams. They make use of different concepts and techniques, notably by employing more binder and with thicker concrete lifts comprising several thin layers compacted into one block. The core of the dam is made of RCD. Traditional concrete surrounds the RCD to a depth of several meters and is designed to improve frost resistance.

- RCC dams with a high paste-content (high proportion of fly ash) and a cement content of 150 kg/m^3 or more. This type of RCC (HCRCC) is impermeable and has mechanical properties similar to traditional concrete. The high pozzolan content, for example, helps reduce thermal stresses.

An intermediary group can also be mentioned—the RCC dam (MCRCC) with a medium cement content of between 100 and 150 kg/m^3.

17.4.1.2 Principal components of RCC

RCC is a mixture of aggregate, cement, a mineral admixture (for example, pozzolan), water, and, when required, occluded air by an air-entraining admixture. RCC differs from traditional concrete through its particle size, consistency, and binder content, which makes it suitable for compaction via vibrating rollers.

Aggregate

A large variety of aggregates may be found in RCC. Among the natural forms of aggregate, the most commonly used in concrete comes from sedimentary, siliceous or limestone rock, metamorphic rock, or igneous rock such as basalt, granite, or porphyry. Aggregate may be sourced from quarries or alluvial deposits, and a mixture of crushed and rounded aggregate is possible. Crushed aggregate has been found to reduce the tendency to segregate. The quality of the aggregate, which, among other criteria, can be characterized by its unit weight and abrasion resistance, must be high if high strength is required. The use of lower quality aggregate is better suited for the internal zone of the dam and can be surrounded by concrete of a higher quality, especially in cases of harsh or moderately harsh external conditions. The dust covering rough aggregate must be restricted, and depending on its quantity, it may be necessary to wash the aggregate.

The maximum diameter of the aggregate affects the volume of voids and the potential for segregation, as well as the cement content. Outside of China and Japan, where 75/80 mm are usually specified (or sometimes 150 mm), in practice a maximum size of between 40 and 60 mm has been recommended in other countries. A continuous particle size is required to obtain a RCC with a minimum of voids. Particle size must be monitored throughout the whole works program. To obtain an high-quality product, each category of aggregate must be uniform and remain within the recommended range. Aggregate is most often separated into 4 categories, frequently into 3 or 5 categories, and rarely into 2 or 1 categories (ICOLD, 2018b, 2020). The maximum diameter of aggregate and the homogeneity of categories has a major impact on the regularity, cohesion, and ease of RCC compaction.

The storing of aggregate and the location of the concrete plant are important factors. It is often necessary to have large stocks of aggregate available before beginning placement of RCC on the dam. Providing a sufficiently large storage area is available, 30 to 40% of the total volume may be required.

Cementitious materials

RCC does not require the use of a special cement, but it must nevertheless generate only a small amount of hydration heat. RCC can be produced from any type of usual cements or a combination of cement and mineral admixtures (siliceous materials with a fine texture). The use of mineral admixtures allows the cement content to be reduced and therefore the cost and generation of hydration heat. They also provide additional fines for the workability of the mixture and the volume of paste.

Generally natural pozzolan (volcanic ash, pumice, tuff, opaline flint, clay schist), fly ash (with either low or high lime content), calcined clay, or slag is added to the cement. The use of one or other of these mineral admixtures depends on their availability, and prior tests are vital. The percentage plays an important role in obtaining a high-quality RCC. Most RCCs contain mineral admixtures added to the cement at a proportion of 15 to 100%. The higher their content, the more the properties of the RCC resemble those of traditional concrete. Long term, its strength can even be greater.

Lastly, the cement must be selected to ensure high resistance to sulphate attacks and the alkali-aggregate reaction. RCC containing mineral admixtures are less sensitive to the alkali-aggregate reaction.

Careful planning is required to ensure a continual supply of cement. The on-site storage capacity for cement will depend on how regularly it can be supplied.

Admixture

Admixtures, principally either water-reducing or air-entraining, have been successfully used in almost 50% of constructed dams. This use of admixtures has enabled the workability and durability of concrete to be improved during freeze/thaw cycles. Retarder admixtures for setting are used to a lesser degree, as a high mineral admixture content already leads to sufficient delay for hardening.

Mixing water

For reference, Table 17.16 gives some moisture content values corresponding to one specific set of aggregates and one workability. Other aggregates and other workabilities will require different moisture contents, however, the relationship between the different maximum sizes will remain the same (ICOLD, 2003).

Table 17.16 Approximate values for moisture content in RCC depending on the maximum particle size (Table 5, from ICOLD, 2003)

Maximum size of aggregate [mm]	150	75	50	38	19
Free moisture content [kg/m^3]	85	95	100	108	117

The use of a VeBe device[4] enables the moisture content to be determined and therefore workability and the best ratio of aggregates.

17.4.1.3 Study into RCC composition

There are several different ways of formulating RCC:
- The composition is set for conventional mass concrete on the basis of consistency tests (the VeBe index must fall within a range of 10 and 40 s; a low value indicates good workability, but lower strength due to the high moisture content). However, new research recommends a lower VeBe value for the composition of LCRCC, with values ranging between 8 and 15 s.
- The composition is set to obtain a lean RCC on the basis of a given particle size. The cement and moisture content are modified until the desired compressive strength is reached. With a cement content between 60 and 120 kg/m^3, compressive strength reaches 6.9 and 10.3 MPa in 1 year.
- The composition is based on methods of soil compaction.

---

[4] The VeBe test developed by Viktor Bährner: a test determining the consistency of fresh concrete by measuring the time it takes for a vibrated sample of concrete (molded into an Abrams cone) to fill a cylindrical container.

As a rule, it is very important to develop a coherent mixture that presents good workability in order to ensure adequate compaction and the lowest possible tendency to segregate. Improving workability and reducing segregation is achieved by lowering the maximum aggregate size, taking into account their shape and particle size, and the use of crushed material. Particle size has an impact on the consistency and workability. Any increase or decrease in fines leads to an increase or decrease in moisture content and workability (the filler acts as a lubricant). A low proportion of sand increases the potential for segregation. Currently, a sand percentage between 38 and 50% of the total weight of aggregates is recommended (previously 34 to 40%). The upper value requires a greater volume of paste, while the lower value is of interest for large structures as it allows the particle size composition and segregation to be carefully controlled (Holderbaum and Roarabaugh, 2001). The proportion of fines for a lean concrete is in the order of 5 to 15% of total weight of aggregates or from 15 to 40% of the weight of the sand.

Pozzolan has similar properties to cement, which means the cement content can be reduced. The amount of pozzolan is between 0–40% of cement for a low content (< 99 kg/m3), between 20–60% cement for a medium content (100–149 kg/m^3) and between 30–80% cement for a high content (> 150 kg/m^3).

No matter the design size of the dam, setting up an initial test program in the laboratory for the composition of the mixture during the preliminary design phase of the dam is recommended. These preliminary tests are essential for large dams or for the first use of RCC in a region.

At the beginning of the construction phase, the initial proportions of the RCC mix are refined in the worksite laboratory using the selected construction aggregates. The in situ properties of the RCC are later confirmed during large-scale tests and during construction.

As a reference, Table 17.17 gives the average cement values for various types of RCC dams (ICOLD, 2018b, 2020).

Table 17.17 Average mixture proportions (kg/m^3) of cementitious paste in various RCC types

Material			Type of RCC		
		LCRCC	MCRCC	RCD	HCRCC
Portland cement	{C}	72	80	87	87
Supplementary cementitious material	{CS}	9	37	35	108
Water	{E}	122	116	96	111
Parameter					
Cementitious materials	{CM} = {C+CS}	81	117	122	195
Water/cementitious ratio	{CE} = {E/CM}	1.51	0.99	0.79	0.57

17.4.1.4 Grout enriched vibratable RCC

The aim is to mix grout on-site and vibrate it through the RCC before compaction. Once this has been done, the RCC is compacted. The resulting RCC is equivalent to a conventional concrete but without the use of an air-entraining agent. This process (which removes the need to manufacture a second concrete) is only used in zones near to the upstream and downstream formworks and to the abutments (§ 17.2.4). Developed in China, the exterior surface obtained by this method is resistant over time and presents reduced permeability.

17.4.2 Typical properties of RCC

As with conventional concrete, the strength and elastic properties of RCC depend on the quality of the aggregate and the cement content. Some guidance (mostly drawn from USACE, 2000, and ICOLD, 2003, 2018b, 2020) is given below, however, it must be noted that both laboratory and field tests provide the most realistic figures for dam design:

- RCC *unit weight* depends on the density of aggregate in the concrete mix, the occluded air, and the degree of compaction. It is of a similar, or even slightly higher quality than that of conventional concrete due to the lower moisture content (> 2.4 t/m^2)[5].
- *Compressive strength* is the parameter best suited to measuring the quality and uniformity of concrete. The design compressive strength is selected based on the level of strength required to cope with compressive stress, tensile stresses, and shear stress under conditions of load. Compressive strength can range from 6.9 MPa to 27.6 MPa, or even more. Commonly accepted values for most projects are between 13.8 MPa and 20.7 MPa at 90 days or even 1 year. A minimal value of 13.8 MPa is recommended primarily for reasons of durability (USACE, 2000).
- *Tensile strength* depends on the age, loading, and cement content. It can be calculated using direct tension tests, the Brazilian test (splitting tensile strength test), or flexural tests (seldom used due to the difficulty in preparing test samples). Depending on the results obtained during project implementation, direct tension tests have given values ranging from 3 to 9% of compressive strength. Most values are between 6 and 8%. Results from the Brazilian test are more consistent than those from the direct test. However, they are overestimated and must therefore be reduced by a factor of 0.75. Dynamic tensile strength is accepted as being 1.5 times direct tensile strength (USACE, 2000).
- *Shear strength* is obtained by direct, bi-axial, and tri-axial tests on samples. Cohesion c and the angle of friction ϕ of RCC are comparable to those of conventional concrete, as long as the aggregate is similar. C depends on the cement content and ϕ on the quality and particle size of the aggregate. The cohesion value C may sit between 0.5 MPa and 4.1 MPa; it is in the order of 20% of compressive strength. The internal angle of friction can vary between 40 and 60°. At the level of the horizontal joints, cohesion equal to 5% of compressive strength (with a layer of mortar between the layers of RCC) and an internal angle of friction of 45° can be accepted for preliminary design.
- The *elasticity modulus* depends on the nature of the aggregate, the water/cement ratio, and the cement content.
- *Poisson's ratio* has a value between 0.17 and 0.22; a value of 0.20 is recommended.
- *Creep* is mainly affected by the elasticity modulus of the aggregate and the proportion of fines. Creep is low for concrete with high elasticity modulus and high strength.
- The *permeability* coefficient of RCC is between 10^{-5} and 10^{-12} m/s; mean permeability is in the order of 10^{-9} m/s. It appears from the observation of constructed dams that permeability is related to the mix content (Figure 17.18). It is accepted that RCC with a high cementitious content has satisfactory impermeability. However, RCC with a low cementitious content is not likely to provide adequate impermeability without specific treatment such as the installation of an upstream facing to reduce seepage (ICOLD, 2003).

---

[5] The value of the density of conventional concrete is in the order of 2.24 to 2.56 t/m^3.

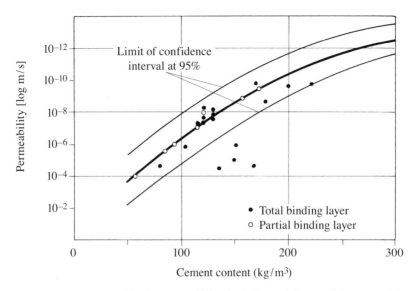

Figure 17.18 Relationship between total in situ permeability (including at joints) and the cementitious content of the concrete from RCC structures (from ICOLD, 2003)

- *The workability* of RCC determines the capacity of the material to be successfully placed and compacted without segregation. Measuring consistency using a VeBe device gives a useful indication of workability. A VeBe of 15 s should be sufficient.
- The RCC should demonstrate the required degree of *durability* in view of ambient conditions and the materials used. The RCC must withstand deterioration resulting from, for example, the alkali-aggregate reaction and must be able to withstand freeze/thaw cycles.

17.4.3 Tests

Just like conventional concrete (see Part V), RCC must be subjected to tests to determine the appropriate mix (aggregate distribution, cement and admixture content) and to establish its mechanical and physical properties. Tests are carried out in the laboratory and on site on both fresh and hardened concrete.

The aggregate must undergo the same quality checks as those used for conventional concrete.

During construction, the testing program focuses on the aggregate properties, the content ratios of RCC components, the characteristics of fresh and hardened concrete, and in situ compaction. The frequency of tests depends on the size of the dam. Table 17.19 provides a list of the main tests.

Table 17.19 (a) Typical constituent materials quality control tests (ACI, 2011), from ICOLD 2020

Material tested	Test procedure	Frequency*
Cement	Physical and chemical properties	Manufacturer's certificate or pre-qualification
Supplementary cementitious material	Physical and chemical properties	Manufacturer's certificate or pre-qualification
Admixture		Manufacturer's certificate
Aggregate	Relative density and absorption	One per month or 50 000 m^3
Aggregate	Flat and elongated particles	Twice per month or 25 000 m^3
Aggregate	Los Angeles abrasion	One per month or 50 000 m^3
Aggregate	Gradation	One per shift or One per day
Aggregate	Moisture content	Before each shift or as required
Aggregate	Compacted bulk density	Twice per month or 25 000 m^3
Aggregate	Sand equivalent (SE)	Twice per month or 25 000 m^3
Aggregate	Efflux	One per month or 50 000 m^3

* Frequency will be dependant upon the size of project and the degree of control required

Table 17.19 (b) Typical fresh RCC quality control tests (from ICOLD 2020)

Test procedure	Frequency*
VeBe consistency and density	500 m^3 or as required
Gradation	1000 m^3 or as required
In situ density and moisture content	1000 m^3 or as required
Setting times	Three times per shift
Oven-dry - Moisture content	1000 m^3 or as required
Temperature	100 m^3 or as required
Variability of mixing procedure	Two per month or 25 000 m^3

* Frequency will be dependant upon the size of project and the degree of control required

Fig 17.19 (c) Typical hardened RCC quality control tests (from ICOLD 2020)

Test procedure	Frequency*
Compressive strength	Two per day or 1000 m^3
Compressive strength on specimens with accelerated curing	Situational, typically approx. 5000 m^3
Tensile strength (direct and/or indirect)	One per day or 2000 m^3
Direct tensile strength on jointed cores	As instructed
Elastic modulus	One per 10 000 m^3
Permeability	One per two months, or 100 000 m^3

* Frequency will be dependant upon the size of project and the degree of control required

With regard to fresh concrete, visual inspections should be carried out to verify the condition of the lift joints (to detect possible soiling and define the method and material to use for cleaning and preparing surfaces) and while spreading and compacting. Field tests enable the density (nuclear densitometry) and moisture content to be established. Checking temperatures during construction is also important.

For tests on hardened concrete, test samples are produced while placement takes place. These are subjected to strength and elasticity modulus tests. The permeability of the placed concrete can be established through pressurized water tests in holes drilled vertically into the body of the dam; these tests are important if no upstream facing has been planned.

17.4.4 Placing RCC

17.4.4.1 Placement tests

Before construction of the dam starts, in situ placement tests must be carried out across a sufficiently large area set aside for this purpose outside of the dam area. In addition to staff training, the aims of these tests are to confirm the appropriateness of the material and methods and to assess the performance of the RCC. They are useful for establishing and verifying the thickness of the layers depending on the effectiveness of the spreading and compacting methods. The test program also includes a study of the binding layer and how the layers will be placed. As such, it is a question of calculating the mix of binding mortar and the acceptable time gap between the placement of two layers, and by possibly analyzing the effects of a setting retarder admixture. Lastly, the best method for cleaning the surfaces is also established.

The test sections also establish the minimum number of passes (passage in one direction) of a vibrating roller to obtain the specified compaction. This depends on the workability of the RCC and the thickness of the layers. Four to eight passes of a 10-tonne vibrating roller is generally sufficient to obtain the desired density for RCC spread in 30-centimeter-thick layers.

The binding and/or direct contact between the RCC and the conventional concrete should also be evaluated.

17.4.4.2 Installation and site equipment for placement

The specific construction methods for RCC dams, which includes the placement of a material over very large surfaces, requires well-designed and efficient logistics and overall organization. All of the various operations (mixing, transportation, unloading, spreading, and compacting) must be carefully programmed in order to complete the process as quickly as possible. By using high-performance installations (concrete manufacturing, transportation, spreading), outputs of up to 1000 m^3 can be reached.

The storing of aggregate and the location of the concrete plant are important factors. Not only should large stocks of aggregate be available, but a considerable amount must already be on site before construction starts. Providing there is sufficient storage space, 30 to 40% of the total volume must be available at the start of the works.

Like any dam, RCC dams are equipped with various measuring instruments to track their behavior (extensometers, thermocouples, piezometers).

Site facilities (Figure 17.20) must take into consideration the following factors:

- Concrete mixing plants are key elements for the rapid manufacture of RCC. They must have sufficient capacity to meet a high rate of placement and be able to correctly mix the components. To ensure sufficient back up, having two plants is recommended. Concrete mixing plants function with little or no downtime, and because of this, maintenance and repair works must be carefully planned.

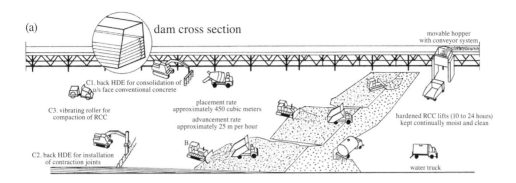

Figure 17.20 General presentation of the plants and site equipment
(from: (a) USACE, 1995; (b) Boes, 1999)

- A separate concrete plant must be set up for the production of conventional concrete (CVC) and mortar.
- The means chosen for transportation of concrete to the work zones must ensure that the least possible segregation occurs (generally conveyor belts and trucks, which are highly flexible methods).
- For spreading, bulldozers with u-shaped blades are used.
- For compaction, vibrating rollers (10-, 15- or even 20-tonne rollers with double or single drums) are generally employed.

17.4.4.3 Treatment of horizontal joints

We have previously seen (§ 17.2.6) that the treatment of joints between layers is an important step to ensure correct binding and ultimately to obtain a monolithic structure. The adherence between the layers of RCC results from the bond created by the binder and the penetration of aggregate into the new layer under the surface of the previously placed layer. Treating joints and the use of binding mixes are usually recommended on the basis of concrete maturity (a product of curing time by temperature). There are two types of binding layers: mortar and concrete binder. Mortar is generally 10 to 20 mm thick, while the depth of concrete binder varies depending on the design.

Three types of joint treatment have been defined (ICOLD, 2003, 2018b, 2020):

- *The fresh (or hot) joint*
  This type of joint (hot joint) is used when the placement of RCC layers occurs quickly and the RCC is still workable when the following layer is applied; the aggregate from the upper layer penetrates the lower layer, which is compacted before it hardens.

- *The intermediate (or warm, or prepared) joint*
  This type of joint (warm joint) is characterized by the fact that the pause in concreting is long enough so that the aggregate from the upper layer does not penetrate the lower layer, but too short to remove the surface mortar without large aggregate coming loose. Furthermore, the compaction of the upper layer can damage the lower layer which is further advanced in the hardening process. However, all debris on the receiving layer must previously have been removed. This type falls between the hot joint and the cold joint.
- *The dry (or cold) joint*
  At this stage, the surface condition of the previously laid layer does not allow (or almost doesn't) the penetration of aggregate from the upper layer into the previously compacted layer. In this case, surface mortar can be removed with water or pressurized air without lifting off aggregate. This type of joint also requires thorough cleaning to remove all debris. Cold lift joints are used when the pause in concreting between RCC placement lasts for several days.

The surface curing of the freshly compacted RCC is an operation that ensures protection from desiccation or frost. To keep the surface continually damp, water trucks spray a fine mist over the concrete. Manual watering must be carried out in places where the trucks cannot go. If a sufficient labor force is available, sprinkler systems and hand-held water hoses can be used. Insulating plastic covers spread over the concrete protect against frost until the concrete has sufficiently matured. However, protection also needs to be put in place for hot temperatures. Thermal protection can be obtained by covering the surface with a layer of damp sand, providing the ambient humidity is relatively low.

17.4.4.4 Stages of construction

Monolithic construction or staged construction of monolithic blocks

RCC dams can be built in a totally monolithic manner or by designing, as for conventional dams, monolithic blocks separated by dilation joints (Figure 17.21). Dividing the dam into several blocks and the construction into stages restricts the work zones but means that layers can be placed rapidly. This method of construction offers a certain degree of freedom for the start of the works, as it is possible to begin placing RCC once the excavations have been fully completed in one particular zone. However, more formwork is required and access to the worksite is more complicated.

Placement using layers

Placing horizontal layers is the simplest method, providing that the layers can be placed quickly. The concrete is placed from one bank to another. It therefore depends on the manufacturing capacity, transportation and distribution systems, and the spreading equipment. Taking these elements into account, a 30-centimeter-deep horizontal layer has gradually become standard. It should be noted that the implementation of thicker layers either in a single layer or in several layers has either been constructed or experimented on.

The process of placing consecutive noncontinuous horizontal layers across the top of a lift has been used on some dams with good results. It is simpler than spreading sloped layers, but before the following layer is placed, it requires the lift surface to be treated like a hot joint.

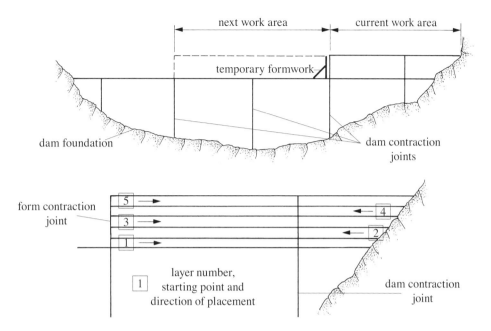

Figure 17.21 Staged construction of an RCC dam using the monolithic block method (from Holderbaum and Roarabaugh, 2001)

Placement in sloped layers

As part of this approach, developed in China, a thick RCC block is made from several individual layers with the slope beginning at the upper part of the block under construction and sloping down toward the upper part of the previous block (Figure 17.22). The slope is determined based on the placement system and the work surface, which must be large enough to allow machinery to work efficiently and allow an acceptable waiting time between two layers. By reducing the time before covering the previous layer, the quality of the joint is improved without having to use a layer of mortar (cold joint or super-cold joint treatment). A steeper slope reduces the interval between the placement of each individual layer. However, this slope should not be too steep, which would make things more difficult for construction machinery. The RCC is placed in layers of 300 mm (after compaction) on a slope smaller than 1:10 (H:V), up to a lift height between 1.2 and 3.0 m, moving from one valley flank to another. On an indicative basis, slopes varying from 15H:1V to 20H:1V were adopted when this process was first employed (Holderbaum and Roarabaugh, 2001).

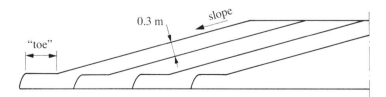

Figure 17.22 Example of placement of RCC in sloped layers

17.5 RCC Arch Dams

17.5.1 Brief historical overview

The first two arched RCC dams were the Knellpoort dam (50 m) and the Wolwedans dam (70 m), both constructed in South Africa in the late 1980s. These dams are both single curvature arch gravity dams, with a vertical upstream face and a sloped downstream face. During the same period, single curvature RCC cofferdams were being built in China. The experience gained from building these structures led to the first RCC arch dam with a thick arch in 1993 (the Puding dam, 93 m) and the first single-center thin arch dam in 1995 (the Xibing dam, 63.5 m). Since then, China has become the world leader in RCC arch dams. By the end of 2017, 60 dams of this type had been constructed, including the Wanjiakouzi double curvature arch dam. With its 168 meters in height, it is the tallest of its kind. RCC arch dams have also been successfully constructed on sites with less-than-ideal geology.

By the second decade of the 21st century, RCC arch dams had not only been completed China, but also in Panama, Puerto Rico, and Turkey, and this Chinese technique has been implemented in Pakistan and Laos. In Algeria, the arched Tabellout gravity dam (121 m), commissioned in 2019, was designed especially to bear seismic loading. And lastly, the Janneh dam (165 m), currently being constructed in Lebanon, will be the highest RCC gravity arch dam once it is completed in 2022.

17.5.2 Key design elements

To make the most of the advantages provided by RCC construction technology, the general project configuration, geometry, and design of RCC arch dams must be as simple as possible. Appurtenant structures should be kept as separate from the dam as is feasible or be designed in such a way that they disrupt the placement of RCC as little as possible. The simpler the construction approach, the more successful the RCC arch dam will be.

Due to their simple shape and low levels of stresses, arch-gravity dam structures are generally well suited to RCC construction. Early arched structures had a simple, circular transversal section with a single center, a vertical or inclined upstream face, and a downstream face with a constant slope of 0.7H:1V and 0.85H:1V, determined by the applied loads, the foundation conditions, and the valley topography.

The design of an RCC arch dam can also follow a similar approach to that of a conventional concrete arch dam, with a vertical section that varies from the crown toward the abutments, the use of a cooling system, and grouting for the transverse contraction joints. Engineers should keep in mind that in principle the dam is built horizontally and that a restricted workspace can have an impact on construction efficiency.

From an economic point of view, the cost of an RCC arch dam will be greater than an RCC gravity dam.

17.5.3 Structural aspects

Geometry and arch sections

Just as with a conventional concrete arch dam, defining the optimal geometry of the arches and transversal sections of an RCC arch dam depends on the topographical and geological conditions of the site.

Most arch dams built in China are thick arch dams. The width:height ratio of the base (B:H) is between 0.20 and 0.25, and the lowest B:H ratio is 0.13. The vast majority of recent arch dams have a B:H ratio lower than 0.25. RCC thick arch dams probably represent the area with greatest potential for development in RCC.

Joints

An analysis of the structure demonstrates that high tensile stresses for arch dams are often located on the upstream face near to the abutments and in the lower sections downstream of the crown cantilever. An innovative solution has been implemented whereby short structural joints were installed in these areas to free up tensile stresses. These short structural joints (from 1 to 4 m long) are only designed for places where tensile stress is high. In principle, short joints enable improved stress distribution in a single curvature arch dam.

Materials

To reduce cracks in arch dams, the aim is to manufacture concrete with relatively high tensile strength, with high tensile deformation capacity, low elasticity modulus, low shrinkage, a reduced increase in adiabatic hydration temperature, and low thermal conductivity.

Depending on the cement quality and the availability of fly ash, cement can generally contain up to a 65% proportion of fly ash (Table 17.23).

Table 17.23 Principal characteristics of RCC arch gravity dams mentioned in the ICOLD Bulletin 177, 2018b, 2020 (Values taken from www.rccdams.co.uk/dams)

Dam	Miel I	Longtan	Gomal Zam	Changuinola 1	Portugues	Kotanli II	Tabellout	Köroglu Kotanli I
Country	Colombia	China	Pakistan	Panama	USA	Turkey	Algeria	Turkey
Year completed	2002	2009	2011	2011	2013	2015	2015	2017
Type	Gravity	Gravity	Arch gravity	Arch gravity	Arch gravity	Arch gravity	Arch gravity	Arch gravity
Height (m)	188	217	133	105	67	70	121	100
Length (m)	345	832	231	595	375	250	366	404
RCC vol. (m^3 × 1000)	1669	4952	390	884	270	245	975	640
Total volume (m^3 × 10^3)	1730	7458	474	910	310	252		680
Reservoir vol. (m^3 × 10^6)	565	29920	1424	347	20	20	294	73
Slope upstream face	V	V/0.25	V	V	V	0.15	V	V
Slope downstream face	0.75/1.00	0.70	0.60	0.5/0.7	0.35	0.70	0.75	0.60
Thickness of layers (mm)	300	300	300	300	300	300	300	300
Thickness of lifts (mm)	300	300	3000	300	300	300	300	300
Cement content (kg/m^3)	85-160	99-86	91	79/65	114	85	85	85
Pozzolan (kg/m^3)	0	121-109	91	1245/150	51	130	45	130
Nature of the pozzolan		Low-lime fly ash	Low-lime fly ash	Low-lime fly ash	Low-lime fly ash	Natural pozzolan		Natural pozzolan

In China, all RCC arch dams have been built using HCRCC mixes, and some of their RCC characteristics are listed below:

- The water/cement ratio is generally between 0.4 and 0.65
- The workability of the RCC is determined by a VeBe test with results situated between 3 and 10 s
- Admixtures are used to increase hardening time, workability, and durability, and using more than one admixture can sometimes be an advantage

- The sand/aggregate ratio is generally between 30 and 38%
- Calcareous aggregates have the best properties for RCC
- For natural aggregate, the quantity of chipped and elongated elements should be as low as possible and always lower than 15%

Examples of types of RCC applied against upstream and downstream formwork of arch dams are presented in Table 17.24.

Table 17.24 Types of formwork for arch dams mentioned in the ICOLD bulletin, 2018b, 2020
(Values taken from www.rccdams.co.uk/dams)

Dam	Changuinola 1	Portugues	Kotanli II	Tabellout	Köroglu Kotanli I
Country	Panama	USA	Turkey	Algeria	Turkey
Year completed	2011	2013	2015	2015	2017
Type	Arch gravity	Arch gravity	Arch gravity	Arch gravity	Arch gravity
Height (m)	105	67	70	121	100
Length (m)	595	375	250	366	404
Type of formwork upstream face	RCC against the formwork GEVR/GE-RCC	RCC against the formwork GEVR/GE-RCC	RCC against the formwork	Conventional reinforced concrete	RCC against the formwork GEVR/GE-RCC
Type of formwork downstream face	RCC against the formwork GEVR/GE-RCC (stepped face)	RCC against the formwork GEVR/GE-RCC (stepped face)	RCC against the formwork GEVR/GE-RCC (stepped face)	Conventional reinforced concrete (stepped face)	RCC against the formwork GEVR/GE-RCC (stepped face)

Refrigeration of RCC (see § 15.7.3)

The use of artificial cooling systems has a major effect on the construction of a RCC arch dam. Consideration must be given to the time and additional costs arising from concrete cooling and joint grouting, which result from the associated thermal effects. This phase requires a substantial supply of electrical energy to reach the necessary temperature for joint closure. It is, however, possible to reduce the amount of artificial cooling required by only placing RCC during the coolest period of the year. This approach has been successfully implemented in a certain number of low-height RCC arch gravity dams of a limited volume.

Artificial cooling has primarily been used for RCC arch dams and RCC gravity dams constructed in China. The Chinese approach of artificial cooling takes place in two or three stages. The first stage occurs immediately after the layer of RCC above the cooling coils has been compacted and continues for a period of 14 days to reduce the maximum hydration temperature. The final cooling stage begins at least one month before grouting the transverse contraction joints in order to bring the concrete temperature down to closure temperature. In more extreme climatic zones, an intermediary cooling stage is introduced to further reduce the temperature gradients in the body of the dam.

Joint grouting (see § 15.7.4)

The aim of grouting transverse contraction joints is to couple the blocks and ensure a monolithic structure. This construction phase is normally planned while considering the stresses and depending on the progression of the works.

Over the course of the development of RCC arch dams, many solutions have been found for injecting grout into transverse contraction joints. For example, the three following methods are generally implemented:
- A single grouting stage after refrigeration of the RCC
- Grouting carried out in two or three stages
- A system of re-groutable joints; the joints have sleeve pipes (see Figure 15.82)

The experience acquired to date has demonstrated that most grouted joints in an RCC arch dam did not open after the first grouting. However, there are exceptions where cracks later developed between the joints, even though the grouted joints themselves did not open.

17.5.4 Monitoring system

The observation of displacement, deformation, and the movement of joints enables dam behavior to be tracked while considering thermal effects and the induced effect of load. This monitoring must be carried out during both construction and operation. As a result, a precise monitoring program must be set up with weekly actions during construction, moving to monthly inspections until the hydration heat has completely dissipated. The installation of long-base sensors in the joints (long-base-strain-gauge-temperature-meter—LBSGTM), which follow the evolution of temperature and movement, enables engineers to establish the precise moment for grouting and to track joint behavior during this operation.[6]

17.6 Hard Embankment Dams

17.6.1 A brief history

Gravity dams have a very high level of safety. Catastrophic consequences can be avoided even if a section of a gravity dam is severely damaged during an earthquake or if dam overtopping occurs during an extreme flood. The Shih Kang dam in Taiwan (Figure 5.15), for example, failed during an earthquake but without resulting in any catastrophic consequences (Chen, 2009). This behavior is what distinguishes gravity dams from other types of dams. However, gravity dams are relatively costly, which is why they account for less than 5% of all large dams higher than 15 m. The American engineer Homer M. Hadley was the first to develop the idea of creating a new type of dam by combining a concrete gravity dam with an embankment dam (Skermer, 1993). The gravity dam with a symmetrical profile, known as the optimum gravity dam, was proposed by Jérôme Raphael in 1970, but no dams were constructed based on this concept. In 1992, French engineers (Londe and Lino, 1992) developed a new concept for RCC dams, called FSHD (Faced Symmetrical Hardfill Dam), which proposed a thicker profile and used hard fill on the symmetrical upstream and downstream faces. The triangular profile (which is the minimum volume profile that does not result in traction at the upstream heel of the dam, according to the concept of pore pressure) was presented in the ICOLD Bulletin 117 in 2000. The Marathia dam in Greece, completed in 1993, was the first hard embankment dam (Figure 17.25).

---

[6] Fibre-optics are used in measuring instrument technology for dam structures; it is possible to measure various parameters along the length of the fibre-optic, in particular temperature.

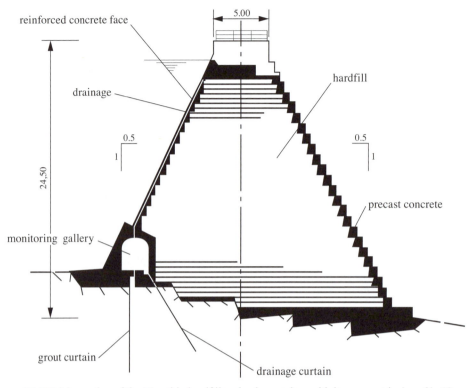

Figure 17.25 Main section of the Marathia hardfill embankment dam with its symmetrical profile (Greece)

Since then, several hard embankment dams have been built in Greece, the Dominican Republic, Peru, Turkey, the Philippines, and Algeria. Figure 17.26 shows the Valsamiotis dam in Greece with a symmetrical profile.

Figure 17.26 Valsamiotis hard embankment dam in Greece, seen from upstream (left) and downstream (right) (Courtesy G. Dounias)

In Japan, this concept was taken up with a trapezoidal profile made of gravel and cemented sand—the Trapezoidal Cemented Sand-Gravel CSG dam. In China, in addition to cement, sand, and gravel, large blocks of up to 300 mm are also used to produce a cement material called CSGR or CSGRD (Cemented Sand Gravel Rock Dam).

17.6.2 The new concept of a symmetrical profile with hardfill

The slope of the downstream face of a traditional concrete gravity dam is in the order of 75 to 80%. According to Figure 14.1, the stresses at the level of the foundation for cases of normal loading vary greatly between an empty and full reservoir with a triangular distribution, with a maximum at the upstream heel with respect to the downstream toe. The symmetrical profile of hardfill dams is characterized by upstream slopes in the order of 50 to 70% (Figure 17.27). Stresses under the foundation are relatively low and are uniformly distributed. They are absolutely constant during dam operation, with a slightly higher compression when the reservoir is full. As a result, a symmetrical profile can be obtained with a hardfill with relatively low strength compared to a gravity dam. Furthermore, the foundation requirements are less stringent; the symmetrical profile can be founded on relatively poor-quality rock. For a symmetrical profile with slopes of 70%, sliding safety is improved by 50% compared to a gravity dam, and stability during earthquakes is excellent. Sensitivity to uplift in the foundation is low. The symmetrical profile is also highly resistant to cases of overloading, for example, due to overtopping of the dam during a flood. To conclude, hard embankment dams can be built on low quality foundations by using local materials of medium quality.

Figure 17.27 Comparison between a traditional concrete gravity dam and a hardfill embankment dam

17.6.3 Characteristics of a hardfill dam

Depending on the height of the hardfill dam with a symmetrical profile, compressive strength at 180 days in the order of 4 to 8 MPa is sufficient. No cohesion in the horizontal construction joints is necessary. The joints do not need any particular treatment; all that is required is for the moving particles to be removed with compressed air cleaning. The segregation of material during placement is not an issue. A hardfill dam does not have to be impermeable if an upstream slab in reinforced concrete is part of the design. By using fly ash or pozzolan, a hardfill dam has a low cement content and low heat release. A hardfill dam is normally less deformable than RCC. As a result, with an upstream slab in reinforced concrete, transverse joints are not necessary.

For the construction of a hardfill dam, relatively low-quality material available near to the dam can be used. Alluvium such as sand and gravel, weathered rock, poor-quality rock (schist, flysch, sandstone, marl) that are easily excavatable are sufficient. During mechanical excavation, transportation, and storage, loose rock is transformed into a granular material and thereafter into a granular cemented mix. The alluvium can even contain a large proportion of non-plastic fines. In Europe, Africa, and America, a maximum diameter of 70 to 80 mm is generally specified for aggregate. In Japan, aggregate size must be smaller than 80 mm for dams and 150 mm for cofferdams that divert rivers. In China, aggregate smaller than 150 mm for dams and less than 250–300 mm for cofferdams have been used.

As mentioned, the lowest possible cement content can be selected in terms of workability. On average, hardfill dams have been constructed with a cement content of 90 kg/m^3 (Lino et al., 2017). Two-thirds of dams built up to 2017 had a cement content lower than 100 kg/m^3. Dams in Greece have a cement content between 60 and 70 kg/m^3, generally without fly ash or pozzolan. In China, the content of cemented materials is lower than 80 kg/m^3, of which a maximum of 40 kg/m^3 cement. Dams in Japan typically have a cement content of 60 to 90 kg/m^3 without the use of fly ash.

17.6.4 Controlling seepage through the hardfill dam

Control of seepage through the hardfill dam is guaranteed by an upstream reinforced concrete slab or a geomembrane preferably installed at the end of the construction phase to limit cracking caused by deformation of the fill during construction (Lino et al., 2017).

In Japan, precast elements of 100 to 150 cm are used as the formwork for traditional concrete for impermeability, carried out at the same time as the hard fill dam. In China at the Shoukoubao dam, a 100-centimeter layer of concrete (C25) was used for impermeability.

The upstream slab must sit on a plinth anchored into hard rock. This plinth, with or without a gallery, enables a curtain to be grouted into the rock. The plinth may possibly be supported by a plastic concrete diaphragm, which makes the rock almost impermeable.

17.7 Rock-filled Concrete Dams

Rock-filled concrete (RFC) is an innovative solution for mass concrete technology proposed in 2003 for the first time by professors Jin and An at the University of Tsinghua in China. RFC technology combines the advantages of masonry with high-performance self-compacting concrete (HSCC). After the rockfill has been placed in several layers, the voids between the blocks are filled with self-compacting concrete, whose key property is to propagate along long distances through the narrow voids and bends. To achieve this, self-protected underwater concrete (SPUC) was developed, which can be applied underwater and which self-compacts without vibration or mixing and without any monitoring of pressure (Jin et al., 2018a, b). This SPUC concrete contains a new agent which guarantees its protection underwater and thus produces a concrete that is superior to traditional nondispersive concrete. More than fifty concrete rockfill dams have been built in China, including the 69-meter-high Baijia dam, which is the tallest arch dam of this type, and the 100-meter-high Songlin dam, which is the tallest RFC gravity dam (Jin et al., 2017, 2018a, b). Typically, the volume of self-compacting concrete is lower than 45% of the total volume of the dam.

17.8 Other Uses for RCC

17.8.1 Protection against dam overtopping

From the mid-1980s, RCC was used in many rehabilitation projects, in particular for the protection of embankment dams. The advantages of RCC facing on the crest and on all or part of the downstream face of a fill dam to protect against erosion in case of overtopping during extreme flood events have been recognized.

17.8.2 Dam strengthening

RCC has been successfully used to strengthen existing dam structures by placing an additional volume of concrete downstream of an existing dam or by filling the open spaces in multiple buttress dams with the aim of improving the static and/or dynamic stability of the structure. Massive RCC structures have also been used to stabilize and protect the rocky versants downstream of a dam against erosion.

17.8.3 Protection against erosion

RCC's resistance to erosion makes it the ideal solution for protecting sensitive zones vulnerable to erosion by the flow of water or scouring from a jet of water. It can be used for the lining of channels, the surfaces of spillways, the creation of the apron of a plunge pool or stilling pool, or for the erodible parts in rock or concrete of the tailrace of a hydropower plant.

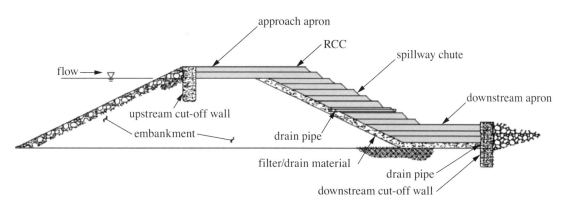

Figure 17.28 RCC protection on the downstream face of an embankment dam

17.8.4 Adapting foundations

RCC is an appropriate material for adapting the surface of dam foundations to create an even base layer. It is important to locally rectify the foundations and avoid stress concentrations, as well as removing zones with weathered surfaces (see § 10.2.4).

17.8.5 Cofferdams

RCC cofferdams have been used as temporary structures or as an integral part of the profile of a proposed dam. RCC has been used in innovative ways for river cofferdams; they can be built quickly in times of low water flow and by using materials from the construction of the dam. Able to resist overtopping and resistant to erosion, RCC is well suited to the construction of cofferdams.

17.8.6 Heightening concrete dams

RCC is also the ideal material for heightening existing concrete gravity dams. All the stages of the works must be carried out with the same care and technical requirements as for a new dam. Specific precautions must be taken to ensure a robust bond between existing and new structures by using GEVR, GERCC, IVRCC, or CVC. The surface of the existing concrete must be treated to remove any damaged or unstable zones (Figure 17.29).

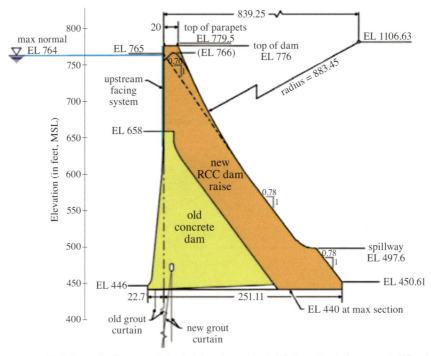

Figure 17.29 Schematic diagram of the heightening of an RCC dam (San Vincente, California)

V. CONCRETE

By definition, concrete is a heterogeneous material resulting from a thorough mixture of aggregate, cement, water, and admixtures. The aggregate forms the skeleton, while the cement, water, and admixtures create a binding paste. The function of the paste is to coat the aggregate and fill the voids between the particles, thus obtaining a compact concrete. Concrete also contains occluded air. For the construction of concrete dams, one of concrete's most important qualities is its workability, a critical factor for correct placement. Furthermore, it is very important that the resulting concrete has a homogeneous structure and performs consistently. This chapter describes the main elements involved in the manufacture and placement of dam concrete.[1]

---

[1] The information in this part is primarily based on the following publications: SwissCoD, 2000a, 2001b; Paillex, 1985; Stucky J.-P., 1956, 1997.

18. Concrete Technology

Dez arch dam in Iran, height 202 m, year commissioned 1962 (Courtesy T. Dietler).

18.1 Concrete Components

18.1.1 Aggregate[2]

Due to the volume of concrete to be placed, there must be a sufficient quantity and quality of aggregate available in the immediate vicinity of the dam. Aggregate may be extracted from gravel pits (alluvial deposits) or quarries. The material from quarries where the parent rock may originally have been crystalline, sedimentary, or metamorphic will have to be crushed to attain the correct dimensions for the manufacture of concrete. A rough surface such as that created by crushing provides improved binding of the hardened paste, but does require, however, a little more wetting water, resulting in less workability than that obtained with smoother rounded aggregate.

In order to ensure that aggregate is consistent, a petrographic test must be carried out. This test enables the lithological and mineralogical identification and classification of the aggregate and determines its physical and chemical characteristics. At this stage, it should already be possible to establish whether there is a risk of an alkali-aggregate reaction.

As a general rule, aggregate from hard, dense rock should be used. It must be stable and resistant to external agents (the freeze-thaw cycle, variations in temperature, chemical actions).

The aggregate must be clean and is therefore put through an intensive mechanical cleaning process to eliminate all the impurities that may have a deleterious effect on the required concrete qualities. Drainage occurs while the aggregate is stored. It is important to ensure that there is no film on the aggregate (clay, for example), which would prevent it from adhering to the cement paste, and to ensure that no organic material is present, which can slow down and prevent cement hardening. Frost-sensitive elements, anhydride, gypsum, pyrite, charcoal, scoria, wood, and organic material are all considered to be harmful. In addition to the risk of low strength in young concrete, negative effects often only appear over the long term.

As its size and shape play a major role in the manufacture and installation of concrete, aggregate is divided into different categories to better control particle size. For example, the distribution chosen for the construction of a dam in Switzerland is as follows:[3]

Categories	Size
filler	< 0.1 mm
sand	0.1–1 mm
sand	1–3.5 mm
chippings	3.5–10 mm
chippings	10–30 mm
gravel	30–70 mm
gravel	70–160 mm

---

[2] Various characteristics and requirements for aggregate are covered in Section 10.3.1 "Aggregate for concrete dams."

[3] There are also other possibilities that also form part of accepted standards.

18.1.2 Binders

The primary function of binder is to ensure not only that voids are filled after concrete placement, but also that the aggregate is sufficiently coated to solidify the hardened conglomerate. The required volume of cement paste depends on the proportion of voids and on the resulting surface, both determined by the aggregate dimensions, shape, and particle size.

Various hydraulic binders are available for the manufacture of dam concrete. Mass concrete requires specific measures to prevent the release of heat from hydration (which also causes a change in volume). A cement with slow and low heat release (for example, pozzolanic cement or blastfurnace cement) is an appropriate solution. The use of normal Portland cement is also suitable. Depending on the size of the concrete blocks, a refrigeration system may need to be envisaged. Portland cement has most commonly been used for the construction of Swiss dams.

18.1.3 Mixing water

Mixing water, as well as the humidity of the aggregate, serves to hydrate the cement. This water must not contain any harmful substances, such as chemical concentrations that attack the hardened concrete, for example. If in doubt, chemical analyses of the water should be carried out. In any event, potable water is suitable to be used as mixing water.

18.1.4 Admixtures

Admixtures are products that are added in only slight quantities (see § 8.2.4). They can modify concrete characteristics either while it is fresh or once hardened. Depending on the desired outcome, there are several types of admixtures:

- Air entrainers have several effects: depending on the type of aggregate (especially when there is flat aggregate), their addition improves concrete workability and facilitates its placement; during transport they lower the risk of segregation; they also improve watertightness and protect the concrete against the effects of the freeze-thaw cycle; however, a lowering of mechanical strength is a possible outcome
- Use of a plasticizer enables the amount of mixing water to be reduced and therefore also the W/C ratio
- Air entraining agents and plasticizers can be combined
- Fluidifiers are plasticizers that enable the greatest reduction in mixing water for the same workability (they are rarely employed for large hydraulic structures but are instead suitable for stilling basins, outlets, and spillways)
- Water reducing agents retain the concrete's plasticity; they increase strength or enable identical strength with less water
- Retarding admixtures allow the setting delay to be extended
- Accelerating admixtures are not advised for mass concrete, as they contribute to the development of heat

18.1.5 Occluded air content (void content)

With the introduction of air entraining agents, measuring the void content of fresh concrete during concrete mixing has become of interest. The value gives an initial indication on the compactness of the mix.

The method is based on the compressibility of the air contained in the fresh concrete. A specific volume of concrete is put into contact with a known volume of compressed air. The air content in the concrete can be read directly on a calibrated manometer depending on the device's functionality. However, for the measurement, aggregate greater than 30 mm is removed and the amount of occluded air or void content is calculated for the total volume. The average value of the void content in concrete 0/30 mm is approximately 3.5% of the volume, which for a D_{max} = 80 to 150 mm corresponds to a void content of approximately 1.8 to 2.3% of the volume (CBS, 2001).

18.2 Formulating Concrete

An important step is to calculate the proportion of aggregate, cement, water, and any eventual admixtures necessary for the concrete composition to meet the required performance. Some principles based on current practice are outlined below.

18.2.1 Particle size

The grading curve of a particular concrete is an important element, as this is what determines the compactness of the concrete. The optimum grading curve is one that, as a % of the weight or volume of each category of aggregate, allows a full, resistant, and durable concrete to be obtained. Particle size can be discontinuous or continuous. Generally, a continuous grading curve is sought. Formulae, often based on the d/D relationship, have been developed, and in Switzerland the following formulae are commonly used:

- Fuller $\qquad$ $P = 100\sqrt{d/D}$ (for round aggregate)[4]
- EMPA (LFEM) $\qquad$ $P = 50\left(d/D + \sqrt{d/D}\right)$
- SIA $\qquad$ $P = 20\left(d/D + \sqrt[4]{d/D}\right)$,

where P represents as a % the path through the sieve or strainer of diameter d. In practice, the SIA formula represents the average curve between that of the Fuller and that of the LFEM. The SIA curve is appropriate for rounded aggregate from 0 to 30 mm. For fine aggregate (smaller or equal to 20 mm), there is a tendency toward the LFEM curve and for mass concrete ($D_{max} \geq 60$ mm), a curve that is higher in sand is sought-after.

For dams, many analyses have been carried out with LFEM with the aim of optimizing particle size. In 1955 this research resulted in the proposition of the theoretical curve known as TS 1955, which has the following shape:

$$P(d) = \frac{10}{9}\left(100\sqrt[3]{\frac{d}{b}} - 10\right),$$

where
  P = percentage of particles with a diameter d
  d = particle diameter [mm]
  D = maximum diameter of the aggregate [mm]

for $D \geq d \geq D/1000$.

Figure 18.1 (a) illustrates different TS 1955 grading curves for D_{max} 80, 100, and 150 mm. In Figure 18.1 (b), the TS 55 curve is compared to other theoretical curves.

---

[4] For crushed aggregate: $P = 100\,(d/D)^{0.4}$.

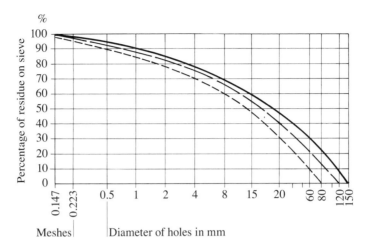

Figure 18.1 (a) TS 1955 grading curves for D_{max} 80, 100, and 150 mm (from Paillex, 1985)

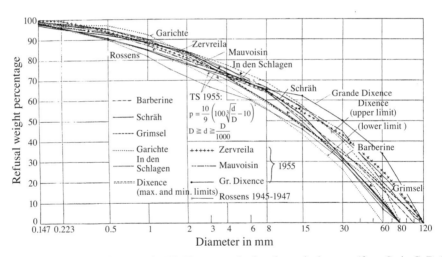

Figure 18.1 (b) Comparison between the TS 55 curve and other theoretical curves (from SwissCoD, 2001b)

In terms of the worksite, the natural particle size of the material used is not consistent and because of this, it is not possible to adhere to an aggregate distribution that is strictly identical to that of the theoretical formula or that proposed by testing. Given the real worksite conditions, it is preferable for the grading curve to sit within the limits of a set zone. It should be noted that the curves inside a zone have practically equivalent workability and mechanical strength when the material is rounded (moraine, alluvions). Furthermore, particle size may change during mixing.

To these considerations can be added some comments drawn from practical experience (Stucky J.-P., 1956, 1997). A lack of fine sand requires a longer period of vibration and must be avoided, especially when a low cement content is being used (160 kg Portland cement/m³). Too much gravel is also detrimental to workability, with a possibility of concrete segregation as the skips open. As for an excess of dust (0–0.1 mm), this has an adverse effect on frost resistance; the quantity of dust should never be greater than 2% of the total aggregate weight. However, a minimum of 1% is necessary to ensure good workability. If the material comes from a quarry (fully crushed aggregate), more sand is generally required to ensure good workability. The amount of dust must be increased to 3% and may even safely reach 4% if it is limestone dust. Lastly, too much sand is not advisable, because the concrete settles, and a mix with a lot of sand and fines requires a lot of wetting water.

18.2.2 Cement content

Cement content is established depending on the constraints to which the dam will be subject and its resistance to weathering (the effect of frost on the faces, abrasion). As such, it can be adapted for the various parts of the dam based on the required strength. However, there is a limit of 2 different cement contents per lift. There must be enough mortar to fill the voids between the particles. Table 18.2 gives, as an example, some values for cement content.

Table 18.2 Cement content for various parts of a concrete dam (from Stucky J.-P., 1997)

Content	Part of the dam
250 kg/m^3	Faces, heavily loaded zones; lifts with galleries, lifts in contact with the rock
200–250 kg/m^3	Mass concrete
180 kg/m^3	A priori minimum content
160 kg/m^3	Minimum content, chosen after many workability tests

As a guide, here is the approach formula established by Paillex (1985) based on constructed dams:

$$C = K / \sqrt[3]{D},$$

where
- C = cement content of full concrete (kg/m^3)
- K = coefficient from 800 to 1000
- D = maximum particle diameter

18.2.3 Moisture content and the W/C ratio

Moisture content is a critical element in terms of the successful workability, durability, and strength of concrete. It is well known that to achieve the best quality, the concrete must be manufactured with the least amount of water that will nevertheless ensure its workability. Firstly, an excess of water should be avoided during mixing, but so should any loss of water during installation. Variations in mixing water are difficult to control. Continuous visual inspections of the appearance of the concrete, frequent checks of its density, the amount of occluded air, and of workability are required so that any necessary corrections can be made.

472   CONCRETE

The W/C ratio is also very important as it determines strength. It is possible to manufacture perfectly workable concrete with W/C values between 0.45 and 0.50; formerly the ratio was between 0.50 and 0.60.

To assess the moisture content based on the largest particle size, Paillex (1985) developed a simplified equation deduced from a formula applied in the laboratory:[5]

$$E = 320/\sqrt[5]{D},$$

where
- E = total amount of moisture in dm^3 per m^3 of mass concrete
- D = max particle diameter in mm

18.2.4 Admixture content

Admixtures are products added in only minimal quantities in the order of 0.02 to 0.5% of cement weight. These low quantities require precise measurement.

18.2.5 Occluded air content

The choice of the optimal occluded air content is decided on by the design engineer. The following values are recommended by Paillex (1985):

Max particle D [mm]	50	75	150
Occluded air [%]	4.0	3.5	3.0

18.3 Properties of Fresh and Hardened Concrete

18.3.1 Fresh concrete[6]

18.3.1.1 Workability and consistency

Workability expresses the aptitude of fresh concrete to transportation, placement, and compaction. It also affects the performance of the hardened concrete. It depends on the size and shape of the aggregate, its particle size distribution, the quality and quantity of binder, the use or not of an admixture (particularly an air entraining agent), and the consistency of the mix. There is a difference between workability and consistency, which reflects the ability of freshly mixed concrete to be poured.

There are several diverse methods for checking the workability and consistency of fresh concrete. For fluid to plastic concrete, workability is determined by a flow test, where a standard cone of concrete is tamped and then lifted for the concrete to flow onto a table. This method is usually employed in the laboratory. The consistency of standard concrete, that is, not too dry nor too fluid, is determined by a slump test, where a standard cone of concrete (Abrams cone) is unmolded and the subsidence measured. This method is well suited for checking concrete on a worksite. VeBe tests (see § 17.4.1.3), as well as the "Walz method," may also be considered.

---

[5] $E = 0.1\,C + K\sqrt[5]{D}$, where K equals 260 for rounded aggregate and 280 for crushed aggregate.
[6] From Paillex, 1985.

18.3.1.2 Bearing capacity
Bearing capacity plays an important role in the placement of mass concrete, which must be able to bear the worksite machinery designed for spreading and compacting concrete. Bearing capacity cannot be assessed prior but can only be verified directly on the worksite.

18.3.1.3 Cohesion
It is accepted that the binding paste, comprising mixing water, cement, and very fine sand, is an essential element for ensuring concrete cohesion. There must be a sufficient quantity of this paste, and it must not be too dry or too fluid, that is, with an appropriate moisture content. It can be reduced down by using an admixture.

18.3.1.4 Sweating
Sweating is a phenomenon during which water moves up to the surface during the placement of fresh concrete. In theory, considerable sweating should be avoided in mass concrete, as it can lead to the formation of pockets under large, flat aggregate and/or small vertical channels in the cement gel, which results in an increase in the concrete's permeability. Sweating can be controlled by the fineness of the binder and filler, partly by the fine bubbles of occluded air and by a careful reduction of mixing water.

18.1.3.5 Premature shrinkage
Premature shrinkage occurs through the rapid evaporation of excess water after the concrete has been installed. It is favored by dry weather, strong sunshine, and the effect of wind. While only initially superficial, the cracks that results from this premature shrinkage soon deepen. This can be avoided by ensuring that the concrete is homogeneous, with a moderate moisture content, and by immediately implementing appropriate setting methods.

18.3.2   Hardened concrete

18.3.2.1 Mechanical strength
For the construction of dams, *compressive strength* is the quality criterion that is most often used (Figure 18.4). It is determined by crushing test samples that have a cubic, prismatic, and/or cylindrical shape. Cube samples have an edge length of 20, 30, or 40 cm. Compared to cubes and prisms, cylindrical samples (generally $\varnothing = 300$ mm and H = 450 mm) have the advantage of not containing any angles, which facilitates the pouring of concrete. Ideally, the diameter of the sample should be $\geq 5D$, where D is the largest aggregate size. It should also be noted that the sample shape affects the strength values (see § 18.6.1). Table 18.3 presents a proposed program for the manufacture of samples during construction. Lastly, it is also possible to get samples by sawing into a block or by drilling a core. Note that sample size is a significant factor for dam concrete with large aggregate size. The results are sensitive to these sizes and to limit conditions.

Table 18.3 Minimum program for the manufacture of samples during construction (Stucky J.-P., 1997)

Minimum program:
- 1–2 times per day: 2–3 cylinders at 7 days
- 1–2 times per week: 2–3 cylinders at 28 days
- 1–2 times per day: 2–3 cylinders at 90 days
- 1–2 times per week: 2–3 cylinders at 365 days

Tests are carried out at 7, 28, 90, and 365 days. Samples are also manufactured and stored for testing at older ages. Tests at 7 days give an initial idea of strength, allow future strength to be evaluated, and ensure that any necessary corrections can be carried out. For dams, compressive strength at 90 days is chosen as a reference value, taking into account the frequency of tests at this age. In mass concrete with a large particle size, the factors that influence strength are the type of rock and the diameter of the largest particle, hardness, the structure and surface of the aggregate, and the W/C ratio. Occluded air has little influence.

Figure 18.4 Compressive strength test on a cube in a worksite laboratory (Courtesy H. Pougatsch)

Tensile strength is deduced using direct tensile tests (affixing perfectly centered tensile heads to the ends of a cylindrical sample), the splitting tensile test (called the Brazilian test, which involves crushing a concrete cylinder between two opposing forces at the disc periphery), and the flexural test (performed on prismatic samples by applying either a concentrated force in the middle of the sample or two concentrated loads on the bearing third) (Figure 18.5 (a), (b)).

It should also be noted that methods for collecting the concrete sample and storing the samples play a role. When the samples are being taken, the concrete must be intact. Unmolding of the samples takes place 24 hours after they were made. They must then be stored in a warm and moist place and frequently sprayed with water.

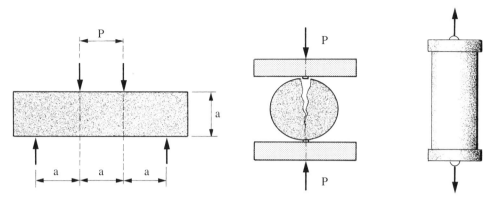

Figure 18.5 (a) Methods for testing tensile strength (test on a prism, splitting test, direct tensile test)

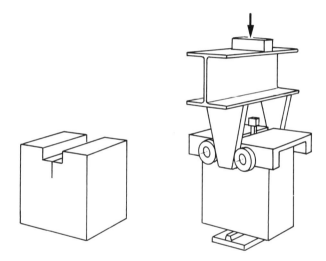

Figure 18.5 (b) Wedge splitting test (from SwissCoD, 2001b)

18.3.2.2 Physical properties

The elasticity modulus E_B, defined as the ratio between the unit stress and relative deformation, is strongly influenced by the nature and size of the aggregate. The elasticity modulus is especially used to calculate deformation.

The order of magnitude of this modulus is in accordance with USACE (1994b):

- E_B = 25–39 GPa to 7 days
- E_B = 30 à 40 GPa to 1 year

and Jansen R. B. (1988):

- E_B = 21–41 GPa at 28 days
- E_B = 28 à 48 GPa at 1 year

Relationships have been established between the instantaneous static elasticity modulus and compressive strength. Standards also provide various formulae. On the basis of values taken from 12 Swiss dams,

Paillex (1985) gave an average value of 30 GPa and an average relationship between the modulus and compressive strength from 830 to 90 days.

Poisson's ratio µ is defined by the relation between the relative transversal and longitudinal deformations. It varies depending on the quality and humidity of the concrete, and its value often lies between 0.16 and 0.20. Average values of 0.17 at 28 days and 0.22 at 1 year are found (Jansen, 1988).

The *thermal expansion coefficient β* depends in particular on the content and type of binder, the nature of the aggregate, the W/C factor, and the age of the concrete. Its value lies between $10 \cdot 10^{-6}$ and $15 \cdot 10^{-6}$. An average value of $12 \cdot 10^{-6}$ can be accepted.

Durability

The watertightness and frost resistance of concrete are important factors in the long-term behavior of concrete dams. Various procedures have been developed to test these. Concrete must also be able to withstand attack from chemical agents and erosion. In these cases, preventive measures can be taken.

Concrete watertightness

In addition to the effect of large aggregate and the cement-aggregate interface, the watertightness of a concrete dam also depends on the method and quality of construction. The care brought to the horizontal concrete lift joints (roughness, cleanliness of surfaces, placement of a mortar layer), the execution of vertical joints, and compaction (honeycombing, which can allow water to find a preferential path, should be avoided) all play an important role regarding watertightness. It should also be noted that the permeability of concrete increases with an increase in the W/C ratio. A low W/C ratio and good placement guarantee low permeability. The use of an air entrainer also lowers the W/C ratio and therefore favors concrete watertightness. Cracking must also be considered.

To assess concrete watertightness, one face of a 60–day-old cube sample is subjected to increasing or constant pressure for a given period of time (for example, 20–30 atm for 10 days). Concrete is considered technically watertight if the amount of water that crosses through the samples, collected in a container, does not exceed, on average, across three cubes (Stucky J.-P., 1997):

- 500 cm^3 for a thin arch dam or buttress dam
- 1000 cm^3 for a thick arch dam
- 1500 cm^3 for a very thick arch dam or gravity dam

Frost resistance[7]

The frost resistance of concrete subject to freeze-thaw cycles is an important factor in preserving the state of the concrete surface. Compared to the substantial thickness of the sections, concrete weathering due to frost manifests as superficial breaking up in only the top few decimeters. To avoid this issue, careful attention must be paid to the concrete in the faces, for example, by increasing the content and using an admixture.

For verifying the behavior of the concrete with regard to freeze-thaw cycles, it is difficult to reproduce in the laboratory the true conditions the dam will be subject to over a number of years. To carry out this test, prisms are exposed to an air freezing phase of approximately -20°C and then to a thawing phase in water at a temperature of +10 to 15°C. Control samples are kept at +18°C. The effect of the frost will modify the elasticity modulus, which is measured during the test (measuring the speed of sound, Figure 18.6),

---

[7] See also Section 19.5.

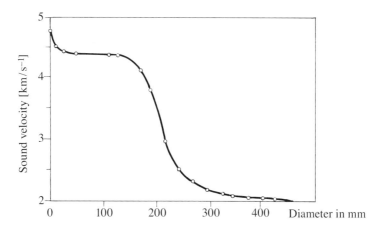

Figure 18.6 Example of a measurement of the gradual deterioration of concrete through freeze-thaw cycles (from SwissCoD, 2001b)

and reduce tensile strength, which is determined at the end of the test. Dam concrete is described as being frost resistant when 60-day-old prisms achieve the following values after 200 freeze-thaw cycles in the laboratory:

- Elasticity modulus (dynamic measurement): $E_{frost}/E_{control} > 0.75$
- Flexural strength: $F_{frost}/F_{control} > 0.75$

Chemical attack
Depending on regions and climates, dam concrete can be exposed to various types of chemical attack: soft water, water laden with organic material, or sulphate waters in ground with gypsum.

Degradation by soft water usually occurs in alpine regions. Soft water issuing from glaciers, snow melt, and springs from areas of granite dissolves the free lime. The effects are negligible when occurring superficially on compact concrete with low permeability. It mostly becomes apparent close to cracks and lift joints.

Attack by organic substances that have been transported or partially dissolved by water can be more dangerous. Protective measures are therefore required.

The greatest attack to which dam concrete can be exposed is from sulphate waters.

Preventing these types of attacks involves analyzing the water that will be in contact with the dam sufficiently in advance. A detailed analysis allows long-term protective measures to be prepared ahead of time.

Erosion by abrasion and cavitation
Erosion is a mechanical action performed by water that dislodges and shifts particles from the concrete elements along which it runs. Erosion by abrasion and cavitation initially concerns surfaces exposed to high-speed flow, for example, from the discharge structures (spillway, bottom outlet, or mid-level outlet) integrated into the dam. Structural measures are necessary to protect these surfaces from major damage, for example, by using epoxy or resin mortar, or concrete with steel fibers.

Abrasion is caused by the impact of transported material and by water itself. In the latter case, it is primarily the speed and volume of moving water that are likely to cause damage.

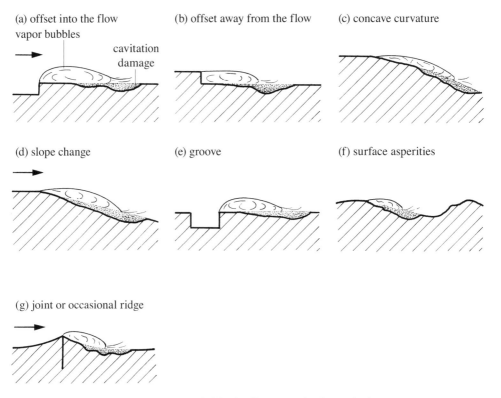

Figure 18.7 Irregularities leading to erosion by cavitation

Cavitation is produced by water flowing at high speed across a surface with even only very slight roughness (Hager et al., 2020).

18.4 Manufacture and Placement

After having carried out intensive exploratory work and numerous preliminary tests, the design team can select the final gravel pit or quarry for the supply of aggregate and choose the types of cement and appropriate admixtures.

First of all, the rock from the front face of the quarry is crushed. Then, the aggregate, no matter its origin, is washed before being graded, sorted into categories, and placed into designated storage areas. From there, it is conveyed to the concrete plant. The aggregate, cement, and water are weighed according to the precise prepared data, before being emptied into the concrete batcher. Any admixtures are also added according to the noted proportions.

In general, the components are mixed for 1.5 to 2 minutes. The correct mixing duration is established by checking the homogeneity of the batches. The concrete must be homogeneous at the start, in the middle, and at the end, when the concrete batcher is emptied. Its consistency must be plastic, that is, with a low moisture content.

The concrete is then transported on site by silobuses and/or blondins (cable crane). As mentioned above, the mix must have the correct consistency—not too little or too much water so as to avoid segregation. The freshly poured concrete is spread into 40- to 50-centimeter-thick horizontal layers by bulldozer. Once spread, it is vibrated in a methodical way so that no areas are left out (Figures 2.27 and 18.8). The concrete is deemed sufficiently vibrated once the hole left by a vibrator that has been removed closes up on its own during the following insertion. The rate of production and placement is between 50 and 100 m³ an hour. For very large dams, this rate can increase to 150, 200, or 250 m³ an hour.

Finally, concrete curing involves protecting it from desiccation from the moment of transport onward. Once the concrete has been installed, the surface of the concrete lift surfaces can be kept humid by spraying an anti-evaporation film, watering, or by placing mats or sand. The water used for curing must not contain any chemical agents or organic material such as tannic acid or iron compounds, which can cause coloration (USACE, 1994b).

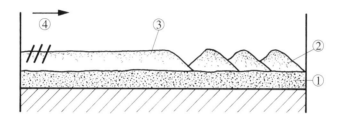

Figure 18.8 Diagram for methodical concrete vibration: ① Already vibrated layer (40 to 50 cm); ② Piles dumped from the wagon; ③ Layer ready to be vibrated; ④ Position of the needle vibrators and direction of travel (from Stucky J.-P., 1997)

18.5   Concrete Testing

This section summarizes the tasks and tests to be carried out at various stages of the design and construction of a dam to guarantee the use of concrete that meets all the requirements. The tests are outlined without a description of how they are conducted—although some are described in preceding sections. It bears repeating that quality must be strictly controlled at each stage of production to ensure the homogeneity and regularity of the concrete.

The various stages of testing are as follows (SwissCoD, 2001b):
- During the design phase:
  - investigation studies and testing
  - preliminary testing
  - suitability tests prior to construction
- During the construction phase:
  - control testing
- During operation:
  - control testing on the structure

18.5.1 Tests during the design phase

18.5.1.1 Investigation studies and testing

This phase takes place during the initial feasibility studies for the dam. The aim is to find a gravel pit or quarry to extract the material for the manufacture of concrete and establish means for transportation, estimate the available quantity, and undertake geological and petrographic examinations. The quality and availability of material will inform the decision-making process for the choice of dam type. A few m^3 of material are extracted (possibly in conjunction with a blasting test) and sent to the nearest laboratory or worksite for qualification testing. Wherever possible, mixing and compressive strength testing is also carried out. An inventory of the nearest cement plants and their products is also established.

In addition to the above points, Table 18.9 (a) summarizes the principal examinations and tests.

Tables 18.9 (a) Examinations and aggregate tests during the investigation phase (from SwissCoD, 2001b)

- Evaluation of aggregate shape
- Qualification of aggregate relative to the alkali-aggregate reaction
- Determination of natural particle size
- Investigation into hardness and abrasion
- Investigation into the compatibility of the cement with the aggregate
- Investigation into possible chemical aggression of water in contact with the dam

18.5.1.2 Preliminary testing

This phase takes places during the preliminary design phase and its aim is not only to outline an initial general approach but also to establish the technical specifications for the project. At this stage, it is a question of defining the guiding principles for concrete manufacture, the necessary strengths, and the targets required in order to meet requirements for the dam structure. This phase lasts for a considerable amount of time and must be started as soon as possible. It encompasses all the materials (aggregate, cement, admixtures), as well as water.

Table 18.9 (b) summarizes the principal tasks to be undertaken as part of preliminary testing.

Table 18.9 (b) Principal tasks required for preliminary testing (from SwissCoD, 2001b)

- Establishing a grading range that offers both good workability and good compactness (various ranges are tested using mixing and placing tests)
- Tests on fresh concrete measuring density, occluded air content, consistency, checking water/cement ratio, including measuring the length of vibration
- Compressive and flexural strength tests at 7 and 28 days
- Checking the density of the hardened concrete
- Dynamic and static elasticity moduli measurements
- Frost resistance tests
- Permeability tests
- Test on the resistance to chemical attack

18.5.1.3 Suitability tests, prior to construction

Before starting the placement of concrete and as soon as the worksite has opened, tests should be carried out to finalize the concrete in the worksite laboratories. These tests include the methods and installation modes that the contractor has chosen.

18.5.2 Control testing during construction

During construction, it is very important to carry out regular control testing as concrete placement progresses, to detect any issues or lowering of production quality as quickly as possible. This includes testing concrete as it is being manufactured (with density and consistency checks carried out on aggregate and fresh concrete) and also testing hardened concrete using samples and cores taken from placed concrete (Table 18.10).

The types of test, as well as their frequency and number, are determined in the technical specifications of the project (quality assurance plan); they depend on the production capacity of the batching plant and are carried out every 300 to 500–600 m^3, or as a function of the number of lifts to be concreted per day or per team, and thus as a function of the dam's dimensions.

Table 18.10 Types of control tests during construction (from SwissCoD, 2001b)

Checking of aggregate	Checking of fresh concrete	Checking of hardened concrete
• Visual inspection of aggregate • Checking the particle size of each load as it comes out of the washing-sorting machine and into intermediary stockpiles • Checking particle size at the entrance to the batching plant • Checking the scales at the batching plant and the actual weight of each particle size load • Checking the amount of fines, whose content and dosing accuracy affect the W/C ratio, workability, and permeability of the concrete • Regular petrographic checking of aggregate to detect any eventual variations	• Visual checking (appearance) • Checking of density • Checking the quantity of occluded air • Checking consistency • Measuring the duration of vibration for placement in samples • Measuring the duration of vibration for placement in blocks • Manufacturing the necessary samples for checking hardened concrete	• Compressive strength testing at 7, 28, 90, and 365 days (for tests at 365 days, only a few series are chosen) • Measuring the density of hardened concrete • Flexural testing at 28 and 90 days, and possibly 365 days • Frost resistance testing; permeability testing • Measuring dynamic and static elasticity moduli • If necessary, control testing of resistance to chemical attack (alkali-aggregate and sulfate reaction)

18.5.3 Control testing during operation

These tests can be subdivided into two categories:
- Control tests upon completion of the dam
- Expertise tests on operating dams

Control tests at the time of completion of the dam test for the structure's acceptance. These concrete tests are run on cores extracted from the structure by drilling. For construction control tests, 300 mm-diameter drillholes are generally used, so that the cores are comparable with the control samples for the follow-up of production. On some dams, prisms and cubes are extracted in order to run bending strength, frost resistance, and permeability tests (this involves extraction by pneumatic hammer, then the sawing of the samples; the samples are generally taken at the end of a yearly concreting campaign).

The objective of expertise tests on operating dams is to track the evolution of its characteristics, remove uncertainty, or identify the explanation for a dam behavioral problem. Sampling for concrete tests is carried out on cores extracted by drilling. It is often combined with drilling related to the dam's monitoring system. The drillhole diameter is determined according to the individual circumstances and the required analyses. The tests can be aimed at checking the mechanical properties (compressive strength on cylinders, splitting, pure tensile strength, density measurement, elasticity modulus measurement, ultrasonic wave velocity, mineralogical and petrographic analyses on thin layers, frost resistance and permeability tests, alkali-aggregate reaction tests).

For structures with major and evolving permanent deformation, creep tests are also of great importance. These are tests of long duration that required the use of costly laboratory equipment.

18.6 Using Test Results on Hardened Concrete

18.6.1 Considerations relative to concrete strength

The information and experience gained concerning concrete properties at the end of construction, and those acquired over time during the construction and operation of concrete dams provide instructive and useful sources for practical use. Without going into detail, it seems appropriate to mention a few of these elements.

Obviously, all properties of hardened concrete are closely connected to those of fresh concrete. However, it must be emphasized that the properties of large aggregate concrete differ significantly from those of standard concrete used for the construction of other engineering works.

In the construction of dams, it is common practice to consider mechanical strength as a primary factor. Compressive strength, in particular, is the most commonly used parameter for checking hardened concrete.

Fracture tests, either compressive or tensile, are run on cubic, prismatic, and cylindrical samples, but also on cores of 30-cm diameter or larger, extracted directly from the dam or from a block manufactured specifically for testing. These cores are highly representative from the age of 90 days, and their strengths are generally slightly greater than those of control samples.

When assessing dam strength, a fundamental aspect concerns the shape and size of samples. It has been found that a cube with a 30-cm edge does not give an accurate idea of true strength, due to parasitic load during the test. As a general rule, cube strength is greater than cylinder strength. After checking the results of several tests, an average value of 0.8 is accepted for cylinder strength compared to cubes. Furthermore, it has been established that strength decreases with an increase in the size of the samples. Tests have proven that in the case of an increasing height/base ratio (H/B), strength decreases until the H/B value reaches 1.5, and then tends towards a constant value above H/B = 2. Ultimately, it appears that the prism, or ideally a cylinder whose height is equal to twice its diameter, is the most appropriate shape for obtaining values closest to true compressive strength (Stucky, 1956).

In the case of optimum workability with more or less constant particle size, a relatively close relationship can be established between the cement content and strength. In Figure 18.11, taken from SwissCoD (2001b), compressive strength at 90 days from eight different Swiss dams was recorded according to cement content. On average, strength was found to increase by approximately 1 N/mm^3 when the cement content increased by 10 kg/m^3, which confirms the linear relationship between the W/C value and strength assuming a constant moisture content (consistency approximately constant):

$$fc(90) = 72 - 60 \cdot E/C, \text{ expressed in N/mm}^2.$$

CONCRETE TECHNOLOGY

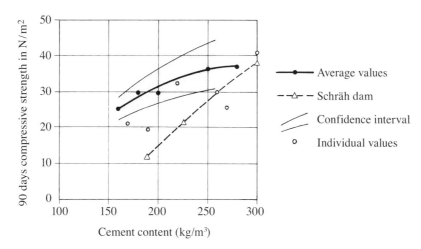

Figure 18.11 Compressive strength at 90 days depending on cement content (from SwissCoD, 2001b)

The increase in strength is less marked for concrete manufactured with crushed aggregate for a constant cement content and invariable consistency:

$$fc(90) = 84 - 72 \cdot E/C, \text{ expressed in N/mm}^2.$$

As an indication of the increase in concrete compressive strength, Table 18.12 presents in % of strength at 90 days that obtained at 7 and 28 days.

Table 18.12 Evolution in % of compressive strength over time (Paillex, 1985)

Parent rock	at 7 days	at 28 days	at 90 days
malm	65 to 69%	84 to 85%	100%
gneiss	60 to 71%	83 to 89%	100%
granite	61 to 74%	81 to 95%	100%

The use of a logarithmic function gives good results for analytically describing the evolution of the compressive strength of construction concrete as a function of time.

As a general rule, compressive strength at 90 days fc(90) is used as a reference value. Figure 18.13, also from SwissCoD, 2001b, groups the values fc(x)/fc(90) from ten different dams built between 1924 and 1985 and in total representing 25 types of concrete, as well as the theoretical relationship:

$$fc(x)/fc(90) = 0{,}34 \cdot (\log x + 1), \text{ where } x \text{ is age expressed in days.}$$

The use of a logarithmic relationship to represent the increase in compressive strength as a function of age thus seems justified.

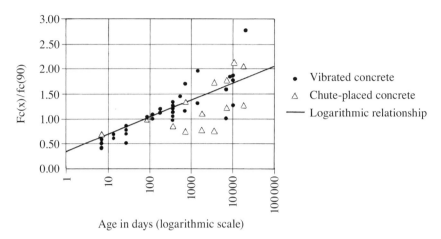

Figure 18.13 Development of compressive strength (from SwissCoD, 2001b)

In order to present the range of values obtained from various dam construction sites, Table 18.14 identifies different compressive strength averages at 90 days calculated throughout the duration of works. As can be observed, the tests were carried out on different types of samples. The dam concrete that was used was manufactured with crushed aggregate whose parent rock was gneiss or granite.

In his book *Le béton de barrages* (1985), Paillex estimated around 380 kg/cm^2 (38 N/mm^2) in the faces compared to 270 kg/cm^2 (27 N/mm^2) in the mass (ratio of 1.4), with the average compressive strength of all types of rocks combined. Furthermore, the homogeneity of the strength results determines the quality of these values, and on the basis of the results from the 10 dams, he cites dispersions of:

- For facing concrete: 7.1 to 13.5% (on average 11.1%)
- For mass concrete: 12.8 to 15.9%

Lastly, tensile strength based on flexural tests on 20 × 20 × 60 cm prisms for concrete graded to 60 mm and a CP content of 250 is in the order of (Stucky J.-P., 1997):

- 30–35 kg/cm^2 (3.0–3.5 N/mm^2) at 7 days
- 35–40 kg/cm^2 (3.5–4.0 N/mm^2) at 28 days
- 40–45 kg/cm^2 (4.0–4.5 N/mm^2) at 90 days

18.6.2 Factor of safety

On the basis of a series of tests, the engineer has, of course, several methods available for assessing the safety factor of a concrete mass. He or she can work off reference values calculated from the fabrication of samples subjected to compressive and tensile fracture tests that were carried out, it must be remembered, during the project design phase and prior to beginning the works. The statistical strength averages obtained are useful parameters.

One simple solution consists in randomly (but cautiously) setting an invariable factor of safety n. It is better to choose a relatively high value, for example, in the order of 4 to 5. The required strength at a given point is therefore equal to the product of the least favorable stress calculated at the place in question by the defined factor of safety.

Table 18.14 Average strength (MPa) at the end of construction of Swiss dams taken from concrete compressive strength tests at 90 days for crushed aggregate from gneiss and granite: P = facing concrete, M = mass concrete (from SwissCoD, 2001b)

Dam	Year	Max D	Type	CP 160 [kg/m³] W20	W30	C30/45	CP 200 [kg/m³] W20	W30	C30/45	CP 250 [kg/m³] W20	W30	C30/45
Albigna	1959	150	P							40.7		
Contra	1965	120	P					28.5				
			M					23.1				
Curnera	1966	80/120	P									25.2
			M					22.1				
Emosson	1974	160	P								41.2	
			M		27.5							
Ferden	1975	100	M								46.3	
Gebidem	1967	80	M					22.7				
Gde Dixence	1961	100	P									24.9
Isola	1960	80	P								35.3	
Mauvoisin	1957	120	P				42					
			M		30.0							
Moiry	1958	150	P					32.5				33.7
			M			25.9						
Nalps	1962	120	P									22.8
			M					20.1				
Oberaar	1953	180	M	24.6								
Sta Maria	1968	100–120	M					20.0				
Vasasca	1967		M									30.9
Zervreilla	1957	120	P	31.7						31.7		
			M				23.6					
Max				31.7	30.0			32.5		40.7	46.3	33.7
Av				17.9	28.8	25.9	23.6	42.0	24.1	36.2	40.9	27.5
Min				24.6	27.5			20.0		31.7	35.3	22.8

The question of whether to apply an average factor of safety or at the minimum value of measured strength can be considered. In addition, by setting this factor of safety value, there is no indication on the probability of obtaining a result that is equal to or lower than the calculated stress, that is, a factor of safety equal to or smaller than 1. It is doubtful whether the concrete will fully attain the expected strength following the results of preliminary testing and checking carried out during construction. Furthermore, we know that the shape and size of samples manufactured for the tests (cube or cylinder) affect the true strength value.

Another approach involves considering average strength R_m and the mean square deviation E, defined in Figure 18.15, to calculate a characteristic strength R_{car} obtained from the following equation:

486   CONCRETE

$$R_{car} = R_m - E.$$

The required strength R_{ex} is equal to

$$R_{ex} = n_o \cdot \sigma,$$

where n_o is the factor of safety.

According to the condition $R_{car} \geq R_{ex}$, we find

$$R_m - E \geq n_o \cdot \sigma,$$

σ: stress obtained by the calculation

R_m: average compressive cube strength

$$R_m = \frac{\Sigma R_i}{N}$$

E: mean square deviation allocated to average strength

$$E = \sqrt{\frac{\Sigma(R_m - R_i)^2}{N-1}}$$

N: number of R_i results

e: dispersion or relative mean square deviation

$$e = \frac{E}{R_m}$$

$n_{0.5}$:   factor of safety corresponding to average strength R_m

$$n_{0,5} = \frac{\text{average strength } R_m}{\text{calculated stress } \sigma} = \frac{R_m}{\sigma}$$

$1 - P$:   probability of obtaining strength lower than given strength R_p

n_p: factor of safety defined by the equation

$$n_p = \frac{R_p}{\sigma}$$

α: given factor as a function of P by the calculation of probabilities

The calculation of probabilities gives

$$R_p = R_m - \alpha E$$

hence the factor of safety n_p having a probability P of being reached or exceeded

$$n_{0,5} = n_p (1 - \alpha E)$$

Figure 18.15 Usual equations for determining the factor of safety (from Stucky, 1956)

A different approach makes the factor of safety n contingent on the dispersion of the test results and also relates it to the probability of occurrence. Dispersion enables the margin of safety to be assessed relative to the actual stresses in the dam. The required strengths are those at 90 days. They are calculated by multiplying the maximum stresses of each dam zone by a factor of safety depending on dispersion. According to the theory of probabilities, the higher the dispersion of a series of values, the greater the probability of finding values that are equal to or even lower than 1. Figure 18.15 summarizes the main equations used for establishing the factor of safety.

Dispersion tends to increase as strength decreases. For a concrete mix between CP 180 and CP 250, dispersion of 0.09 is considered low and a value of 0.15 excessive. For such mixes, a value of 0.12 is a normal average. For lower mixes than those mentioned above, for example, CP 140, the normal average is 0.15 (Stucky, 1956).

Stucky (1956) proposed a practical rule founded on three conditions, on the basis of test results from a 30-cm cube (Figure 18.16):

- If dispersion is medium, that is, e = 0.12, we use the factor of safety n = 4.2, which is the usually accepted value, and the probability of a strength lower than calculated stress σ is $1 - P = 10^{-10}$
- If dispersion is high, that is, e = 0.15, we require $1 - P = 10^{-7}$ hence 4.45
- If dispersion is very low, that is, e ≤ 0.0533, we admit a minimum value n = 3.2 and probability $1 - P$ is very low

By imposing these three conditions, a relationship can be established between the factor of safety n in conjunction with the average of the results of concrete tests and with their dispersion. Figure 18.16 gives the required factors of safety n, taking into account average strength tests on 30-cm cubes and 30/45-cm cylinders. Practice shows that cube strength is 20% greater than cylinder strength, hence $R_{cyl\,30/45} = 0.80\,R_{cube\,30}$.

Another method introduces the fracture diagram while taking into account a biaxial load (Figure 18.17). This involves searching for pairs of the least favorable principal stresses on the surfaces of the concrete, deducted from all cases of load considered. In this case, the factor of safety is expressed by the relationship between point η_{rupt}, located at the limit of the biaxial fracture and η_{eff}, corresponding to

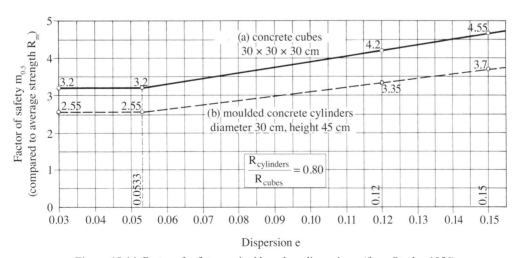

Figure 18.16 Factor of safety required based on dispersion e (from Stucky, 1956)

488   CONCRETE

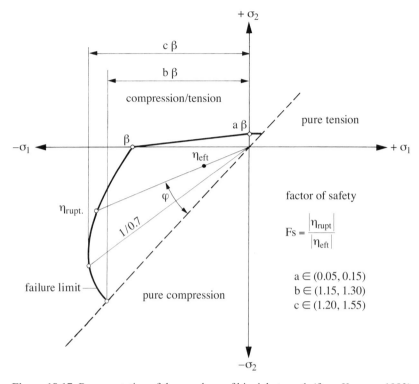

Figure 18.17 Representation of the envelope of biaxial strength (from Kreuzer, 1983)

the extreme edge of the vector of principal stresses at a given point of the dam. The limit of compressive strength is based on strength obtained at 90 days. The pure tensile section can be defined on the basis of cube or prism compressive strength[8] or determined using tensile tests. In the compressive/tensile section, a rectilinear transition can be accepted.

As with the diagram in Figure 18.17, it is possible to establish a failure diagram for several nominal concrete strengths (Figure 18.18). The latter can be reduced to principal acceptable stresses by setting a factor of safety firstly for pure compression and then for pure tensile stresses. The factors of safety are chosen based on a dispersion in percentage of average concrete compressive strength. For example, for dams built in Switzerland, on the basis of measured cube compressive strength, a factor n was accepted equal to 4 for compressive strength and 2 for tensile stresses, taking into account a dispersion of 12%. For other dispersion values, the factor of safety is adapted by introducing a correction coefficient for average strength.

By referring to the literature, other factor of safety values can be found. For example, Kreuzer (1983) introduced the following values by distinguishing between loads, as well as cube or cylinder compressive strengths:[9]

---

[8] Using prism compressive strength β_w as a basis, tensile strength is, according to the sources, in the order of (2.5 to 3) $\sqrt{\beta_w}$, or on average $0.82\, \beta_w^{2/3}$.

[9] It was previously mentioned that cube strength is 20% greater than cylinder strength.

	Case of normal loads	Case of exceptional loads
Cube strength	3.5	2.25
Cylinder strength	3.0	2.0

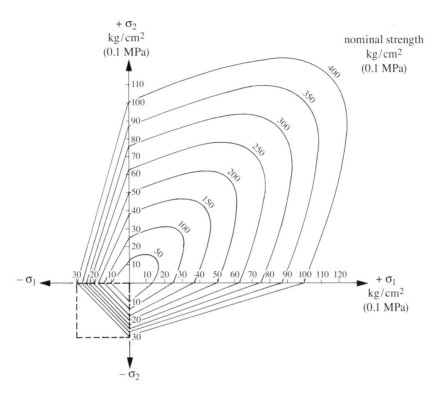

Figure 18.18 Biaxial failure diagram with nominal strengths

18.6.3 Determining cement content

Cement content is chosen so that the required coefficients are respected. To do this, tests are run on the various types of planned concrete. The test results, with their dispersions, are ordered according to the type of concrete (facing concrete, mass concrete) and contents, which enables the factor of safety to be calculated (Table 18.19). It is then possible to calculate the required strengths at various block levels in the dam by multiplying the least favorable principal stresses by the factor of safety. In cases where for the same loading the principal combined stress σ_{II} is a tensile stress, a penalizing effect must be introduced. This amounts to imposing a greater quality than would be necessary if only the compressive stress σ_I was considered. A fictitious compressive stress σ^*, proposed by Stucky (1956) can then be introduced, which has the value

$$\sigma^* = \sigma_I - 1.4\,\sigma_{II}.$$

Table 18.19 Example of the calculation of the factor of safety compared to average cylinder compressive strength

Categories	Content in kg/m³	Number of cylinders	Strength at 90 days in kg/cm² [MPa]	Dispersion in %	Factor of safety according to Figure 18.16
Mass concrete	200	52	220 (22.0)	13.3	3.48
	220	66	227 (22.7)	16.7	3.87
	250	54	250 (25.0)	12.2	3.36
Facing concrete	250	62	248 (24.8)	16.1	3.80
	280	30	264 (26.4)	15.5	3.73

It should also be kept in mind that a block can be comprised of facing concrete on either side of the mass concrete (Figure 18.20). To determine the content of the mass concrete, the principal stresses are brought to the edges, generally by linear interpolation, in order to find out the maximum load at each specific point.

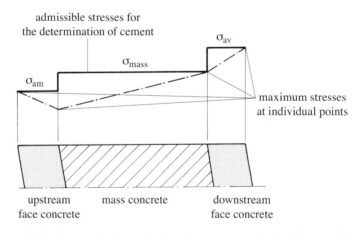

Figure 18.20 Stresses that determine how the concrete mix should be established

18.6.4 Worksite laboratory

A worksite laboratory is an essential installation on a large construction site, to be able to systematically check the manufacture of concrete and quickly assess its homogeneity and quality. During this process, tests on aggregate, cement, and fresh concrete, as well as compressive strength tests (already at 7 days) are of paramount importance in identifying any defect or drop in production quality as soon as possible. If necessary, the content can quickly be adjusted to correct the mechanical strength of the concrete.

Sampling the aggregate categories enables the moisture content, particle size composition, and dust content to be determined.

Cement is subject to mechanical (compressive and tensile strength tests at 7 and 28 days, for example), physical, and chemical assessments.

Lastly, checks carried out on the fresh and hardened concrete must be such that correct placement on the dam is guaranteed, according to the required conditions. These conditions notably include workability, mechanical strength, and durability. In order to carry out tests on fresh concrete (workability, occluded air content), sampling is carried out as the concrete comes out of the concrete batchers, and samples are manufactured. It is common practice to prepare a series of three samples per defined age per test; an average value is calculated for each series of three. The samples are stored and subjected to laboratory tests. To carry out these mechanical tests, the surfaces of the samples in contact with the press must be flat. If not, they must be rectified or undergo a finishing process before the test. The laboratory is also responsible for carrying out permeability and frost resistance tests.

18.7 Research on Concrete

The SFWG has supported various research studies on concrete. The conclusions drawn are briefly summarized in the following sections.

18.7.1 Behavior of unreinforced concrete under dynamic forces

Overview

During an earthquake, concrete dams can be subject to high levels of loading alternating between compressive stresses and tensile stresses. It is therefore very important to analyze the creation and propagation of cracks caused by this loading using computational models based on fracture mechanics. Initially, however, the work consists in experimentally determining the fracture characteristics of unreinforced concrete, which can then be directly applied in the development of computational models. Specific research studies have been undertaken with the aim of further understanding the behavior of unreinforced concrete under dynamic forces. Studies are also planned to establish models better reflecting the true behavior of materials by introducing a damage law. Tests carried out on different concrete samples, including some from dams (Mauvoisin, Solis, Zervreilla), should also take into account their past history of loading and the impact of the speed of deformation (Chappuis, 1987; Brühwiler, 1988).

Tests

Research studies should enable, with the help of several tests on samples, the characteristic values of both static and dynamic concrete fracture under loading to be determined. To determine the specific fracture energy G_F of mass concrete, the most appropriate sample shape had to be found, one with the best possible ratio between its volume and the fracture surface. Comparatively to other types of tests (direct tensile test, flexural test, compact tensile test), the corner fracture test has turned out to be the most favorable, particularly due to its stability in the face of high deformation speeds (Figure 18.25 (a) and (b)).

Several series of tests were carried out firstly on standard concrete and then on concrete from three dams: Mauvoisin, Zervreilla, and Solis. The general characteristics of these concrete samples were:

Dams	D_{max} grain Φ [mm]	Cement content [kg/m^3]	W/C	Sample dia. Φ [mm]	Age of concrete [years]
Solis	80	250	0.50	200	2
Zervreilla	120	200	0.59	200	30
Mauvoisin	120	250	0.47	300	30

Tests under almost static loading have demonstrated that the specific energy of the fracture G_F does not depend on the shape of the samples or the testing method. All types of tests result in comparable values. Furthermore, the G_F value of dam concrete is substantially higher than that of standard concrete, due to the maximum diameter and characteristics of the aggregate used in dam concrete (Figure 18.21). In addition, the specific fracture energy G_F is affected by the ligament length; it increases with the length until a limit value is reached.

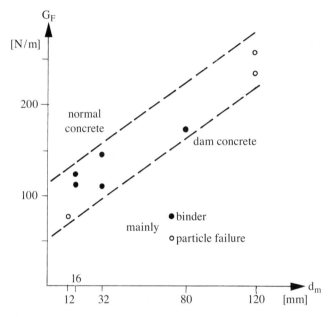

Figure 18.21 Effect of maximum particle diameter on the specific fracture energy (from Brühwiler, 1988)

Dam behavior under seismic loading has been studied using dynamic tests. Research has been carried out on the influence of dynamic preloading on fracture characteristics with a high speed of deformation. The results show that this deformation speed acts strongly on both compressive and tensile concrete strength. It should also be noted that the past history of compressive loading has a decisive influence on concrete tensile strength (Figure 18.22). The specific fracture energy G_F of concrete that has been subjected to pre-loading is lower than that of concrete that has not previously undergone loading.

Computational model of the propagation of cracks
A simple method involves considering primarily tensile strength using a linear or nonlinear model, while admitting that the formation of a crack begins at the moment that tensile strength is exceeded (Rankine's material strength criterion).

In the course of work on a thesis, a program of finite elements was developed to model discrete cracks and their influence on the dynamic behavior of a concrete dam. The discrete crack (Figure 18.23) was modeled by the disconnection of nodes of finite elements on each side of the lip of the crack. Crack propagation required a redefining of the mesh at each step in the calculation.

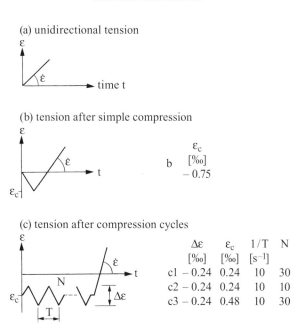

Figure 18.22 Presentation of the timeline of loading (from Brühwiler, 1988)

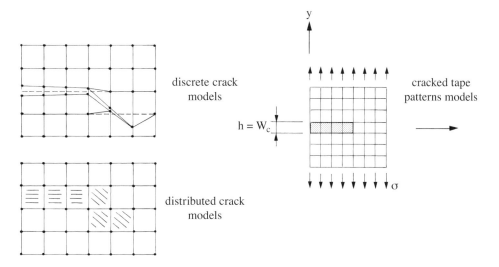

Figure 18.23 Crack models

Initially, this program did not include a model employing fracture mechanics but applied the criterion of tensile stress to reproduce the propagation of cracks (Skrikerud, 1983). Later, the program was supplemented with an improved model of the material, which in particular included the effect of deformation speed and the past history of loading. With regard to the behavior of the material once tensile strength has been exceeded, the model includes the damage based on the fictitious crack model (Chappuis, 1987). The fictitious crack model can not only be applied when describing the tensile test (see inset box) but also when describing the stress state. It is therefore of value in the research into the formation and stability of cracks. The graphs in Figure 18.24 show the distribution of stress along a crack in a bent beam. As the deformation increases, the stresses at the front of the crack also increase. It is admitted that the stress cannot be greater than that of tensile strength. As soon as this level of stress is reached, any increase in deformation creates a lengthening in the fictitious crack. The stress at the front of the fictitious crack is equal to tensile strength, for the duration that the fictitious crack continues to grow. Energy G_F has been dissipated along the fictitious crack.

Another finite element program simulates the growth of a crack in a gravity dam after an earthquake. The program is based on elastic linear fracture mechanics combined with a modeling of discrete cracks (Chapuis et al., 1985; Droz, 1987).

As with the fictitious crack model, the crack band model (where the crack propagates within a zone) takes into account the band where the softening process of the tensile stresses occurs and can be used to simulate the development of the crack as well as damage to the material. The crack band model stipulates that the failure of a heterogeneous material can be modeled by a band of dense and parallel microcracks.

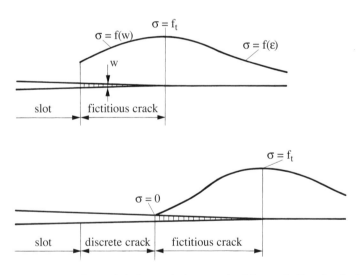

Figure 18.24 Stress around a crack before and after growth of the crack (from Brühwiler, 1988)

"Fictitious crack" model

In order to consider, for the cracking process of concrete, the cracked zone ahead of the crack, in 1976 Hillerborg and his colleagues proposed the "fictitious crack model."

A sample is subjected to direct tensile stress looking at its deformation in order to follow the descending branch of the stress-deformation curve right down to zero (Figure 18.25). The deformation is measured by gauges placed in the two zones A and B. As the sample is presumed to be homogeneous with a constant section, the ascending parts of the A and B curves are identical until maximum load is reached. Beyond that, somewhere in the sample (zone A, for example), a crack forms; its width is limited in the direction of tensile stresses. As it develops, deformation of the sample becomes localized, the force of the tensile stresses decreases, and deformation in the adjacent zone B decreases. The graph in Figure 18.25 shows the deformation in zones A and B. Deformation in zone B is represented by a traditional stress-deformation curve with loading and unloading. Deformation in zone A also includes the lengthening of the cracked part. These are the differences between the abscissa of the descending branches A and B. Ultimately, there are two curves:

- Outside of the cracked zone, specific stress/deformation curve, a = f(z), including unloading
- In the cracked zone, stress/lengthening curve $\sigma = f(w)$

With these two curves, it is possible to calculate deformation M of any section l_0:

- in an uncracked zone $\Delta l = \varepsilon \, l_0$
- in a cracked zone $\Delta l = \varepsilon \, l_0 + w$

It should be noted that the width of the cracked zone is not of relevance in the abovementioned equations. The hypothesis can be put forward that originally this width was zero. During the cracking process, the total width of the cracked zone is therefore w. Stresses σ can still be transmitted within this width, according to the law $\sigma = f(w)$. This simplified representation of the cracked zone is called the *fictitious crack*.

The energy dissipated during the test, per surface unit of the sample section, is represented by the surface of the curve $\sigma = f(\varepsilon)$. For a perfectly elastic material, this energy is zero.

Similarly, the energy dissipated in the fictitious crack zone is represented by the surface under the curve $\sigma = f(w)$. This is the specific crack energy G_F.

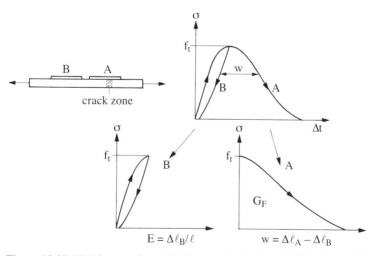

Figure 18.25 Fictitious crack model (Sources: Brühwiler, 1988; Lafitte, 1989)

18.7.2 Size of samples and scaling effect[10]

Sample size is a major problem for dam concrete due to the large aggregate size. Test results are very sensitive to these sizes, as well as to limit conditions.

Knowledge of scaling laws is a vital element for relating test results to the behavior of the actual structures. For concrete, this is a domain that is in full expansion, and intensive research is underway in many countries, as is much debate.

It is an important field of research for dams for two main reasons:

- The exceptional size of the structure comprising the actual dam
- The fact that the aggregate used in dam concrete has a maximum diameter that is much greater than that of standard concrete; it can reach 150 mm

Scaling effects are generally affected by limit conditions. Representative elementary volume (REV) is defined as being a volume whose size does not affect the test results. It is accepted that representative volume is obtained for sample sizes between five and ten times the maximum size of the largest heterogeneity.

For real laboratory samples, a more or less large dispersion is always observed. This means that the real samples have never reached the REV size. As a general rule, samples should be limited to twice or even three times the maximum size of the aggregate.

In order to better understand this scaling effect, splitting tests on variable size elements were carried out in 1995 and 1996 on samples manufactured on site during heightening works on the Mauvoisin and Luzzone dams. Tables 18.26 and 18.27 clearly demonstrate that the tensile strength and fracture energy of a concrete dam depend on the sample size.

Table 18.26 Influence of sample size on concrete characteristics in the Luzzone dam; elasticity modulus $E = 19{,}000$ N/mm^2; content 250 kg/m^3 (CP and 50 kg of fly ash); particle size 0–63 mm

Samples [cm]	Number	Tensile strength [N/mm^2]	Fracture energy [N/m]
20 × 20 × 20	6	1.82	271
40 × 40 × 40	3	1.27	334
80 × 80 × 40	3	1.43	442

Table 18.27 Influence of sample size on concrete characteristics in the Mauvoisin dam; elasticity modulus $E = 28{,}300$ N/mm^2 (CP250), $E = 24{,}300$ N/mm^2 (CP175)

Size [cm]	Samples Number CP 250	CP 175	Tensile strength [N/mm^2] CP 250	CP 175	Fracture energy [N/m] CP 250	CP 175
20 × 20 × 20	16	14	2.64	2.10	266	219
40 × 40 × 40	8	7	2.27	2.51	300	259
80 × 80 × 40	7	7	2.12	2.00	373	392
160 × 160 × 40	7	3	2.11	1.80	482	418
320 × 320 × 40	3	0	2.27	–	480	–

---

[10] From SwissCoD, 2000a.

19. Concrete Behavior and Observed Phenomena

Moiry arch dam in Switzerland, height 148 m, year commissioned 1958 (Courtesy J.-C. Dufour, IDEALP)

19.1 Development of Deformation Over Time

As dams (particularly seasonal storage dams) age, a gradual increase in maximum and minimum displacement is often observed over the years, both with a full and empty reservoir. This range in annual maximum and minimum deformation can add up to a total of several centimeters and is often still detectable after decades of operation. Over a longer period, the movement can in some cases have completed its irreversible delayed phase (sometimes called "plastic") and occurs annually, in near elasticity, between the limit positions reached. However, until this stabilization phase has been directly observed, it is difficult to predict whether it will occur or how long it will take.

The delayed deformation that affects dams can have various causes and can manifest in very different ways from one dam to another, depending on the dam shape and, in particular, its static system. This delayed deformation is due to the combination of several phenomena and is added to seasonal or cyclical deformation. These different types of deformation can sometimes cancel one another out or, quite the opposite, add up. It is often difficult to isolate the various delayed components of the total displacement. However, after an exceptional drawdown, it may be possible to carry out geodesic measurements providing useful information. Delayed deformation occurs in both the horizontal and vertical directions. The dam monitoring system must be capable of measuring in both directions.

19.2 Cracking

19.2.1 How cracks form

Concrete cracking is a complex process, due in part to the material's own properties and also to the geometry of the structure or part of the structure. A distinction can be made between crack origin and propagation. Cracking is often caused by tension that is generated internally. We know that the cracks originate from microcracking in the concrete caused by the unsuccessful shrinkage of cement paste. The ensuing formation of macroscopic cracks occurs for several reasons.

During concreting, thermal gradients develop in massive elements and cause cracks. These appear in the first few days, or even weeks, after concreting. In general, they are surface cracks and often propagate in the vertical plane or perpendicular to the foundation (Figure 19.1). During dam operation, considerable variations in ambient temperature can also provoke thermal gradients leading to cracks, especially for parts of the dam with a thin section (for example, buttresses).

Figure 19.1 Example of cracks perpendicular to the foundation of an arch dam (Courtesy H. Pougatsch)

After formwork removal, the concrete loses part of the mixing water and a water gradient is established. If the resulting shrinkage is excessive, cracks may begin to appear. In massive elements, self-desiccation can also create a stress state that may lead to internal cracks in random directions.

Under loading (compressive or tensile stresses), the concentration of stress around the aggregate causes these microcracks to get bigger. With an increase in load, true cracks begin to appear following the development of connecting microcracks between the microcracks that touch the aggregate.

19.2.2 Development of cracks

The question then arises around the stability of a crack that has developed, that is, will it remain the same length or will it continue to grow. The theory of linear elastic fracture mechanics may help answer this question, and even more so for dams that have large dimensions compared to the crack zone. This theory introduces the notion of the stress intensity factor at the tip of the crack and compares it to concrete fragility or resistance to cracking. Although there are more and more tests available, this particular approach is continuing to develop for dam concrete. The same applies for our understanding of the speed of crack propagation.

There are two separate cases for cracks filled with water that is under pressure (cracking in the upstream face of a dam):

- The crack opens slowly, and the hydrostatic pressure acts like a wedge separating the lips of the crack
- The crack opens at a speed at which the water can no longer follow, and the slight hollow that appears tends to hold back the lips of the crack, thus checking its propagation

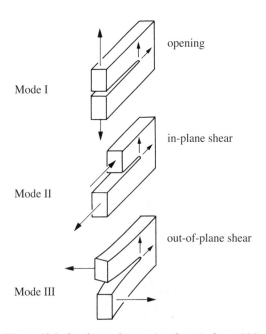

Figure 19.2 Crack opening modes (from Lafitte, 1989)

Crack opening can be categorized into three basic mechanisms: normal opening of the lips (mode I), in-plane shearing (mode II), and out-of-plane shearing (mode III) (Figure 19.2). Modeling of crack formation has become possible thanks to the implementation of discrete crack models in the two-dimensional domain (plane deformation models). Regarding the overall behavior of the structure, exterior actions emerging either from normal operation or from accidental events can lead to cracking in some parts of the dam.

19.2.3 Observations made concerning concrete dams

During normal operation, the actions in question include hydrostatic loading and temperature. For massive structures such as gravity dams, the principal factor is that of water pressure, while for thinner structures (arch and buttress dams), thermal variations also play a major role. Cracking during normal operation may occur for a number of reasons: unfavorable geometry, poorly executed concrete lift joints, the dam's thermal state at different phases (at the end of construction/during joint closure/during impoundment), foundation movement, etc. Concrete cracking often appears after construction and then develops over time, with its influence on the dam only becoming significant after several years. Although it has a different origin, this type of cracking can occur in conjunction with that caused by concrete shrinkage.

Depending on the type of dam, the following behavior can be observed:

a) Gravity dams

Most loading is due to hydrostatic pressure and can cause tensile stresses near the upstream face that gradually overcome the material's resistance, leading to horizontal cracking or the opening of concrete lift joints. Drawing down the reservoir tends to worsen cracking and increase the rate of leaks.

b) Arch Dams

The appearance of cracks differs on the two faces. These cracks are naturally produced by the concrete's resistance to tensile stresses being overcome. Given the high degree of hyperstatic behavior in arch dams, cracking is not really dangerous, as load can be taken up by the arch-wall distribution. Cracks may be highly varied: horizontal cracks at the concrete lift surfaces, cracks perpendicular to the foundation near the downstream abutments, cracks parallel to the foundation at the upstream heel.

c) Buttress dams

Cracking in these structures (whether they have buttresses or closed downstream facing) is common. It is first of all related to the dissipation of hydration heat and then to concrete shrinkage. During the operating phase, hydrostatic and thermal cycles cause high bending stress in the buttresses, leading to a worsening of the cracks. However, it is primarily variations in temperature that affect the development of cracks.

Indeed, the presence of cracks is frequently observed during gallery inspections. This observation mostly affects upper galleries. Some cracks may already have been recorded once operation began, while others may be noticed later on. These may be vertical or subvertical cracks that initiate at the top of the gallery, or horizontal or sloped cracks located to the sides of the galleries. These cracks may be almost closed but may also be clearly open. Sometimes, these cracks are caused by a thermal action related to the process of concrete hydration or by temperature gradients related to exterior conditions. A combination of the two causes is also possible. In some cases, the emergence of these cracks is associated with the phenomenon of swelling.

As an accidental action that cause cracks to form, the alkali-aggregate reaction can also be mentioned, as well as the erratic settlement of one of the abutments, a sudden movement of the dam walls under the effect of a tectonic or hydrogeological shift, the action of a high-intensity earthquake, or hydraulic overpressure that may even result in the overtopping of the reservoir should the weir sections be blocked.

19.3 Creep and Shrinkage

19.3.1 Description and characteristics of creep and shrinkage

Shrinkage, in the generally accepted sense of the word, is a loss of volume caused by the evaporation of non-chemically bound water (drying shrinkage). This initially depends on concrete temperature but is especially related to the water gradient between the dam core and its surface. Although shrinkage in thin structures generally stops after a few years, it can continue for many years (although on a limited scale) in massive structures such as dams.

Concrete is a viscoelastic material, that is, material loading is accompanied by almost instantaneous deformation and by delayed deformation (creep). *Creep* is defined as being permanent deformation under the effect of a load, usually constant over time. As the load varies, the resulting creep is obtained by applying the principle of superposition. If, however, the imposed deformation is admitted as being constant, the same phenomenon leads to a relaxation of stresses. Creep develops relatively quickly at the beginning and then slows down. It continues for several decades after the loading. Young concrete undergoes a specific type of creep that is considerably greater than for old concrete. After fifteen to twenty years, the deformation associated with creep is in the order of two to four times greater than instantaneous deformation. Creep in concrete is considerable and is one of the main causes of the increase in nonreversible deformation.

The respective influences of shrinkage and creep on a dam are not easy to separate.

19.3.2 Implications for concrete dams

The effects of shrinkage and creep on different types of dams are the following:

a) Gravity dams

For the above-mentioned reasons, the effects remain generally limited for this type of isostatic dam. A contraction in the concrete mass manifests in crest settlement, which, depending on geometry and the zone most affected by shrinkage, may be accompanied by slight upstream or downstream crest displacement.

b) Arch dam

Crest settlement, accompanied by a displacement of the arches and downstream tilting of the cantilevers, is generally observed.

This mechanism leads to decompression in the foundation at the upstream heel of the dam, and may even result in unfavorable consequences for the bearing capacity of the foundation from a structural and hydraulic viewpoint:

- Opening of discontinuities in the rock and the uplift of the upstream heel of the dam, accompanied by an increase in compressive stress and shearing in the concrete downstream
- Increase in hydrostatic pressure in cracks in the rock and concrete upstream (to the point that it competes with the full pressure of the reservoir), thus amplifying cantilever tilting and increasing the upstream-downstream hydraulic gradient (an increase in seepage flow)

c) Buttress dams or multiple arch dams

Cracking can occur where relative displacement is hindered. The cracks generally develop in the buttress webs, departing vertically from the foundation and curving upstream as they climb.

19.3.3 Laboratory testing

Provision should be made for laboratory testing to obtain more information concerning creep. Creep tests are conducted by seeking to best reproduce normal operating conditions, which for some dams means annual loading-unloading cycles (the reservoir is full in summer-autumn and empty in spring).

To establish the concrete's potential for creep, extracted samples are subjected to compressive axial loading for the number of days corresponding to the duration of loading, during which time the shortening is measured to 1/1000 mm accuracy. The samples are then unloaded and deformation measurements are taken over several months. The unloading enables the elastic deformation to be estimated (ε_{elast}) and, therefore, the plastic deformation (ε_{plast}) to be re-established under loading. Similarly, viscoelastic deformation (ε_{ve} or "reversible" creep) and actual creep deformation (ε_{fl}) can be deduced from this. It goes without saying that the categorization of these different types of deformation is not always easy; it remains affected by not inconsiderable dispersion. The illustration of a test, with the different effects identified, is presented in Figure 19.3.

To define concrete's potential for creep, tests on samples extracted from operating dams (Emosson, Luzzone, Les Toules) were carried out (Gunn and Bossoney, 1996; LMC EPFL, 1996). It is interesting to note that the effect of creep is still significant even though the loading tests were carried out practically 30 years after the dams were constructed. In one case, it was observed that the total deformation after 3.5 years of loading reached more than twice the instantaneous deformation.

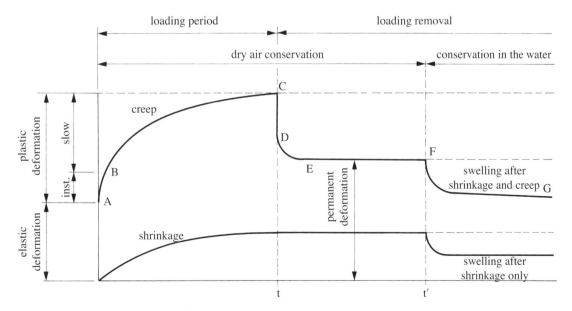

Figure 19.3 Creep test under constant loading

19.4. Swelling

19.4.1 Description and characteristics of swelling

The phenomenon of concrete swelling is always associated with the presence of water. Its development is very slow. The relatively recent understanding of swelling is also due to the fact the creep in concrete dams under constant loading obscured the effect of swelling. Furthermore, it is extremely difficult to estimate its duration. Concrete expansion introduces nonreversible deformation and cracks can initiate, which further favors the circulation of water.

Concrete swelling in a dam can result from different causes (Amberg et al., 2018; SwissCoD, 2000a, 2017a):

- An increase in temperature (thermal swelling)
- An increase in water content (moisture swelling)
- The hydration process (hydration swelling)
- Other chemical reactions, such as, for example, the alkali-aggregate reaction (chemical swelling)

Thermal swelling follows annual temperature cycles and is practically reversible (effect of frost near the faces).

Moisture swelling, unlike thermal swelling, continues to develop over a relatively long time. Fresh concrete and young concrete only use part of the total mixing water for cement hydration (hardening). The result is a porous material partially saturated with water. If the concrete enters into contact with water during its liquid stage, it tends to resaturate. This tendency is all the more pronounced when the water is under considerable hydrostatic pressure, as is the case in the deep waters of a reservoir. Over the years, this resaturation occurs in conjunction with concrete swelling.

Different ***chemical reactions*** leading to swelling can occur in hardened concrete. If the concrete is in contact with selenitic water (water containing sulphates), the sulphates will react firstly with the aluminates in the cement and form ettringite or thaumasite. This swelling, caused by the sulphates, can eventually destroy the concrete from the surface, or, in less severe cases, contribute to the overall swelling of the dam. The sulphates in the concrete do not cause generalized swelling if the temperature does not reach values of more than 50°C. In massive elements, this temperature may temporarily be exceeded during setting. There may then be a delayed risk of ettringite forming.

19.4.2 Alkali-aggregate reaction (AAR)

The damaging process of the alkali-aggregate reaction (AAR) was first discovered in 1940 by T. Stanton on concrete roads, bridge structures, and retaining walls in California. He attributed this damage to the results of a chemical reaction that had occurred within the concrete. Later, reports of similar damage came from elsewhere in the USA, Denmark, the Netherlands, Germany, the United Kingdom, Iceland, South Africa, France, and Japan. In Switzerland, damage due to AAR was reported for the first time in 1988; it had occurred on a dam (SwissCoD, 2017a).

Some minerals comprising amorphous silica are liable to be attacked by a solution containing alkalis. These are known as reactive. However, cement always contains a certain amount of alkalis (sodium and potassium). In their presence, these reactive minerals create a gel that incorporates large quantities of water into the structure. This gel produces local swelling that on the surface can lead to a network of cracks or radial cracks (Figure 19.4). In cases of high compressive loading, the direction of the cracks is parallel to the pressure trajectory. Gel droplets form, which become milky when exposed to air. They turn white after dehydration. Shattering of particles on the surface can also be observed.

Figure 19.4 Surface damage due to the alkali-aggregate reaction (Courtesy H. Pougatsch)

If there is a considerable amount of these reactive minerals, and they are spread throughout the concrete mass, the alkali-aggregate reaction causes concrete swelling. Depending on the case, this swelling can lead to intense cracking and severe damage or remain below a critical level.

The most spectacular swelling has been recorded in cases of chemical reaction. The alkali-aggregate reaction, when it occurs, often affects a large part or even all of the structure. This deformation can be distinguished from deformation due to creep or an increase in temperature by the following indications:

- A different initiation (the deformation manifests after a delay, which may range from a few years to a few decades), the time that the gel fills the concrete pores and that thermal contraction due to the dissipation of hydration heat falls below that of swelling
- An acceleration in deformation over time
- A deformation pattern that expresses an expansion of volume into plane and elevation

19.4.3 Identification and analysis of swelling

With regard to concrete dams built in Switzerland, early explanations attributed the increase in volume to increases in concrete temperature and ambient temperature. It was only after more substantial deformation was observed that AAR was integrated into the research on the origin of long-term deformation.

Identifying one effect of swelling requires long periods of observation with measurement systems based on stable points of reference. It is vitally important to dissociate thermal effects from those of chemical swelling.

The first element to identify in the swelling phenomenon is above all the demonstration of a vertical raising deformation, as this vector for deformation contains fewer induced side effects or components due to other forms of loading than horizontal deformation does. However, even though the appearance of uprising is a hint that swelling may be occurring, its observation is not sufficient; an analysis of overall deformation is necessary.

As part of this analysis, it should be remembered that deformation can be affected by a nonhomogeneous distribution of expansion within the dam structure. Furthermore, the development of zones with swelling is related to local conditions inside the dam (for example, the reactivity of aggregate itself (see § 19.4.5), water seepage made possible by porosity, in joints, cracks, or honeycombing). As for porosity, initially pores can act as an expansion space for gel to swell and thus delay the appearance of deformation. It is only when some of the pores start to seal that the first indications of swelling appear, without all of these expansion spaces having been completely mobilized. Concrete expansion also manifests through the development of a slight detachment between the cement paste and the aggregate, and "fusiform" lenticular cracks (sometimes containing silica gel), which can traverse the aggregate and cement paste.

19.4.4 Measuring displacement and visual observations

Displacement measurements are the most appropriate for detecting swelling. Leveling along the dam crest is particularly sensitive to this phenomenon. Difficulties appearing in gate operation due to deformation at their site location can also be an early sign of swelling.

As dams have a monitoring system, it is therefore possible to closely track their behavior (see § 32.4). This essential activity enables the early detection of any abnormal behavior, before swelling becomes apparent and before it presents a safety problem for the dam. As previously mentioned, horizontal and vertical displacement are determining parameters. Table 19.5 gives guidance on possible methods for measuring displacement.

Table 19.5 Methods for measuring displacement

Type of displacement	Methods of measurement
Horizontal displacement	Pendulums Geodesic measurements
Vertical displacement	Levelling Invar wire, extensometers

Meanwhile, visual inspection, which is also very important, enables the engineer to:
- Identify cracks that may contain gel or moisture
- Track the opening/closing of vertical joints between blocks on the upstream and downstream faces
- Check the functionality of the mechanical elements (gates) of the water-release system

19.4.5 Laboratory analyses and tests

As soon as an anomaly has been detected, the exact diagnoses must be made by testing samples in the laboratory. Investigations can be carried out using an optical microscope, electronic microscope, and spectroscopy. For example, the analysis program that was established for the Illsee dam (VS) included:
- Mineralogical analyses of thin sections to identify possible expansive formations due to the formation of ettringite (Candlot's salt) or thaumasite, or the alkali-aggregate reaction
- Chemical analyses on concrete to measure levels of SiO_2 and
- SO_3 and alkalis (Na, K)
- Chemical analyses of the water
- Mineralogical analyses of rock surrounding the quarry (levels of sulphur-pyrite FeS_2) (Figure 19.6)
- Mineralogical analyses using x-ray diffractometry on the concrete and on the crust taken from the downstream face of the dam

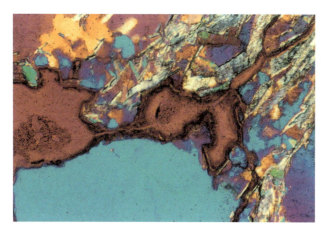

Figure 19.6 Illsee dam, mineralogical analyses of thin sections: structure of an aggregate along a string of alveoli containing silica gel (from SwissCoD, 2000a)

These tests should demonstrate that the origin of dam deformation is due to concrete swelling. The alkali-aggregate reaction is the main reason behind expansion, with the formation of ettringite playing a secondary role.

19.4.6 Implications for concrete dams

Concrete expansion due to the alkali-aggregate reaction leads to structural movement, major cracking (Figure 19.7), a modification of concrete strength, differential movement, and a deterioration in the joints.

Figure 19.7 Cracking following an AAR observed in a gallery (from SwissCoD, 2017a)

For concrete structures, AAR leads to an annual elongation of the concrete of 20 to 150 μm/m (the upper value applies to structures in the tropics). As a result, annual expansion in the order of 2 to 15 mm is observed for a dam with a height of 100 m, and annual elongation in the order of 5 to 40 mm is noted for a crest with a width of 250 m (SwissCoD, 2017a).

In cases where expansion has been prevented, compressive forces develop, leading to cracking. Following the formation of cracks, the mechanical properties of the concrete are reduced. For a highly developed AAR, laboratory tests indicate the following values:

- Compressive strength decreases by 25 to 60%
- Tensile strength decreases by 50 to 70%
- Elasticity modulus decreases by 60 to 70%

It should be noted that issues below the foundation and/or the abutments and, as has been mentioned previously, the impossibility of actioning gates, following the swelling of the piers or walls and thus a reduction in the width of the openings in which they are installed, can also occur.

In addition, differential swelling across the width of a dam can lead to cracking in the faces in the form of crazing (Figure 19.4). Lastly, in the case of attack by sulphates, the crystallization of complex salts (especially ettringite and thaumasite) is accompanied by considerable swelling and a loss of binder cohesion. This phenomenon advances quickly, is generally localized (no generalized swelling), and results in substantial damage to the concrete.

19.4.7 Effects on different types of dams

The effects of swelling on different types of dams are as follows:

a) Gravity dams

It has been demonstrated that in gravity dams, the horizontal tendency is primarily due to the differential expansion of the upstream and downstream faces. In the Alps, the temperature distribution is mainly influenced by sunshine and therefore the direction in which the dam is oriented.

If we consider the hypothetical case of homogeneous swelling affecting a single block in a rectilinear dam, upward and upstream crest displacement will take place. Because swelling occurs more quickly when the concrete is wet and when water circulation is more active, the upstream zone of the dam is more exposed, and crest displacement is generally directed upward and upstream. Cracking can affect both the upstream and downstream faces; upstream cracks favor water penetration and uplift action. Ultimately, crest displacement will always be directed upward (free expansion), while its upstream or downstream orientation will depend on local conditions.

With regard to the gravity dam as a whole, swelling in blocks will lead to joint closure between these blocks and the appearance of confining stresses. This lateral confinement may have side effects, notably:

- The slowing down or even stopping of swelling in the direction of the confinement
- Appearance of tensile stresses and cracks along the access shaft or drainage wells located in the field of compressive stresses

b) Arch dams

The uprising of the crest due to concrete swelling is generally accompanied by an overall upstream displacement. The stresses resulting from swelling are not as great as those that occur in a rectilinear dam, but the following consequences may arise:

- Opening of vertical joints, upstream in the zone of the crown and the downstream face along the dam's abutment
- Appearance of cracks on the downstream face, parallel to the foundation (these cracks normally have no impact on stability, as swelling tends to increase the arch effect, especially on the banks; however, the development of these cracks should nevertheless be carefully monitored)

c) Buttress dams

Swelling develops more quickly in the heads of the buttresses than in the web, which is not in contact with the reservoir water. Differential swelling causes upward and downstream crest displacement, as well as the appearance of zones with tensile stresses.

19.4.8 Preventive measures

To assess the impact of swelling caused by AAR on the structural safety of dams, simple yet sophisticated computational models have been developed. To successfully complete this assessment, it is a good idea to begin by interpreting the data from the monitoring system and then progressing on to in situ and laboratory tests. The computational analysis concerns the observed behavior, structural safety, and the impact of defensive measures.

The origin of the AAR depends on 3 parameters: reactive aggregate, humidity, and the presence of alkali. These parameters will determine the nature of the preventive measures. During the construction of a new dam, the use of nonreactive aggregate is one way to avoid the problem of swelling (see Table 10.19, which summarizes the main reactive stone materials). However, for the construction of dams, it is important to remember that priority should first be given to finding material that is available near the dam site, due to issues of transportation and cost. To alleviate this phenomenon, the alkali content of the cement must be as low as possible. The solution consists in replacing part of the cement with fly ash, pozzolan, or blast furnace slag.

As for the aggregate, many tests on the materials must be carried out for each new dam. The main tests include:

- The *petrographic investigation of aggregate* (in particular, microscope analysis of thin sections)
- The *calculation of aggregate reactivity through microbar testing (according to AFNOR P 18-588)*; the test is carried out on 10 × 10 × 40 mm prisms comprising crushed and ground stone particles; these are then subjected to four hours of water vapor and six hours in a potassium solution at 150°C; after which time the increase in length is measured; the mix can be considered nonreactive if the increase is lower than 0.11%; the microbar test cannot be used to assess the reactivity of aggregate containing primarily metamorphic or crystalline rocks[1]
- *Verification of concrete evolution* (using samples made from the concrete mix planned for the new construction)

To protect against moisture, water seepage into the concrete should be prevented. Sealing and drainage should theoretically suffice. For dams, however, this measure is generally insufficient to maintain the moisture content below the critical 80% for AAR.

It is also possible to introduce stress relaxation measures such as cuts or grooves in the concrete. Although it is still under development, this method is one solution for prolonging the service life of the dam.

---

[1] See § 10.3.1.

19.4.9　In cases of rehabilitation

Once the presence of a swelling phenomenon has been confirmed, its effects on the dam should be assessed by clearly distinguishing between operational aspects and safety aspects. Identifying possible measures to mitigate the effects of concrete swelling on dam behavior is a step that occurs at a later stage.[2] The effect of some measures is limited in time and is not permanent.

For Swiss dams, the demonstration of nonreversible deformation, the observation of the general condition of the concrete, and a crack analysis have often been early indications. The assumption that AAR is present must then be confirmed by targeted investigations.

Some of the measures taken have included installing a membrane on the upstream face with the aim of preventing seepage into the Illsee dam as a first step. Vertical grooves were sawn in an upstream-downstream direction at the Salanfe and Illsee dams.

As for the solution involved in the rehabilitation of the Serra dam, it required the total demolition of the existing dam. An entirely new dam was built downstream of the old dam. In view of the experience gained around AAR, the project design and construction were carried out with all the necessary precautionary measures.

19.5　The Effects of Frost

As a general rule, the temperature of mass concrete remains relatively stable and does not drop sufficiently for frost damage to occur. The surface zones, however, are subject to daily and annual temperature cycles. Depending on the location of the dam and the climatic conditions (temperature variations, frequency of precipitation), the faces may undergo several frost-thaw cycles. Due to the presence of mineral salts in the pores, concrete is considered to freeze at –2°C and to thaw at 0°C. Frost is, of course, accompanied by a change in volume (increase) depending on the moisture content of the affected area. Damage due to frost first only manifests locally, initially as cracking, before gradually the concrete disintegrates and finally shatters at the surface. Damage occurs only very slowly and does not affect the safety and long-term life of the dam (arch dams may comprise an exception).

We have seen that modern techniques in the manufacture of concrete can produce a frost-resistant material that is suited to local climatic conditions (for example, through the use of an air entraining agent, water repellent impregnation, or by lowering the water/cement ratio). Current technology adapts to fit the local environment of the dam (whether it is located at high altitudes or on the plain).

Dam behavior in response to frost depends on the characteristics of the cement paste and aggregate, primarily their porosity and the distribution of pore size. It is clear that sensitivity to frost is much more visible on less compact, older concrete, and those manufactured without air entraining agents.

Depending on the dam zone in question, the consequences of frost action are not the same:

- On the upstream face, damage is usually limited to the drawdown zone. The speed at which the concrete disintegrates varies; although for a frost-resistant concrete this speed is almost zero, for frost-sensitive concrete (for example, old poured concrete), it can reach up to several millimeters per year (Figure 19.8).
- On the downstream face, concrete that has been attacked by frost is very common, as there is both seepage water and water present in the atmosphere. Seepage water can lead to the total saturation of the concrete; although surface water tends to penetrate from the surface of the face, it only has a superficial effect.

2　See text box: Monitoring phases for a dam affected by swelling.

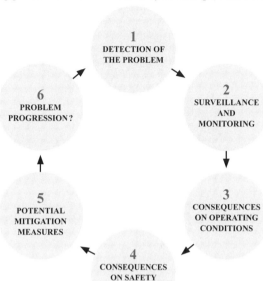

Monitoring phases for a dam affected by swelling (from SwissCoD, 2017a)

DETECTION OF THE PROBLEM: The initial detection or the slightest sign of concrete swelling is linked to the demonstration of nonreversible deformation.

SURVEILLANCE AND MONITORING: The hypothesis usually needs to be strengthened with further measurements.

CONSEQUENCES ON OPERATING CONDITIONS: Once the consequences for the dam have been identified, specific inspections may need to be carried out, which will enable the detailed monitoring of each specific aspect.

CONSEQUENCES ON SAFETY CONDITIONS: The consequences of the progression of swelling on dam safety are more difficult to establish. Indeed, safety depends on a large number of factors and requires, among other things, the (partial or total) modeling of failure scenarios for the dam and/or its appurtenant structures.

POTENTIAL MITIGATION MEASURES: After the previous analyses, measures mitigating the effects of swelling may be envisaged. The reasons for these interventions and their objectives may be multiple, but it is important to identify whether the measures aim to improve operating conditions or dam safety.

PROBLEM PROGRESSION: The interventions will generally act on the consequences of swelling, and only in exceptional cases on the actual causes. Mitigation interventions are therefore generally temporary, and their effectiveness gradually lessens over time.

Although the surface of concrete that has been damaged by frost often looks rough and unappealing, the depth of damage is only in the order of a few centimeters (sometimes it may reach 20 cm or more), and the mass concrete is not affected. In fact, the layer affected by frost acts as a kind of thermal shield that insulates the concrete mass.

Figure 19.8 Frost damage on the upstream face of the Cleuson dam (from SwissCoD, 2000a)

VI. EMBANKMENT DAMS

Embankment dams are some of the oldest engineering works. Traces of structures more than 2,000 years old have been found, especially in Asia. Today, this type of dam accounts for a considerable percentage of dams built around the world. Its development has been based on experience gained, improvements in equipment and methods for installing materials, as well as progress made in soil mechanics and means of analysis. The current highest built embankment dam is the Nurek dam (300 m), in Tajikistan. The Rogun embankment dam is being constructed and with its projected 335 m, will be the highest dam in the world.

As with concrete dams, there are several types of embankment dams depending on the materials available and the sealing system that is chosen (sealed cut-off), including the homogeneous (without a sealing system) and zoned (internal sealing system) embankment dams, and the embankment dam with an upstream facing. Embankment dams have a larger footprint than concrete dams.

Chapters 20 to 24 aim to describe their principal characteristics and how the different types of embankment dams are built.

Homogeneous embankment dams, which are covered in Chapter 21, have a simple profile, as they are made of just one, tight material; the slopes of this type of embankment dam have a gentle incline. They may have a drainage system.

The materials in zoned embankment dams, described in Chapter 22, differ in the manner they ensure firstly impermeability through their clay core (vertical or sloped) and secondly stability through upstream and downstream shells made of soil or rockfill. These elements are separated by filters to avoid the migration of particles.

Chapter 23 presents another variant that, should there be a lack of clay material available, has an internal sealing system using a concrete or asphalt core, or even sheet piling.

Lastly, embankment dams with a primarily asphalt or reinforced concrete upstream facing are dealt with in Chapter 24. This type of dam is also useful in cases where there is insufficient clay material. This watertight slab cover on the face ensures that the body of the dam is not subject to the effects of fluctuations in the reservoir water level. Problems relating to a loss of water through joints have affected embankment dams with an upstream concrete facing, due to a poor support system and the design of these joints. The use of well-graded rockfill and a clear improvement in compaction techniques have since 1960 led to an upsurge in interest for this type of dam, which has translated into a sharp increase in both numbers constructed and their height. However, for dams higher than 200 m, issues have arisen during impoundment.

The asphalt slab is a tried-and-tested procedure that is well known. This type of facing can be installed both at high altitudes where climatic conditions are harsh and in very sunny regions with high temperatures.

Other kinds of upstream facings include geomembranes.

To establish the best way of placing the material, it must be rigorously tested. Chapter 25 gives indications on the tests that must be carried out during both the construction and operation of the dam.

It is also necessary to verify the overall stability of an embankment dam. Several analysis methods have been developed to achieve this, and these are explained in Chapter 26. Verification of seismic loading is an important step, and the Swiss guidelines are provided as an example. This chapter also addresses the subject of internal erosion in embankment dams and foundations, which is a real danger with regard to dam safety.

Finally, some examples of structural details are provided in Chapter 27.

20. Overview

Marmorera (Castiletto) rockfill embankment dam in Switzerland, height 91 m, year commissioned 1954
(Courtesy © Swiss Air Force)

20.1 Background

Embankment dams, used as flood-protection structures along rivers or as water-retaining structures, are some of the oldest civil engineering works in the history of humankind. Traces of these types of dams dating back some 2,000 years have been discovered in India, Sri Lanka, Yemen, and China (Figure 20.1) (Schnitter, 1988, 1994).

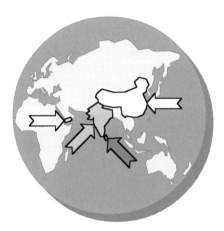

Figure 20.1 The location of dam structures more than 2,000 years old

Although thousands of embankment dams were built throughout the centuries prior to industrialization, analytical methods for dealing with problems of stability are only very recent.

Table 20.2 illustrates by way of some key dates the historical development of construction techniques for embankment dams (see also § 2.1.4).

Table 20.2 The development in construction techniques for embankment dams

1907 Bassel:	Proposed that the base of a dike must be sufficiently wide to mobilize friction; he proposed a safety factor of 10.
1926 von Terzaghi	Published a book on the principles of soil mechanics; he was the first to introduce the effects of pore pressure and criteria for internal erosion.
1926 Fellenius	Proposed a calculation method for slope stability based on sliding circles (and Terzaghi's theories).
1933 Proctor	Suggested that dike slopes should vary between 1:2 to 1:4 depending on foundation conditions; he proposed the concept of optimum moisture content for compaction.
1954/55 Janbu and Bishop	Published analytical methods for calculating dike stability, which are still in use today.
1965 Morgenstern and Price	Published an analytical method for various failure surfaces.

Most dams built around the world are embankment dams (Table 20.3).

Table 20.3 Number of embankment dams as a proportion of total number of dams (in %)

	Height from the foundation	
	H > 15 m	H > 100 m
World	80–85%	45%
Switzerland	37%	8%

As with all water-retaining structures, the heights of embankment dams have gradually increased thanks to previous experience, scientific developments, and improvements in material installation. Today almost half of all embankment dams are very high. Structures of more than 100 m height were first built in the 1930s, and then in the late 1970s/early 1980s, the 200-m barrier was crossed. Table 20.4 lists some very high constructions in Switzerland and around the world. The largest is the Nurek embankment dam, in Tajikistan. In the same country, the initial project design of the Rogun embankment dam with a core was supposed to reach 335 m in three construction stages, making it the highest embankment dam in the world. The preparatory works began in 1976 and the main activities in 1982. With the collapse of the Soviet Union, the construction works slowed down. In 2004, a study was undertaken looking to revise the project. It recommended building the dam in two stages, with an initial height of 225 m and a final height of 285 m (Schmidt, 2007). In October 2016, construction of the initial project for an embankment dam with a core and a height of 335 m began again. This structure is being built in stages. Two turbines in the underground power plant have been in operation since 2019. The end of construction is planned for 2028.

Table 20.4 Maximum height of embankment dams from the foundation

	Name (year of construction)	Height
World	Nurek (1980)	300 m
	Rogun (under construction)	335 m
Switzerland	Göscheneralp (1960)	155 m
	Mattmark (1967)	120 m
	Marmorera (1956)	91 m

20.2 Criteria for Choosing a Site

If we make a general comparison between concrete dams and embankment dams, the latter have the following advantages:
- Geological and topographical conditions are less critical
- They require less installation and equipment, but require the same care in construction
- They are not especially sensitive to settlement and earthquakes

However, they have the major disadvantage of generally having no tolerance for overtopping; sufficient freeboard is therefore essential.

The conditions related to the construction site for dams was outlined in chapter 10. It may be useful to recall a few general principles concerning geological and topographical conditions that affect the choice of site for an embankment dam.

First of all, embankment dams adapt to practically any topographical shape. However, in addition to valley shape, geology and availability of fill material are criteria that must be assessed conjointly.

With regard to geology, the precautions to be taken and the necessary investigations for an embankment dam are the same as those for a concrete dam, even though the requirements for subsurface quality are not as high.

When the geology is not favorable for the construction of a concrete structure, an embankment dam may be an alternative. In this case, thorough knowledge of the geology is vital to guarantee the feasibility of the embankment dam.

Furthermore, its construction requires a large volume of fill material. Detailed prospecting work in the surrounding region, in addition to a geotechnical investigation, is vital.

The feasibility study is thus the fruit of the joint work of the engineer and the geologist.

The geologist is responsible for detailing all the geological conditions pertaining to the reservoir, the site itself, and to the availability of material.[1]

20.3 Types of Embankment Dams

As mentioned in section 3.3, embankment dams can be divided into two categories: earth dams and rockfill dams (Figure 3.18). The rockfill embankment dam consists mainly of rocks that are crushed or simply dumped. As with all dams, hydrostatic pressure must be taken up (dam shell), and they must be built with a material that is sufficiently impermeable, or else fitted with an internal (core, central membrane) or external (slab, geomembrane) sealing system. Depending on the method of construction and the sealing system (Figure 20.5), we can further categorize embankment dams into the following types:

- Homogeneous embankment dam (comprising one single material)
- Zoned embankment dam made up of several types of materials, each carrying out functions of impermeability or general stability (internal sealing system: vertical or sloped clay core, asphalt core, or a diaphragm wall)
- Embankment dam with an upstream facing (external sealing system: asphalt or reinforced concrete slab, upstream membrane)

Figure 20.6 presents different options for the sealing system. The sealing system element in dams with a core is made from loose material with a high clay content. This core can be thin, wide, vertical, or sloped in an upstream direction.

The static behavior of this core is similar to that of the dam shell. However, membranes and asphalt or concrete facings behave completely differently depending on the type of fill in the dam shell. The membrane is made of a thin impermeable layer of a synthetic material that provides a watertight facing.

Downstream of the sealing system, it is important to have sufficient drainage to evacuate seepage flow coming from the subsurface or that may have crossed through the sealing system. This seepage can be due to the natural permeability of the sealing system or the subsurface. Consideration should also be given to possible events that may result from a defect in the sealing system (cracking in the core, membrane, or slab).

The drainage system must capture this seepage flow while avoiding any loading in:

- The upstream and downstream dam shell
- The downstream dam foundation
- The lateral abutments downstream of the dam

---

[1] For details, see § 10.2.1.

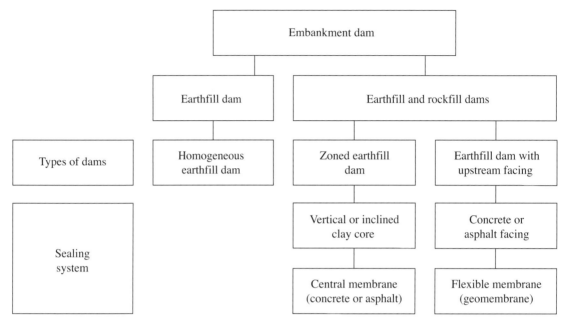

Figure 20.5 Principal types of embankment dam

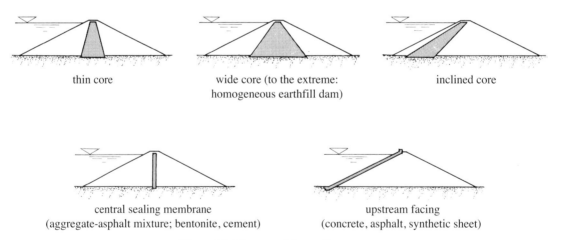

Figure 20.6 Layout of the sealing system

Several drainage systems are possible (Figure 20.7):
- Highly permeable downstream dam shell (rockfill)
- Draining layer downstream of the core and along the foundation
- Chimney drain and drainage blankets integrated into a homogeneous earthfill dam
- Partial drainage of the downstream toe (reasonably permeable dam shell)
- Drainage gallery at the base of the dam combined with drainage layers/chimney drains
- Drainage gallery in the rock with drainage holes combined with drainage blankets

Layers of filter must be installed between the sealing system and drainage zones to avoid the migration of fine material from the most watertight zone to the most permeable zone. Such phenomena can result in piping, which can eventually lead to dam failure (see § 26.8.2.1).

All drainage systems must be equipped with measuring and monitoring systems (pressure and flow).

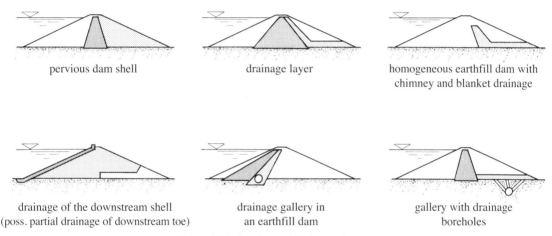

Figure 20.7 Sealing and drainage elements

20.4 Foundation

There are few civil engineering structures where the interdependence between the structure and its foundation is as clear as it is in embankment dams. Construction conditions for the foundations have a major effect on:
- The shape and depth of the excavation
- The layout and dimensions of the sealing system
- The upstream and downstream slopes (stability)
- Future deformation
- Seepage

To best define the true characteristics of the foundation, the following points must be investigated with particular care:

a) Rocky subsurface

- Faults crossing the foundation (active or not)
- Orientation and nature of any discontinuities (cracking and stratification)
- Groundwater table
- Foundation permeability (loss of water)
- Long-term behavior under the effect of seepage (internal erosion, dissolution)
- Deformability and shear strength in different directions

b) Loose terrain

- Deformability and shear strength in different layers
- Horizontal and vertical permeability in different materials in the foundation
- Nature of the heterogeneity of the soil (layers, lens)
- Moisture content and groundwater table

Figures 20.8 (a) and 20.8 (b) demonstrate the various possible foundation layouts. Figure 20.9 illustrates the creation of an upstream blanket made from clay material whose aim is to increase the seepage path and thus reduce uplift under the dam.

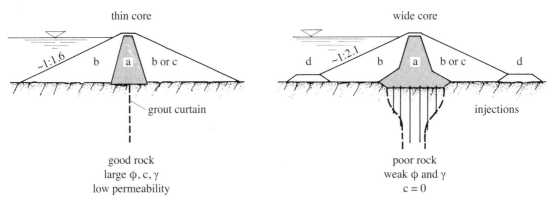

Figure 20.8 (a) Types of foundations for embankment dams on rock: (a) core; (b, c, d) dam shell of varying quality

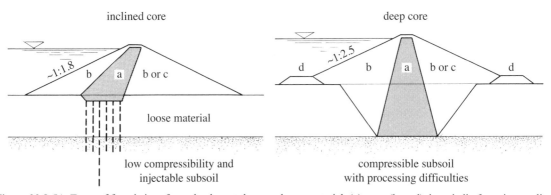

Figure 20.8 (b) Types of foundations for embankment dams on loose material: (a) core; (b, c, d) dam shell of varying quality

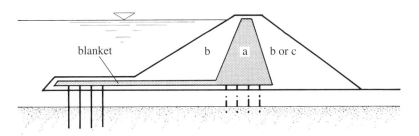

Figure 20.9 Core combined with upstream blanket: (a) core; (b, c) dam shell of varying quality

20.5 Behavior of Embankment Dams

There are three main phases in analyzing the behavior of an embankment dam:

Construction phase
The material is placed in consecutive layers. The following effects are immediately observed:
- Major vertical settlement due to the increase in weight (Figure 20.10)
- Lateral deformation due to vertical compression

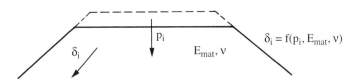

Figure 20.10 Settlement in an embankment dam

Settlement must be compensated for as each layer of fill is placed; this is not the case for the faces. The selection of the initial area must take this effect into account. This deformation can be affected by compaction, which is necessary to avoid internal shearing of the material. The appearance of cracks in the core may be observed.

Consolidation phase[2]
This commences alongside placement, but if the material has low permeability ($k < 10^{-6}$ m/s), it finishes many years after construction has ended. To compensate for settlement due to consolidation, the crest must be heightened proportionally to the theoretical dimension (Figure 20.11).

---

[2] See also § 25.4.1

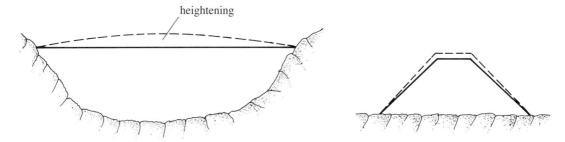

Figure 20.11 Heightening fill during construction

Operating phase with variations in water level

The first impoundment of the reservoir creates settlement due to loading from the water and, when impermeability is provided by a central core, due to the submersion of the upstream dam shell.

Water pressure from the reservoir also causes horizontal deformation of the dam at the level of the crest (Figure 20.12).

Each time the reservoir is filled this deformation is repeated, but to a far lesser degree, as plastic deformation is mostly produced during the first impoundment.

It is therefore wise to curve the axis of the embankment dam upstream, especially if located in a narrow valley.

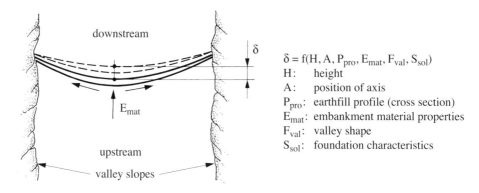

$\delta = f(H, A, P_{pro}, E_{mat}, F_{val}, S_{sol})$
H: height
A: position of axis
P_{pro}: earthfill profile (cross section)
E_{mat}: embankment material properties
F_{val}: valley shape
S_{sol}: foundation characteristics

Figure 20.12 Downstream crest displacement

20.6 Appurtenant Structures

Appurtenant structures such as the

- diversion tunnel
- bottom outlet
- water intake and
- spillway

are usually made from concrete, and, due to their rigidity, are therefore vulnerable to differential settlement.

As a rule, concrete structures should be placed outside of the fill, and at the very least, outside the core. The various possibilities for their installation have a major impact on the choice of site.

The only exceptions that may sometimes be found are galleries that cross the dam longitudinally or transversally. Some embankment dams have a longitudinal gallery to carry out grouting activities, as well as a drainage gallery. The Mattmark embankment dam, in Switzerland, is an example of this. Longitudinal galleries combined with vertical towers can often be found. The lower parts of the towers are encased in the upstream dam shell. These towers may be used for

- Access shafts
- Spillways (for example, a bellmouth spillway)
- Water intakes
- Bottom outlets

The fill-concrete structure interface can be the site of localized seepage, which may increase the risk of internal erosion (see § 11.4.5).

This is therefore a fundamental difference with concrete dams, whose appurtenant structures are often integrated into the body of the dam. This combination is not practically possible in embankment dams.

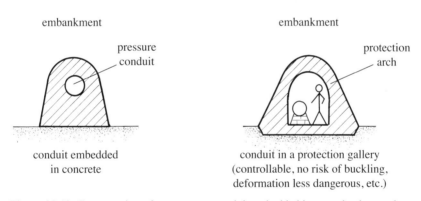

Figure 20.13 Cross-section of a transverse conduit embedded in an embankment dam

When the foundation is comprised of loose material (moraine), the gallery must adapt to foundation settlement through provisions that prevent or avoid cracking.

As a result, ideally these transverse galleries should be placed directly on a rocky foundation.

If the gallery is used for access or for free surface flow, the installation of deformation joints is recommended. However, for conduit flow under pressure, the lining of the entire gallery is the safest solution (Figure 20.13).

Any seepage of water from the gallery toward the main body of the embankment creates a risk of washouts or internal erosion and therefore compromises the stability of the dam.

The transverse section of the gallery should not include any abrupt changes in slope nor should any vertical walls in the section touch the watertight core. Figure 20.4 illustrates several possible solutions.

The surface of the concrete constitutes a preferential seepage path both at the contact point with the foundation and with the fill (in the watertight zone).

Figure 20.14 Foundation methods for a longitudinal or transverse conduit in an embankment dam

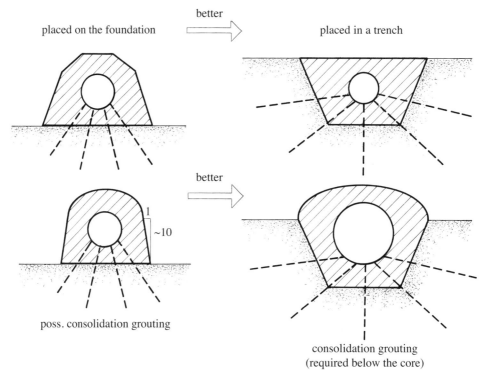

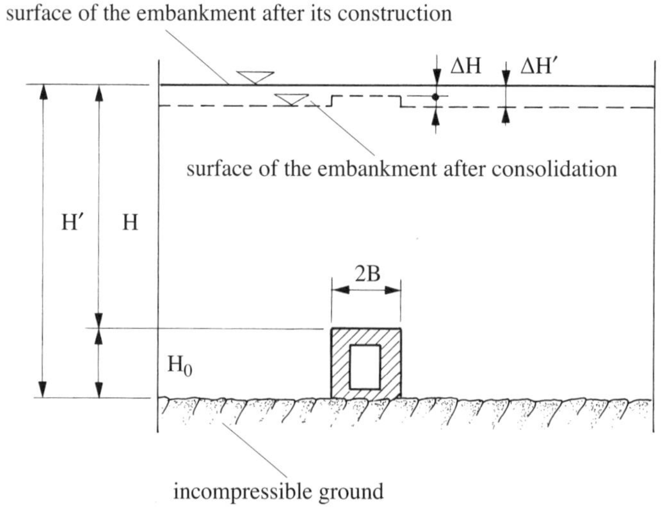

Figure 20.15 Pressure exerted on a gallery inside the fill

In the past, engineers attempted to avoid preferential seepage paths by using concrete peripherical rings along the conduit. Experience has proven that the use of such peripherical rings is not recommended as it becomes difficult to adequately compact the core around the galleries. A surface treatment for reducing friction (for example, with asphalt) and especially careful placement of the core material is more effective.

Galleries placed in foundation trenches are most certainly the best solution. The parts of the galleries that are surrounded by permeable material, that is, situated in the dam shells, do not require any particular surface treatment.

Note that a conduit placed on a rigid ground surface (foundation) will alter fill settlement (Figure 20.15). This settlement above the conduit is theoretically smaller than in other parts of the fill. The total pressure applied on the conduit is equal to the sum of the weight of the fill and a secondary force caused by the various deformations of the fill.

According to Guerrin [1960, 1961], we have the following equation:

$$P = 1{,}5 \cdot \gamma \cdot H,$$

where γ is the density of fill and H the height of the fill above the conduit.

According to Pruska (1963), and by considering secondary force, we have:

$$P = \gamma \cdot H + P1 = \gamma \cdot H + \frac{1}{2} \frac{\gamma \cdot \pi \cdot B(H'^2 - H^2)}{H \cdot \operatorname{arccotg}\frac{H}{B} + B \cdot \ln\frac{B^2 + H^2}{B^2}},$$

where (cf. Figure 20.15)
- γ = fill density
- H = fill height above the gallery
- H' = fill height above the ground
- $2B$ = gallery width

Figure 20.16 Example of the construction of a gallery underneath an embankment dam (Courtesy G. Azzolini)

21. Homogeneous Earthfill Dams

Bannalp embankment dam with central core in Switzerland, height 32 m, year commissioned 1937 (Courtesy EW Nidwalden).

21.1 Overall Layout

A homogeneous earthfill dam is an embankment dam primarily made from fine compacted material. It has the simplest profile among all embankment dams. One specificity of this type of structure is that it comprises only one material uniformly spread throughout the section. This material alone must fulfil the functions of both impermeability and dam shell. The most permeable materials are placed near the faces. In order to achieve stability, the slopes are gentle, and because of this the dam covers a large surface area and requires the placement of a large cubic volume of material. This is why permanent homogeneous earthfill dams are only rarely higher than 15 m. They are often found along waterways as flood protection (especially as dikes, some of which are very old). For greater heights, homogeneous earthfill dams are not a cost-effective option. However, they may be considered for use as temporary dams, for example, as cofferdams up to 30 m high.

Some large dams of this type do, however, exist and have quite specific functions:

- Protection dikes (against floods, avalanches, or rockfalls)
- Dikes for mining waste
- Dikes for dredge material (hydraulic fill)

With regard to protective dikes against avalanches, it should be noted that these are generally installed in difficult topographical and geological conditions. Because of this, the option of the embankment dam is usually selected, especially as the material is generally available nearby. Care should of course be taken that these structures meet the usual safety criteria and that their construction is undertaken to a high standard. These structures can hold back water, sediment, and mud, so the evacuation of water and floodwaters must also be allowed for.

21.2 Main Characteristics of the Materials

For reasons of economy and to reduce transportation requirements, the borrow areas for the material used for fill should be located near the worksite and should ensure a sufficient quantity. Site investigation by way of test pits and/or exploratory trenches should be carried out to assess the quality of the material and then identify it in the laboratory. Depending on its characteristics, excavated material can also be used.

Material used for the fill must be resistant to weathering and contain little organic matter. The material must have high shear strength and be sufficiently impermeable ($k < 10^{-5}$-10^{-6} m/s) to create a barrier for water by preventing unwanted seepage through the dam. The use of soil with low variation in permeability is recommended. Lastly, it must be suitable for compaction.

21.3 Factors Concerning the Design and Operation of the Dam

Homogeneous earthfill dams do not require many special conditions with regard to the foundation. They can adapt to deformable foundations. However, it is important to seal cracks in the foundation, which may cause water seepage and erosion within the dike (this phenomenon has been observed in Canada in particular).

As for the characteristics of the material used, it must have good shear strength and low compressibility. During placement and compaction, it should not have a tendency to fragment.

The fill is mechanically placed and compacted in consecutive horizontal layers. The depth of the layers depends on the nature of the material and is determined from in situ compaction tests. These tests also establish the number of passes to be made by the compaction equipment in order to obtain optimum density and while taking into the account an appropriate moisture content (see § 25.1.1.1). However, homogeneity in terms of permeability is not easy to achieve for compacted fill, as the method of construction always introduces greater permeability in the direction of the layers (Figure 21.1).

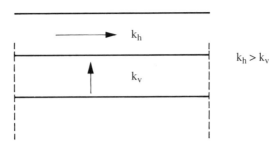

Figure 21.1. Difference in permeability between the layers

Seepage zones on the downstream surface often result from this kind of fill. To avoid internal erosion (piping)[1] and instability, drainage zones are vital.

The risk of piping is related to the potential gradient. As height H of the reservoir is a given, the total length of the traverse is a major factor. For small structures, seepage water can be forced along a reasonably long traverse so that it loses its full head at the downstream end. One or two cut offs can be installed to this end. The following empirical equation by Lane, reviewed by Bligh, can also be used for assessment. It first takes into consideration the sum of vertical traversing along the cut-off (L_v) and then the horizontal distance traveled by the water under the dam (L_h):

$$cH = L_v + 1/3\, L_h,$$

where c is a coefficient dependent on the nature of the terrain, whose average values are given in Table 21.2 as an example.

Table 21.2 Average values of coefficient c from Lane and Bligh's equation (from Post and Londe, 1953)

Nature of the ground	c
Fine sand and silt	8.5
Fine sand	7
Medium-grained sand	6
Large-grained sand	5
Small gravel	4
Large gravel	3
Plastic clay	3

After impoundment, seepage paths develop in the body of the dam, and the saturation line reaches the toe of the downstream slope. It is therefore inevitable that water will emerge at the downstream face. A drainage system lowers the saturation line, which inevitably develops in the body of the dam and delineates the saturated part. For dikes of a low height (< 10 m), drainage in the form of blankets or a drain at the downstream toe is sufficient (Figure 21.3).

---

[1] See Section 7.3. Piping is a type of internal erosion that is very dangerous. It is caused by downstream to upstream backward erosion along a conduit.

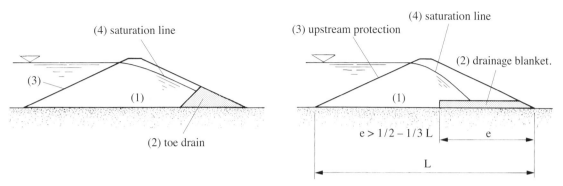

Figure 21.3 Drainage zones in a homogeneous dam: (1) body of the dam in a homogeneous material; (2) downstream drainage; (3) upstream slope protection (riprap); (4) saturation line

For permanent dams higher than 10 or 15 m, it is preferable to install a vertical or slightly inclined drain in the central part of the dam (Figure 21.4). This type of drain accelerates the dissipation of pore pressure in the fill. The criteria for the filter between the drain and the rest of the fill must of course be met.

Lastly, in terms of operation, this type of dam is highly sensitive to rapid variations in reservoir water level.

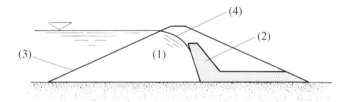

Figure 21.4 Drain in the body of a homogeneous earthfill dam: (1) dam body in a homogeneous material; (2) chimney and drainage blanket; (3) upstream slope protection (riprap); (4) saturation line

It should be reiterated that when the dam foundation sits on ground with untreated, loose material, there is a risk of the formation of piping or internal erosion if no measures are taken. Installing cut offs and/or a diaphragm wall into the ground, which are then embedded into near impermeable rock are possible options.

22. Zoned Embankment Dams

El Makhazzine embankment dam with a central core in Morocco, height 65 m, year commissioned 1974
(Courtesy R. Sinniger)

22.1 Overall Layout

If the volume of material available for constructing a homogeneous earthfill dam is clearly insufficient, another possibility is to design a dam that uses the various material that is available nearby or perhaps comes from the excavation of appurtenant structures (spillway, galleries, tunnels) to best effect. This material is categorized depending on its characteristics so that impermeability and then dam stability is guaranteed. The traditional profile of the zoned embankment dam includes a watertight clay core (vertical or inclined) surrounded by an upstream and downstream shell made from compacted and relatively impermeable rockfill or random fill. Depending on the characteristics of the placed material, the slope of the shell faces may vary between 2:1 and 4:1. Other specific elements, such as filters, drains, and slope protection are also incorporated (Figure 22.1). Compared to a homogeneous earthfill dam, the volume to be placed and therefore the duration of the works can be less.

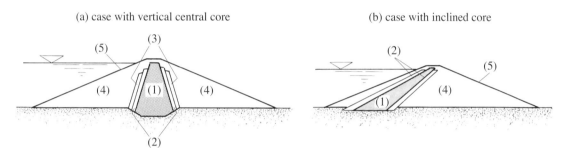

Figure 22.1 Embankment dam with clay core: (1) clay core; (2) filters; (3) transition zones; (4) dam shell; (5) slope protection

22.2 Embankment Dam with Central Core

This type of embankment dam is a common option. It consists of two dam shells, upstream and downstream, that guarantee the stability on either side of the central core, which is the impermeable element. The dam shells are generally made of rockfill, but may also be built out of coarse, relatively impermeable material. The number and layout of the types of material used may vary depending on the design chosen by the engineer. Due to its permeability, the upstream shell should readily tolerate fluctuations in the reservoir water level. Protection of the upstream face is necessary against the erosive effect of waves and rain. Protection of the downstream face is also required against storms.

The clay core option may be chosen if natural material with very low permeability is available near the site: either clay silt ($k = 10^{-8}$ m/s) or moraine soil ($k = 10^{-6}$ à 10^{-7} m/s).

If the foundation rock is suitable, that is, impermeable, resistant to internal erosion, and not highly deformable, the core can be relatively thin. To start with, the seepage gradient should not be greater than 3. These thin cores have little negative impact on dike stability, which is an advantage. The upstream and downstream slopes depend on the material used for the dam shells.

Due to the large discontinuity between their particle size, the core and dam shell must be separated by filters and, if necessary, transition zones (Figure 22.2). Filters are of paramount importance:

- They prevent the migration of the core toward the dam shell in the case of a rapid drawdown of the reservoir
- They create a sealing effect should preferential seepage paths occur through the core

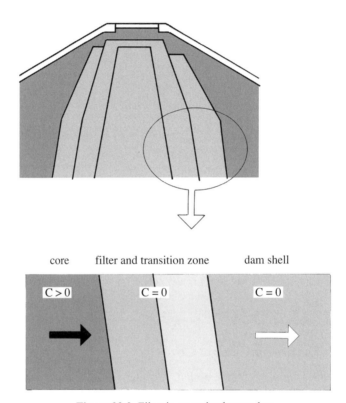

Figure 22.2 Filter in an embankment dam

Although the core is made from a cohesive material with low permeability, the filter is noncohesive and is relatively more permeable. These differences in characteristics must meet filter criteria (see § 22.4.3). In addition, a drain can be installed between the core and downstream face if the latter is not sufficiently permeable to evacuate without pressure the seepage that has reached the downstream face of the core. A system for capturing seepage water is also crucial.

The core and dam shell have very different behaviors with regard to settlement. Excellent compaction of the core is required to prevent hanging at the dam shells during the consolidation process. Should this happen, the phenomenon of hydraulic cracking could occur in the core.

After impoundment, a pattern of pore pressure due to seepage forms in the core. This pressure affects shear strength and stability.

The advantage of a dam with a central core is its ability to tolerate ground settlement. Also, the foundation of these dams can be built on ground with loose material. Excavating down to healthy rock is not always necessary. If the dam foundations are on loose ground, the core must be thicker to lower the seepage gradient. Moreover, the loose ground beneath the core must be made watertight by grouting down to rock that has low permeability (Figure 20.8 (b)).

22.3 Embankment Dam with Inclined Core

The watertight (or highly impermeable) element, such as, for example, a core in clay silt, can in some cases be shifted upstream, which in effect eliminates the upstream shell entirely. This is known as an embankment with an inclined core.

The more the core is inclined, the more the slope of the upstream face is determined by the mechanical properties of the core material. Potential sliding surfaces cross the core, and the slope of the upstream face must therefore be gentler in order to guarantee stability. The mass of the downstream dam shell also increases, which provides a certain advantage for the dam.

An inclined core has the three following main advantages:

- The core rests against the dam shell and is compressed by water pressure
- The dam shell can be placed independently of the core; this means that work can still take place during periods of heavy rain, when placing the core is difficult or even impossible
- The possible heightening of the dam is easier to achieve

Lastly, water pressure is exerted in a favorable direction, with a vertical component, which increases the overall stability of the dam

22.4 Description and Characteristics of the Materials[1]

22.4.1 Dam shell

Noncohesive material should be used, such as rockfill and also gravel (earthfill) from

- Alluvial formations
- Moraines
- Quarries (limestone, gneiss, granite, dolomite)
- Gallery excavation

As for its characteristics, the material used must be resistant to water and external conditions (weather). It must have good compressive and shear strength, and low compressibility. During transportation, placement (dumping), and compaction, it must not have a tendency to fragment, which may affect particle size. It must not contain any mineral particles, which, as they are unstable, may cause mechanical or chemical disintegration of the rock. A wide-ranging particle size from silt to gravel is ideal. The maximum diameter (Figure 22.4) is limited by the thickness of the compaction layers:

- $d_{max} \leq 2/3$ of the layer height
- $d_{max} \leq 60$ to 120 cm

By increasing the height of the layer, compaction becomes more complex. According to Terzaghi and Beck (1965), the material must meet the following criteria:

---

[1] See also § 10.3.2.

- No more than 10% of the material must be smaller than 0.2 mm
- Max particle size is between 200 and 300 mm
- The uniformity coefficient $C_u = D_{60}/D_{10}$ must be equal to or greater than 15
- The particles must have a strong contact force

As for the angle of friction for the rockfill, it can vary from 35 to 40° (lowest values) to 50 to 60° (highest values). Face slopes between 1:2 and 1:4 may be considered.

Table 22.3 indicates the exploratory work and tests required for qualifying the rockfill.

Table 22.3 Summary of investigation for rockfill (established by R. W. Müller)
Key: XX = necessary test; X = test not always necessary

	Exploratory works	Feasibility study	Final design
Field reconnaissance			
Geological study	XX	XX	
Firing, mining test			XX
Crushing test			X
Fill/backfill test			X
Collecting samples	X	XX	XX
Laboratory testing			
Petrography	X	X	X
Aging		XX	XX
Abrasion		XX	XX
Unit weight		XX	XX
Compaction			X
Compressibility			X
Permeability			X
Compressive strength			X
Shear strength			X

22.4.2 Clay core

The particle size of the material comprising the core must be continuous. It generally ranges from clay (< 0.002 mm) to sand (< 2 mm). Wide-ranging and regular particle size distribution favors the self-sealing of the core, however, the risk of liquefaction increases at the same time. Figure 22.4 compares the particle size distribution of the material in core A (cohesive) with the material in the dam shell B (noncohesive).

It is a good idea to choose a material with a high plasticity index $I_P = w_L - w_P$. Bear in the mind that the plasticity index is the difference between the liquidity limit w_L (moisture content at which point the ground shifts from a plastic state to a liquid state) and the plasticity limit w_P (moisture content at which point the ground shifts from a plastic state to a solid state).

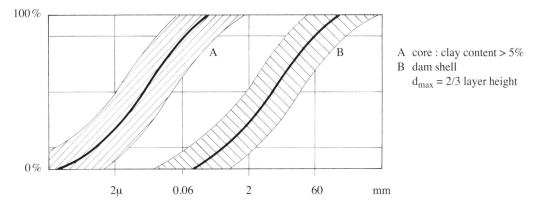

Figure 22.4 Particle size distribution in the construction materials

The core functions as a barrier against seepage, so it must therefore be made from impermeable materials. The permeability coefficient must be low; it should be at 10^{-6} m/s or less.

As mentioned above, clay silt ($k = 10^{-8}$ m/s) or moraine soil ($k = 10^{-6}$ à 10^{-7} m/s) should be chosen. Clay silt is characterized by low to medium plasticity. Its liquidity limit is lower than 50%. At least half of its components have a diameter smaller than 0.06 mm. To guarantee low permeability, the clay content must be at least 5%. As for moraine material, this is suitable when it comprises more than 15% silt and more than 2% clay.

22.4.3 Filter

The filter is designed to hold in place particles from the core and foundation that might be shifted through the action of seepage (Figure 22.5). The filter therefore prevents

- Suffusion through the transportation of material within the filter
- Erosion through the exchange of materials at the filter boundary
- Sealing at the filter surface

A filter, which is a protective element, is always necessary as a transition zone between cohesive and non-cohesive materials. The filter most often comprises a granular material (sand, gravel), whose particle size is carefully designed to block the finest particles. The filter must be made of a material with a particle size greater than that of the material it is protecting. The most commonly used criteria for the composition of a filter were established by Terzaghi and Peck (1965) and are as follows:

- D_{15}, filter $\geq 5\ D_{15}$, core
- D_{15}, filter $\leq 5\ D_{85}$, core
- D_{50}, filter $\leq 25\ D_{50}$, core

The graph in Figure 22.6 illustrates these criteria. It is, of course, possible to find other equations and criteria in the literature (USBR, 1987; USACE, 1994a).

At the interface between two zones with a different particle size, such as between the core and filter, specific criteria known as *filter criteria* must be applied to avoid the migration of particles from the fine zone through the coarse zone. In small dams only, the use of a geotextile may be envisaged.

Figure 22.5 Placement of a clay core and filters at the Taham embankment dam in Iran (Courtesy Mahab Ghodss)

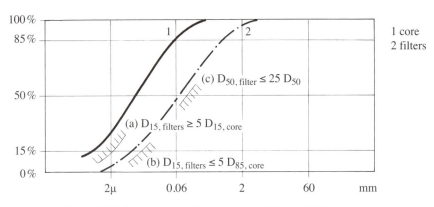

Figure 22.6 Grading curve for the core and filter, and filter criteria

23. Embankment Dam with Asphalt Concrete Core

Schmalwasser rockfill embankment dam in Germany, height 81 m, with an asphalt concrete core

23.1 Overall Layout

The embankment dam with an asphalt concrete core is an attractive alternative when there is not enough material in the vicinity of the construction site for a clay core. However, its design does depend on the availability of material for the dam shell. The main methods for installing this internal core (Figure 23.1) are:

- An asphalt core (e_{min} = 50 cm; H/e ≤ 120)
- A dry diaphragm wall (e_{min} = 80 cm)
- Sheet piles (especially for cofferdams and lateral dikes along a waterway)

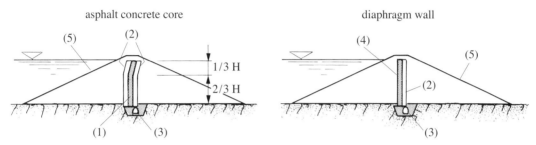

Figure 23.1 Embankment dam with an asphalt core: (1) asphalt concrete core; (2) transition zones: (3) grouting gallery; (4) membrane: mixture of bentonite and cement (poss. concrete wall); (5) slope protection (downstream and upstream)

This type of structure has several advantages (ICOLD, 1992, 2018d). First of all, it is suitable for dams that are located in sites where adverse weather conditions can make the placement and compaction of a highly impermeable, cohesive core difficult or even impossible. As the installation of the core is independent of climatic conditions, rockfill or earthfill shells can be constructed without stopping for poor weather. Furthermore, the internal core, which is located between the shells, is protected against external actions due, for example, to atmospheric conditions (the sun's rays, rain, run-off, frost). Lastly, partial impounding can be envisaged before the completion of the works.

A thin core implies very high gradients, which requires particular care in construction, and moreover, considerable grouting. The base of the core is the most complex element. A monitoring and grouting gallery is one possible option for checking watertightness, measuring seepage flow, and, if necessary, grouting the foundation.

Of the options mentioned above, the asphalt core is the most widespread due to the great flexibility of the seal, allowing it to adapt to deformation in the dam shell. The asphalt mix must be workable and compactable in order to ensure proper installation. Moreover, the asphalt is insoluble in water and does not negatively affect potable water. Lastly, it is cost-effective compared to the option with an upstream asphalt facing. It is possible to construct dams with an asphalt central core up to a height of around 150 m. Due to its adaptability to the freeze-thaw cycle, several of these dams have been built in Norway.

Dry or concrete diaphragm walls are only able to be used in dams of a low height, and their installation can present a challenge. These walls, approximately 80 cm thick, are very rigid and cause issues in cases of settlement and earthquakes. It is therefore preferable to install a flexible asphaltic core.

23.2 Description of the Asphalt Core

23.2.1 Main characteristics

The core is located in the central part and may be vertical, inclined along its whole height (in the literature, values between 1.0:0.05 and 1.0:0.8 can be found), or only inclined in the upper part (Figure 23.2).

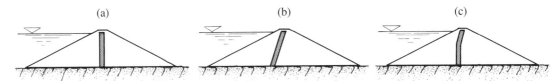

Figure 23.2 Position and shape of the asphalt core: (a) vertical core wall; (b) inclined core wall; (c) vertical/inclined wall

Transition zones must be planned between the core and the dam shell. The bond between them must be homogeneous and intensive compaction must prevent any differential deformation from occurring. The asphalt concrete does not offer any resistance to deformation caused by the dam; due to this fact, dam deformation is independent of asphalt properties. In addition to its flexibility and workability, asphalt concrete withstands erosion and ageing. Some uncertainty does remain, however, with regard to the long-term behavior of asphalt, particularly concerning creep and drying. The core is made from a viscoelastic material and can be installed without joints. Thanks to its ductile properties, it can be relied on to self-seal should any cracks form; the ability to self-seal depends on the type and content of the asphalt.

First and foremost, the core must be watertight. This watertightness is said to be reached if the volume of voids is lower than 4%. Furthermore, it must be resistant to cracking under deformation and have good shear strength.

The thickness of the asphalt core must be at least 50 cm and can vary depending on the height to reach a value that may be greater than 100 cm (it can be admitted that the thickness of the core is equal to 1% of the height of the water). Some constructions have the following max/min core values: 140/50, 130/60, 120/80.

The downstream transition zone plays the role of a chimney drain, collecting seepage flow as it crosses through the core. Seepage flow is generally directed toward a gallery. Moreover, the particle size of the upstream transition zone must enable grouting should the core lose its watertightness.

Large dams of this type have been constructed in Austria, among other countries. Examples include the Finstertal dam, 150 m high, completed in 1998, which was one of the first of these dams to be built. The core wall is inclined up its entire 96-meter height, is 70 cm wide at the base, and 50 cm wide at the crest. Topographical issues determined the position of the sealing system (Pircher and Schwab, 1998; Schober, 2003). The Feistritzbach dam, 85 m high and completed in 1990, is unusual in that the first 70 meters of the core wall is vertical before becoming sloped in the upper part (Tschernutter and Nackler, 1992; Schober, 2003). The highest core wall is in the Storvatn dam in Norway, completed in 1990. It is 125 m high and is inclined along its entire height (Høeg, 1992).

23.2.2 Components of asphalt concrete

The mix used for manufacturing the core comprises asphalt, filler, sand, and gravel. It is common practice for the particle size to be continuous and equal to that proposed by Fuller-Thompson, that is:

$$d(\%) = (d/d_{max})^{1/2}.$$

Broadly speaking, the standard maximum particle diameter (D_{max}) is 16 or 22 mm. It must, however, remain lower than 40 mm. D_{max} should not exceed 1/5 of the thickness of the layers. If D_{max} is small, placement is made easier and the risk of segregation decreases.

An asphalt content of around 6% of the weight is sufficient to fill the voids. It can be increased slightly (0.2 to 0.5%) to favor workability and flexibility. The choice of content is based on void content tests and triaxial tests verifying the behavior of the mix.

With regard to filler, its function is to fill the voids between the large particles. It also acts as a binding agent and enables the viscosity of the binder to be increased and the risk of segregation to be decreased. Lastly, a filler with consistent properties is required.

The transition zones are created by following the filter criteria and by using a rocky, hard material whose maximum particle diameter must not exceed 100 mm. The same material can be used both upstream and downstream. The upstream filter must contain enough fines to enable a sealing effect. The upstream particle size must be such that grouting is possible should a watertightness defect in the core be observed.

23.2.3 Testing and monitoring

Tests are carried out on the aggregate, asphalt, and the mix. Monitoring is carried out on the core once it has been installed.

As for the aggregate, the process begins with the calculation of particle size for the components in the mix. The asphalt undergoes traditional testing, including penetration and viscosity tests, tests assessing the presence of insoluble and volatile materials, tests evaluating the modification of properties after "heating," and density measurements. The mix is tested for asphalt content, particle size, void percentage, and permeability.

On site, visual checks and temperature measurements are carried out. To verify in situ compaction, samples are collected (in the form of cores with a diameter of 10 cm and height of 40 cm) to determine density and the proportion of voids. The holes are then filled with compacted asphalt or bituminous mastic.

23.2.4 Placement

The construction of embankment dams with an asphalt concrete core must be entrusted to specialized construction companies with appropriate equipment for the placement of the core and transition zones. In addition, the methods of construction and monitoring must be of high quality.

After having placed the bottom layer comprising the bituminous mastic (the only operation that is sensitive to weather conditions), the placement of the asphalt and transition zones occurs simultaneously in layers (lifts) of equal thickness, generally limited to 20 cm, using an engine specially equipped with a finishing machine (Figure 23.3). Note that the asphalt is placed and compacted at a temperature of 160 to 180°C. It is placed first to prevent it from penetrating the filters. Furthermore, it then becomes possible to use a lateral support for the adjacent sections. Placement near the shell is done manually. During construction, the level of the core and transition zones is always higher than the dam shell.

It is important to avoid contaminating the surface of the core, and it is a good idea to plan temporary access paths for machinery.

Lastly, the worksite does not need to be paused during bad weather (rain).

Figure 23.3 Placement of an asphalt core during the construction of the Maoping embankment dam, an accessory dam in the Three Gorges system in China (Courtesy A. Schleiss)

24. Embankment Dam with an Upstream Facing

New Spicer Meadow rockfill embankment dam with a concrete upstream facing in California, height 81 m, year commissioned 1989 (Courtesy Elektrowatt)

24.1 Overall Layout

If there is not enough material available to install a clay core, the option of a watertight element in the form of a facing placed on the upstream face may be envisaged (Figure 24.1). With this option, the fill is entirely protected from the influence of the reservoir water. Major changes in water level during operation, or even a rapid drawdown, are thus tolerated. For an equal height, this type of embankment dam requires a lesser volume of fill than other options. It should also be noted that the direction of the resultant of water pressure is favorable and increases the vertical component that acts on the foundation, which increases sliding stability.

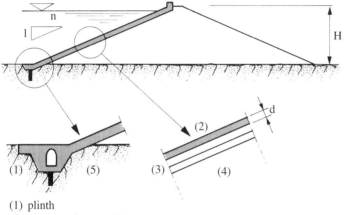

(1) plinth
(2) upstream facing, thickness d
(3) support and transition zones
(4) dam body
(5) bedrock

Asphalt facing
- n_{max} = 1.70-1.75 for execution reasons
- $d \cong H/300$ for H > 30 m,
- $H_{max} \approx 75$ m

Concrete facing
- n = 1.35-1.40 for unweathered rocky subsoils
- $d \approx 35$ to 45 cm,
- reinforcement 0.3-0.4 %

Figure 24.1 Embankment dam with an upstream facing

For medium to high dams (H > 50 m), the materials currently used for the facing are mainly:
- asphalt
- reinforced concrete

For low dams, synthetic flexible membranes (geomembrane)[1] are also used.

Dams with an upstream facing, at least those of a certain height, require adequate and homogeneous foundation rock, as the seepage gradient is considerable at the upstream heel of the dam, in the interface between the facing and the subsurface. At the upstream heel, a plinth or toe wall, which is a continuous concrete element, ensures the connection between the facing and the rock foundation. Depending on the nature of the foundation, this element is extended vertically by a cut-off wall or grouting.

---
[1] Exceptionally, in cases of stable ground.

Lastly, the subsurface and fill must only settle slightly under the effect of consolidation and water pressure. Concrete facing is more rigid than asphalt facing.

24.2 Embankment Dam with a Concrete Upstream Facing[2]

24.2.1 Main characteristics

Embankment dams with an upstream facing higher than 80 m built during the first half of the 20th century have proven to have numerous issues related to water loss. Most leaks were observed through the joints due primarily to the excessive movement of the slabs, resulting from a poor supporting layer and joint design. In the past, roughly graded rockfill was dumped in thick layers and then compacted after having been soaked by a high-speed water jet. Hard blocks were placed by a crane to support the facing. The slabs were relatively thick and included a tight network of horizontal and vertical joints. A cut-off was also included at the upstream heel. Later on, Terzaghi's comments were taken into account. He proposed using rockfill that was heavily compacted by vibrating rollers to reduce deformation and observed damage. From 1960, there was renewed interest in this type of dam structure due to significant improvement in rockfill compaction techniques. Using well-graded rockfill ensures a low volume of voids and a high compressibility modulus. It then became possible to consider the embankment dam with upstream facing as an alternative to other types of dams, especially when one or other types prove impracticable. A sharp rise was therefore observed in not only the construction but also the height of this type of dam. This was particularly pronounced in Australia, and then in South America and China (Xu, 2008). The most recent structures have been built to heights greater than 200 m. However, it should be noted that the behavior of these types of large dams have demonstrated significant issues and caused serious damage to the concrete facing (see § 24.2.2.7). This mainly concerns the Tianshengquiao dam (H = 178 m), China; the Barra Grande dam (H = 185 m), Brazil; the Campos Novos dam (H = 202 m), Brazil; and the Mohale dam (H = 145 m), Lesotho (Cruz and Freitas Jr, 2007).

The body of modern embankment dams with a concrete facing is made of compacted and, in theory, highly permeable rockfill. To achieve high-quality rockfill, the upstream and downstream slopes must be in the order of 1.3H/1.0V to 1.4H/1.0V. However, if the material is of lesser quality or the foundation is less resistant, the slopes will be gentler (approx. 1.5H/1.0V). When hard rockfill is used, drainage of all seepage automatically occurs. This type of dam has zones of increasing permeability in the downstream direction. The aim of this arrangement is to maintain stability under the effect of water pressure upstream of the dam during the construction phase, before the facing is placed or should the facing become damaged. The dam shell must be strong enough to support the concrete facing. Due to the effect of load (self-weight, hydrostatic pressure), it must be made in such a way that deformation is as low as possible. Measurements carried out on built dams demonstrate that this deformation is inversely proportional to the volumetric deformation modulus of the rockfill and increases with the square of the height. Deformation is primarily caused by the failure of rock elements at their contact point, followed by their reorganization. Adequate prior compaction is therefore recommended. Observations of the behavior of some dams after impoundment reveals that settlement occurs primarily in the lower third of the fill in the vicinity of the upstream face. It then gradually decreases in the downstream direction, which means that compaction can be less intense toward the downstream toe.

---

[2] Also known as a CFRD (*Concrete Face Rockfill Dam*).

Figure 24.2 presents cross-section diagrams of an embankment dam with a concrete upstream facing; it is completed by Table 24.3, which provides some of the characteristics of each zone.

Zone 1 is primarily designed to bear pressure from the facing prior to impoundment. The higher the dam, the thicker the zone becomes.

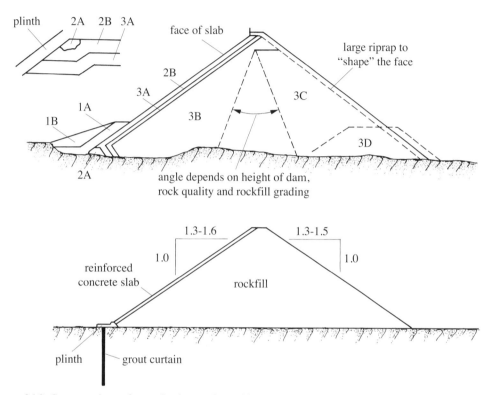

Figure 24.2 Cross-sections of an embankment dam with a concrete upstream facing, with the common names for each zone.

Zone 2, which is a transition zone, supports the upstream facing. It is subdivided into two parts: the first is situated just below the perimetric joint (zone 2A) and the second (zone 2B) is semi-permeable and located behind the facing so as to provide a barrier to any seepage flow.

With regard to the zone supporting the facing (zone 2B, 3A), it usually has a horizontal width of 4 to 5 m at the level of the crest. For dams higher than 100 m, this width may increase moderately down to the toe of the dam. The facing is supported by a layer of material that is as incompressible as possible. This layer is followed by a layer of the same size made up of crushed random fill. The transition between these two layers is important, and the filter criteria must be respected to avoid any loss of support to the facing in the case of seepage through the upstream lining. These two zones are placed in layers of 30 to 50 cm and compacted horizontally with a vibrating roller. Before placing the concrete facing, it is vital to protect the external face against erosion due to heavy rain by guniting or with an asphalt emulsion.

Table 24.3 Principal characteristics of the various zones in an embankment dam with a concrete upstream facing (from Cooke, 1997; Materón, 2007)

Zone label	Material description	Layer height (cm)	Placement and compaction method
1A	Dumped fill (silt)		Trucking and dumping
1B	Dumped semi-permeable material (silty sand)		
2A	Fine filter graded to 3.8 cm (incompressible material)		
2B	Crushed rock material from 7.5 to 10 cm	30, 40, or 50 cm	4–6 passes of a 10 t vibrating roller
3A	Small, crushed rock	40 to 50 cm	Compacted as per zone 2B
3B	Quarry rockfill or possibly random fill	80 to 100 cm	4–6 passes of a 10–12 t vibrating roller
3C	Same material as adjacent layer 3B	150 to 200 cm	
3D	Dumped rockfill		

During the construction of a dam in Brazil, a solution was developed that consisted of placing prefabricated concrete elements onto the transition layer before placing the following layer (Materón, 1992, 2007).

Upstream, zone 3B is subject to hydrostatic load and transfers this load to the foundations. It is important that all deformation is restricted. To do this, rockfill or gravel (random fill) can be used and placed in layers of 80 to 100 cm. It is compacted by a vibrating roller and sometimes sprayed with water. For rockfill of poor quality, compaction must be denser and the layers less thick. In the downstream zone 3C, which is less critical in terms of the behavior of the concrete facing, it is possible to use the same material as upstream, however, the layers must be thicker (1.5–2.0 m). Lastly, zone 3A ensures that there are fewer voids adjacent to zone 3B.

With regard to the facing, its function is to form a watertight barrier that does not deteriorate, especially in the drawdown zone. The facing must also be able to tolerate fill deformation under the effect of hydrostatic pressure. Temperature variations will of course also lead to deformation in the facing. The facing generally works in compression across its surface, but most especially in the center. High compressive stress has at times led to concrete bursting, especially in high dams constructed in narrow valleys (Pinto, 2007). Slight tensile stresses may occur near to the dam toe, crest, and periphery.

24.2.2 Constructing a concrete facing

24.2.2.1 Facing thickness

Different empirical formulae have been proposed to establish the depth of the facing. These formulae, which are not applied to the letter, mostly take into account the water level. With the progress that has been made in the placement of fill, it has become possible to decrease this depth.

Previously, the depth of concrete facing was calculated using the following equation:

$$d = 0.3 + 0.0067H \text{ [m]}.$$

Today, facing or slab thickness on built dams ranges between

$$d = (0.3 + 0.002H) \text{ and } (0.3 + 0.004H),$$

that is, a reduction of 40 to 70%. Table 24.4 includes some formulae based on the type of fill and the dam height (from Materón, 2007).

Table 24.4 Formulae for calculating facing thickness d (from Materón, 2007)

	Height	Facing thickness d
Well-graded rockfill	H < 120 m	d = 0.3 + 0.002 H [m]
		d = 0.3 + 0.003 H [m] if modulus < 100 MPa
	H > 120 m	d = 0.0045 H [m]
Rockfill with uniform particle size	H < 120 m	d = 0.3 + 0.003 H [m]
		d = 0.4 + 0.003 H [m] if modulus < 100 MPa
	H > 120 m	d = 0.0045 H [m]
		d = 0.0063 H [m] if modulus < 100 MPa

24.2.2.2 Concrete

Concrete with an aggregate of D_{max} in the order of 38 mm is generally used. Resistance after 28 days reaches 20 to 24 MPa. Using an air entraining agent reinforces impermeability and durability. Pozzolan, fly ash, and a plastifying agent are also recommended to reduce the water-cement content as much as possible. Concrete curing must sometimes be maintained until the first impoundment so as to reduce the effects of drying shrinkage. Experience has also proven that CFRD tolerate cold climates where severe freeze-thaw cycles occur.

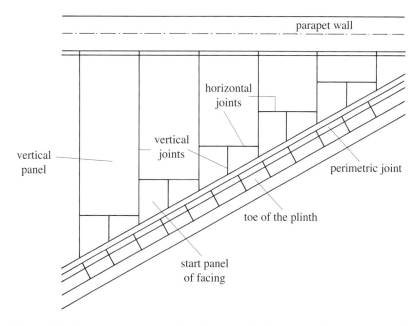

Figure 24.5 Construction layout of the slab along the foundation (from Schewe, 1990)

Concrete is placed using mechanical sliding formwork, which reduces the network of joints. During concreting, correct placement and appropriate concrete tightening using vibrators must be ensured. To facilitate placement of the main slabs, triangular or trapezoidal slabs are used to form a flat surface from which to begin placement of the sliding formwork and to provide a junction with the plinth (Figure 24.5).

24.2.2.3 Steel reinforcement
It is common practice to reinforce the slabs with steel bars. Today, reinforcement is deemed sufficient if it reduces cracking from thermal effects and shrinkage. The percentage of reinforcement lies between 0.3 and 0.4% It is distributed in both directions and laid out 10–15 cm from the surface (Pinto, 2001).

24.2.2.4 The plinth
Purpose
The watertight connection between the facing and the rock foundation is achieved thanks to the plinth, which is a continuous concrete base. By increasing the seepage path, the plinth limits the seepage gradient. The face of the plinth also provides a surface for carrying out grouting work (cut-off or strengthening). The plinth can be fitted with anchoring rods to counter uplift force and grouting pressure.

Geometric characteristics
The width of the plinth is from 0.4 to 0.5 times the height of the water; however, it has a minimum width of 3 m, or possibly 2 m for a dam lower than 40 m. It is not thicker than 1 m; generally 0.3 to 0.5 m is enough. Reinforcement is provided to distribute the temperatures and limit cracking due to possible tensile stresses (Figure 24.6).

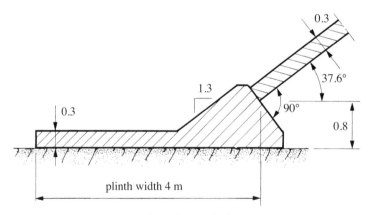

Figure 24.6 Details of the plinth (from Cooke, 2000)

24.2.2.5 Joints
Vertical and horizontal joints
Prior to 1960, the facing included a system of flexible vertical and horizontal joints that allowed it to adapt to deformation of the dam shell, which was made of dumped rockfill. The joints were filled with an asphalt material or fitted with wood.

In today's modern design, thanks to improved compaction methods (which in turn have led to lower deformation of the fill), the horizontal joints have been removed, thus facilitating the use of sliding formwork. Horizontal joints are still present for the connection between the initial slab and the facing. Current design only includes vertical joints and the perimetric joint. The vertical joint is now designed as an open joint, fitted with a waterstop.

As most of the facing is compressed, it is possible to reduce the number of vertical joints. In practice, these joints are spaced by 12, 15, or 18 m—15 m being the usual distance. It is important to note that the values are determined by the size of the formwork used.

In terms of the design of the joints, a copper plate placed on a bed of mortar is located under the slab, beneath the vertical joints. An elastic filler protected by a membrane completes the system on the surface (Figure 24.7).

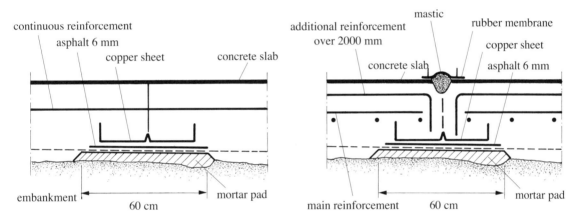

Figure 24.7 Construction of joints

Peripheral joint
The peripheral joint (Figure 24.8) is the most delicate part of the facing, as this is where major leakage problems can occur. The peripheral joint opens and closes depending on the reservoir operation. To avoid the risk of seepage, a multiple sealing system is installed with grout covered by a membrane on the surface and a PVC or rubber waterstop inside, as well as a copper sheet on the inside of the facing. An asphalt felt or plywood element is used at the interface with the plinth.

Following the experience gained from observing the major deformation occurring on embankment dams with an upstream facing, and its effects (see § 24.2.2.7), a new type of peripheral joint was developed in China (Figure 24.9). Compared to traditional joints, the principal idea is to remove the waterstop in the middle of the slab and reinforce the sealing joint at the surface. Laboratory tests and experiments have indicated that this type of joint can tolerate a load of 200 m of water. A new type of mastic, similar to rubber, has also been developed, which swells once submerged in water, thus achieving a sort of self-seal (Xu, 2008).

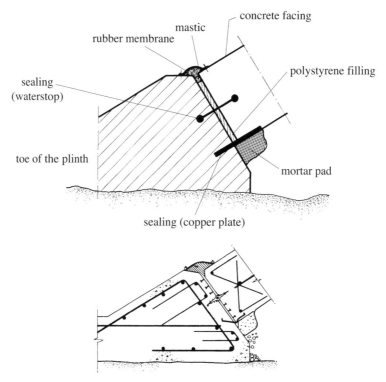

Figure 24.8 Detail of the peripheral joint with reinforcement below (from Marulanda et al., 1991)

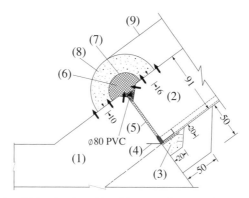

Figure 24.9 Detail of the peripheral joint developed in China: (1) plinth; (2) facing slab; (3) asphalt sand; (4) copper waterstop; (5) hard wood; (6) swelling mastic; (7) rubber waterstop; (8) fly ash; (9) stainless steel cap (from Xu, 2008)

24.2.2.6 Crest

It is usual practice to construct a parapet upstream of the crest to ensure a good connection with the facing. A flexible joint is designed in this zone. The parapet is shaped like a retaining wall with a large foundation slab, stabilized by fill material (Figure 24.10).

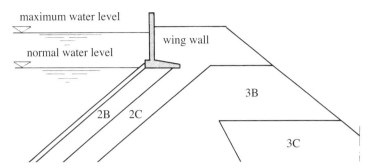

Figure 24.10 Detail of the crest of an embankment dam with upstream facing (from Hein, 1986)

24.2.2.7 Behavior of very tall dams

The construction of embankment dams with an upstream concrete facing is largely based on empirical criteria. Indeed, tests to characterize the fill material are not conducted with the same diligence as with zoned embankment dams. Serious problems have emerged in the construction of dams with a height of around 200 m, especially regarding the upstream facing—primarily during impoundment.

The principal mechanism for the transfer of load from the rockfill to the concrete slab is the development of a friction force at the concrete-rockfill interface when the rockfill deforms under the effect of hydrostatic load. The facing tends to slide if the force required is greater than the force available.

Furthermore, the settlement of the rockfill creates differential deformation between the facing and the rockfill, with the resulting deformation in the rockfill causing a compression effect in most of the facing. The ensuing excessive compressive stress can lead to cracks in the concrete slab along the vertical joints in the center part of the facing. Horizontal cracks in the lower part can also form (Pinto et al., 1998).

The zone of the central vertical joints becomes severely damaged, with the emergence of discontinuities, concrete shattering, and bent and visible reinforcing bars. For the dams in question, the total rehabilitation of the vertical joints was undertaken. Joint design was reviewed to better absorb the compressive stress due to horizontal displacement and avoid any fracturing of the facing. At the Campos Novos dam (Brazil), an open joint filled with mastic was installed (Sobrinho et al., 2007).

Seepage flow is a good indicator of the existence of cracks or fractures along the facing joints, or even of defects in the perimetric joint. It should be noted, though, that seepage transits without much risk of causing instability.

The major issue is to assess whether there is a risk of high compressive stress developing in the plane of the facing in the case of hydrostatic load. One simple approach for gaining an initial idea is to compare the conditions of the site and the deformation modulus of embankment dams with concrete facings that have not suffered from any issues. In the case of narrow valleys ($A/H^2 < 4$, where H is dam height and A is the slab surface), a dam facing with a vertical deformation modulus E lower than that given by the equation $E = 160 - 27\, A/H^2$ is likely to sustain damage. For wider valleys, a horizontal limit of $E = 40$ MPa is set (Figure 24.11).

Following these events, it seemed necessary to develop design criteria and methods based on an analysis of dam behavior, as has been done for other types of embankment dams. For example, thanks to FEM models, it should be possible to reproduce the friction transfer mechanism of the rockfill on the surface of the facing by including the effects of the rockfill-concrete interface. The results should enable the effects of the structural arrangement to be evaluated in order to reduce compressive load by planning flexible horizontal and vertical joints capable of absorbing the movement and by increasing the depth of the slab. Rockfill testing should also be undertaken systematically. Lastly, 3D calculations should enable deformation to be estimated prior to the filling of the reservoir.

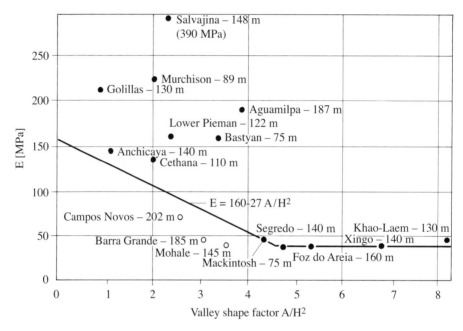

Figure 24.11 Vertical deformation modulus for rockfill related to the valley shape (from Pinto, 2997)

24.3 Constructing an Asphalt Concrete Facing

24.3.1 General information

In Switzerland, this type of facing is often used for afterbay reservoirs integrated into hydroelectric schemes where there are frequent variations in water level. The interest of such an installation lies in the fact that not only can it be used at a high altitude, where climatic conditions (especially in winter) are harsh, but it can also be used in hot and sunny regions. This type of facing has the advantage of being a tried-and-tested solution whose durability can be described as good. There are other benefits: the facing can be installed once the fill has been completed, and it adapts to the inevitable deformation of the shell. The surface of the upstream face can be totally cleared to allow inspection and, if necessary, maintenance works can be carried out. Weathering often occurs in the drawdown zone.

The bond at the upstream heel is provided by a toe wall possibly equipped with a gallery. A cut-off wall or grout curtain ensures watertightness under the upstream heel.

24.3.2 Technical characteristics

In terms of requirements, asphalt concrete must have low permeability (10^{-2} cm/s to 10^{-5} cm/s) and be flexible in order to absorb deformation from the fill. It must be able to resist erosion, and any creep along the slope must be avoided. This is why the upstream slope is between 1:1.6 and 1:1.75.

To ensure a good layer of facing, the upstream slope of the fill comprises a permeable material with graded and well-compacted particles with a D$_{max}$ of 100 mm. Its horizontal width is around 4.0 m.

The facing is a multilayered structure placed onto a supporting layer that ensures the transition.

Asphalt facing can have different structures, each comprising several layers. In the following example (double lining facing, Figure 24.12,), we can observe, from the contact point with the body of the fill, a surface treatment by penetration of an asphalt binder, a binding layer made of coated chippings or asphalt concrete a few centimeters thick, an impermeable layer of dense asphalt concrete (4 to 7.5 cm), a draining layer of asphalt concrete (5 to 15 cm), a binding layer of asphalt concrete or coated chippings (3 to 10 cm), a second layer of impermeable facing consisting of one or two layers of dense asphalt concrete, and a closing facing made from a layer of fine asphalt mastic.

Another structure (single lining facing, Figure 24.12), is composed of a surface stabilizing treatment, an equalizing surface layer, a drainage layer (3 to 15 cm), a binding layer (3 to 12 cm), one or several layers of dense asphalt concrete, and a closing layer.

Previously, the impermeable part of the dense asphalt concrete facing was installed in layers. In many cases, the formation of gas bubbles was observed between the two layers. This is why today the facings are placed in a single layer of minimum 10 cm or 1/300 of the height, for large dams (H > 30 m).

Asphalt concrete is a mixture of aggregate (D$_{max}$ 20-30 mm), filler (2 to 5%), and asphalt (7 to 8.5%). It includes approximately 3% voids. It is placed at a temperature of 180°C. After spreading it is compacted by vibrating rollers.

It is also common practice to plan a closing layer to protect the surface of the facing against atmospheric agents and aging. This is made of an asphalt emulsion applied cold or a mastic applied warm.

Table 24.13 summarizes the main structural aspects in the use of a watertight facing.

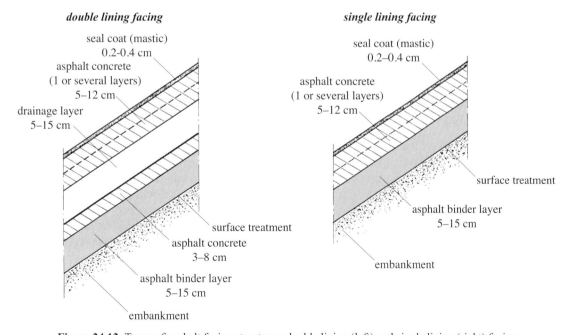

Figure 24.12 Types of asphalt facing structures: double lining (left) and single lining (right) facing

Table 24.13 Main structural elements in the use of a watertight facing (from Pougatsch and Müller, 2000)

Construction and operational aspects		Purpose/Characteristics
Slope of upstream face		depends on: • the characteristics of the materials • the adopted sealing system
Components of the sealing system	Type of sealing	• influences bearing capacity over time (durability)
	Supporting layer	• evens out roughness in the fill • ensures the transition between the body of the embankment dam and the sealing element
	Drainage	• collects losses and leaks in the case of damage to the sealing system • protects the body of the embankment dam in the case of leaks • reduces the effect of uplift from groundwater
	Protective layer	protects against the effects of: • drawdown • ice • waves • rain • wind • vandalism
Placement		requires: • quality checks of the material • careful installation • careful treatment of the joints in the sealing system and connections to other elements in the dam (generally in concrete)
Maintenance		requires: • regular condition inspections • repair work, if necessary

24.4 Other Types of Upstream Facing

24.4.1 Geomembrane

Geomembranes, thick synthetic films made of polyvinyl chloride (PVC) or polyethylene (PE), are generally used for low- to medium-height dams. They can be used in conjunction with (nonwoven) geotextiles, which affords them better resistance to impact. The ageing of geomembranes depends on the quality of the water, solar radiation and UV rays, heat, frost, and the movement of ice. According to information in the literature, experience using this type of facing spans 25 years on average.

Geomembranes are installed on a supporting layer. They are often protected against impact, waves, uplift by wind, and vandalism by rockfill or concrete slabs. The drawback to a protective layer is that it prevents close inspection of the condition of the geomembrane. Installation of a drainage system is recommended. If this protective layer is not used, anchoring methods must be used. The effects of the repeated movement of wind and waves leaves undulations in the membrane. In some cases, a specific water level must be maintained in order to counter the effects of wind (Pougatsch and Müller, 2000).

24.4.2 Lining based on stabilized ground

This type of lining is commonly used in roadworks, warehouse surfacing, carparks, and even landfills. It has also been used in dam projects, especially for sealing the upstream slope of embankment dams, without, however, constituting the most appropriate option. It consists in using a mixture of cohesive soil with cement or another additive such as bentonite, a clay mineral, or organic agents to improve the characteristics of these soils, including the permeability coefficient and compaction properties. In this system, the supporting layer, drainage zone, and protective layer are important elements.

Surface damage (cracking, the formation of grooves) is inevitable. Without protection, it is difficult to keep the surfaces intact. This type of lining also suffers from changes in water level (Pougatsch et Müller, 2000).

24.5 Comparison Between an Upstream Facing and a Central Core

The option comprising an upstream facing has the following advantages compared to a central core (Figure 24.14):

- The dam shell is out of the water (no weathering due to the effect of water, no variation in pore pressure due to changes in reservoir water level)
- Water pressure acts in a more favorable direction; it increases the vertical component E_v, which acts on the foundation, reinforcing the dam's stability against sliding
- The dam shell is not subject to seepage, and therefore the slope of the faces can be steeper

However, it also has the following disadvantages:

- The seepage gradient at the toe of the facing is very high, requiring a special construction (the plinth) and specific treatment of the foundation
- Placement of the facing is a complex operation, requiring specific equipment and skills
- The facing is more sensitive to differential settlement and dynamic forces due to earthquakes

The plinth sits directly on the rock and is often fixed with anchors. To limit the effect of seepage gradient, grouting, or even a grout curtain leading to sound, highly impermeable rock, is very important.

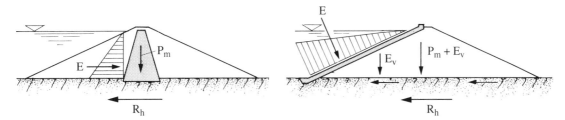

Figure 24.14 Comparison between an upstream facing and a central core

25. Construction and Behavior of Embankment Dams

Shuibuya rockfill dam in China, height 233 m, year commissioned 2008 (Courtesy M. Wieland)

25.1 Placement and Compaction of Rockfill

25.1.1 Cohesive materials

25.1.1.1 Finding the optimum moisture content
For cohesive materials, moisture content w plays a vital role for:

- Utilization of the borrow zone
- Fill placement
- Compaction
- Shear strength
- Consolidation (settlement)

The optimum moisture content w_{opt} is defined as being that which leads to the maximum apparent dry unit density γ_d for a given compaction energy. For each cohesive material, there is an optimum moisture content for which the apparent dry unit weight reaches a maximum value.

The optimum moisture content for a specific material is determined using compaction tests (the Proctor test). This test involves compacting soil into a cylindrical mold either with low compaction energy (standard Proctor test) or with greater compaction energy (modified Proctor test) for various moisture contents. The aim of the Proctor compaction test is to establish the highest possible dry density. Table 25.1 summarizes the principal data from the Proctor test.

Table 25.1 Characteristics of the Proctor test (AASHO mold)

	Standard test	Modified test
Mold diameter	10.16 cm	10.16 cm
Sample height	11.7 cm	11.7 cm
Sample volume	948 cm^3	948 cm^3
Tamp diameter	5.1 cm	5.1 cm
Tamp mass	2.49 kg	4.54 kg
Fall height	30.5 cm	45.7 cm
Number of layers	3	5
Blows per layer	25	25
Compaction energy	0.6 J/cm^3	2.7 J/cm^3

After each compaction, the apparent dry unit weight is measured. Using the obtained results, the Proctor curve is established by plotting the moisture content along the horizontal axis in relation to the apparent dry unit weight. This curve shows the maximum value for the optimum moisture content (Figure 25.2):

$$\gamma_d = \frac{w}{V(1+w_e)} \quad \text{en } [g/cm^3],$$

where

w: wet weight of the sample
V: volume of the sample
w_e: final moisture content

The apparent dry unit weight γ_d influences:

- The shear property $\phi' = f(\gamma_d)$
- Permeability $k = f(\gamma_d)$
- Settlement

We have the following comments to make about the Proctor curve (Figure 25.2):

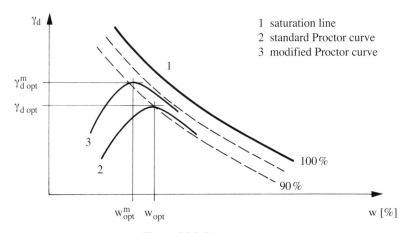

Figure 25.2 Proctor curve

- For a specific, slightly wet moisture content, the increase in compaction energy only increases unit weight by a small margin; the material nears saturation
- For a slightly dry moisture content, high compaction energy noticeably increases the unit weight γ_d and therefore ϕ'

The required moisture content in the various zone of the core is depicted in Figure 25.3.

In the lower part of the core, the moisture content must remain below optimum moisture content, as pore pressure, as well as moisture content, will increase as construction progresses through the overload created by the upper layers (w_{opt} – 1 to 2%).

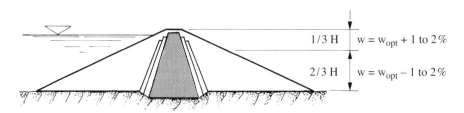

Figure 25.3 Required moisture content

When the void index lowers, the degree of saturation increases. If the degree of saturation reaches 100%, pore pressure u occurs (see also § 25.1.1.3), decreasing shear strength in line with actual stress calculations.

To avoid an increase in the degree of saturation due to settlement, a moisture content that is slightly drier than optimum ($w_{opt} - 1$ to 2%) is used for the lower zones. The disadvantage is that after compaction, the materials become slightly more friable.

However, in the upper part of the core (1/3 H), greater flexibility (plasticity) is favored, while requiring a moisture content that is increased by 1 to 2% compared to the optimum ($w_{opt} + 1$ to 2%). This flexibility is advisable in the upper part of the core to tolerate settlement without cracking. The risk of pore pressure due to the higher degree of saturation is low, as the overload from the layers above is also low. The reduced density because of the increased moisture content results in slightly higher settlement, but this is tolerable in the upper part of the core.

25.1.1.2 Placement

The clay content in the material used for the core is normally greater than 5%, which guarantees low permeability.

The materials are placed in layers that are 25 to 30 cm thick, or 40 cm at the very most. The maximum particle diameter is limited to 10 cm regardless (for cores in natural moraine). Notwithstanding this, the maximum tolerable particle diameter primarily depends on the grading curve. The proportion of coarse particles must be limited so that contact between the largest particles is avoided. This requirement guarantees that no compressive arch forms in the core under the effect of consolidation (Figure 25.4).

The Proctor test ensures that the optimum moisture content w_{opt} can be determined prior, depending on compaction energy. Monitoring moisture content is vital throughout the whole placement and compaction process (see section 25.3).

If the moisture content in one layer is higher than that prescribed, drying time should be allowed before beginning compaction. If the moisture content is lower than prescribed, watering may be required, although the effect of this action is low and confined to the surface.

Compaction is most often carried out by:

- A pneumatic roller (6–8 tonnes, 4 to 6 passes)
- A vibrating roller (8–10 tonnes)

To reduce pore pressure in the layers, rollers with large surface areas are also used, such as

- Grid and disk rollers
- Sheepsfoot rollers

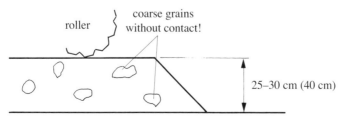

Figure 25.4 Placing cohesive materials

These also prevent the material from sticking to the rollers through the effect of suction.

Flat-drum rollers, however, have the advantage of protecting the placed layer from water penetration in case of rain. Before placing the next layer, the surface is textured by using a set of harrows.

25.1.1.3 Development of pore pressure

Thanks to compaction and the weight of the upper layers, the lower layers are more compressed and their volume decreases. As a result, the apparent density and degree of saturation increases, leading to the development of pore pressure (Figure 25.5). Once construction has been completed, this pressure is of particular interest, as it forms part of dam stability calculations.

Final pore pressure is therefore not a constant value up the entire height of the dam core. Typically, at the base of the core, pore pressure can reach 40% of the pressure of weight of the material above. As previously mentioned, pore pressure is limited by applying a reduced moisture content that is kept between 1 and 2% below the optimum content.

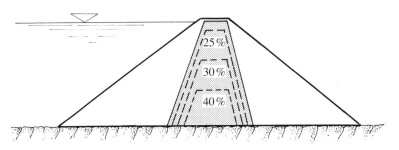

Figure 25.5 Diagram illustrating pore pressure after construction

Pore pressure begins to dissipate as soon as it appears. In addition to permeability, the speed of construction also affects pore pressure once construction has finished.

Depending on the permeability of the material, different degrees of dissipation can be observed in the central part of the core:

$K > 5 \cdot 10^{-7}$ cm/s	no dissipation during construction
$K > 5 \cdot 10^{-6}$ cm/s	partial dissipation
$K > 5 \cdot 10^{-5}$ cm/s	major dissipation
$K > 10^{-5}$ cm/s	complete dissipation

Careful placement of the core near the contact with the foundation rock is essential to avoid seepage through this point. The surface of the rock must be thoroughly cleaned, and the core compacted with a high moisture content close to the rocky foundation.

25.1.2 Noncohesive materials

Noncohesive materials ($c = 0$) (Figure 25.6) are used for the construction of:
- Filters
- Transition zones
- Dam shells

- Drainage
- Riprap

Typical thicknesses for the placed layers range from:

- Approximately 60 cm for the filters
- Between 90 and 120 cm for alluvium and talus (depending on permeability)
- Approximately 150 cm for quarry rockfill

The maximum particle diameter is equal to ¾ of the thickness of layers for the dam shell. With regard to fines content (d [mm] ≤ 0.08), this should be at a maximum of 6%, or possibly from 8 to 10%.

Compaction is carried out with vibrating rollers (8–10 tonnes). The type of roller is selected based on prior tests.

Rockfill is generally placed using compaction. It is watered during this operation, as the presence of water both facilitates the movement of the blocks against one another and absorbs most of the compaction energy. Settlement during compaction and operation results notably in the crushing of the edges of the blocks.

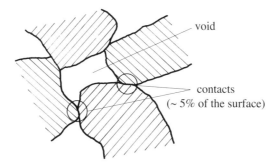

Figure 25.6 Macroscopic structure of rockfill with noncohesive material

The presence of water can therefore also affect the compaction of noncohesive material. Relative density is used as an indicator and is defined as follows:

$$D = \frac{e_{max} - e}{e_{max} - e_{min}} \quad [en\ \%]$$

$$e = \frac{n}{1-n},$$

where e = void index and n = percentage of voids [%].

A relative density greater than 70% is satisfactory. Relative density is not easy to calculate.

In conclusion, it should be noted that before the works begin, compaction tests (see Section 25.2) must be carried out for the various materials, with the aim of determining

- optimum layer thickness
- optimum moisture content and
- most appropriate compaction methods

to guarantee the most cost-effective placement that complies with the specifications.

25.2 Trial Areas

Before constructing the fill, a trial area of around 30 to 40 m long and 4 to 6 m wide should be established. This will allow the conditions for the placement of material to be determined and the most effective compaction methods to be chosen. The thickness of the layers and the number of compactor passes are also determined during these trials.

25.3 Inspections and Measurements During Construction

During construction several inspections and measurements must be carried out before, during, and after the placement of material. The inspections take place based on a predetermined program and are adjusted for each category of material (core, dam shell, filter). The frequency and number of tests vary and can be established depending on the volume of placed material, also taking into account the size of the dam. The aim of these inspections and measurements is to monitor the homogeneity and compliance of the materials, as well as construction quality (in particular, to inspect compaction). Tests may be carried out in situ or in a worksite laboratory. Visual observations also enable any specificities to be detected and the appropriate actions to be taken. The permanent on-site presence of a senior geotechnical engineer is vital to ensure adequate inspections.

Although this list is not exhaustive, the following inspections should be undertaken:

Tests

- Particle size — Every 3000 m^3
- Apparent unit weight (γ_d) — Compaction inspection: at the beginning of the works, every 500 m^3, then every 2000 m^3
- Moisture content (w%)
- Plasticity (Atterberg limits)
- Normal Proctor
- Mechanical strength — Triaxial and/or direct shear tests

Measurements

- Pore pressure u
- Horizontal deformation — For example, by using deformation gauges, rod extensometers
- Vertical deformation (settlement) — For example, by using fixed settlement guides on a telescopic tube, with induction sensor measurement

25.4 Behavior During and After Construction

25.4.1 Vertical deformation

With regard to vertical deformation, the following phases have been identified:
- Settlement due to an increase in weight during the placement of fill, which is an instantaneous deformation, and
- Consolidation after construction has finished

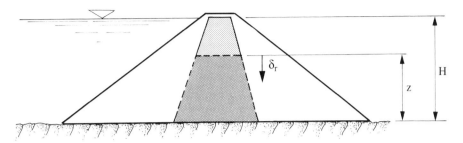

Figure 25.7 Behavior of the core during construction

It is interesting to trace the behavior of the core during construction by using a simplified model (Herzog, 1993) (Figure 25.7).

The part of height z already placed settles under the effect of overloading $(H-z)\gamma$. Deformation in one layer is given by the equation

$$\varepsilon = \frac{\sigma}{E} \cdot \ell,$$

where ε: deformation, σ: normal stress, E: deformation modulus and ℓ: sample length.

Therefore, settlement is deduced from

$$\delta_r = \frac{1}{E_r} \underbrace{z}_{\text{compacted zone}} \cdot \underbrace{(H-z)\gamma}_{\sigma}$$

$$\delta_r = \frac{\gamma}{E_r}(H \cdot z - z^2)$$

where γ: fill unit weight and E_r: fill deformation modulus (unknown).

As illustrated in Figure 25.8, the distribution of settlement as a function of dam height is parabolic.

This relation is theoretical, as in reality the deformation modulus is not constant along the height of the embankment dam, and overloading is triangular.

Settlement after construction (consolidation) is calculated using the equation given in Figure 25.9. The distribution of consolidation is therefore a half-parabolic function along the dam height (Figure 25.9). Theoretically, deformation modulus is not known and can only be estimated from deformation measurements.

Figure 25.10 illustrates the measured settlement and consolidation at the Göscheneralp embankment dam. The measurements have been compared with theoretical equations while assuming a deformation modulus.

During construction the material settles and expands slightly. This expansion is partly transferred to the subsurface. Ultimately, the volume is therefore greater than the theoretical sum of the placed layers.

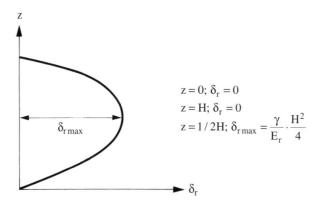

Figure 25.8 Distribution of settlement as a function of embankment dam height

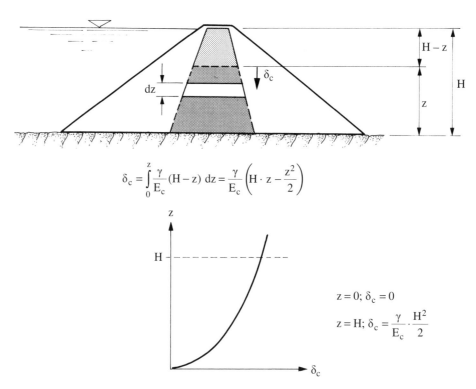

Figure 25.9 Distribution of vertical deformation due to post-construction consolidation as a function of embankment dam height

CONSTRUCTION AND BEHAVIOR OF EMBANKMENT DAMS

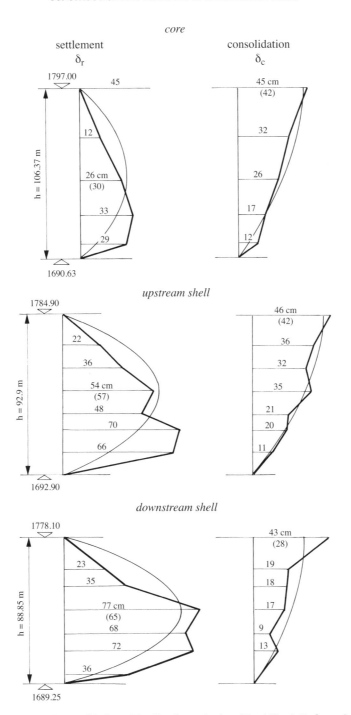

Figure 25.10 Settlement and consolidation of the Göscheneralp dam (H = 155 m). Deformation moduli according to adjusted theoretical curves (Herzog, 1993): core: $E_r = 207$ MN/m^2, $E_c = 303$ MN/m^2 (top), upstream shell: $E_r = 82$ MN/m^2, $E_c = 232$ MN/m^2 (middle), downstream shell: $E_r = 66$ MN/m^2, $E_c = 323$ MN/m^2 (bottom).

Here are some typical values for settlement during construction (for an approximate height of 30 m):
- Gravel and sand   0.9–1.4 %   of total height
- Sandy silt   1.3–2.1 %   of total height
- Gravel and clayey sand 1.9–3.3 %   of total height
- Clayey silt   2.8–4.2 %   of total height

One per cent of the total height may be a first approximation for the consolidation settlement of the crest of an embankment dam, measured after it has been completed. It is therefore necessary to compensate for this settlement by heightening the crest during its construction, with a view to also ensuring the determined freeboard. Settlement decreases toward zero near the abutments (see Figure 20.11).

25.4.2 Conditions with a full reservoir

For an embankment dam, the presence of water in the reservoir creates the following effects:
- Change in effective stresses (upstream of the sealing system)
- Water pressure against the dam
- Seepage through the dam and its foundation
- Change in compressive strength (upstream)
- Loss of cohesion in the zones with saturated cohesive materials

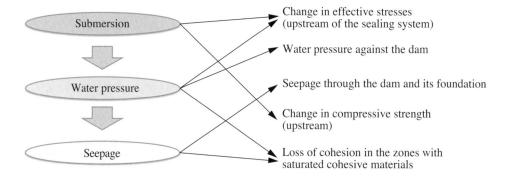

25.4.2.1 Submersion

Submersion of the upstream shell causes a change in stress in the fill material. In the case of an embankment dam with a central core, the upstream shoulder is firstly subject to Archimedes' uplift which decreases the effective stresses. Despite this decrease, settlement is often observed, especially in an upstream shell made of rockfill. This is the result of a possible reduction in compressive strength in the particles once they are saturated with water, as the contact surfaces are therefore smoother. This settlement can be reduced by appropriate compaction and by adding sufficient water during placement (20% of the volume of rockfill).

25.4.2.2 Water pressure

Water pressure on the embankment dam causes a change in stresses in the part of the dam upstream of the sealing system, including the foundation subsurface (effective stresses).

Water pressure always acts against the sealing system (Figure 25.11); this is why the behavior of a central core is completely different from that of an upstream facing (cf. Section 24.5).

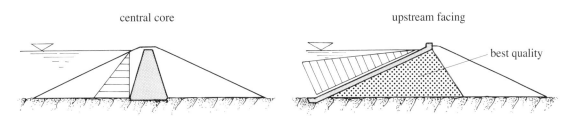

Figure 25.11 Effect of water pressure on the sealing system in an embankment dam

In the case of an upstream facing, a high vertical component of water pressure can be observed, which is maximal in the upstream heel zone of the dam. Maximum settlement due to water pressure is therefore located about half-way up the dam and not at the crest. The upstream two-thirds of the fill require high quality material and compaction. This type of embankment dam has the advantage that the entire fill-foundation interface is available to resist the horizontal component of water pressure, which is not the case for an embankment dam with a central core.

The embankment dam with a central core is only subject to a horizontal component of water pressure, which also causes horizontal deformation (shear stresses).

25.4.2.3 Seepage

Seepage through the embankment dam and its foundation has a profound effect on dam stability. An analysis of accidents occurring in all embankment dams built after 1900 demonstrated that the most frequent cause, that is, 38% of the accidents, was due to internal erosion or the formation of piping following seepage.

The rules to follow to avoid such consequences are:

- A wise choice of materials
- Meeting the filter criteria between the various zones of the embankment dam and the subsurface interface
- The correct treatment of embankment dam contact zones with the subsurface and concrete structures
- The appropriate treatment of the subsurface with grouting and/or drainage
- Placement of material in accordance with best practice
- Adequate compaction of material near to conduits, test pits, and other elements in direct contact with the fill

Seepage through the embankment dam is characterized by:

- Flow and its volume
- Distribution of pore pressure

Seepage flow is an excellent indicator of the correct functioning of the dam; every effort should therefore be made to ensure its continuous and permanent monitoring.

Periodic analysis of the water includes:
- Electrical conductivity
- Chemical composition
- Temperature
- Solids content

This analysis then enables
- The source of the water to be identified (reservoir or subsurface)
- Understanding of any eventual modifications in material characteristics affected by the seepage (washouts, movement of fines, internal erosion, chemical reactions)

The distribution and intensity of pore pressure due to seepage are also vitally important for the stability and safety of the dam. Monitoring with frequent measurements and immediately interpreting the results is essential.

25.4.3 Reservoir operation

For many water storage schemes comprising a reservoir created by a dam, the operation of large reservoirs creates annual variations with a major effect on the water level. The rhythm of these fluctuations can also be monthly or weekly in the case of reservoirs with a medium or small volume of water.

The core, and most especially its upstream side, are affected by movements in water level, as the water level in the impermeable material does not follow that of the reservoir. This means that the transient seepage regime has a particularly adverse effect on the stability of the upstream part of the embankment dam.

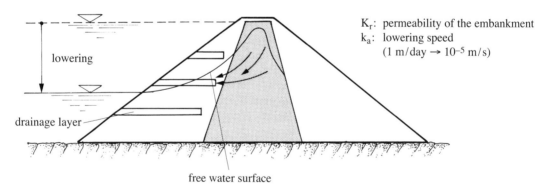

Figure 25.12 Effect of drawdown on water level

To avoid high water levels in the upstream body of embankment dams, it is preferable to use materials with high permeability ($K_r/k_a > 10$). If this is not possible, drainage layers can also be included (Figure 25.12).

26. Stability Analysis

Mattmark earthfill embankment dam with an inclined core in Switzerland, height 120 m,
year commissioned 1969 (Courtesy T. Andenmatten, Mattmark AG)

26.1 Principles of Analysis

Sliding safety is calculated for various failure surfaces that may cross several zones of an embankment dam and possibly the subsurface. The aim is to find the least stable, that is, the most critical of the potential failure surfaces. The hypothesis of a circular slip surface is generally made (Figure 16.1). However, depending on dam configuration, the calculation can also be made by considering any slip surface.

Several slip surfaces oriented either upstream or downstream are checked, cutting across:

- The surface protection (riprap)
- The upstream or downstream dam shell
- The core
- The transition zones (filter)

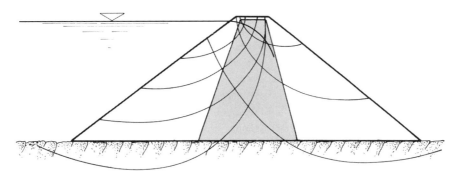

Figure 26.1 Position of various failure circles considered for stability analysis

26.2 Methods of Analysis

For heterogeneous masses such as zoned embankment dams, well-known analytical methods from soil mechanics are used. The unstable mass is usually divided into vertical slices. The equilibrium of all the slices is then examined. The slip surface depends greatly on the profile of the zoned embankment.

The following analytical methods are most often used:

- Circular slip surface: – Fellenius (1948)
  – Bishop (1955)
- Any slip surface: – Janbu (1954)
  – Morgenstern-Price (1965)

Critical slip surfaces can be found relatively easily using available software.

26.3 Load Cases

To verify the safety of embankment dams, different load cases and the combinations thereof have been described in sections 11.1 and 11.2. As a general rule, the following types of loading are described:

- Type 1   normal loads
- Type 2   exceptional loads
- Type 3   extreme loads

For embankment dams, it should be noted that verifications also concern the work stages, that is, during and after completion of construction. Table 26.2 summarizes the different types that may occur.[1] It is nevertheless up to the engineer to establish, while considering the type of dam, the least favorable type of loading that must be checked.

Table 26.2 Summary of different types of loading concerning embankment dams

Types of loading	Phases in question	Description of loading
Type 1 (normal)	Operation	Normal water level
		Normal drawdown of water level (depending on reservoir use)
		Water level reached during a design flood
Type 2 (exceptional)	During construction	Loading of the embankment dam during floods (final height of the dam is not yet reached)
	End of construction	Empty reservoir
		Empty reservoir + earthquake (pore pressure in the core is at its maximum just after the fill has been finished)
	Operation	Rapid drawdown of water level
		Water level reached during a design flood
		Avalanche, debris flow
Type 3 (extreme)	Operation	Reservoir with normal water level + verification earthquake
		Empty reservoir or normal drawdown + verification earthquake (optional)
		Water level reached during a safety flood

In comparison with Table 26.2, we can make the following remarks:
- Pore pressure can vary with time, permeability, hydrostatic loading, and drainage conditions. Depending on the cases envisaged (prior to consolidation, intermediary water level, effect of an earthquake), pore pressure must be adapted accordingly. A network of flow lines and equipotential lines indicates the intensity and distribution of pore pressure.
- In the case of an earthquake and with high pore pressure, there is a risk of core liquefaction. Lastly, with regard to stability, pore pressure can reduce shear strength.

Rapid drawdown is a type of loading for which we assume that the water must be evacuated by all available water-release structures, open to a maximum, and without considering any natural water intake. This type of loading may occur through
- A voluntary action to avoid another risk (bank instability, bombing, etc.)
- An operating error
- Sabotage of water-release structures

It is not usual to combine two types of extreme loading, for example, an earthquake with a rapid drawdown or exceptional flood.

---

[1] Also refer to paragraph 6.3.2 (Table 6.3) in the publication "Sécurité des ouvrages d'accumulation," SFOE guidelines (SFWG), 2002b.

26.4 Factor of Safety

In most methods using vertical slices, the factor of safety is defined by the equation:

$$FS = \frac{\text{stabilising forces}}{\text{driving forces}}.$$

The values for materials at the end of construction must be used while taking into account test variations (±standard deviation).

In the standards or in the literature, different factor of safety values can be found. As an example, the factor of safety values proposed by the DIN 4084 are as follows:

- Normal loads: $FS \geq 1.40$
- Exceptional loads: $FS \geq 1.30$
- Extreme loads: $FS \geq 1.20$

In the case of small dams, the following factors are proposed (from Pougatsch and Müller, 2000):

Type of loading			Factor of safety
End of construction	empty reservoir		1.3
	empty reservoir	with earthquake	1.1
Operation	full reservoir		1.5
	rapid drawdown		1.3
	full reservoir	with earthquake	1.1
	empty reservoir	with earthquake	1.1

26.5 Slope Angles

Stability calculations determine the required slope angles. They are above all a function of the material used and the sealing system chosen. Table 26.3 gives some guidance for the design of embankment dams.

Table 26.3 Indicative slope angle values for the design of embankment dams

Fill material (dam shell)	Sealing element	Upstream slope	Downstream slope
Rockfill	Center core	1:1.80	1:1.80
	Inclined core	1:2.10	1:1.80
	Upstream facing	1:1.50	1:1.40
Permeable alluvium	Center core	1:2.00	1:2.00
	Inclined core	1:2.30	1:2.00
Fines alluvium	Center core	1:3.00	1:2.50
	Inclined core	1:3.30	1:2.50
Homogeneous dam		1:3.00	1:3.00

26.6 Safety in the Case of an Earthquake Based on a Pseudo-static Analysis

The effect of an earthquake is taken into account by introducing horizontal forces onto the sections. Horizontal force depends on the acceleration of the mass in question, that is, the response, in a horizontal direction, of the earthquake. This is known as the pseudo-static method.

With this method, it is not usual to guarantee sufficient safety (FS = 1.20) for the sliding of critical failure surfaces. In practice, limited deformation and settlement due to instability during earthquakes are accepted.

The analysis follows the approach proposed by Makdisi and Seed (1978) and is based on Newmark's theories (1965):

- Definition of the accelerograph for the site of the embankment dam (Figure 26.4)
- Calculation of critical acceleration for which the critical failure surface is still stable (FS = 1.0)
- For all acceleration greater than critical acceleration, the unstable mass slides a little for a limited time (plastic deformation)[2]
- The addition of all these partial slidings thus gives a total plastic deformation, whose amount must remain acceptable compared to the freeboard; this latter must remain sufficiently high to avoid overtopping of the embankment dam[3]

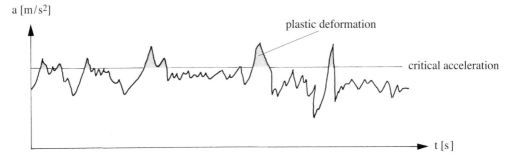

Figure 26.4 Accepted accelerograph at the site of the embankment dam

This deformation in the case of an earthquake leads to a lowering of the level of the crest. If such plastic deformation in the case of an earthquake is to be accepted, the crest must be heightened (see Figure 20.11) or a larger freeboard must be allowed for.

26.7 Verification of Embankment Dams with Regard to Earthquakes, Based on Swiss Guidelines[4]

26.7.1 Verification basis and requirements

According to the guidelines of the Office fédéral de l'énergie (SFOE (SFWG), 2003), dams are grouped into three categories defined in Section 11.5.2 and are subject to different requirements. As an example,

---

[2] See also § 26.7.5 and § 26.7.6.
[3] Section 26.7.6 gives guidance on acceptable plastic deformation.
[4] The notes from the guidelines have been retained.

the verification approach for category II embankment dams is presented below; this category includes retaining structures with a height below 40 m and a volume of reservoir water smaller than 1 mio m^3. Larger dams (category I) are subject to stricter requirements. The simplified method presented below does not apply to category I dams.

The basis for verification is as follows:

- Static, characteristic values of the materials obtained through investigation (new constructions and existing dams)
- Empirical calculation of sensitivity of material in the embankment dam to an increase in pore pressure following cyclical loading
- Stability analysis of potential sliding blocks after dynamic horizontal acceleration determined empirically (1 mode) and under pseudo-static vertical seismic loading
- Otherwise, calculation of nonreversible sliding displacement (while considering in an empirical manner the dynamic behavior of the embankment dam)

These principles lead to the requirements outlined in Table 26.5.

Table 26.5 Requirements for category II dams

Return time for a verification earthquake	5,000 years, intensity values as per the Swiss seismic hazard map[5]
Material properties and investigation methods	*Properties necessary for all material zones:* • Shear strength parameters (static): φ′ (effective angle of friction), c (cohesion), s_u (shear strength in undrained conditions) • Grading curves • Compactness • Possibly SPT tests or similar (cf. § 26.7.4) • Possibly typical dynamic values of the materials (§ 26.7.7) *Investigation methods:* • For new constructions: standard static tests, dynamic tests advisable • For existing dams: depends on construction documents, supplemented by testing if necessary • For sensitive materials, analysis of the increase in pore pressure due to the earthquake, using cyclical laboratory tests or equivalent monitoring methods
Modeling	Geometry, geotechnical model of the foundation and body of the dam (piezometric line included) Bi-dimensional model of the dam Negligible hydrodynamic pressure
Calculation methods	Simple assessment of the increase in pore pressure Simplified analysis of stability during the earthquake (spectrum response method with one mode), horizontal and vertical loading Possibly a simple calculation of displacement through sliding, should sliding be possible (cf. § 26.7.5 and § 26.7.6) Possibly calculation of sliding displacement, on the basis of the temporal development of acceleration (cf. § 26.7.6)

---

[5] The I_{MSK} intensities for a return period of 1,000 and 10,000 years are taken from intensity maps (Figures 11.24 and 11.25). Values for a return period of 5,000 years are interpolated as follows: $I_{5000} = 0.3 \cdot I_{1000} + 0.7 \cdot I_{10,000}$.

26.7.2 Flowchart for the calculation process

Calculations for category II dams are carried out in line with Figure 26.6.

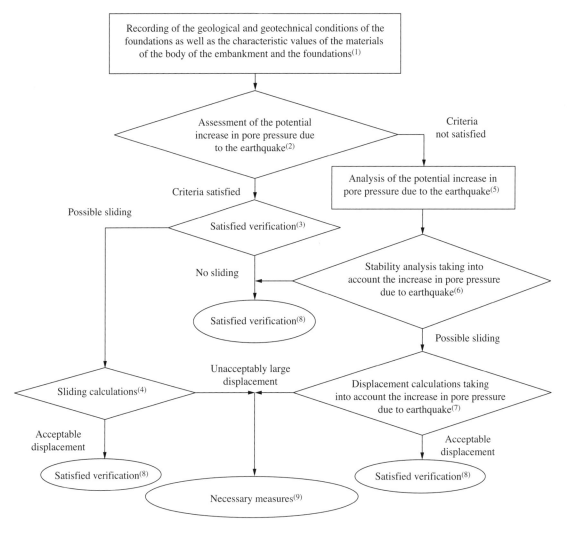

Figure 26.6 Calculation flowchart for category II dams. Key: (1) standard static geotechnical investigations and laboratory tests (§ 26.7.3); (2) as per Section 26.7.4; (3) for example, based on Bishop with substitution seismic forces (§ 26.7.5); (4) simplified calculation of sliding displacement (§ 26.7.6); (5) cyclical tests in the laboratory (§ 26.7.7); (6) for example, based on Bishop with substitution seismic forces, pore overpressure, and reduced shear strength (§ 26.7.8); (7) simplified calculation of sliding displacement with pore overpressure and reduced shear strength (§ 26.7.8); (8) earthquake safety verification is deemed to be provided; (9) earthquake safety verification is not deemed to be provided. Some measures may be necessary (for example, a more detailed calculation with less conservative parameter values obtained in a more precise manner, structural measures, drawdown of the reservoir).

26.7.3 Summary of geological and geotechnical conditions of the foundation as well as typical values of materials comprising the body of the embankment dam and the foundation soil

The geological and geotechnical conditions of the foundation, as well as the typical values of the materials are determined through standard geotechnical investigations and laboratory testing. Dynamic tests for category II dams are not necessary but are recommended.

26.7.4 Assessment of the potential increase in pore pressure due to an earthquake

The increase in pore pressure due to an earthquake is estimated based on the grading curve and compactness of materials.

An analysis of the increase in pore pressure due to an earthquake, based on Section 26.7.7, becomes necessary if the criteria mentioned below are all met across widespread zones of the dam or in continuous layers in the foundation:

- The grading curve lies in the critical zone presented in Figure 26.7 (based on § 26.7.4.1)
- Low density placement (relative density < 0.5) (based on § 26.7.4.2)
- Saturated materials

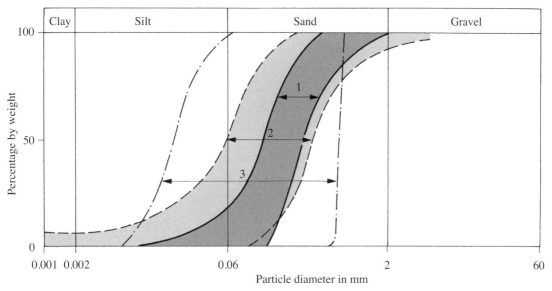

Figure 26.7 The particle size fields for liquefiable soil: (a) Niigata sand; (2) grouping of 10 Japanese sands that liquefied under seismic loading; (3) indicated by laboratory tests by Lee and Focht

26.7.4.1 Grading curve analysis

Figure 26.7 illustrates the types of soil that experience a substantial increase in pore pressure when subjected to cyclical loading in a saturated state. This may lead, in extreme cases, to a complete loss of shear strength. The part of the grading curve between 10 and 90% (% weight) well as the uniformity coefficient in this range are vital.

The uniformity coefficient C_u of a grading curve is defined as being the relationship between the spacing in the mesh of a sieve allowing 60% of particles to pass through and the spacing in the mesh of a sieve allowing 10% of particles to pass through, that is, $C_u = d_{60/d10}$.

An increase in pore pressure needs to be studied in more detail (with the help of STP tests or cyclical tests in a laboratory) if the grading curve of a material lies inside the fields in Figure 26.6 (especially in the field "2" between 10 and 90% and if the uniformity coefficient is lower than approximately 2.

This simplified method cannot be used for category I dams.

26.7.4.2 Analysis of compactness, low compactness

The compactness of the soil is represented by the relative density D_3 and equals:

$$D_r = \frac{n_{max} - n}{n_{max} - n_{min}} = \frac{\gamma_d - \gamma_{dmin}}{\gamma_{dmax} - \gamma_{dmin}} \cdot \frac{\gamma_{dmax}}{\gamma_d},$$

where

n:	porosity of the soil in situ
n_{max}:	porosity at maximum compactness
n_{min}:	porosity at minimum compactness
γ_d:	unit weight of dry soil
$\gamma_{d\,max}$:	unit weight of dry soil at maximum compactness
$\gamma_{d\,min}$:	unit weight of dry soil at minimum compactness

$\gamma_{d\,min}$, as well as $\gamma_{d\,max}$ are calculated according to the standards outlined in USBR 5525, with respect to USBR 5530 of the Bureau of Reclamation (USBR, 1990).

Compactness is low if the relative density D_r is lower than 0.5.

26.7.4.3 Assessing liquefaction potential using SPT tests

Alternatively, empirical correlations between the soil liquefaction potential and penetration test results, such as SPT (Standard Penetration Test) or CPTU can be used to estimate the increase in pore pressure due to an earthquake.

The behavior of sand liquefaction is assessed on the basis of SPT tests in line with Figure 26.8.

26.7.4.4 Calculation of SPT penetration resistance $(N_1)_{60\text{-cs}}$

The number of blows measured during SPT tests is designated by N_{60}.

An initial correlation according to the below equation enables us to obtain $(N_1)_{60}$ (corresponding to penetration resistance with a normal effective vertical stress of 100 kPa):

$$(N_1)_{60} = \frac{N_{60}}{\sqrt{\sigma'_{v0}}},$$

where σ'_{v0}: normal effective stress at the corresponding depth prior to the test.

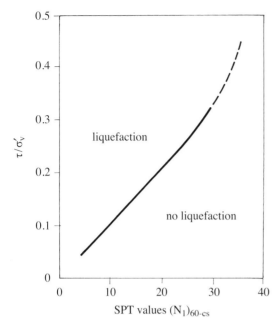

Figure 26.8 Correlation between the liquefaction of in situ sand and standardized SPT penetration resistance $(N_1)_{60\text{-cs}}$ (modified from Seed and Idriss, 1982)

Both N_{60} and $(N_1)_{60}$ are adimensional sizes (number of blows), σ'_{v0} must be introduced in [kg/cm²] in the equation below.

Then, a second correction enables $(N_1)_{60\text{-cs}}$ to be obtained from $(N_1)_{60}$ using the equation below:

$$(N_1)_{60-CS} = (N_1)_{60} + \Delta(N_1)_{60},$$

where $\Delta(N_1)_{60}$ depends on the fines content (percentage < 0.06 mm of the grading curve) and is calculated using the below table:

Fines content in % (fraction < 0.06 mm)	$\Delta(N_1)_{60}$
10	1
25	2
35	3
50	4
75	5

26.7.4.5 Calculating shear stress due to cyclical loading

Cyclical shear stress τ in the foundation

Cyclical shear stress τ is calculated at each depth (Figure 26.9) using the following equation:

$$\tau = 0.65 \cdot \frac{a_h}{g} \cdot \sigma_v \cdot r_d,$$

where

- a_h: peak acceleration according to the (Swiss) seismic hazard map
- g: gravity acceleration
- σ_v: normal vertical stress at depth z
- z: depth
- r_d: reduction factor according to Figure 26.9

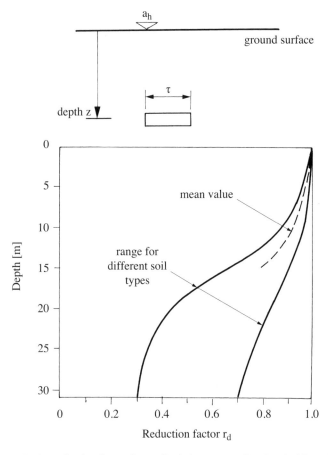

Figure 26.9 Reduction factor for cyclical shear stress (Seed and Idriss, 1982)

Cyclical shear stress τ in the dam body

Similarly, cyclical shear stress τ at the center of gravity of a slip surface is calculated in the profile across the dam according to the following equation:

$$\tau = 0.65 \cdot \frac{a_G}{g} \cdot \sigma_v,$$

where
- a_G: maximum acceleration at the center of gravity (cf. below)
- σ_v: total normal vertical stress at the center of gravity in question

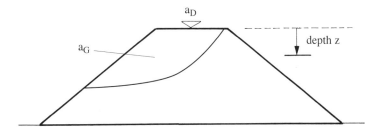

Maximum acceleration a_G at the center of gravity

The value of acceleration a_G is determined using Figure 26.10 depending on the position of the slip surface.

Figure 26.10 enables the relationship between a_G and maximum acceleration at the crest a_D to be determined based on depth y of the sliding block. The field of values in Figure 26.10 was determined by Makdisi and Seed (1978) on the basis of several calculations. The values for the shear modulus G and damping D of these calculations are located in the usual range of values in practice.

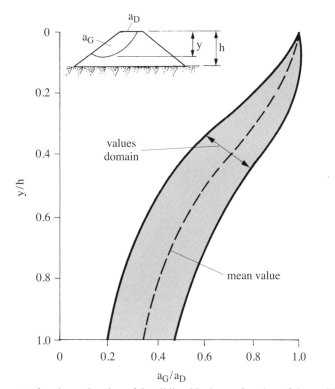

Figure 26.10 Development of peak acceleration of the sliding block as a function of the position of the slip surface (Makdisi and Seed, 1978)

Maximum acceleration a_D at the crest

The maximum value of acceleration at the crest a_D is determined using the following formula (Makdisi and Seed, 1978):

$$a_D = \sqrt{(1.60 \cdot a_1)^2 + (1.06 \cdot a_2)^2 + (0.86 \cdot a_3)^2},$$

a_1, a_2 and a_3 are the spectral acceleration values according to the response spectra for various types of material (with damping of 15%) for the three first self-frequencies ω_1, ω_2, with respect to ω_3.

The influence of local foundation material conditions on loading due to an earthquake must be considered (different response spectra for A, B, and C types of material).

Table 26.11 Estimation of shear wave speed for different types of material

Soil type	v_s [m/s]
Loose material	
Layers of overburden material with low compactness, broken up, unsaturated (depth 3 to 6 m)	110…480
Ballast (sandy gravel), unsaturated	220…450
Ballast, saturated by underground water	400…600
Cemented ballast	1000…1500
Silt from the reservoir bottom, not completely saturated	290…540
Silt from the reservoir bottom, saturated	390…530
Silt from the banks, unsaturated	120…400
Moraine	500…1150
Lœss	150…300
Rock	
Soft marl, broken up	520…1050
Soft molasse sandstone, broken up	520…1050
Marl, not broken up	1000…1900
Molasse sandstone, hard	1100…2200
Molasse from the plateau	600…2500
Schist	1100…3100
Limestone	1800…3700
Gneiss	1900…3500
Granite	2500…3900

ω_1, ω_2, and ω_3 can be calculated using the following formulae:

$$\omega_1 = 2.40 \cdot \frac{v_s}{h} \quad ; \quad \omega_2 = 5.52 \cdot \frac{v_s}{h} \quad ; \quad \omega_3 = 8.65 \cdot \frac{v_s}{h},$$

where
- h: dam height
- v_s: mean speed of the shear wave in the dam materials (Table 26.11).

These values correspond with the self-frequencies of a homogeneous dam on a rigid foundation. Periods T_1 to T_3 corresponding to values ω_1, ω_2 and ω_3 are calculated using:

$$T = \frac{2 \cdot \pi}{\omega}.$$

26.7.5 Simplified analysis of seismic stability

The verification of seismic safety using a simplified stability analysis includes the following calculation steps:
- Calculation of the base period T_0 of the dam in the direction perpendicular to the axis of the dam close to the highest section, according to Figure 26.12
- Calculation of horizontal and vertical substitution seismic forces for various sliding blocks, based on Section 26.7.5.2

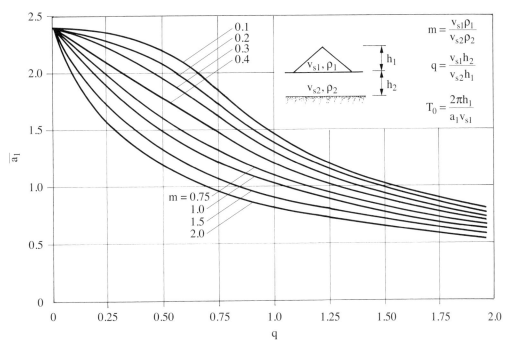

Figure 26.12 Base period T_0 for a dam on an elastic layer of foundation (Sharma, 1979).
v_s: speed of shear wave; ρ: density of materials; $\bar{a}_1$: determined using the figure from the values calculated for m and q; T_0 can finally be calculated.

- Calculation of sliding safety for the selected sliding blocks while taking into account self weight and horizontal and vertical substitution seismic forces, based on Section 26.7.5.3. The least favorable combination of directions of replacement seismic forces is a determining factor in verification.

If the conclusion of the calculation is that the sliding block is not stable, displacement by sliding must be calculated using Section 26.7.6. A sliding block is assumed to be stable if the factor of safety based on Section 26.7.5.3 is greater than 1.0.

26.7.5.1 Determining the dam's base period (perpendicular to the dam axis)

The dam's base period T_0 is estimated using Figure 26.12.
This simplified method cannot be used for category I dams.

26.7.5.2 Calculating substitution seismic forces for a sliding block

The horizontal substitution seismic force E_h for a potential sliding block is calculated using the following equation:

$$E_h = a_G \cdot m,$$

where

a_G: mean acceleration at the center of gravity of the sliding block, according to Figure 26.10
m: mass of the sliding block.

The vertical substitution seismic force E_v is calculated similarly by assuming that dynamic behavior in the vertical direction is approximately rigid. Vertical acceleration in the dam body therefore corresponds with that at the base:

$$E_v = a_v \cdot m,$$

a_v being the vertical component of seismic loading according to a_h.
Peak vertical acceleration can be calculated from the horizontal component a_h with $a_v = 2/3 a_h$.
Acceleration values are determined on the basis of intensity, according to Swiss seismic hazard maps for a given return period (Figure 11.25 and 11.26) through the following transformation (category II: 5,000 years):

$$\log a_h = 0.26 \cdot I_{MSK} + 0.19.$$

26.7.5.3 Calculating safety of a specific sliding block

The factor of safety F for a selected block is calculated using the usual static methods (Section 26.2) while considering the given horizontal and vertical seismic loading (cf. above).
The applicable static methods are, for example, Bishop's slice method, respectively Janbu or other simplified methods.
The contribution of each slice to total strength is calculated using Coulomb's shear strength equation:

$$\tau_f = \sigma' \tan \varphi' + c'.$$

STABILITY ANALYSIS

The factor of safety is then calculated using the following equation on the basis of the shear strength τ_f and stress τ due to seismic loading and the overall static loading:

$$F = \frac{\sum \tau_f}{\sum \tau},$$

where Σ: sum of all the slices along the interface of the potential sliding.

(For category I dams (H > 40 m), this simplified method can only be applied to the verification of the condition after the earthquake.)

26.7.6 Simplified calculation for sliding displacement

Determining sliding displacement using a simplified calculation is carried out using the following steps:

- Calculation of the base period T_0 of the dam (in the direction perpendicular to the axis of the dam) close to the highest section, according to Figure 26.12
- The critical acceleration of a potential slip surface is that which leads to a factor of safety F of 1.0 in the corresponding sliding block

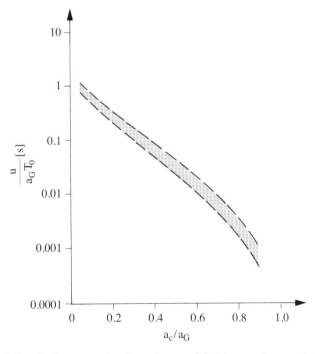

Figure 26.13 Residual sliding displacement of a slip surface, modified in accordance with Makdisi and Seed (1978) a_c: critical acceleration of a slip surface; a_G: according to Figure 26.10; T_0: according to Figure 26.12.

- The factor of sliding safety F is calculated using the usual static methods (for example, those of Bishop or Janbu) while considering the static substitution forces for each slice. The horizontal static substitution force I for a slice is equal to:

$$I = m \cdot a_c$$

where m: mass of the slice
- Calculation of acceleration values a_G (mean acceleration at the center of gravity of the sliding block) according to Figure 26.10 and a_c (critical acceleration)
- Determining total residual sliding displacement u

Total residual sliding displacement u is determined using Figure 26.13.

Comments

According to Makdisi and Seed (1978), total displacement u thus corresponds to the total equivalent deformation along a horizontal slip surface following seismic loading, after critical acceleration leading to the sliding of the block has been surpassed. On the basis of investigations, displacement along an inclined slip surface is approximately 20% greater.

Due to the fact firstly that Figure 26.13 indicates a deformation range and that secondly, the scale is logarithmic, u corresponds in an initial approximation both to the total vertical settlement and to the total horizontal displacement of the sliding block.

For category II dams, determining sliding displacement using a simplified method such as this is only acceptable if the calculated displacement is lower than the following limit values:
- 0.3 m for deep sliding blocks
- 0.15 m for surface sliding blocks

Otherwise, the sliding displacement must be calculated with the development over time of acceleration. This requires a step-by-step calculation over time.

Total displacement along a slip surface is obtained in this case using a double integral of acceleration due to seismic loading exceeding the critical acceleration value (the difference between acceleration due to seismic loading and critical acceleration is a determining factor). The calculation is based on Newmark's sliding block method (1965).[6]

Acceptable displacement is therefore determined as follows:

Objectives	The general aim when determining sliding displacement is to verify the long-term safety of the dam. This implies the following objectives: • Avoiding any overtopping of the dam • Preventing any risk of internal erosion • Ensuring stability of the sliding block in a deformed state

---

[6] See also Section 26.6.

Criteria	These objectives are generally reached if the sliding displacement is lower than the following limit values: • 0.5 m for deep sliding blocks • 0.2 m for surface sliding blocks The depth of the sliding block is determined using one parallel to the dam face: 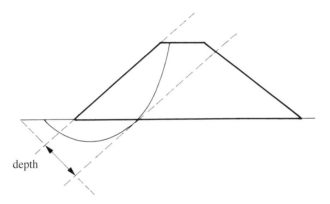 Sliding blocks with a height not greater than 10 ÷ 20% of dam height are considered to be "surface." Otherwise, they are considered to be "deep." For dams with an upstream facing, the entire sealing system must be verified after an earthquake.

Actions	If these criteria are not met, specific investigations are necessary to reach the objectives listed above.
	In particular, the following verifications must be carried out:
	• Even if the embankment dam is deformed, there must be sufficient freeboard to prevent overtopping at any moment.
	• The risk of internal erosion must be excluded by ensuring that the following conditions are met:
	  – the filter criteria between the different zones in the embankment dam are fulfilled
	  – the core of the embankment dam has sufficient self-regenerating overburden material, that is, binding material that can adapt to the required deformation without substantial change in its permeability properties
	  – the residual thickness of filter and drainage layers in a deformed state must be at least equal to half the initial thickness in an undeformed state
	• Stability verification of the sliding block in a deformed state is made on the basis of strength related solely to the residual angle of friction φ'_r.

26.7.7 Analysis of the increase in pore pressure due to an earthquake

The increase in pore pressure due to an earthquake must be analyzed using cyclical testing in a laboratory.

During laboratory tests, the verification earthquake is modeled with a mean shear stress τ_m and an equivalent number of loading cycles.

Mean shear stress at a given depth in the foundation or body of an embankment dam is calculated in accordance with Section 26.7.4.5 (category II and III dams), respectively using a finite element calculation (category I dam).

At least 15 loading cycles should be applied.

The following results are of particular interest:

- Increase in pore pressure due to cyclical loading
- Nonelastic deformation under cyclical loading

26.7.8 Simplified analysis of seismic stability or calculation of sliding displacement by considering the increase in pore pressure due to an earthquake

The process for a simplified analysis of seismic stability and calculation of sliding displacement, taking into account the possible increase in pore pressure due to an earthquake, differs from the same analysis and corresponding calculations not considering an increase in pore pressure resulting only from the presence of pore overpressure due to the earthquake. This overpressure is determined using physical tests in the laboratory. It leads to a reduction in effective stress (based on $\sigma' = \sigma - U$) and, as a result, in shear strength.

Apart from this, the methods described in Sections 26.7.5 and 26.7.6 are applicable without any modification.

26.7.9 Stability analysis after an earthquake while considering an increase in pore pressure due to the earthquake

In addition to the above verifications, sliding safety must also be guaranteed after the earthquake (and, therefore, without inertia forces) while taking into account the persistence of a possible increase in pore pressure due to the earthquake. A sliding block is regarded as sufficiently stable if the factor of safety F, as outlined in Section 26.7.5.3, is greater than 1.2 (the minimum static factor of safety without pore overpressure is 1.5 for normal operation).

26.8 The Processes of Internal Erosion and Their Consequences[7]

26.8.1 Characteristics of the phenomena[8]

Internal erosion is a complex phenomenon that may be the source of major disorders both in the foundation and the embankment dam itself. It has a direct influence on dam safety and can even lead to the dam's total failure. Most incidents concerning internal erosion appear during the first impoundment, but many take place during operation, years after the first impoundment (see Section 7.3, Failure of the Teton embankment dam, USA).

Internal erosion occurs when particles of material in an embankment dam or in its foundation are carried downstream by seepage flow. It begins when the erosive forces caused by hydraulic loading exceed the resistance to erosion of the material in the dam and/or the foundation. The erosive forces are directly related to the water level of the reservoir (ICOLD, 2017). The potential initiation points for internal erosion are illustrated in Figure 26.14.

---

[7] This section is primarily based on the bulletin B164 *Internal Erosion in Dams, Levees and Dikes* published by ICOLD (2017); this bulletin covers the process of internal erosion, dam vulnerability assessments, and protective measures.

[8] Some specific cases have already been noted in previous chapters.

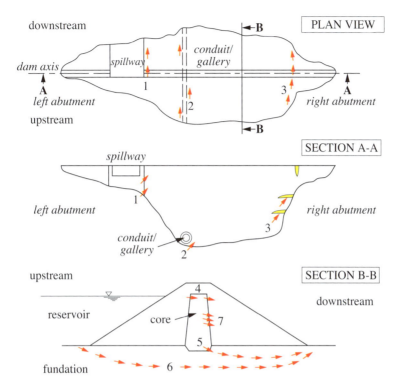

Figure 26.14 Possible locations of initiation of internal erosion (based on Fell and Fry in ICOLD, 2017): 1 spillway wall interface; 2 adjacent to conduit; 3 cracks associated with steep abutment profile; 4 desiccation on top of core; 5 embankment to foundation; 6 foundation (if the foundation is soil or erodible rock); 7 embankment through poorly compacted layer, crack (or erosion in the case of a cohesionless core).

The propagation of this phenomenon takes the four following preferential paths:
- Through the fill
- Through the foundation
- Along or in the conduits crossing the fill
- Touching the adjoining structures

The seepage flow pattern provides an image of seepage paths through the dam and foundations, as well as the distribution of equipotential lines. Flow lines and equipotential lines form an orthogonal network that enables calculation of seepage flow and hydraulic pore pressure at a given point. Determining this network can be done manually or by using computational or analog models. Figure 11.19 provides the image of a network.

Internal erosion can be identified through resurgence, an increase in seepage and leaks, the presence of water in eroded particles, the appearance of surface ground collapse, and settlement. It can be detected during visual inspections, as well as by measuring the discharge of seepage flow and pore pressure (piezometers). Flow can also be detected and localized by measuring temperature with fiber optics (see § 26.8.4, § 32.4.6.4).

26.8.2 Description of the process of internal erosion

The process of internal erosion includes four phases that are illustrated in Figure 26.15; it involves:

- Initiation of the erosion
- Continuation of the erosion
- Progression to form a pipe, surface instability
- Initiation of a breach or failure through sliding

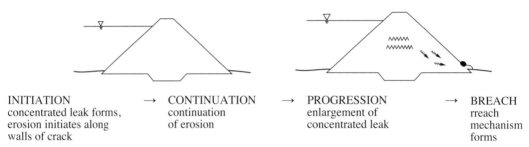

(a) Internal erosion in the embankment initiated by erosion in a concentrated leak

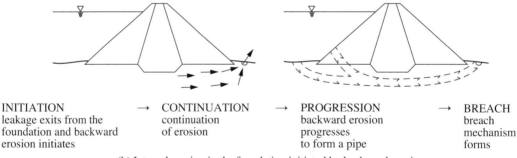

(b) Internal erosion in the foundation initiated by backward erosion

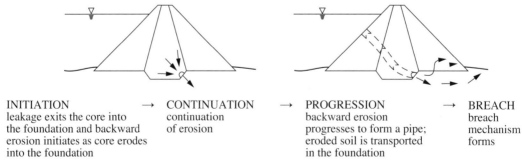

(c) Internal erosion from the embankment to foundation initiated by backward erosion

Figure 26.15 Failure mechanisms through internal erosion (based on Foster and Fell in ICOLD, 2017)

26.8.2.1 Mechanisms involved in the initiation of internal erosion

Piping, backward erosion, contact erosion, and suffusion are the four mechanisms involved in the initiation of internal erosion.

a) Piping

This type of erosion depends on the presence of a crack creating a preferential flow path. This crack (or weak point) may result from differential settlement, a hydraulic fracture, a seismic shock, or desiccation. It appears in highly permeable zones with low compaction is always be located below the level of the reservoir.

Furthermore, piping may also occur due to the presence of animal dens and the effects of vegetation (in addition to obstructing visual observation, roots may block drains or initiate seepage paths due to their decomposition). One particular risk for piping exists in foundations comprising residual soils with preferential paths filled with fine sand resulting from the decomposition of old roots, also known as *canaliculus*.

This process may also begin at the contact with a conduit crossing through the fill (faulty compaction, use of an anti-piping system, crack resulting from differential settlement) and from the detachment of the embankment adjoining a guide wall, due to the movement of this element or to a poorly compacted zone.

b) Backward erosion

Backward erosion is a particular erosion phenomenon that occurs in nonplastic fine soils at a downstream point and continues its progress in the upstream direction. It develops primarily in the foundation, but also occurs inside the embankment. This phenomenon begins at a free surface (e.g., a ditch, the seepage resurgence at the downstream face, or the riverbed at the foot of the dam).

Three different types can be distinguished. Backward erosion in a sandy foundation which generally develops in a downstream direction. Its presence is often identified through the appearance of a sand mound downstream. The total backward erosion of a dam core occurs when the surrounding material cannot hold up the roofs of initial erosion pipes and it collapses. Backward erosion of the downstream face concerns dams built with silt, sand, or gravel, as well as rockfill dams prior to placement of the upstream facing. Particles can often detach from the upstream face under the action of gravity or the seepage gradient at the start of the resurgence zone.

The approaches outlined by Bligh and Lane (see Section. 21.3), as well as those of other researchers are widely used to verify if the seepage path length under dams is sufficient to prevent erosion.[9]

c) Contact erosion

This occurs when coarse material is in contact with a fine material and when a seepage flow parallel to the contact surface in the coarse material erodes the fine material. The particles from the layer of fine material can be detached by the seepage flow and transported through the voids in the layer of coarse material parallel to the interface. The voids must be sufficiently large to allow the passage of fine particles. Moreover, the speed of flow must be sufficient to detach and transport the particles.

---

[9] The ICOLD bulletin 164 (2017) contains many equations that have recently been developed by various authors following research and laboratory tests.

All the interfaces between materials in the dams or their foundation are potential sources of contact erosion. It may be triggered between any particle layer (filter, drain, riprap) and a fine material in contact with this layer.

d) Suffusion

This concerns the transportation of fine particles by seepage through pores in powdery and self-filtering materials. It occurs particularly when water flows through noncohesive soil with a wide or discontinuous range of particle sizes (alluvium, colluvium, core with material from glacial deposits), filters with a wide, discontinuous or excessive range of fines particle size.

Suffusion leads to an increase in permeability, greater seepage velocities, and, depending on the seepage gradient, mass transportation of fines leading to possible clogging or even hydraulic fracturing. Suffusion occurs inside the core or foundation of a dam; it can cause settlement in the embankment (ICOLD, 2017).

26.8.2.2 Continuation or interruption of erosion by filtering

If erosion continues, it can become a real threat for the dam. Zoning is an important factor in preventing erosion. It is possible to interrupt erosion by filtering, where the eroded particles are held back. In most cases, filters are present downstream of the core or in the form of a filtering blanket on the foundation or downstream (see Figure 26.17). In cases of homogeneous dams, erosion can develop due to the absence of a filter. Lastly, some methods for assessing the continuation or interruption of erosion exist.

A filter may have different functions; these are briefly described below (ICOLD, 2017).

a) Retention

The filter stops the transportation of base soil particles; the filter voids are sufficiently small to prevent the erosion of base soil. To guarantee this function, methods corresponding to geometrical and hydraulic criteria are available. As an example, we can cite the method based on particle size, where D_{15F} characterizes the size of voids in the filter and D_{85B} the size of the base soil (Terzaghi and Peck, 1965; Sherard and Dunniga, 1989).

b) Self-filtering or stability

The filter has its own internal stability, and its particles are not subject to erosion. Particle size instability (the possibility that the fines are transported under the effect of flow) may be avoided.

c) Lack of cohesion

The filter is powdery and fills up the cracks. It does not hold the cracks open.

d) Drainage

The filter is sufficiently permeable to transport water flowing through the base soil toward the drainage layers or the toe of the dam; the proportion of fines D_{15F}/D_{85B} is limited to ensure that the filter is permeable enough.

e) Strength

The filter transfers the stress to the interior of the dam without being crushed and therefore without becoming finer.

26.8.2.3 Development of erosion

Erosion that is not stopped by the action of a filter can develop if the flow has enough speed to transport particles continuously and carry them out of the dam. Development stops if the flow is limited by a head loss between the upstream and downstream zones (hydraulic condition). Furthermore, erosion develops if the flow through the dam continues due to the fact that the cracks are maintained open and that the pipe or cavity through which the eroded particles travel does not collapse (mechanic condition). Erosion will develop in the absence of clogging (filter obstruction), flow limitation, or reservoir drawdown (no reduction in the flow gradient).

At this stage, the question to be asked is whether the dam may fail or stabilize itself. An enlargement of the pipe, an overflow, a sliding or unraveling of the downstream slope, or even a downstream static liquefaction in the embankment itself or a strong resurgence at the toe with liquefaction of the foundation followed by deep sliding, may lead to the formation of a breach. Failure during impoundment can occur very quickly as the critical threshold may be reached and exceeded during the initial rise in water level. For existing dams that have already withstood high water levels, extreme levels (e.g., during a 10,000-year flood) may exceed the critical threshold and lead to rapid failure. Figure 26.16 illustrates some of these failure mechanisms.

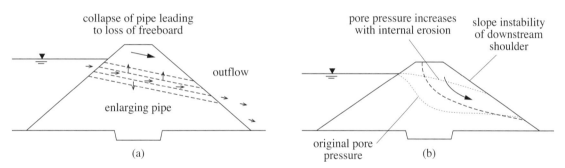

Potential breach (failure) phenomena–pipe enlargement and slope instability

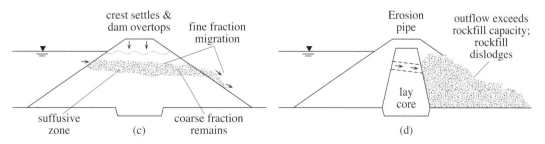

Potential breach (failure) phenomena–overtopping by settlement and unraveling of the downstream face

Figure 26.16 Types of potential breaches (based on Fell and Fry in ICOLD, 2017): (a) enlargement of the pipe; (b) breach due to slope instability; (c) breach through overflow due to settlement; (d) breach due to unraveling of the downstream slope

26.8.3 Repair and available methods

To avoid problems resulting from internal erosion, two approaches can be applied as preventive solutions:
- Barriers to prevent flow through the foundation and dam
- Filters to intercept eroded particles

In the foundation, use of a cut-off curtain is one approach that has several options: sheet piles, diaphragm wall, secant piles, grouting (Chapters 29 and 30). It is vital that the cut-off be total, watertight, and continuous. As they require binding with the impermeable layers, these curtains or walls may go down a substantial depth (sometimes over 100 m). The interconnection with the impermeable zones of the banks must also be ensured (Figure 29.12 (a) and (b)).

Different types of filters can be envisaged: filtering cut-off wall and filtering shell (for dams without a filter or with an inadequate filter), drainage walls (filter installed in a trench), filtering decompression shafts (to unload pore pressure at great depths or under impermeable layers), peripheral pipe and spillway filters, multilayer and drainage capacity filters (Figure 26.17).

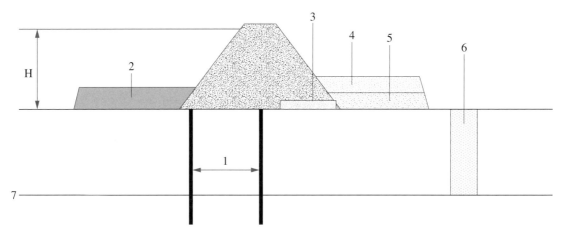

Figure 26.17 Layout of various approaches for reducing the risk of internal erosion:
1 impermeable vertical water barrier; 2 impermeable material blanket; 3 drainage blanket; 4 reloading;
5 filter; 6 discharge well; 7 bedrock, i.e., impermeable rock

26.8.4 Surveillance and monitoring

A phenomenon of internal erosion may appear quite quickly and lead to failure. Most accidents occur when the water level is high. This is enough reason, as with any water-retaining structure, to ensure, first of all, intensive surveillance, especially through visual inspections, and secondly, a monitoring system to highlight specific behavior. An action list must be established, and depending on the situation, may be more or less intensive.

The organization of the implemented surveillance measures must ensure that the initialization process of internal erosion is identified and that the appropriate actions are undertaken. Depending on the observations, several interventions are possible:
- Lower the reservoir water level
- Install discharge well (foundation and embankment)

- Construct a reverse filter
- Install a shell to reduce floating or sliding
- Place granular material on the upstream face

Visual inspection, however, remains essential. This includes the observation of any increase in seepage flow or the presence of dirty water on the downstream face of the dam or in the foundation, as well as cracks in the embankment and downstream of the dam. However, it is important to note that in a normal situation, monitoring is difficult during the night and at weekends. The presence of vegetation can hide the existence of leaks or, if it appears rapidly in localized spots, it can indicate leaks.

Given that the phenomenon of internal erosion already occurs during the first impoundment of the reservoir, a highly experienced team must be permanently present on site for surveillance.

As for measuring equipment, this must enable water pressure to be recorded (piezometers) in the foundation and embankment, which will then be correlated with the level of the reservoir. Settlement and deformation can be observed through geodesic measurements. Fiber-optic temperature sensors distributed in the dam enable water circulation to be monitored.

Specific features in the context of surveillance and detection

If the development time for the initiation process from the first appearance of pipe erosion (piping) to the breach is 6 hours and the dam is only inspected or monitored weekly, it is highly unlikely that the pipe erosion incident will be detected before the breach occurs. However, if the dam can be seen by members of the public, it is possible that the leak will be detected before this.

During the first impoundment of the reservoir, permanent surveillance must be ensured by an experienced team.

Rapid detection of the process of internal erosion is generally difficult, especially for erosion beginning in cracks or for backward erosion, because the seepage flow is initially very low. It has been noted that most piping incidents, initiated by pipe erosion, were detected during the development phase. Suffusion is easier to detect using piezometers, as the development process is slower. The presence of conditions potentially leading to uplift and backward erosion in the foundation may also be detected by correctly positioned piezometers and read as a function of the increase in the reservoir water level. Abrupt settlement and even slight drops in the piezometric level during first impoundment may be a warning sign of the development of internal erosion. Resurgence of water at the downstream toe, even slight, is also a worrying sign.

27. Structural Details

Upper Gotvand earthfill dam with a vertical core in Iran, height 180 m, under construction (Courtesy A. Schleiss)

27.1 Determining Freeboard

27.1.1 Definition of freeboard

The maximum level of operation for a reservoir is established on the basis of economic considerations. Very often an upper limit is imposed by the topography or geology. Due to the vulnerability of an embankment dam to overtopping, it is vital to include a freeboard in the design, which, by definition, is the vertical measured distance between the maximum operating level and the crest level (after consolidation) (Figure 10.36; Table 10.38). Determination of the freeboard must take into account a reasonable combination of the various following effects:

- A rise in water level during the maximum flood (safety flood) considered for the design (see § 10.6.2 and § 10.6.3)
- The malfunctioning of mobile water-release structures (gates)
- The maximum height of waves caused by the extreme wind considered
- Wave run-up on the surface of the upstream face
- The rise in water level caused by the extreme wind considered (wind set-up)
- Additional safety requirements (for example, for waves caused by landslides, avalanches, glacier break-up, earthquakes) (see § 5.4.2)

27.1.2 Effects of wind and waves

27.1.2.1 Height of waves

Wind of a certain duration and intensity causes waves of various heights on the surface of a reservoir.

Molitor's equation may be mentioned among the many empirical formulae based on tests or in situ observations (Figure 27.1):

$$F < 30 \text{ km}: h_{wave} = 0.76 + 0.032 \cdot (v_v \cdot F)^{1/2} - 0.26 \cdot F^{1/ \cdot}$$

$$F > 30 \text{ km}: h_{wave} = 0.032 \cdot (v_v \cdot F)^{1/2},$$

where
- h_{wave}: wave height [m]
- v_v: wind speed [km/h]
- F: fetch [km]

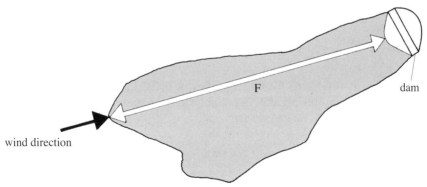

Figure 27.1 Definition of fetch

27.1.2.2 Wave run-up

The height R of wave run-up depends not only on the height of the wave h, but also on the wave length, the slope n of the surface, and the type of surface on the upstream face (Figure 17.2).

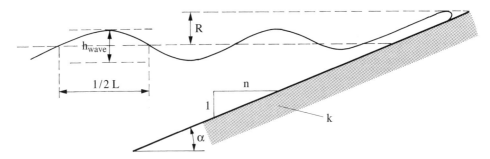

Figure 27.2 Wave run-up

R can be estimated by using the Kâlal formula

$$R_{wave} = 3.2 \cdot h_{wave} \cdot k \cdot \frac{1}{n} \qquad \frac{1}{n} = \tan \alpha,$$

where
- $k = 0.72$ riprap
- $k = 1.00$ paved
- $k = 1.25$ paved with blocks of prefabricated concrete
- $k = 1.40$ smooth surfaces

Should there be a surface protection with blocks (riprap), the USBR (USBR, 1981) proposes the following equation:

$$R_{wave} = \frac{h_{wave}}{0.4 + (h_{wave}/L_{wave})^{0.5} \cot \alpha} \quad \text{(applicable for } \cot \alpha < 5\text{),}$$

where $L_{wave} = 1.56\, T^2$ and $T = 0.556 \cdot v_v^{0.41} \cdot F^{1/3}$; T in [s]; v_v in [m/s]; L_{wave} in [m]; F in [m] and R_{wave} in [m].

For inclined surfaces, R_{wave} only depends to a slight degree on the wave length. However, for vertical faces, the wave length must be taken into account:

$$R_{wave} = h_v \left(1.5 + \frac{\pi \cdot h_{wave}^3}{32 \cdot L_{wave}}\right),$$

where L_{wave} = wave length = $0.152 \cdot v_v \cdot F^{1/2}$; L_{wave}, h_{wave}, R_{wave} in [m]; v_v in [km/h]; F in [km].

27.1.2.3 Wind set-up

The rise in water level caused by extreme wind is very low and does not exceed 5 to 10 cm, even for very large reservoirs.

27.1.2.4 Waves caused by landslides, rockfalls, avalanches, glacier break-up, etc.

Impulse waves result from unstable masses (avalanches, rockfalls, glacier falls) falling into the lake behind a natural or artificial retaining structure. The masses set in motion and their speed of impact on these bodies of water may be considerable. They are capable of causing dangerous waves that can reach a dam crest and damage it or overtop it. Wave height can be estimated through special calculations or laboratory tests (Figures 27.3 and 27.4).

Figure 27.3 The effect of a rock mass arriving into a body of water

To assess the risks, the first step is to determine the unstable slopes and zones, quantify the volume in question, and define their possible trajectory in the case of a slip. This step is done with the participation of geological, geotechnical, and snow experts.

Scale-models tests are a simple way of determining the extent and process of the decline in impulse waves, as well as their impact along the banks. Test such as these have been carried out by the Laboratory of Hydraulics, Hydrology and Glaciology at the ETH Zurich (VAW/ETHZ).[1] These tests enabled the laboratory to develop a calculation process to assess "impulse waves generated by a landslide into reservoirs" (Heller et al., 2009).

---

[1] Fuchs (2013), Heller et al. (2009), Heller (2008).

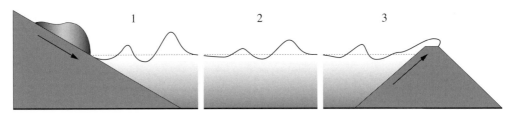

Figure 27.4 Illustration of the propagation of impulse waves on a body of water (from Heller et al., 2009): 1 slide impact with wave generation; 2 wave propagation with wave transformation; 3 impact and run-up of the impulse wave

After an avalanche fell into a reservoir and the wave reached the dam crest, the Swiss supervising authority set winter restrictions for winter water level from early December (possibly November) to late April for various large dams in the Alps.

27.1.3 Required freeboard

Swiss guidelines have established the danger level that may not be exceeded during a safety flood. This level is equal to the crest for homogeneous embankment dams or the upper level of the sealing system (clay core, membrane). The amount of freeboard required is determined in relation to the theoretical crest, that is, its level after consolidation and after exceptional settlement due to an earthquake (see § 25.4.1).

The following criteria are given as an example and are taken from Minor (1988). They are essentially based on hydraulic considerations.

Freeboard criteria for embankment dams with a core (Figure 27.5)

- Design flood HQ_{1000} (thousand-year flood)
    - rule n-1 for water-release structures (gates)
    - wave height and run-up are determined for a wind with a return period of 100 years
    - these heights are added to the rise in water level in the case of a 1,000-year flood
    - minimum freeboard is $f_D \geq 2.0$ m
- 10,000-year flood $HQ_{10,000}$
    - all water-release systems are assumed to be in use (gates)
    - wave height and run-up are determined for a wind with a return period of 10 years

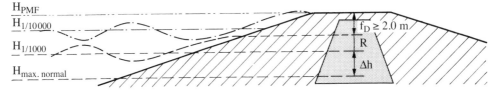

Figure 27.5 Freeboard required for embankment dams with a core (from Minor, 1988)

STRUCTURAL DETAILS 613

- these heights are added to the rise in water level in the case of a 10,000-year flood
- water level (with waves) must be lower than the crest
- water level (without waves) must be lower than the crown of the core
• Safety flood (deluge) (PMF)
  - water level (without waves) may reach the crest
  - wave run-up is assumed

Freeboard criteria for embankment dams with upstream facing (Figure 27.6)
• Design flood HQ_{1000} (thousand-year flood)
  - as with the embankment dam with a core but $f_D > 1.5$ m is sufficient.
• 10,000-year flood $HQ_{10,000}$
  - all water-release systems are assumed to be in use (gates)
  - wave height and run-up are determined for a wind with a return period of 100 years
  - the increase in water level from a 10,000-year flood combined with waves must not exceed the crest; a wave wall of max. 1.5 m in height can be used against the effect of waves, however, the level without waves must not be higher than the crest.
• Safety flood (deluge) (PMF)
  - water level (without waves) must be lower than the crest
  - the run-up of waves is assumed

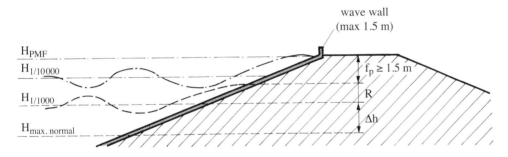

Figure 27.6 Freeboard required for embankment dams with upstream facing (from Minor, 1988)

27.2 Crest

The dam crest is particularly important for the structure and requires special care (Figure 27.7). The following points must be considered:
• There is limited space for the various zones, which often means a reduction in the thickness of filters, drains, and transition zones
• The crest width affects the total volume of the dam and also the core width at the crown
• During an earthquake, acceleration is maximum at the crest
• Deformation after consolidation and during operation are maximum (zones with high risk of cracking)

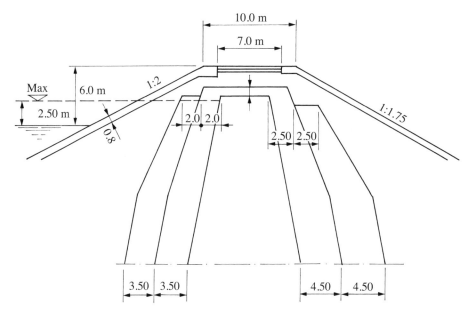

Figure 27.7 Example of the crest zone of a dam with a central core

27.3 Berms

Designing berms on the upstream and downstream faces is always recommended. This allows the passage of trucks. The minimum width is 3.5 to 4.0 m. Berms facilitate monitoring and protection of the surface. They also prevent small local slips. The stability calculation can be based on the average slope.

27.4 Grouting and Monitoring Galleries

For large embankment dams, it is always wise to include grouting and monitoring galleries along the foundation in the area of contact with the sealing system (see also Section 20.6).

These present the following advantages:

- Grouting works can take place independently of fill works
- Possibility of monitoring post-construction
- Possibility of rehabilitation by additional grouting
- Enables high-pressure grouting to be carried out thanks to the overload of the fill
- Drainage holes can be drilled independently of the fill
- Monitoring during dam operation is possible

Galleries can be excavated either into the rock below the foundation or constructed in concrete in a trench or excavation founded on rock. Galleries excavated within the rock are more costly but can be done at the same time as the fill (Figure 27.8).

For a concrete gallery, the interface between the core and the concrete structure is delicate and must be carried out with great care, according to current procedures and practices.

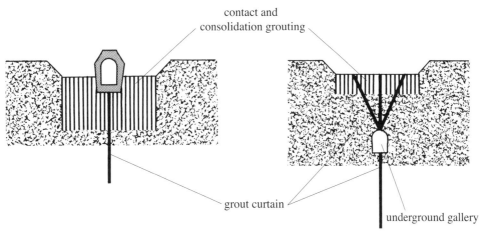

Figure 27.8 Grouting and monitoring galleries

27.5 Contact between Sealing Elements and the Subsurface

The connection between the subsurface and sealing elements such as a central core or an upstream facing is always a complex operation, as high seepage gradients or flow must be avoided in the interface zone (Figures 27.9 and 27.10).

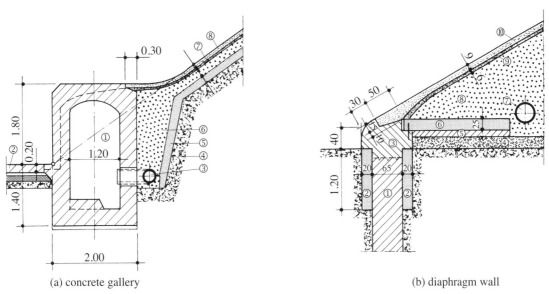

(a) concrete gallery (b) diaphragm wall

Figure 27.9 Interface between sealing elements and the subsurface for a dam with an asphalt upstream facing. (a) Val d'Ambra, connection between the upstream facing and the lining at the bottom of the compensation basin: ① monitoring and drainage gallery; ② concrete lining; ③ drain, diameter 200 mm; ④ moraine; ⑤ filtering layer; ⑥ drainage layer; ⑦asphalt layer; ⑧watertight asphalt lining. (b) Godey, connecting structure between the diaphragm wall and the upstream facing: ① diaphragm wall; ② guiding walls; ③head of the diaphragm wall; ④ leveling layer ⑤ leveling concrete; ⑥ transition slab; ⑦drain, diameter 400 mm; ⑧ drainage layer; ⑨porous asphalt; ⑩ watertight facing.

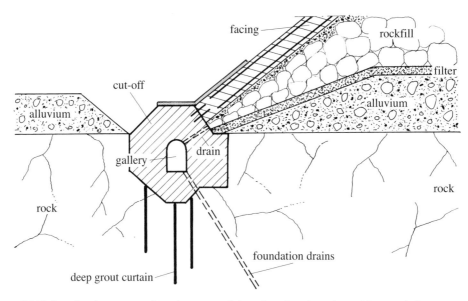

Figure 27.10 Interface between sealing elements and the subsurface for a dam with an asphalt upstream facing

Where possible, the earth core must have its foundation directly on a layer of rock. If the area of loose overburden material is not too large (< 30 m), it can be removed from the core zone (see also Figure 20.10 (b)).

To guarantee a faultless contact, the rock surface must be clean and as regular as possible (see also § 10.2.4, Figures 10.16 and 10.17, Figure 28.1). Overhangs, bumps with sharp angles, and steps are to be avoided or removed. Holes can be filled with concrete sealing. It should be noted that irregular contact surfaces can cause local cracking in the core near the rock, thus forming preferential seepage paths. This could in turn lead to internal erosion of the core.

The rock next to the contact point must be sound. Cracks and joints must be treated with contact and consolidation grouting. This grouting covers the entire width of the core and guarantees the connection with a deep cut-off curtain.

27.6 Slope Protection

Depending on the nature of the material in the dam shell, the face of the upstream facing must be protected against the erosion that large waves can cause. The type of protection depends on the operation regime, the size and shape of the reservoir, the climate, and wind. The protection zone must cover at least the drawdown zone. The downstream face must be protected against damage that may be caused by the effect of wind and the flow of rainwater.

27.6.1 Protection of the upstream slope

The upstream slope can be covered by a layer of rockfill or riprap (Figure 27.11). Riprap is often preferred due to its durability and lower cost. The roughness and porosity of rocks that are dumped are very effective at dissipating wave energy and alleviating the wave run-up. Riprap stability must ensure stability against sliding on its supporting structure and resistance against displacements of blocks.

The effectiveness of riprap depends on the quality of the rock, its size or the block size (it must not shift under the action of waves), its shape, the thickness of the layer, the incline of the slope, its stability, and the effectiveness of the filter. Riprap is made up of blocks of hard, dense, and resistant rocks. Riprap is a coarse, angular material resulting from the crushing process or blasting. Igneous and metamorphic rocks can be used, and possibly limestone. The material may also come from talus. It is dumped or placed manually. Dumped riprap is preferable as the blocks of rock find their own position and maintenance costs are low.

Rock size and layer depth are a question of experience. The thickness of the layer can be from 75 to 90 cm. The riprap is placed on connecting layers that fulfil filter criteria.

Lastly, riprap rock placed manually, in a thickness of 30 cm minimum, must be of high quality. A contact filter must also be included.

Figure 27.11 Protection of the upstream and downstream slopes for an embankment dam in Norway
(Courtesy A. Schleiss)

27.6.2 Protection of the downstream slope

It is important to ensure that the downstream slope does not become covered with dense vegetation (trees, shrubs, and thick bushes) that compromises visual observation. Trees can cause major damage to slopes should they be ripped out in high winds.

It is important to be able to identify deformation in the dam body, especially damp zones and areas of seepage. If necessary, action must be taken to prevent roots from disturbing drainage and sealing systems, so as not to negatively affect slope stability (an overprofile may be implemented). It is possible to cover the downstream slope with a layer of topsoil with grass. Animals should also be prevented from digging holes and tunnels and thus creating preferential seepage paths.

Vegetation is only permitted for plants with short root systems. They should be arranged sparsely, as long as a fill overprofile is present (no roots in the static profile).

27.6.3 Construction near to fill

Slopes, the crest, and any ground in the immediate vicinity of the embankment structure (a 10-meter strip) must remain free of any construction that may hinder the correct implementation of the necessary visual checks and measurements (particularly geodesic measurements) (WRFA, 2010).

VII. TREATMENT OF FOUNDATIONS

The foundations of water-retaining structures are an essential element, as they must bear the substantial forces that are transferred to them.

Chapter 28 gives guidance on the execution of excavation works, which is the first phase to be undertaken. Most importantly, these must be carried out in such a way as to create an appropriate base for the dam.

Due to the massive geological structure of the foundation, the presence of seepage and uplift cannot be excluded, the intensity of which will vary depending on fluctuations in reservoir water level. To offset these, grouting is employed, which is the subject of Chapter 29, as well as a drainage system, described in Chapter 30. Beneath the dam, and extending out to the banks, a grout curtain is created by injecting grout into boreholes, which lengthens seepage paths, therefore reducing seepage. Drainage is one way of reducing uplift.

Grouting works take place only after tests to determine the best spacing between the boreholes and the most effective grout.

There are other types of works required for foundation treatment, for example, the installation of a diaphragm wall to act as a core wall or the installation of anchors to strengthen the abutments.

28. Excavations

Rossens arch dam in Switzerland, height 83 m, year commissioned 1948 (Courtesy © Swiss Air Force)

Excavations are an important stage in the construction works. They must be carried out with great care to prepare an appropriate base for the dam by going down to sound rock. In addition, overhangs must be removed and slope irregularities corrected to prepare a foundation surface with a regular topography (Figures 10.16 and 28.1). Lastly, any unstable elements present at the surface must be removed.

The first thing to do is clear the dam zone and remove trees and bushes, overburden, and organic soils, as well as any detritus and debris.

It is also important to develop a blasting plan and use any explosives with care. Near the final excavation area, only small quantities of explosive are used so as not to break up the rock. The risk becomes greater when the rock is layered and fractured. If, at a given depth, the bedrock is mediocre, excavation must go below this depth. Open discontinuities at the surface (for example, open joints, mylonite faults) must be cleaned, possibly by hand depending on their size, and then sealed with a concrete filling. The same applies for cavities. In the case of the Grande Dixence dam, a canyon discovered once excavation had begun was filled with concrete.

The presplitting method can be used on some compact rocks. This involves prior creation of a separation plane between the mass to be excavated and that to remain in place. A series of parallel blast holes is detonated simultaneously. The abutment surface for the future concrete is then highly regular and has few fragments.

Water from any springs is collected and piped, if possible by gravity, outside the excavation zone. Prior to concreting, the rock surface must be thoroughly cleaned with water and/or compressed air.

All debris must be removed. It is also important to carry out a geological survey of the excavation footprint, supplemented by photographs. This precious information should be filed in a geology file for reference.

For worker safety, nets to catch falling rocks must be set up, and any slopes liable to slip must be monitored. If required, any unstable blocks of rock should be fastened by anchors or bolts. The walls can also be protected by shotcrete.

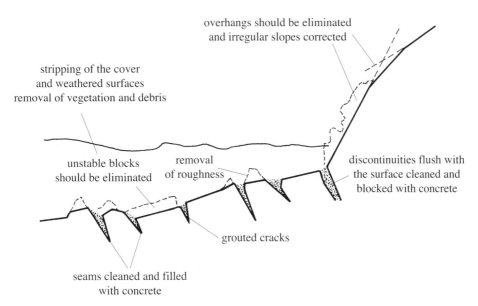

Figure 28.1 Treatment of the foundation surface prior to concreting (from Jansen, 1988)

29. Grouting of Rock Foundations

Vieux-Emosson arch dam in Switzerland prior to heightening (height 45 m, year commissioned 1955)
(Courtesy Straub AG)

During the operation of a water-retaining dam, the geological structure of the rock mass in the foundation (faults, diaclases, cracks, joints) largely determines the development of seepage and uplift. Their intensities may vary depending on fluctuations in the reservoir water level, but may also tend to increase or decrease, which means that the situation in the subsurface may change over time. For example, a decrease in seepage may indicate a blockage in seepage paths, with a simultaneous increase in uplift.

Grouting and drainage are appropriate measures for combating seepage and uplift. With grouting, a grout mix is injected into a more or less permeable zone through a borehole. The grout mix is a liquid or suspension that hardens. Depending on the quality of the subsurface, a grout cut-off should be installed beneath the dam, and should, as a general rule, continue out to the banks. Care should be taken that the grout cut-off is able to continue functioning adequately over the long term. In addition, the upper part of the foundation must be treated with contact and consolidation grouting.

The design plans for the grout cut-off must include the choice of grout mix, the spacing and depth of boreholes, the grouting sequences, the grouting process, and monitoring during construction.

A drainage system (galleries, drainage holes) is also a vital and efficient element in the treatment of foundations, which effectively strengthens safety in relation to uplift (see Section 30.1). This is not always the case when grouting alone is carried out. If the initial permeability of the ground is low, uplift develops in a similar way as to permeable ground. Efforts should therefore be made in terms of drainage, as a grout cut-off will have little impact. In permeable ground, drainage on its own could lead to high seepage flow; in this case, the cut-off is capable of reducing ground permeability (Sabarly, 1968). These are extreme cases, and many intermediary situations may well occur.

The drainage system also allows the bearing capacity of the cut-off to be monitored.

29.1 Objectives

The aim of grouting is to strengthen a cracked rock mass or make it watertight. Pressurized grout fills the voids by being sent through boreholes that cross through the cracks and discontinuities in the rock mass. Installing boreholes requires in-depth and additional knowledge of the subsurface, particularly regarding discontinuities and permeabilities.

By installing a deep cut-off, grouting notably allows:

- Seepage flow to be reduced across the dam foundation
- Seepage paths to be lengthened
- Uplift to be reduced (in association with drainage holes)
- Shallow foundation deformation to be reduced, thus limiting differential deformation
- An increase in the strength of the rock mass

29.2 Geological Knowledge

In order to establish an appropriate program and optimize the grouting process, good knowledge of the site's geological conditions is imperative. The structure of the subsurface and the various characteristics of the rocks present will dictate the method and extent of grouting treatment of the rock. Some rocks are highly fractured and highly permeable. Joints, too, create preferential seepage paths.

All discontinuities (joints, cracks, faults, and other accidents) should be recorded. The various discontinuity systems and their orientation can be established by drawing up a polar diagram. On this basis, the planning of grout boreholes can then best be established by seeking to intercept a maximum of vertical or inclined cracks. Furthermore, it is important to complete information on whether to fill discontinuities with clay or a similar material and the frequency of cracks, as well as their degree of opening (large or narrow). The RQD factor gives a good indication of the quality of the rock mass (see § 10.2.3).

Rock permeability is determined by using Lugeon tests (see § 10.2.3). Note that as the drilling progresses, a double packer is used to carry out a water test every 5 m at a pressure of 10 bar (1 MPa). The unit of measurement is defined as per Lugeon:

$$1 \text{ Lugeon} = 1 \text{ [l/min} \cdot \text{m] at 10 bar.}$$

According to Darcy, an approximate and simple relationship is established between the Lugeon values and permeability:

$$1 \text{ Lugeon} \approx 1.7 \cdot 10^{-7} \text{ [m/s]}.$$

Based on the Lugeon test results, the rock can be roughly qualified as follows:

- 0 to 1 Lugeon   excellent rock
- 1 to 5 Lugeon   good rock
- 5 to 10 Lugeon   average rock
- > 10 Lugeon   mediocre rock

In the so-called "traditional" method, the decision to carry out grouting works was made as soon as a certain set, a priori value of the Lugeon test had been exceeded. A sector was considered impermeable if the Lugeon unit was equal to or lower than 1; values between 2 and 5 Lugeon were also accepted (CNSGB, 1992). Ewert (1992) notes that a Lugeon value lower than 5 Lugeon corresponds to a sufficiently impermeable and not easily groutable rock while a value greater than 20 Lugeon indicates a permeable and easily groutable rock.[1]

However, it should be noted that this criterion is far from sufficient, because, as has been mentioned, the importance of the foundation's geological nature should not be overlooked while developing the grouting design. The type of dam structure should also be considered.

Ultimately, only in situ tests can provide reliable guidance regarding how groutable the rock is (see § 29.5.8).

In current grouting design, it is agreed that the Lugeon test does not allow a numerical relationship to be established between the values issuing from the test and the volume of grout that can be absorbed during grouting. As such, water tests can be avoided during the works. The tests do, however, provide a general overview of the permeability of foundations prior to treatment, and at the end of the works they give an indication regarding the reduction in permeability. Lastly, it is important to check the groutability of the rock through direct grouting tests.

29.3 Construction Method for Rock Grouting

29.3.1 Drilling methods

Rotary and percussion are the most common means for drilling holes into rock.

---

[1] Some publications (Lombardi, 1996, 2003; Ewert, 1992; Balissat, 1985) provide interesting considerations on water tests.

Rotary drilling uses a metal or diamond bit mounted on the bottom of a drill string and powered by rotation. It requires rinsing with water and leaves a hole that is suitable for grout injection. Core samples can be taken with a single or double corer.

Figure 29.1 illustrates some examples of bits and corers used for drilling.

Percussion drilling (using a cutting tool at the head, high rotation speeds, and intense washing) breaks the rock into fine particles that are taken back up to the surface in a fluid (mud, air, or water). Circulation of the fluid is said to be direct when the fluid is injected inside the drill string, sending the material up the annulus between the casing and the string; it is reverse when the fluid goes down this annulus and sends the material back up the inside of the drill string. Washing the borehole to eliminate any residual debris is recommended. This results in a hole that is ready for grouting in hard rock; in soft rock, the mud and other debris that is hard to remove with washing may block some of the cracks.

The choice of hole diameter depends on the type and quality of rock. It normally ranges from 38 to 76 millimeters. Small diameters can be employed for hard rock, while a larger diameter is necessary for drilling in poor-quality rock.

Compared to rotary drilling, percussion drilling is less costly and quicker.

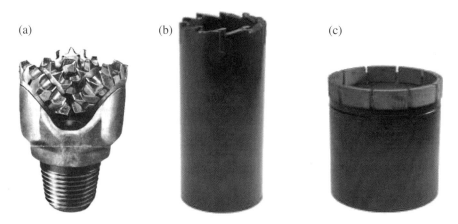

Figure 29.1 Examples of bits and corers: (a) a rotary drilling tricone; (b) metal bit; (c) diamond bit

29.3.2 Grouting method

Grouting in rock mass may be either downstage or upstage. Upstage grouting in sections involves drilling the hole along the total length. The rock is then grouted in intervals going up from bottom to top, using a single, inflatable packer. This method can be difficult if the quality of the hole wall does not allow adequate placement of the packer. Downstage grouting in intervals involves drilling an interval that is immediately grouted, also with a packer. The interval that has just been grouted must be redrilled before continuing down to the next interval. This method has the advantage of creating a cap that allows increasing pressure when grouting lower intervals. In highly cracked rock mass, downstage grouting is preferred to avoid grout emerging at the surface, especially if the crack network is diagonal compared to the drilling axis. This is a relatively costly option as it requires each grouted interval to be redrilled.

630   TREATMENT OF FOUNDATIONS

In both these methods, the length of the grout interval may vary depending on the nature of the terrain to be grouted. The length may range from 3 to 8 m; a standard length of 5 m is generally used. A set value should not be strictly applied along the entire height of the grout borehole or cut-off. Prior washing of the hole, before grouting, is generally recommended. This operation saturates the rock mass and furthermore ensures the presence of water for hydrating the particles of cement grout.

Figure 29.2 illustrates the two types of grouting procedures.

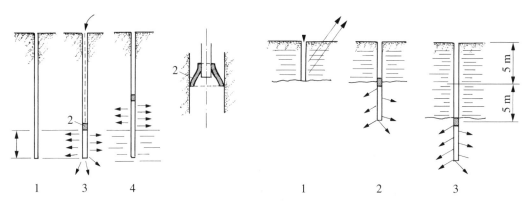

Rock grouting from the bottom to the top
1. Drilling of the hole
2. Packer with flexible walls that are pushed against those of the borehole by internal pressure
3. Grouting of the lower interval
4. Grouting of the next interval, etc.

Rock grouting from the top to the bottom
1. Drilling by 5 m intervals
2. Grouting of this 1st interval (if necessary, under very low pressure) to form a cap, then drilling of the following 5 m interval
3. Grouting of the 2nd interval, then drilling of the following 5 m interval

Figure 29.2 Illustration of the two grouting methods in foundation rock

29.3.3 Types of grouting material

A certain number of materials may be used for consolidating cracked rock mass, including cement grout (with or without the addition of fly ash, fillers, or silica), resins, silicates, and other chemical products (Lombardi, 2003).

Cement grout is a mix of cement and water, whose proportions are defined by the weight ratio W/C. Various types of cement, differentiated by their fineness and chemical resistance, may be used. Moreover, using small amounts of additives (admixtures or chemical products) enables its properties to be improved. To ensure high-quality grout, its selection must be based on laboratory tests on different cements and mixes, with or without the use of additives (Bremen, 1997). Flow, sedimentation, setting time, and mechanical properties are all measured, among other things.

There are two types of grout:

- Unstable grout (cement, water)
- Stable grout (cement, clay, water, and admixture)

29.3.3.1 Unstable grout

Unstable grout is made from a combination of water and cement whose weight ratio may vary during grouting. The cement particles remain suspended in water during the initial phase, before quickly separating from the water. One to two per cent bentonite can be added to the grout to stabilize the mix, reduce sedimentation, and absorb any excess water.

The sealing of cracks is obtained by the hydraulic filling of cement particles. The amount of pressure required for grouting can be very high. It is advisable to always begin the traditional method by using a fluid grout. Grouting takes place in intervals with increasingly thick grout. Grouting ends with a dense grout. This method assumes that a fluid grout with low cohesion will penetrate fine cracks better than a dense grout. There are different schools of thought: some recommend beginning with a grout with a 5:1 W/C ratio (or even 10:1). However, current practice does not exceed a ratio of 3:1.

The quantity of water will have an impact on mechanical strength and the chemical durability of the grout.

29.3.3.2 Stable grout

Stable grout is a dense grout with limited settling, whose final mechanical properties and resistance to washout are high.

Lombardi (2003) recommends the systematic use of this type of grout, as *"due to the good results obtained on many dam sites, there is no doubt that stable grout should be preferred. In order to simplify the process, one single type of grout should be used for each dam (or part of the dam) taken individually. This is the most appropriate grout, or the inherently best."*

Lombardi and Deere (1993) note that a mix with a W/C weight ratio between 0.67:1 and 0.8:1 results in the required densities and mechanical strength of hard grout. Furthermore, the quantity of water is sufficient to ensure hydration of the cement, facilitate propagation within discontinuities, and compensate for water loss during grouting. In addition, the use of a plastifying agent, which reduces cohesion and viscosity, facilitates penetration of the grout into the discontinuities. A finer grout may possibly be used if this can be justified on a cost basis.

29.3.3.3 Grout properties

Lombardi (2003) specifies that to judge a grout, two types of properties should be considered, which are not entirely independent of one another. The first relates to fresh grout, which is actually a suspension of cement particles in water, and which behaves like a Bingham body requiring the application of a force to begin moving in order to overcome the resistance brought about by cohesion. As for water, this is a Newtonian body. Density, settling (bleeding), viscosity, cohesion, and setting time are primarily included.

The properties required for fresh grout, in order to facilitate penetration into the rock mass (that is, cohesion, viscosity, setting time) can be obtained by using appropriate additives without compromising the final strength properties of the grout and therefore the quality of the completed grouting works, as happens when the W/C ratio is increased.

The second group of properties relates instead to grout that has set and concerns mechanical strength, resistance to chemical agents and washout, as well as permeability.

29.3.4 Grouting pressure

Grouting pressure must ensure that cracks open and the grout can penetrate.

There is a traditional procedure that has been applied for many years, which relies on very simple rules. These consist in defining the grouting pressure limit simply as a function of depth below the surface. However, it is important to remember that not only the type of rock and its geological characteristics should be considered, but also the type of grout to be injected. There is a pressure limit known as "refusal pressure," beyond which the rock mass will not absorb any further volume of grout. Moreover, related to grouting pressure, the worry is firstly that hydraulic cracking (filling fracture) may occur, which results in the brusque opening of a new discontinuity, and secondly that uprising, perceptible or not, of the surface of the grouted rock mass may occur, as a result of the opening of discontinuities.

For Swiss dams that are higher than 100 m, maximum pressure is at 5 MPa in limestone and 7 MPa in gneiss (CNSGB, 1992).

The GIN method presented in the next section is a modern approach that is increasingly common. It enables pressure to be managed by controlling the intensity of grout injection.

29.4 The GIN Method[2]

29.4.1 Principle

With the aim of firstly limiting the volume of grouting (in cases where grout penetrates easily at low pressure) and secondly limiting pressure (in cases where grout does not penetrate easily), or even using graduated pressure (grouting of fine cracks), Lombardi and Deere (1993) developed in the early 1980s a modern practice that introduced the notion of known intensity, under the name of the GIN method (*grouting intensity number*). This method can be used for easily grouted cracks as well as for fine cracks that require high pressure.

The method proposes the use of single stable grout with a low W/C ratio to avoid injecting a large quantity of water. In addition, a superplasticizer is added to reduce cohesion and viscosity, as well as to increase grout penetrability. The main idea is to use, during the works, a constant grouting intensity defined by the equation:

$$GIN = p \cdot V \; [bar \cdot l/m],$$

where p is the final grouting pressure (in bar) and V the final volume of injected grout (in l/m). The GIN value is calculated for a zero flow, that is, with the injecting pumps at a standstill. This value qualifies, as it were, the energy mobilized by the grouting process.

By strictly controlling the operation, this method allows the grouting process to be controlled independently of crack opening. For each type of rock, two maximum values (grouting pressure P and absorption V) are attributed; grouting stops when the product p · V reached a set limit intensity. The GIN is therefore used to limit the grouting process by avoiding the concomitance of elevated pressure and an elevated absorbed volume. This principle allows a reduction in grouting pressure in view of the volume of grout already injected (Figure 29.3). The GIN method prevents the rock from degrading through hydraulic fracturing (or filling fracture).

---

[2] The main points in this chapter have been drawn from several articles written on this subject by Dr. G. Lombardi, who was the promoter of the GIN method, along with D. U. Deere. Table 29.7 summarizes the main elements of the method.

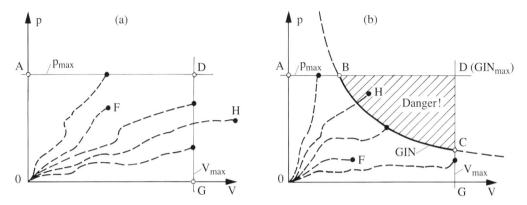

Figure 29.3 The limits for the grouting procedure (from Lombardi, 2003): (a) traditional method: ADG limit; (b) GIN method: ABCG limit

A limiting GIN curve (Figure 29.4), a kind of warning curve, is defined by considering the three following factors:

- Maximum pressure (p_{max})
- Maximum absorption (V_{max})
- Limit intensity (GIN)

The optimum GIN value must be established beforehand by grouting tests on site in order to adapt it to the real conditions of each homogeneous field in the rock mass. Figure 29.5 refers to practical values dictated by the experience of many grouting worksites where fundamentally sound rock mass has been treated. These are standard limit functions or practical guidelines. An average GIN of 1500 bar · l/m can be accepted at the beginning and may be modified depending on the circumstances. Note that the GIN value for the abutment is lower than that accepted in the valley, to account for the difference in hydrostatic pressure.

Furthermore, GIN intensity is also an indicator of the mean distance reached by the grout.

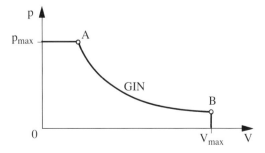

Figure 29.4 Limiting GIN curve

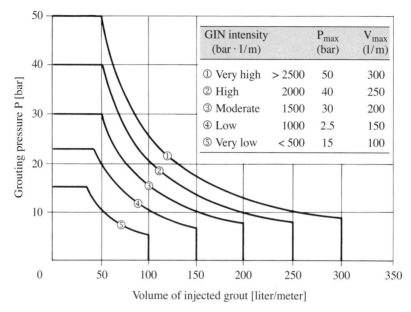

Figure 29.5 Series of standard limiting GIN curves that may be useful as an initial guidance (from Lombardi, 2003)

29.4.2 Monitoring grouting works

Given the high standards of today's grouting equipment, no major grouting activity should be undertaken without the complete recording in real time of the main parameters for each grouting interval. Computer processing of all data is vital.[3]

The principal parameters to be analyzed, recorded, and then turned into graphs are:

- Pressure p as a function of time
- Grout flow q (l/m) as a function of time
- Volumes (l/m) absorbed as a function of time
- Pressure in relation to volume absorbed
- Penetrability (q/p) as a function of the volume of grout absorbed

The last two parameters are by far the most important. They affect the management of grouting pumps. With regard to penetrability, this is defined as the relationship between the instantaneous values of flow and grouting pressure (Figure 29.6).

Interpreting these functions in real time means that the grouting process can be characterized and any hydro-fracturation phenomenon or excessive opening of discontinuities can be detected immediately, and the process can therefore be adapted to the actual conditions in the rock mass. Several software programs that have demonstrated their effectiveness for use in this field are currently available.

---

[3] The development of software by companies specializing in grouting means that both grouting work and monitoring can now be computerized. It is therefore possible to manage instructions, collect data, and produce graphs, notably in 3D. This in turn allows the quality of the works to be evaluated in real time, with any interventions being carried out rapidly if needs be.

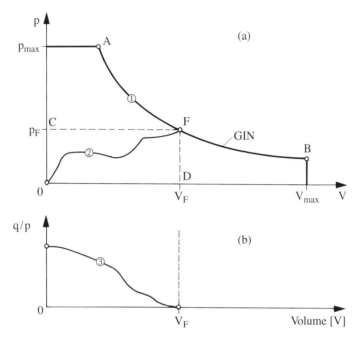

Figure 29.6 Typical grouting process for one grouting stage: (a) GIN curve and grouting process; (b) penetrability curve (from Lombardi, 2003) ① = limiting grout curve: pressure as a function of absorption; ② = real-time progression of grouting: pressure as a function of absorbed volume; ③ = penetrability (q/p) as a function of absorption. F = final grouting point; pF = final grouting pressure; V_F = total absorption. The OCFDO rectangle is defined as the "grouting intensity."

29.4.3 Summary of essential aspects of the GIN method

The essential aspects are:

- Use of a single grout mix that is the most appropriate for the site, defined on the basis of laboratory tests, particularly with a view to guaranteeing durability of the dam
- Use of a GIN limit function (with the three parameters p_{max}, V_{max}, GIN), correct distance between drilled intervals, interval lengths, and the correct fineness of the cement, which must best adapt to the properties of the rock and project requirements on the basis of site tests and mechanical considerations of the rock
- Tracking of the grouting process on an electronic screen and recording of actual data
- Continuous analysis of recorded data to optimize grouting works
- Elimination of water pressure tests in zones that have already been grouted
- Saturation of the rock mass immediately prior to grouting

It bears repeating that compared to the GIN method, the traditional method sets grouting pressure limits and uses different grouts by decreasing water content (for example W/C: 4:1, 3:1, 2:1, 1:1, etc.). As cohesion increases, the penetrability of the grout can be stopped.

Table 29.7 Generally accepted steps in the "GIN method" (from Lombardi, 2003)

GIN method principles

1. Define the exact purpose of the grouting works.
2. Establish a comprehensive grouting project and do not rely on "general grouting conditions," which may be outdated.
3. Use the "best grout mix" for the project in question, through the use of laboratory tests, and from an economical as well as technical point of view. Only stable mixes, with the addition of a superplasticizer, should be used. (Note that dense and stable grouts have been preferred by European grouting experts for several decades.)
4. Use one type of grout, "the best possible," for all grouting intervals in order to ensure consistent results and to simplify procedures, as well as to reduce grout loss.
5. Define the parameters of the limiting GIN curve: p_{max}, v_{max}, and the GIN = $p \cdot V$ value, while considering all the geological and mechanical factors of the rock that may have an impact, as well as the aims of the project and project budget.
6. Confirm this study through in situ tests and monitor the completed works using appropriately grouted additional perforations.
7. Avoid water pressure tests during grouting works, as these are unnecessary and can even be dangerous.
8. The split-spacing method is not a new technique but is used as part of the GIN method as a self-regulating process.
9. The use of interval lengths that increase with depth below surface is one way of speeding up grouting works and introducing some (modest) cost savings.
10. Saturate by injecting water into dry rock formations that may absorb the grout water. Do so immediately prior to grouting to avoid having to stop the process due to the creation of internal friction in the grout from loss of water.
11. The need for new drilled intervals and their length (for example, in a grout cut-off) are decided by taking into account observed absorption on neighboring drilled sections.
12. Computer controlled grouting is a requisite condition for achieving high-quality grouting. The graphs produced from this monitoring provide much helpful information, enabling the creation of statistical relations between various parameters and various grouting stages. Unfortunately, it has been observed that worksites often draw up these graphs, but that the correct conclusions are not drawn.

29.5 Structural Arrangement of the Grout Cut-off

29.5.1 Overall layout

With its position and function, the grout cut-off is an extension of the concrete dam into the rock mass of the foundation; the same applies for an embankment dam built on loose material (see Figures 20.8 and 20.9). As it behaves like a relatively watertight element upon which hydrostatic pressure acts, it must be treated with the same care as the dam itself. Poor execution may mean that the reservoir cannot be filled due to excessive seepage.

The grout curtain may be installed from the surface, a base gallery in the dam, or galleries in the rock. Grouting works from a gallery normally take place when the dam is practically finished, which means that the overload created by the dam can be capitalized on for high-pressure grouting. Use of perimetric and lateral galleries enable works to be carried out entirely independently of the concreting works taking place on the surface, thus allowing grouting to be picked up again or completed later (Figure 29.8).

Lastly, it is important that the grout cut-off remains effective throughout the operating life of the dam.

With regard to consolidation grouting, short boreholes are distributed throughout the foundation zone. Contact grouting, carried out after concreting, ensures the contact between the dam and the surface of the foundation. They only go down to a certain depth, in the order of 10 to 20 m (Figure 29.9). Drainage holes (see Section 30.1) are only installed once grouting has finished.

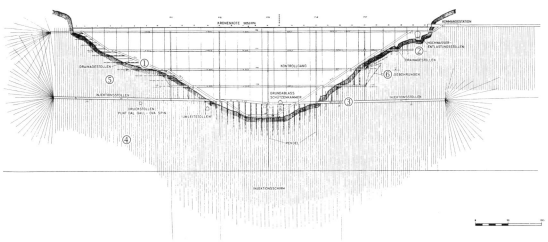

Figure 29.8 Grout curtain at the Punt dal Gall dam installed from galleries: ① bottom gallery; ② drainage gallery; ③ grouting gallery; ④ boreholes drilled from the grouting gallery; ⑤ boreholes drilled from the bottom gallery; ⑥ drainage holes

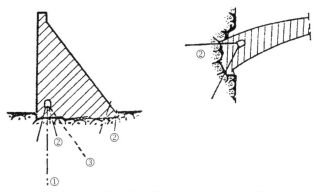

Figure 29.9 Binding and consolidation grouting. Key: ① main grout curtain; ② contact and consolidation grouting (at a limited depth); ③ drainage holes

29.5.2 Number of grouting lines

A deep grout curtain comprises one or several grouting lines. The cut-off is double or triple when the ground does not tolerate high pressure and there is a risk that the boreholes will only have a small active range. The boreholes are then distributed in a quincunx pattern.

As a general rule, a single grout line is sufficient for concrete dams. Grout cut-offs beneath an embankment dam, however, generally require several lines. For embankment dams with a core, the grout cut-off should be installed before the placement of fill, possibly from a concrete plate. If the dam has an upstream facing, grouting can be carried out from a gallery located at the upstream plinth (see Figure 27.8).

29.5.3 Position of the grout cut-off

For concrete dams, the grout cut-off is generally situated as far upstream as possible. Its exact location depends to a large extent on the specific geological conditions at the site.

For embankment dams, the grout cut-off is usually a continuation of the sealing system (core, facing).

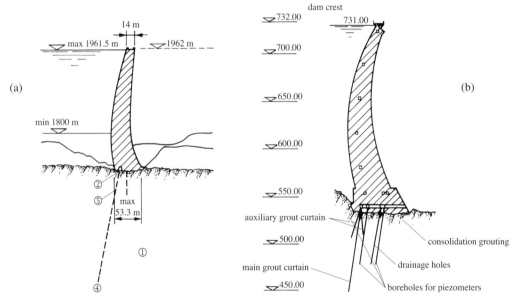

Figure 29.10 Position of the grout cut-off for an arch dam: (a) Mauvoisin dam in Switzerland; (b) Almendra dam in Spain

29.5.4 Spacing between boreholes

The grout cut-off is installed in different stages, with the distance between the boreholes being reduced in each stage (Figure 29.11). The first holes (primary holes) are generally longer and spaced at an interval of 10 to 12 m. Samples are taken to provide supplementary information relative to the geology, and permeability tests can be carried out. Less deep intermediary holes (secondary, tertiary holes, etc.) are carried out, with the space reduced by half for each stage. In most cases, the final distance, which depends on the rock and the degree of impermeability that is sought, varies between 2.5 and 3 m. For dams with a lower height, the space between holes can be lower than 3 m; grouting is then carried out with lower pressure.

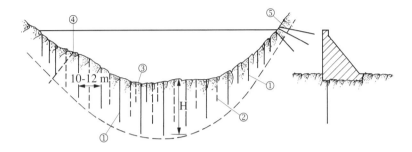

Figure 29.11 Diagram of a grout cut-off: ① primary holes; ② secondary holes; ③ tertiary holes; ④ oblique control holes; ⑤ wing cut-off

29.5.5 Orientation of holes

The direction of holes can be either vertical or inclined upstream, depending on the geomechanical properties of the foundation. As the objective is to fill cracks with a material, the line or plane that includes the most fractures and vertical and inclined cracks must be found. Drawing up a polar diagram enables the various existing discontinuity systems to be determined and the correct orientation of the holes to be established.

Sloping the grout cut-off in a concrete dam in an upstream direction ensures good separation between the grout cut-off and the drainage.

29.5.6 Depth of grout cut-off

The depth and geometry of the grout curtain depend on geology and the height of the dam. It must be sufficiently deep to minimize seepage and contribute to a reduction in uplift. The curtain can be continued down to a zone where permeability is less than in the grouted zone. Generally, the height of the curtain is 35 to 100% of dam height, with a minimum depth of 50 m.

In relation to dam height H, the USBR proposes a cut-off depth equal to

$$H_{cut\text{-}off} = 1/3\ H + C, \text{ where } C = 8 \text{ to } 25 \text{ m.}$$

For dams built in Switzerland, a depth equal to 2/3 of dam height is the average for dams with a height greater than 50 m and built on gneiss foundations. For limestone, this is the case for 80% of dams (SNCOLD, 1992).

29.5.7 Extension past the dam

It is common practice to extend the grout curtain a substantial distance into the banks. Moreover, there have been cases where it was also necessary to extend grouting works along entire slopes of the reservoir in order to avoid loss (Limmern dam (GL/1963/H = 146 m), Salanfe dam (VS/1952/H = 52 m), or Sanetsch dam (VS/1965/H = 42 m)).

As an example, Figures 29.12 (a) and 29.12 (b) illustrate axonometric views of dams and their grout curtains.

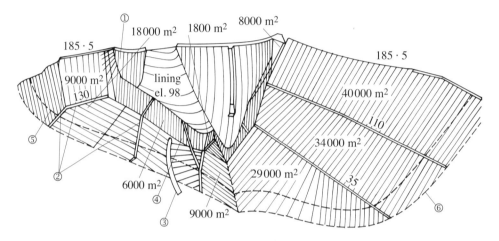

Figure 29.12 (a) Grout curtain at the Oymapinar arch dam in Greece with a total surface of 180,000 m² (from Balissat, 1985): ① auxiliary dam; ② grouting gallery; ③ diversion gallery; ④ grouting boreholes; ⑤ intersection of the grout curtain with the schist; ⑥ limit of the grout curtain

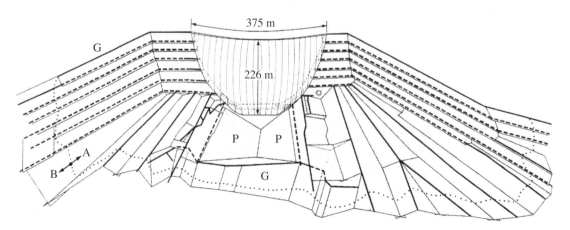

Figure 29.12 (b) Grout curtain at the El Cajón arch dam in Honduras (from Balissat, 1985): A limestone section; B volcanite section; G grouting gallery; P plane of the grout curtain

29.5.8 Test zones

Before beginning full-scale grouting, it is a good idea to test the rock's suitability to the procedure. By setting up one or several test zones, it is possible to check the quality and penetrability of the grout, verify that the grouting sequences are correct, and test the planned drilling and grouting equipment.

The test zone can correspond to one section of the cut-off for which the distribution of boreholes (in one or several lines) is the same as for the dam itself. The primary boreholes can also be laid out in a triangle (Figure 29.13).

In both cases, the subsequent boreholes are split-spaced while reducing the distance by half. Lastly, control boreholes ensure that the effectiveness of the grout can be verified by permeability tests.

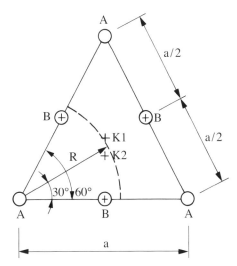

Figure 29.13 Example of a triangular test zone: A: primary boreholes; B: secondary boreholes; K: control boreholes; R: grout penetration; a: distance between primary boreholes

29.5.9 Depiction of grouting results

For the monitoring of works, it is common practice to record the quantity of grout injected in each interval, as well as results from water tests when these are carried out, on a general arrangement drawing that shows the depth of the borehole. Figure 29.14 provides an illustration of this.

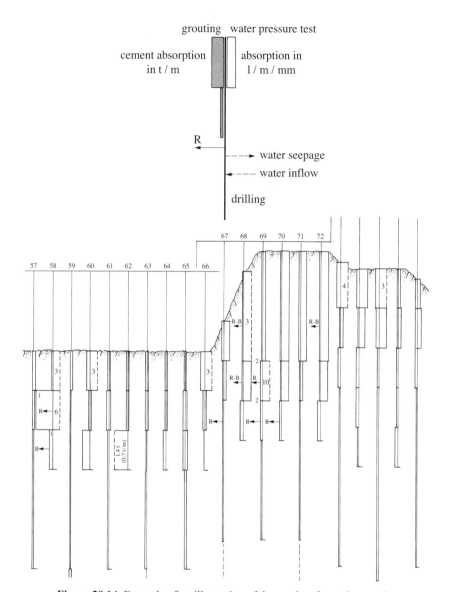

Figure 29.14 Example of an illustration of the results of grouting works

30. Other Methods for Treating Foundations

Limmernboden arch dam in Switzerland, height 146 m, year commissioned 1968
(Courtesy © Swiss Air Force)

30.1 Drainage Systems for Concrete Dams

The importance of the drainage system was emphasized in Chapter 29. In particular, its aim is to diminish the effect of uplift and capture seepage. It thus reinforces the safety of the dam and its foundation. The drainage system also allows the tightening capacity of the cut-off to be monitored.

The two most commonly used methods are:
- Drainage holes
- Drainage galleries

30.1.1 Drainage holes

Drainage holes are generally installed downstream of a grout curtain. They can be vertical or inclined, depending on the geological structure of the rock. Sometimes, they are spread out along the transversal section of the dam.

Drainage holes are excavated from the dam's base gallery once grouting works have been completed. Their walls are "free," but, depending on the bearing capacity of the rock, they may be fitted with a partially or totally perforated sleeve. It is not unusual for drainage holes to become blocked by calcite deposits over time. It is then necessary to redrill them or to drill new ones.

30.1.2 Drainage galleries

These are galleries within the rock (lined or not) that may have been former exploratory galleries and/or grouting galleries. They cross the rock mass of the abutments; some are located under the dam (Figure 30.1). They are oriented either as an extension of or in parallel with the dam's plane of reference, or in an upstream direction.

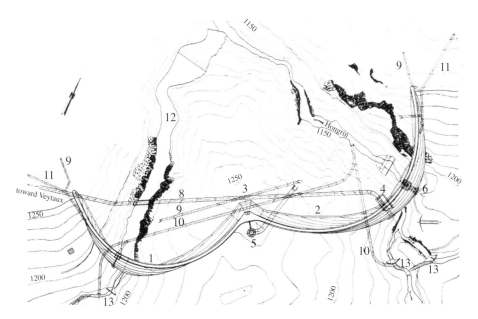

Figure 30.1 The Hongrin dams: drainage galleries on the wings and in the central section (nos. 9 and 11)

Seepage is captured in order to measure its discharge. It is also collected in a gutter and directed to an outlet. If it ends up in a low point, it is pumped out. A siphon system enables permanent flow and prevents the orifice from becoming blocked by a build-up of calcite.

30.2 Jet Grouting

Jet grouting, which was developed in the United Kingdom in the late 1950s, is a method that consolidates the ground and creates a watertight barrier. It uses a jet of high-speed fluid to break up the rock material and mix it with liquid grout. The removal of part of the material up to the surface is done with jetting fluid. The effectiveness of this method obviously depends on the nature of the terrain, which is difficult to break up if it is dense and compact. This technique has been applied in the field of dams to construct sealing systems, notably in the case of some highly fractured rocks with low mechanical strength (Balissat, 1985). Boreholes can be more spaced out than in conventional grouting. By using a system of neighboring panels, it is possible to create a continuous wall. The method is more cost-effective compared to the construction of a diaphragm wall.

With regard to construction, the works begin with the creation of a small-diameter borehole (100 to 200 mm) up the entire height to be treated with grout. In the next stage, material in the lower intervals is broken up thanks to a powerful jet of fluid from a high-pressure pump. At the same time as the material is being eroded, grout, sometimes with the addition of bentonite, is injected. The grout is incorporated into the material to form a uniform mix. Lastly, the rotating rods are slowly brought the surface, thus forming a column of soil-concrete (Figure 30.2).

There are different methods for jet grouting (single fluid, double fluid, triple fluid). In the single-fluid system, the fluid is grout only. It performs the actions of breaking up the material, extraction, and incorporation. In the double-fluid system, the jet of high-speed grout is surrounded by a ring of high-pressure air, which improves its range and ability to erode. The jet of grout simultaneously cuts into and mixes the material. And lastly, in the triple-fluid system, the breaking up and extraction of the material are achieved by a double jet of water and air. The grout is injected simultaneously through an additional nozzle lower down.

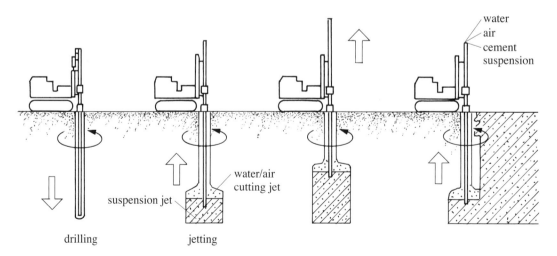

Figure 30.2 Construction phases for jet grouting

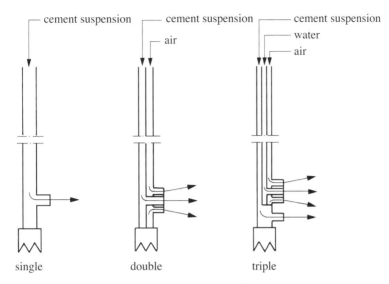

Figure 30.3 Three methods of jet grouting

30.3 Alluvial Grouting

The most commonly used method is the sleeve pipe (or valve pipe) method. As the ground may collapse, the borehole is sleeved as it is drilled.

Once the drilling is complete, a metal or plastic pipe, with a diameter that depends on its use (in the order of 50 mm), is inserted (Figure 30.4). The pipe is perforated at regular intervals (between 0.30 and 0.50 m) with holes protected by sleeves (distance between them from 0.25 to 0.50 m). The sleeve is made of a rubber membrane, which acts as a "check valve." This allows grout to pass through into the surrounding terrain but prevents it from returning inside the pipe.

The pipe adheres to the surrounding material thanks to a kind of grout called *annular grout*. The placement of this grout takes place during drilling or from the bottom of the borehole as the supporting pipe

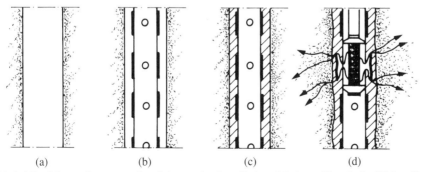

Figure 30.4 Alluvial grouting: example of sleeve pipe installation: (a) sleeved borehole; (b) installation of sleeve pipe; (c) filling of the annulus with annular grout; (d) grouting with a double packer

slowly ascends. The grout must have adequate setting time and be sufficiently hardened before use, while nevertheless ensuring it does not become too hard.

Grouting with stable grout (clay-cement, gel, colloidal solution) occurs in short bursts, usually while ascending (30 to 50 cm) and while using a double packer for one or several of the sleeves. The pressurized grout is able to lift up the sleeve and pass through to the exterior. The pressure of the injection is enough to fracture the cement sheath in which the set of sleeve pipes is sealed. Grouting pressure and flow remain low in order not to fracture the terrain, but rather to permeate it. In fact, the grout fills the soil pores by forcing out the interstitial water that is present.

Once the operation is complete, the packer and pipe are removed and the borehole is cleaned with water to remove any of the grout that may still be in the pipe and may cause a blockage for the next grouting.

30.4 Vertical Diaphragm

Vertical cut-offs may be constructed like a watertight curtain in the foundation, as either a diaphragm wall or as secant piles that extend the toe of the upstream facing of an embankment dam (Figure 30.5).

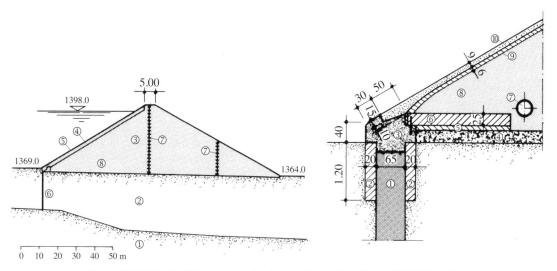

Figure 30.5 Detail of the upstream heel at the Godey dike (from SNCOLD, 1985)

This option can be used in cases where grouting works would be too costly and as long as the ground can be excavated.

Before beginning actual excavation of the diaphragm wall, a shallow starting trench must first be constructed. It is used to feed the excavation with drilling slurry and guide the excavation machinery. Excavation of the wall takes place in panels, with a clamshell bucket or by drilling. Should the bucket be used, the ground is broken up with a heavy suspended drill bit. The drilling tool is alternately dropped down and brought back up. With the other option, a cutting drill is used. The cutting excavator sinks continuously into the ground with a reverse circulation rotational drilling technique. It is equipped with rolling cutter drill bits that crush and grind the rock, which is continuously sucked back up (Bauer system). No time is wasted in withdrawing the tool. Furthermore, the system enables continual surveillance of any wall deviation. This option is therefore used when the ground is very hard (Figure 30.6). During drilling, the extracted material must be replaced by a mixture of bentonite and water that functions as a prop before the concrete is placed.

Figure 30.6 Drill with a rolling cutter bit (Bauer system)

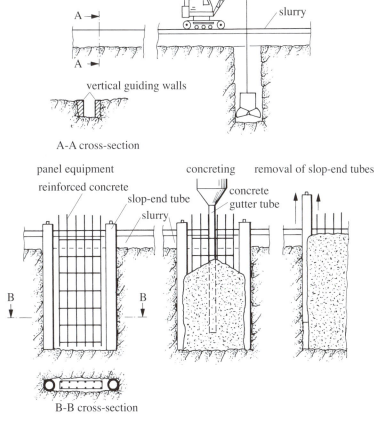

Figure 30.7 Layout of elements in a diaphragm wall (from Schneebeli, 1971)

Panels can be placed one after the other, or a series of primary panels can be placed, followed by a series of secondary panels. Drilling slurry ensures trench stability. Once a panel is ready, reinforcement is inserted and concreting can occur (Figure 30.7). For a diaphragm wall project, certain limit values must be taken into account: trench width: ≤ 1.2 m; length of elements: ≤ 4.2 m; depth: ≤ 80 m.

With regard to concrete, as these are watertight cut-offs designed to reduce seepage loss in loose material, it is recommended to construct the wall in plastic concrete (concrete that slumps in an Abrams cone by 5 to 9 cm). Plastic concrete is made up of sand, gravel, cement, and colloidal clay. It has strong plasticity, which enables it to adapt to the deformation and settlement of the ground.

A wall made from secant piles is constructed when the ground is too hard to construct a diaphragm wall. It has a series of overlapping piles with intersections (Figure 30.8). There is also another option with piles that are simply adjacent.

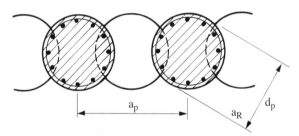

Figure 30.8 Cross-section of a secant pile wall

30.5 Abutment Consolidation

There may be cases in which specific geological conditions (orientation of layers, fracture system of the rock) raises fear of abutment instability during impoundment of the reservoir or after an earthquake. It is therefore necessary to strengthen this element by the placement of anchors (rock bolts, rods). The execution of these strengthening measures may occur as construction begins or after excavation.

Strengthening works may also be undertaken during operation, should any movement in the rock mass be observed. The observation of the behavior of at-risk zones must be integrated into the monitoring system (see Part VIII).

To illustrate this, some concrete cases are described in the following sections.

At the Gebidem dam, advanced cracking of the upper part of the two abutments required consolidation. It should also be mentioned that blasting work had led to a modification in topography due to falling rocks and the removal of the disturbed elements. The placement of a network of prestressed anchors was designed and constructed. The aim of this consolidation was also to enable efficient grouting without the risk of disturbing boulders. Cables with a tension of between 60 and 160 t were laid out in a mesh arrangement of approximately 4 m by 4 m (Figure 30.9).

At the Lessoc dam (FR/1973/H = 33 m), the planned excavations for construction of a stilling basin required cutting into the base of rock layers that were steeply inclined downstream. To avoid any movement during and after the works, strengthening works had to be carried out along a thirty-five-meter length. To achieve this, as the excavation progressed, a series of 37 prestressed cable anchors were installed in 5

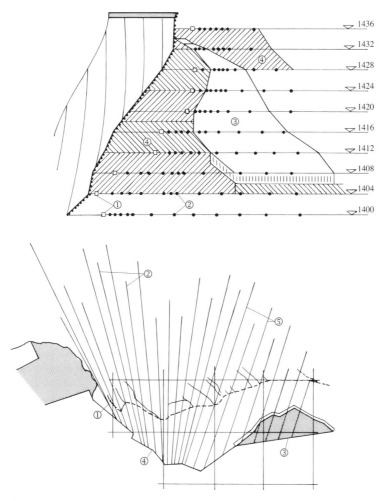

Figure 30.9 Gebidem dam: consolidation of the left-bank abutment; elevation and horizontal sections (from Gicot, 1970). Key: ① coupled anchors; ② anchors; ③ formwork concrete; ④ local filling of cracks or shotcrete; ⑤ grouting injections.

stages. Horizontally inclined by 20° to 35°, the force in the rods varies between 150 and 170 tonnes. The excavations were carried out in benched sections of 5 m. Braces and concrete distribution beams guarantee the transfer of load. The cables were grouted after construction in 1972 (Pougatsch, 1975). As, firstly, geodesic measurements revealed deformation in the direction of the waterway, and, secondly, it was not easy to establish the condition of the anchors, which were subject to stray currents due to the presence of a railway track nearby, the project owner decided, for reasons of safety, to reinforce consolidation of the rock mass. A total of 33 fifteen-meter sheathed bars (passive anchorage) were placed at the level of the distribution beams (Figure 30.10). They were slid into the borehole and were then grouted. Moreover, uplift is present in the rock mass and so grouting to strengthen the grout curtain was carried out on the left bank (Fern et al., 2019).

Figure 30.10 Lessoc hydroelectric dam: consolidation of the left-bank abutment beneath the stilling basin: (a) during construction in 1972; (b) during consolidation works in spring 2019 (Courtesy H. Pougatsch)

Built between 1966 and 1971, the north and south Hongrin arch dams (Figures 2.19 and 30.1) create a double-arch structure. The two arches rest on a central abutment.

The rock on the right bank of the north dam is tectonized, and its stratification is not favorable. During construction, this zone was strengthened by anchors. During impoundment, the appearance of small quantities of seepage flow were observed, which raised fears of imperfections in the grout curtain and/or the drainage. The seepage flow was therefore monitored as part of the usual surveillance system.

Following the results observed over the years, in-depth studies were undertaken. These related to the study of underground flow and the assessment of stability in the zone. There ended up being a significant correlation between the reservoir water level and the seepage flow. The installation of pressure sensors in boreholes confirmed this correlation. In fact, the chemical basis of the water is similar. A gradual increase in pressure just downstream of the grout curtain is a sign of its deterioration and/or of drains becoming blocked.

On the basis of these results, an action plan was developed. In the short term, it proposed strengthening surveillance, in particular through the installation of two extensometers. In the medium term, grouting works in the weak zone of the grout curtain and the construction of additional drainage is planned. Lastly, in the long term, an intervention on the anchors or even their replacement is envisaged (Bussard et al., 2015).

As previously mentioned in Section 2.2.3 (Part I), various measurements (geodesic, pendulums, extensometers) have demonstrated weak movement in the downstream part of the left-bank abutment of the Montsalvens dam (FR/1920/H = 52 m), a movement that had already occurred during impoundment in 1921. A rockfall also occurred in 1944. Analyses carried out have shown that the state of equilibrium of the rock mass lies at the limit of elastic behavior. It was therefore decided that strengthening works should be undertaken. The strengthening works were achieved using passive anchors and drainage, and a surface protection with rock bolts and gunite to prevent or at least limit the increase in observed plastic displacement (Figure 30.11). The completed works have increased stability of the rock mass and should prevent a new rockfall (Lazaro and Golliard, 1999).

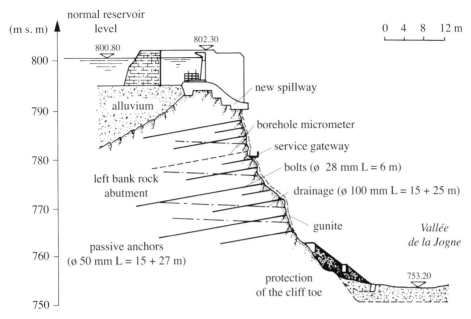

Figure 30.11 Montsalvens dam: strengthening measure on the left-bank abutment, general view and cross-section (from Lazaro and Golliard, 1999)

In a formerly used technique, it was common to grout prestressed rods in order to protect them against corrosion. This technique has the disadvantage that the rods cannot be monitored during operation, and the required stability cannot always be guaranteed. The current technique not only provides appropriate anti-corrosion protection of the steel anchors, but also allows monitoring of rod tension. The SIA 267 standards recommend that the number of measuring and monitoring rods should depend on the specific conditions at the dam. The total number of measuring and monitoring rods together should be at least 5% of the total anchors, or three rods in each element of the dam.

In the case of the Pfaffensprung dam (UR/1921/H = 32 m), also mentioned in Section 2.2.3, additional anchors were installed and fitted with a tension monitoring system to supplement existing anchors, whose bearing capacity is unknown.

VIII. OPERATION

A constructed, operating water-retaining structure cannot be left without surveillance, nor regular maintenance. As such, the dam operator must first appoint staff members to be responsible for the various missions of surveillance and maintenance. The aim is to highlight in a timely manner any anomaly concerning the behavior and condition of the dam, in order to take the necessary measures to safeguard against all possible hazards.

As an example, Chapter 31 describes the creation of a set of guidelines clearly outlining the various tasks and who is responsible for them. This document must be regularly updated. This chapter also highlights the role of documentation and archiving. All documents produced during construction and while operating must be kept in a safe place. Given the quantity of these documents, it is a good idea to compile a dam monograph that covers the most important elements of the project, as well as significant events that took place during construction and operation. Thanks to this document, basic information concerning the dam can be accessed quickly. It can also be updated regularly, for example every five years.

Chapter 32 focuses on the method for and the main tasks involved in organizing the monitoring and maintenance of the dam, which must result in the recording of the behavior of the dam and its general condition. Achieving the objectives relating to surveillance and maintenance is made possible by installing a monitoring system and by carrying out visual inspections and operating tests.

The monitoring system, which not only includes the dam itself, but also the surrounding environment, is a vital and compulsory element in the detailed tracking of the behavior of water-retaining structures. It allows representative parameters (deformation, seepage, and pressure) to be measured. It is also of use during the reporting of any anomaly and must allow its interpretation.

Part of this chapter gives several guidelines concerning the general overview of a monitoring system (why measure, what to measure, how to measure, and when to measure) and the many and varied measurement instruments available.

Thanks to rapid advances in technology (electronic, metrology, computer sciences, material sciences), the possibilities and options for measurement and recording are also continually evolving. Namely, this includes means of measurement (measuring instruments), means of data transmission (cables), and automatic means of data acquisition and storage (databases), as well as the means for processing and presenting data (analysis of measurement results, creation of graphics and reports). Although the list of measuring instruments has expanded, it should be noted that the parameters measured remain largely the same.

International experience has shown that two thirds of specific events are brought to light through visual inspections. Lastly, by means of testing, the correct functioning of various elements (water-release structures, in particular) and equipment can be guaranteed.

In order for the surveillance system to be effective, the various measurements and observations must be interpreted rapidly and then integrated into visit reports or annual reports.

Of course, this is not to rule out or ignore the possibility of an extraordinary event, such as abnormal behavior or even hazards relating to the risk of rockfall into the reservoir, floods, or earthquakes. In view of such possibilities, adequate methods (alarm and warning systems, emergency organization) for managing these specific events must be put in place to protect and inform members of the public residing downstream from a dam. Planning and current measures are described in Chapter 33. It is nonetheless clear that such a system and the provisions taken may evolve over time as technology and experience gained from operation continue to develop.

31. Necessary Operating Documents

Gelmer gravity dam in Switzerland, height 35 m, year commissioned 1929 (Courtesy © Swiss Air Force)

31.1 Surveillance Guidelines

The dam operator must take all necessary actions to guarantee the safety of the water-retaining structure at all times. How surveillance and maintenance is organized must be clearly defined. To achieve this, the dam operator should prepare a file containing maintenance and surveillance guidelines for the water-retaining structure during normal operation as well as during extraordinary events. This important document should notably contain the following elements:

- The list and description of all tasks falling to the dam operator (listing all managers and their replacements). These tasks include:
  - Visual inspections
  - Operating tests
  - Monitoring system measurements
  - Instructions for the preliminary analysis and transmission of measurement data
  - Instructions for actions in the case of an anomaly or extraordinary event (flood, earthquake)
  - Updating the dam file
  - Measurement program
- The organization chart, as well as the list of all relevant staff members and their contact details (address, telephone number, email), must always be kept up to date

To achieve these various tasks, "Regulations for the operation and monitoring of water-release structures" and "Regulations relating to the surveillance of the water-retaining structure" should be drawn up.

31.2 Dam File

The Swiss legal framework requires the dam operator to create a "Dam File"[1] (WRFO, 2012). This file includes all documents produced during planning, design, and construction, as well as all results from observations made during operation (Table 31.1). The file, which includes the main dam archives, must be regularly updated. The file contents must be stored in an appropriate, protected location (safe from water and fire) and must be easily accessible for all managers.

In its publication about the "dam file", the Swiss Committee on Dams (SwissCoD) puts forward a proposed structure and content outline for this document and establishes a detailed list of all elements that should be contained in it (SwissCoD, 2005). In the first part, the file contains all documents relating to surveillance and maintenance of the water-retaining structure. This part should be continually updated and added to with new measurements and observations, as well as with annual reports, expert opinions, test protocols, etc. The second part includes the design documents for the construction and operation of the water-retaining structure. This is the basis upon which dam safety can be assessed. It is only added to if the structure is modified or extended due to alterations, significant rehabilitation, new static verifications, or safety verifications in the case of a flood, etc.

---

[1] Formerly called the "Dam Journal."

Table 31.1 Content of the dam file according to the Swiss legal framework

The dam operator must create a file concerning the water-retaining structure and keep it up to date. It must be available at all times for the surveillance authority to access. The file must include:
- The most important as-built drawings and data on construction works
- Calculations and reports on static, hydrological, and hydraulic behavior
- Geological appraisal
- Commissioning report
- Annual measurement reports
- Minutes from annual monitoring
- Minutes from tests with water release
- Five-yearly reports
- Reports on safety assessments
- Reports on geodesic deformation measurements
- Surveillance, gate operation, and emergency regulations
- Reports on incidents and operating anomalies

31.3 Dam Monograph

Another very useful document is the dam monograph, which gathers together information concerning the life of the structure. All key events should be mentioned. This document gives the opportunity for each new manager to get an overview of dam operation, without having to consult the voluminous dam file. This document must also be kept up to date.

The dam monograph should include a description of local conditions, and the condition of the dam and its appurtenant structures, including any modification to the surveillance concept. It should also contain a bibliography and an index of the documents contained in the dam file. The "Dam Monograph" established by a working group of the Swiss Committee of Dams provides a list of points that must be included (SwissCoD, 2001a).

32. Surveillance and Maintenance

Ova Spin arch dam in Switzerland, height 73 m, year commissioned 1968 (Courtesy B. Kirchen)

32.1 Aim and Organization

In order to take all necessary measures to protect against possible hazards or even to decide on rehabilitation works, the aim of surveillance is to detect, as soon as possible:

- Damage to the dam and appurtenant structures
- Any failures in structural safety
- Any threats to safety by external phenomena

Surveillance firstly involves regular monitoring of the condition and behavior of the water-retaining structure and, secondly, periodic safety inspections. The aim of this regular monitoring is to track behavior from day to day. The aim of periodic inspections is to monitor long-term behavior and assess structural safety.

To thoroughly monitor the dam's condition and behavior, the following actions must be carried out:

- Visual inspections
- The measurement of relevant parameters using various instruments
- Operating tests for water-release structures (gates), instruments, and means of communication

As a general rule, the minimum monitoring frequency for large dams is the following:

- Measurements taken using monitoring instruments should be carried out at least once a month
- The most important parameters are continuously recorded, and some results are transferred via computer data link
- Visual inspections are carried out at least once a week
- Operating tests for water-release structures (bottom outlet) are carried out a minimum of once a year with the reservoir near full and with wet testing (Figure 32.2).

In Switzerland, surveillance is organized as follows (§ 5.3.3):

1st level: Staff employed by the dam operator (or possibly the dam owner) carry out visual inspections, measurements and preliminary analyses, and operating tests, and supervise maintenance work

Table 32.1 summarizes the various tasks to be carried out by personnel in the first three levels

What	By whom	Duties
a) visual checks	1) staff employed by the dam operator	regular inspections (at least 1/week)
	2) experienced dam engineer	annual inspection (with report)
	3) senior experts	one inspection every 5 years
b) measurements	1) staff employed by the dam operator	record measurement data first analysis of data
	2) experienced dam engineer	analysis of dam monitoring data annual report on dam behavior
	3) senior experts	analysis of dam behavior special safety inspection
c) operating tests	1) staff employed by the dam operator	wet operating tests of gates and valves (at least 1/year)

2nd level: Results are transferred within a week to a dam engineer (experienced engineer), who analyzes the measurements and annual checks; the engineer drafts an annual report on the behavior and condition of the dam

3rd level: Every five years various experts (civil engineering, geology) carry out an in-depth study of the condition and behavior of the dam, as well as a safety inspection, and draw up a five-year report containing any recommendations; a surveyor may also be called upon to contribute

4th level: Official surveillance authority

32.2 Visual Inspections

These inspections are essential and effective. In fact, more than two thirds of extraordinary events are brought to light through visual observations. Even with current technological means and possibilities for automation, visual inspections must be carried out regularly. Measurements alone do not guarantee that changes in condition may be detected. For example, the appearance of a wet zone or spring downstream of a dam may indicate a change in underground flow before this is detected by measuring instruments (Pougatsch et al., 2001).

As tasks are diverse and are carried out at different frequencies, careful planning is needed. This planning must establish:

- What is to be monitored
- How the monitoring is to be carried out
- How often monitoring must occur
- Who is responsible for monitoring

Monitoring includes ensuring that there are no major variations, primarily with regard to:

- Dam condition:
  - for concrete dams: new cracks or a continuation of existing cracks, shattering, relative displacement of blocks, swelling and sweating, drainage water flow and quality, etc.
  - for embankment dams: local settlement, local sliding zones, humidity on the downstream face, drainage water flow and quality, etc.
- The dam abutments and galleries in rock
- The immediate surroundings and banks of the reservoir[1]
  - Rockfalls
  - Indications of slope instability (breaks in the plant cover, swelling, sliding) and any indication of slips (rock color, outcrop curvature)
  - Settlement of roads, paths, fences, or trees, pipe failure, etc.

This visual inspection may be carried out using an existing form that is filled out by the person doing the inspection. The form can then be used as an inspection report and is filed in the dam file, or even included in an annual report (SNCOLD, 1997a).[2] It can also be useful to record observations on drawings (for example, cracks, sweating, seepage, etc.).

---

[1] See Schneider, 1987; Sinniger, 1987.
[2] The document "Surveillance of dam condition and checklists for visual inspections," which was developed by a working group from the Comité national suisse des grands barrages, provides examples of visual inspections for different types of dams (SNCOLD, 1997a).

32.3 Testing of Water-Release Structures

All mobile elements (gates) must be tested regularly. These operating tests are generally carried out at least once a year. Wet testing of gates is carried out when the reservoir is almost full.

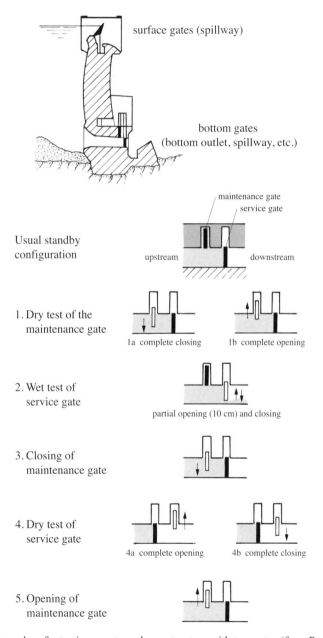

Figure 32.2 Procedure for testing a water-release structure with two gates (from Pougatsch, 1983)

32.4 General Overview of the Monitoring System

32.4.1 Purpose of a monitoring system

The monitoring system is a key component for the appropriate surveillance of a water-retaining structure. It is essential and should be developed during the design phase. It must be able to reliably measure (that is, produce measurements that are verifiable) the parameters required to assess the behavior of the dam and foundations, given the loads they are under. Reliably means using various methods to independently verify measurements. This is known as monitoring system redundancy.

The design of the monitoring system obviously depends on the specific characteristics of the dam: the type of dam, but also its geometric and geological specificities, the materials used, the period in which it was constructed, and the technical means available (Pougatsch, 2002; Pougatsch and Sonderegger, 2003).

The monitoring system is designed to fulfil these objectives:

- Carry out inspections during construction and first impoundment
- Carry out inspections during operation
- Detect any behavioral anomalies in time
- Provide additional information should a behavioral anomaly occur (or implement additional measuring instrumentation should this be necessary)
- Add to and improve information for the engineer (technical or scientific research, analytical forecasts, risk analysis)

32.4.2 Parameters to be tracked

The monitoring system is a carefully designed system that measures both forces exerted on the dam (causes) and the various parameters (dimensions) that characterize the behavior of a water-retaining structure (consequences) (Figure 32.3).

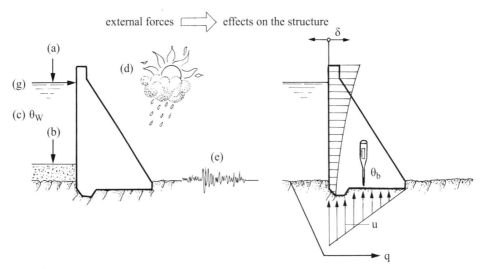

Figure 32.3 Effects of external forces on a structure. External forces: (a) water pressure; (b) sediment pressure; (c) water temperature θ_W; (d) weather conditions (air temperature, rainfall, snow); (e) seismic conditions; (g) ice load. Effects of forces: deformation (δ); concrete temperature variation (θ_b); seepage (q); uplift (u), and pore pressure.

With regard to external forces, these include:
- Water pressure (by measuring the reservoir water level)
- Sedimentation pressure
- Water temperature
- Internal temperature
- Weather conditions (solar radiation, air temperature, rainfall, depth of snow)
- Seismic conditions
- Ice load

These direct loads and external conditions lead to:
- Deformation and displacement
- Variation in temperature, especially in the body of a concrete dam
- Pressure (uplift, pore pressure)
- Leaks and seepage through the dam and foundations

A difference should be drawn between parameters indicating an intermittent effect (internal stresses), a local effect across a broad field (temperatures, water pressure), and overall behavior resulting from several local effects (deformation, seepage flow).

Table 32.4 summarizes the primary relevant parameters that must be recorded and the inspections that must be carried out to adequately track the behavior of concrete and embankment dams, as well as their foundations.

Table 32.4 Relevant parameters and inspections for monitoring a dam and its foundations

Concrete dam	Embankment dam	Foundations
Structural deformation	Deformation in the dam body	Deformation Abutment movements
Local movements (cracks, joints)	Specific movements (connection with a concrete structure)	Specific movement (cracks, diaclases)
Dam body temperature	Dam body temperature to detect seepage (possible)	
Uplift (at the concrete-foundation interface and in the rock)	Pore pressure in embankment dam body and piezometric level	Pore pressure Deep body uplift pressure Piezometric level Groundwater level
Seepage and drainage flow	Seepage and drainage flow	Seepage and drainage flow, resurgence (springs)
Chemical analysis of seepage water Turbidity (possible)	Chemical analysis of seepage water Turbidity	Chemical analysis of seepage water Turbidity

32.4.3 Some key principles[3]

When designing a monitoring system, it is important to understand some key principles, including:
- The dam and its foundations create one combined structure. However, the monitoring system must be able to clearly differentiate between the behavior of the dam and the behavior of its foundations and surrounding environment.

---
[3] See also Biedermann (1987a).

- The monitoring system must be adapted to the specificities and size of the dam in question.
- There is no rule for defining the number of measuring devices required to ensure satisfactory monitoring of behavior; it is better to use a limited number of reliable instruments, which also makes interpreting measurements easier.
- There is a difference between instruments that track the general behavior of the dam and indicate an anomaly and complementary instruments that help to explain the reason behind a particular phenomenon.
- The monitoring system is not fixed. Indeed, it is a good idea to regularly assess whether it still satisfies all the needs and requirements; if necessary, it should be adapted, extended, or modernized. Even though instruments on the market continue to evolve, the parameters to be measured remain the same.

And lastly, should measuring instruments be changed or replaced, care should be taken to ensure that data is recorded continuously.

32.4.4 Choice and characteristics of measuring devices

32.4.4.1 Criteria for selecting devices

The choice of measuring device depends on the parameters to be observed, the method of dam construction, and possibilities for installation. The choice must be determined based on each specific case. However, priority must be given to instruments that meet the following criteria:

- Simple in their design and use (measurements are usually taken by staff employed by the dam operator)
- Robust
- Not sensitive to external conditions: temperature, humidity, electrical surge
- Long-lasting (device longevity must be guaranteed, especially for those integrated into the body of the dam during construction)
- Precise and reliable
- Easy to read

For those that are not integrated into the body of the dam, they must be

- Accessible
- Replaceable while ensuring guaranteed continuity of measurements

To prevent breakdowns or failures, redundancy measurements should be planned for certain parameters (for example, deformation measurement). With regard to the reliability of measuring devices, the failure rate is variable and depends on the type of instrument. It should also be noted that instrument longevity is less than the lifetime of the dam. A faulty device may need to be replaced. Appropriate back-up instruments should therefore be readily available (SwissCoD, 2013).

32.4.4.2 Types of instruments to be considered

The Tables 32.5, 32.6, and 32.7 provide lists of the various instruments and general measurement methods that form part of a monitoring system. Further information and descriptions can be found in the document published by the Comité suisse des barrages "Dispositif d'auscultation des barrages. Concept, fiabilité et redundance" (SwissCoD, 2005a, 2006).

Table 32.5 Instruments for measuring load and external conditions

Type of measurements	Equipment
Reservoir level	Depth gauge
	Pressure balance
	Manometer
	Cable with indicator (audio or light)
Sedimentation level	Bathymetry
Water temperature	Normal thermometer
Weather conditions	Thermograph, thermometer
	Rain gauge
Seismic conditions	Seismometer, accelerometer

Table 32.6 Instruments and inspections for concrete dams

Type of measurements	Equipment
Structural deformation	Direct pendulum
	Inverted pendulum
	Inclinometer
	Extensometer
	Fiber-optic sensor and cable
	Geodesy
	Geodetic survey (terrestrial measurements and GPS)
	Leveling
	Polygonal
	Vertical line of sight
	Simple angular measurements
	Alignment
Local movements (cracks, joints)	Jointmeter
	Micrometer
	Fiber-optic sensor and cable
	Dilatometer
	Deformeter
Dam body temperature	Normal thermometer
	Electronic thermometer
	Fiber-optic sensor and cable
Uplift at the concrete-foundation interface	Manometer
	Pressure cell
Leaks, seepage, and drainage flow	Weir, venturi
	Volumetric measurements
Chemical analysis of seepage water	
Tension of anchors (in the body of the dam, in the foundation)	Load cell (hydraulic or electrical system)

Table 32.7 Instruments and inspections for embankment dams

Type of measurements	Equipment
Deformation along horizontal and vertical lines (settlement)	Geodesy Geodetic survey (terrestrial measurements and GPS) Leveling Polygonal Alignment Angle and distance measurements Settlement gauge
Saturation curve	Piezometer Fiber-optic sensor and cable
Pore pressure	Manometer Pressure cell Fiber-optic sensor and cable
Seepage and drainage flow	Weir, venturi Volumetric measurement
Chemical analysis of seepage water	
Dam body temperature	Electronic thermometer Fiber-optic sensor and cable

In recent years, technological advances in instrumentation and surveillance techniques have led to improved monitoring accuracy and data assessment (ICOLD, 2018c). The following equipment and applications are given as examples of these improvements (Table 32.8).

Table 32.8 Recent developments in instrumentation and their applications

Equipment	Applications
Laser scanning and digital imagery	Accurate distance measurements using laser with high spatial resolution on surfaces (3D geometry of dam)
Fiber-optic sensor and cable	Temperature and strain measurements (in optical fiber using laser light)
GPS (Global Positioning System)	Space measurements by satellite (accurate distance measurements between orbits and sensor)
Ground Survey Aperture Radar (GBInSAR)	Photogrammetry method using ground station images
Ground Penetrating Radar (GPR)	Detect changes in properties of near-surface ground layers, localization of defects or voids in concrete structures
Borehole micrometer	3D measurements of deformation in a borehole
Resistivity	Active electrical method to detect changes in material properties
Multibeam bathymetry	Echo sounding

Fiber-optic sensors and cables

In the context of new measurement technology, it should be noted that fiber-optics have been used in measuring instruments in dams since the 1990s, primarily as (SwissCoD, 2005a):

- A device in which the optical fiber is itself the measuring instrument (such as the extensometer)
- A device in which various phenomena are measured along the fiber-optic cable
- A device in which the optical fiber provides a means for transporting data (pressure, temperature, difference in length)

Indeed, compared to normal electrical cable, fiber-optic cables are less liable to wear and other effects. The measurement cells behave in the same way as an electronic instrument – undergoing a small deformation due to external influences (temperature, deformation, pressure).

Measuring instruments using fiber-optic technology include:

- Earth pressure sensors
- Settlement indicators
- Extensometers in loose material
- Extensometer in concrete
- Jointmeter, displacement measurement
- Piezometer
- Temperature sensor

3D measurements of deformation in a borehole

The borehole micrometer (developed at the ISETH, which is affiliated with the ETH Zurich) is a mobile measuring device that enables measurement in successive 1-meter sections of, for example, differential variations in length along the borehole. This instrument is equipped with an inclinometer that determines displacement in 3 orthogonal directions along a vertical borehole (SwissCoD, 2005a).

Face observation using a drone

A hi-res photographic recording of the downstream face of the dam can be produced by using a drone, which can then form part of an inspection report. This type of inspection was used for the Zeuzier dam in Switzerland in 2016.

32.4.5 Automation and transmission of measurements

As a result of the development in electronics and computer processing, the capacities and value in automating monitoring systems have only increased. A direct liaison with the user is possible. Such systems comprise means for measurement (measuring instruments), means for transmission of data, and automatic means for the acquisition and storage of data (databases), as well as means of treatment and presentation of data (data analysis, preparation of graphs, and writing of reports).[4]

The use of automatic measurements in carefully selected locations, as well as computer data links, enable near permanent surveillance. An automatic measurement system can be useful for dam sites with difficult access (due to unfavorable external conditions, for example, in winter), as well as for hard-to-reach measurement points.

---

[4] See publications SNCOLD, 1993, and SNCOLD, 1994.

Even though this type of installation may be added to a traditional manual arrangement, it does not in any way replace it. Continuity of measurements must be guaranteed, even in cases where the automation fails.

The automatic logging of all measurement points is not essential; instead, limiting measurements to only some characteristic parameters and points is recommended (Table 32.9).

Table 32.9 Parameters recommended for continuous measurements

For all types of dams	For concrete dams	For embankment dams
Reservoir level	Deformation characteristics (for example, total deformation of the section)	Seepage and drainage flow rate characteristics (for example, total discharge)
Weather conditions (precipitation)	Concrete temperature	Pore pressure
Impermeability	Seepage and drainage flow rates at specific points	Turbidity of seepage water

For the automated measurement of variables, the following points must be considered:
- Simple and robust sensors
- Electromagnetic compatibility
- Protection against electrical surge
- Protection against humidity
- Adequate range of operating temperature

Finally, the results from automatic measurements (for example, pendulums) must be checked if possible at least once a year using manual measurements. This approach ensures periodic and regular staff presence at the dam and means that visual observations and the monitoring of the various systems can also be carried out.

Furthermore, it is important to regularly verify the operation of all automatic devices (site inspection, remote testing). Any dysfunction in part of the installation could lead to the loss of data.

At least one value should be saved each day (instantaneous or average) for the automatically measured parameters (reservoir level, air temperature, behavioral parameters such as deformation, pressure, seepage rates). Care must be taken to ensure that the information contained in automatically produced reports does not contain any errors.

32.4.6 Description of instruments

32.4.6.1 Measurement of external conditions and load

A summary of instruments measuring load and exterior conditions can be found in Table 32.5.

Reservoir level

Measuring water level in the reservoir is vital to allow analysis of dam behavior. This measurement is often managed by the reservoir operator, for whom the volume stored behind the dam is an important piece of information.

There are several systems for measuring water level. The most reliable is undoubtedly the pressure balance, whose measurement range can cover all variations in reservoir level with an accuracy of around ± 10 cm. Calibration and verification of the pressure balance are vital (redundancy, with other measurement modes). Other methods include the depth gauge, which is a graded ruler in wood or metal with decimetric and centimetric marked increments, and an ultrasound sensor that detects the surface of the water through reflecting ultrasound waves.

During a flood event, staff must be able to visually track the increase in water level, which, if nothing is done to prevent this, may reach a level higher than the crest of the dam. It is therefore important to install reference marks and/or depth gauges with visible increment markings. This redundancy of tracking systems also provides a back-up should the automatic measuring system malfunction.

Sedimentation level
Measuring the level of sediment deposits in a reservoir can be done from a small boat, either by using a sounding rod or weighted line, or by ultrasound. In order to draw the profiles and topography of the deposits, the position of the boat when carrying out measurements must be calculated. It is obtained from the bank using an angle and distance measurement or by using a GPS positioning measurement. An even simpler option involves stretching cords with distance markers from one bank to the other.

Figure 32.10 Mapping the bottom of a reservoir using ultrasound and GPS positioning measurements

Temperatures
Air and reservoir water temperatures are easy to measure. These parameters are measured indirectly during the verification of dam safety. For concrete dams, for example, the distribution and variation of temperatures inside the dam should be recorded (see § 32.4.6.4).

Air temperature can be measured using an industrial capillary thermometer. A maximum/minimum thermometer is especially recommended, because as well as indicating the current temperature, it also enables the lowest and highest temperatures from the measurement period to be recorded. A thermograph enables the continuous measurement and recording of ambient temperatures. The information is logged on a roll of paper attached to a drum.

For measuring water temperature, a thermometer is placed in a protective metal tube with a perforated water compartment. The idea is to fill the compartment with water at a certain temperature. The thermometer is then brought to the surface with the compartment full of water from a given depth. It is then possible to read the temperature.

Rainfall
Standard rain gauges and cumulative rain gauges provide information on how much precipitation has fallen in the region of the dam. The results from these measurements mean that weather conditions can be considered when interpreting seepage and drainage rates, as well as water pressure in the foundation, which are all influenced by the level of the groundwater table. The recorded measurements also provide data for hydrological studies.

Seismology
The distribution of accelerometers with three components inside and around the dam is generally established based on the type of dam and the desired objectives. The equipment must be designed so that after a large earthquake, information relative to the dynamic behavior of the structure and to the effective movement along the abutments can be accessed. External devices provide information from a zone situated near the dam that is not affected by it (Darbre and Pougatsch, 1993; Pougatsch, 1993; Darbre, 2004). As a minimum, the SFOE guidelines recommend placing sensors on the crest (if possible in the main section), at the toe of the dam, and on the foundation rock at a certain distance (SFOE (SFWG), 2003). The installation of accelerometers also serves as a sort of alarm system.

As an example, Figure 32.11 shows the equipment installed at the Mauvoisin dam as part of the seismic network of accelerometers implemented in 1992 in Switzerland.

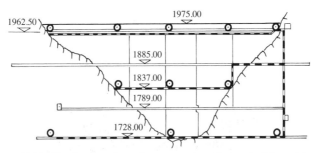

Figure 32.11 Distribution of accelerometers installed at the Mauvoisin dam

32.4.6.2 Measuring deformation along horizontal and vertical lines (concrete and embankment dams)

A difference must be drawn between measurements that provide information on the overall deformation of the dam and those that measure particular points (for example, joint or crack movement).

Appropriate surveillance of deformation generally requires a spatial measuring device that collects information on the altimetric and planimetric displacement of the selected points. The monitoring or measuring points should be located on the crest and in the galleries, on the faces or slopes, as well as in the surrounding vicinity. As part of the regular surveillance of a small or medium-height dam, only the movement of points located at the level of the crest are tracked.

Concrete dam

If a concrete dam has monitoring galleries and shafts, an orthogonal system can be generated for measurements along vertical and horizontal lines. Where possible, measurement axes must be extended into the abutments and banks (Figure 32.12). If there are no shafts, geodesic sighting targets can be installed and distributed across the downstream face.

Deformation along a vertical line can be obtained by measurements from direct and inverted pendulums (Figure 32.13). Inverted pendulums can be installed in vertical boreholes drilled into the rock; the deepest anchorage point is generally considered to be a fixed point (only geodesic measurements can indicate whether this point moves). If there are no shafts, a borehole can be drilled, providing it is perfectly vertical. A self-centering guidewire can also be slid down a grooved tube, which gives fixed points at different elevations.

In RCC dams, pendulums are installed in boreholes that are drilled after construction.

The required accuracy for measurements is ± 0.2 mm.[5] Manual measurements can be carried out by one person alone. They can also be automated and sent by computer link in real time to a monitoring center.

Horizontal deformation at the level of the crest and horizontal galleries can be determined by polygonal measurements. For concrete dams that have not been fitted with pendulums, angle and distance measurements (vector measurement) on outside targets affixed at the level of the crest and on the downstream face

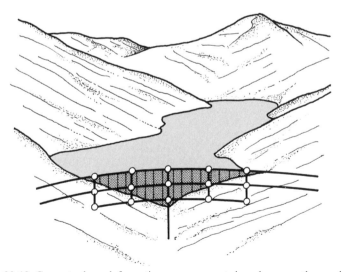

Figure 32.12 Concrete dam: deformation measurements based on an orthogonal network

---

[5] The required accuracy is related to a value that should be reached, based on the type of measurement. It does not necessarily correspond to the accuracy of the device.

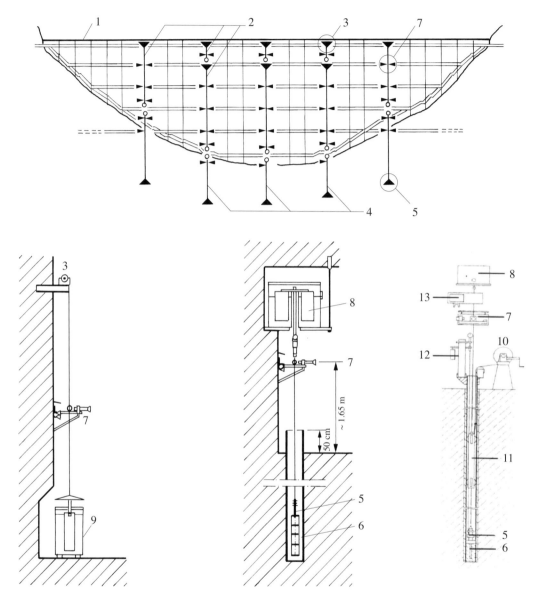

Figure 32.13 Layout of pendulums and inverted pendulums (SwissCoD, 2005a): 1 dam crest; 2 direct pendulum; 3 suspension point of a pendulum; 4 inverted pendulum; 5 fixation of inverted pendulum; 6 hardened cement grout; 7 measurement point (for coordiscope); 8 float/float vessel; 9 damping vessel; 10 hoist; 11 mobile centering system; 12 automatic vertical distance measurements; 13 automatic measurements (Huggenberger SA system)

are a simple geodesic method for measuring deformation; this method is appropriate for small or medium-height dams. Sighting a point deemed to be fixed requires good weather conditions. Accuracy depends on the network installed. It is less reliable than that obtained from pendulum measurements.

Other methods for calculating horizontal deformation exist: these involve optical alignment measurements and wire alignment. For optical alignment, sighting from a stationary point toward a fixed reference mark constitutes a vertical reference plane, along which the monitoring points are found. Deformation corresponds to the variations in distance compared to the reference plane. As for wire alignment, this works by stretching a wire from one bank to the other along rectilinear walls in galleries or along the parapet at the level of the crest. This wire defines a vertical reference plane. Deformation compared to the plane that corresponds to variations in the distance of the wire to the wall are calculated using a measuring instrument attached to the wall. Measuring the position of the wire is carried out in a similar way to that of a pendulum.

For all types of dams, leveling at the level of the crest and galleries provides information on vertical movement. Where possible, extending the leveling path into the banks is recommended (Figure 32.15). Accuracy depends on the instruments and the sighting length chosen.

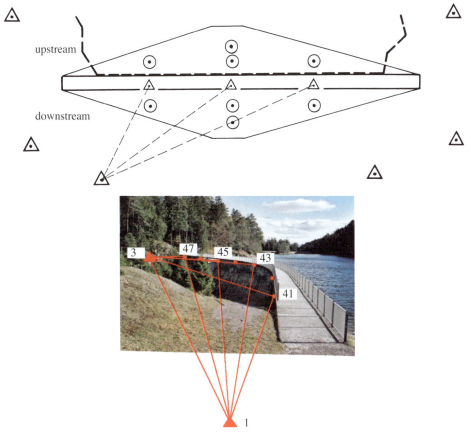

Figure 32.14 Concrete and embankment dams: deformation measurement using angle and distance measurements (Courtesy Schneider Ingénieurs SA, Chur)

Once displacement is known, it is then a simple matter to obtain the deformation along the horizontal and vertical lines.

Lastly, geodesic measurements are also used to determine deformation in RCC dams.

Embankment dam

For embankment dams, the aim is firstly to understand the evolution of vertical (settlement) and horizontal deformation of the dam at the crown, but also, if possible, at other levels (in particular, foundation settlement). The horizontal displacement of points is generally determined using geodesic measurements such as angle and distance measurements (vector measurement), alignment, and polygonals (Figure 32.14). As for vertical displacement (settlement or heaving), this can be determined by leveling measurements (Figure 32.15), as well as by settlement gauges or hydraulic settlement markers. Leveling can be carried out at the level of the crest and possibly from a berm.

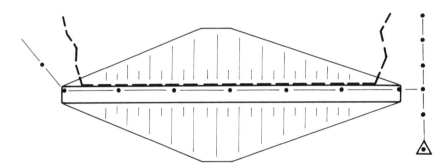

Figure 32.15 Concrete and embankment dams: measuring vertical deformation through leveling along the crest and on the banks

Vertical settlement mark (settlement gauge)

The measuring device for embankment dams is made up of a vertical plastic tube placed in the embankment as construction progresses (Figure 32.16 (a) and (b)). Steel plates (or metallic rings) are placed at regular intervals outside the tube, connected to the embankment and able to slide along the tube. An induction sensor introduced into the tube from the surface captures the electromagnetic interference created by the presence of the steel plates. The distance is measured compared to the edge of the tube's upper rim. Required accuracy is in the order of ± 1 cm during operation, for lengths of up to 100 m. This measurement is always combined with crown leveling.

Measuring local incline

Different types of devices have been developed to measure slopes from the base of a concrete dam or along a vertical section. In the case of the latter, it is possible to draw the deformation if measurements of the various elevations are available (Deinum, 1987). Devices can be separated into two types: systems with liquid (Figure 32.17) and those with a pendulum mass. Attention should be paid, however, to the fact that variations in slope near to niches or galleries can be influenced by the local movement of forces.

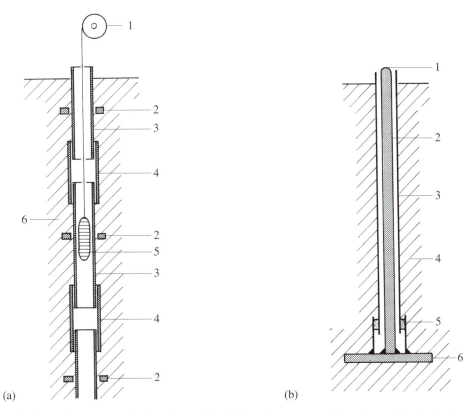

Figure 32.16 (a) Settlement mark (from SwissCoD, 2005a): 1 measuring cable and recording device; 2 metal plates; 3 tubes; 4 sleeve; 5 sensor; 6 embankment; (b) Settlement mark with base plate: 1 measuring point (leveling); 2 measuring rod; 3 protective tube; 4 embankment; 5 sliding sleeve; 6 base plate

Figure 32.17 Installation of an inclinometer (from SwissCoD, 2005a)

Crack and joint movement

Concrete dams are not exempt from cracks. Although a visual survey can sometimes be enough, in some instances it is a good idea to track the lip movement of certain cracks. Reference marks can also be installed to measure the movement of joints in the dam. The instruments used are generally comparing devices. For example, the most commonly used devices are micrometers (Figure 32,18), jointmeters, and dilatometers/deformeters.

Figure 32.18 3D joint deformation measurements in a dam (from SwissCoD, 2005a)

32.4.6.3 Geodesic measurements

a) Geodesic networks

Geodesic measuring is an integral part of dam surveillance. It is used alone or in conjunction with other measurement techniques to carry out the three goals of surveillance, that is:

- To continuously assess forces on the dam, as well as the state and behavior of the dam
- To rapidly assess the state and behavior of the dam in the case of an extraordinary event
- To support research into the cause of anomalies, when they occur (SwissCoD, 2018c)

Geodesic measurements support other measuring systems (redundancy). It is important to underline that geodesic measuring is a specialist activity and that it is dependent on weather conditions.[6]

Thanks to geodesy, it is possible to measure absolute displacement from reference marks compared against a network of stations presumed to be fixed. It should be noted that the measurement of points located inside the dam only enables relative deformation to be determined, which is sufficient in the context of regular surveillance. However, to obtain information on the development of permanent deformation or to detect, explain, or track abnormal behavior compared to projected deformation, absolute deformation values are needed. To achieve this, measurements from inside the dam (that is, using pendulums, borehole micrometers, leveling, invar wire) are coupled to a triangulation network based on fixed marks outside the dam (optical geodesy or GPS).

---

[6] The publication by the Swiss Committee on Large Dams (SNCOLD, 1997a) provides useful information concerning geodesic measurements in the context of dam surveillance. See also Biedermann, 1981; Egger, 1982.

The plan for the geodesic network must take into account the fact that the water-storage structure includes not only the dam, but also its foundations, the reservoir and its banks, and any part of the surrounding terrain that might have an influence on the dam or reservoir. The measuring network is thus divided into regular networks and extended networks. Regular networks cover the immediate surroundings of the water-retaining structure and enable all the tasks of local surveillance to be carried out. Extended networks cover a larger perimeter and integrate the regular networks into a broader framework (SwissCoD, 2013).

Accordingly, the system specific to the dam is complemented by the reference space within which it is located. It can then be further connected to the extended reference network, whose point locations can be measured by satellites (GPS). The *Global Positioning System* (GPS) provides an elegant way of integrating geologically stable points into the surveillance network, outside of the deformation zone. Coupling to existing geodesic networks can be achieved with the conventional terrestrial method or with GPS. The combination of GPS and terrestrial geodesic measurements forms a hybrid network. The surveillance of deformation can thus be based on a spatial measuring device (see point b, below).

Although tests have been carried out, it should be noted that currently it is not commonplace to set up GPS stations directly on dams to measure displacement.

To create a network of geodesic measurements (Figure 32.20), the fixed points of the external network must be, among other things:

- Set up sufficiently far from the dam so as not to be influenced by movement in the foundation
- Situated both upstream and downstream (at least 4 points)
- Made of concrete pillars

Figure 32.19 Various networks for geodesic measurements (Courtesy KW Vorderheim AG)

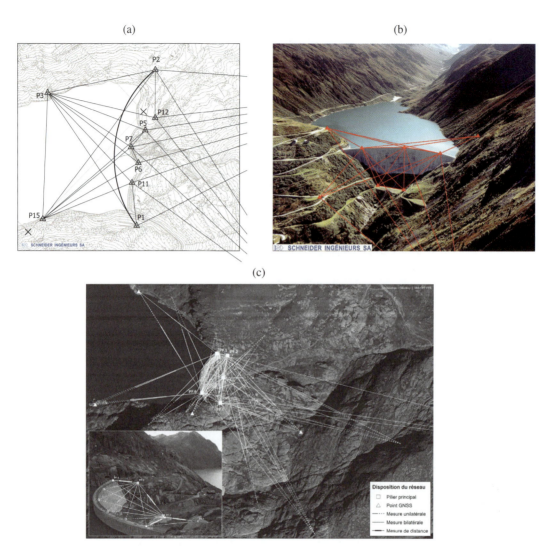

Figure 32.20 Network of geodesic measurements: (a) plan of the geodesic network at the Zeuzier arch dam (plan established by Schneider Ingénieurs SA); (b) arrangement of points in the geodesic network at the Nalps arch dam (photographic montage Schneider Ingénieurs SA); (c) aerial view of the geodesic network of the Vieux-Emosson arch gravity dam (photographic montage Swisstopo)

Setting up points for the various networks is the result of collaboration between the surveyor, the dam engineer, and the engineering geologist. Lastly, the other measurement systems located in the body of the dam and the surrounding area must be coupled to the geodesic network.

b) Geodesic measurements and use of GPS (Global Positioning System)

The US Department of Defense has developed and created a spatial navigation system comprising a constellation of 24 satellites rotating twice daily on an orbit situated approximately 20,000 km from the Earth. A European consortium has also developed and implemented the GALILEO system. Russia and China also have their own systems.

These systems can be used to determine absolute or relative (3D Cartesian coordinates) positions by measuring the distances between the receiver antenna and the satellites. For geometric resolution, at least 3 distance measurements are required. The position of the antenna is obtained by the intersection of 3 spheres whose centers are positioned at the satellite and for which the 3 measured distances are the radii.

Provided that the measurement method is appropriate, this spatial navigation system can be used to accurately determine a position relative to a given point. To achieve this, there must be 2 GPS receivers simultaneously measuring the distance to the visible satellites (at least 3).

Compared to traditional geodesic methods (for example, angle and distance measurements), measurements using GPS are largely independent of weather conditions. A visual liaison between the various stations is not required (SwissCoD, 2005a).

c) Laser scanning and digital imagery

Laser scanning uses the principle of distance measurement without reflectors. It is combined with a system of horizontal and vertical scans made by a laser beam at high speed; depending on the device this can be from 1,000 to 500,000 measurements per second in the direction of the object to scan (dam, building, tunnel, ground surface, etc.).

The raw data from a scanner are in the form of a point cloud. Through the intermediary of reference points, such as national coordinates, and the use of sophisticated software, drawings, ground surveys, 2- and 3-dimensional drawings of 3D objects, etc. can be obtained.

Any object can be scanned using laser scanning, and it has now replaced terrestrial photogrammetry. Distance measurement without reflectors between the laser scanning station and the object ranges from 50 to 200 m maximum. The relative accuracy of the points from land laser scanning is less than 1 centimeter but depends on the distance and quality of the reflective surface of the object (SwissCoD, 2005a).

Laser scanning can be used to scan the surfaces of concrete dams and, for example, to locate damage. It can also provide information on the abutments, appurtenant structures, or unstable zones upstream or downstream (Figure 32.21 (a)).

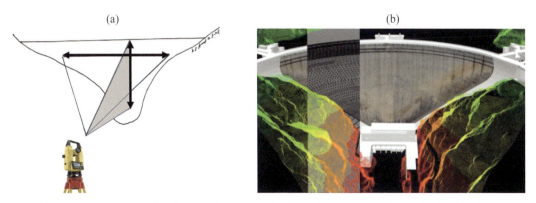

Figure 32.21 (a) Diagram illustrating the scanning method for the surface of a concrete dam (SwissCoD, 2005a); (b) example of a laser scan from the Cabril arch dam in Portugal (ICOLD, 2018c)

d) Surveillance of critical ground zones

Surveillance of landslides focuses on the size, direction, and speed of movement of critical ground zones. As displacement can often reach several centimeters, the biggest challenge is generally not to achieve accurate measurements but to ensure reliable measurements.

32.4.6.4 Temperature measurement

Concrete dams

During the construction of conventional concrete and RCC dams, changes in the temperature of the concrete mass must be tracked during the hydration and hardening phases. After that, during operation, it is useful to track variations in the internal temperature of the concrete to determine how they affect deformation.

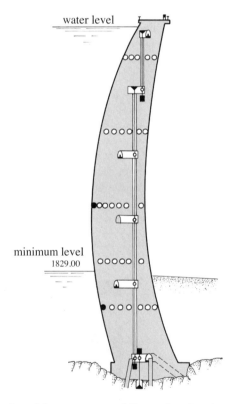

Figure 32.22 Distribution of thermometers at different elevations in one section of an arch dam (from SwissCoD, 2005a)

In a conventional concrete dam, concrete temperature is measured using electronic thermometers (resistance thermometer, thermoelement, vibrating wire thermometer) embedded in the mass during construction. These are installed at various elevations and are spread throughout the depth of the dam wall. The number at each elevation depends on the thickness.

The thermometers are integrated directly into the mass during concreting. In order to ensure adequate data redundancy, a large number of sensors must be installed (Figure 32.22). The thermometers located near the surface are strongly influenced by local external conditions (air and water temperatures).

With regard to the accessibility of instruments, electronic thermometers can be slid into vertical or inclined boreholes in a tube that has been encased in the concrete, so that they can be read at the required depth. They are insulated so as not to be affected by outside temperatures or by the gallery temperature. In the case of a failure, they can be removed and replaced.

Measuring temperature using fiber-optic cables (concrete and embankment dams)
Temperature measurements using a fiber-optic cable are carried out with a laser beam. Temperatures are determined by spectroscopy applied to the reflected light. The location of temperatures is determined by measuring the distance between the travel time of the emitted light and that of the reflected light, while taking into account the propagation and speed of light in the fiber-optic cable. This method is particularly suitable for tracking temperature variations in conventional concrete or RCC during the hardening and cooling phases in one or several sections.[7]

This system can also be used for detecting water infiltration in embankment dams. The temperature of an embankment dam is subject to the influence of external factors such as rain and air, but also internal factors such as the presence of a leak or groundwater.

Temperature measurements can be used indirectly to determine the presence, location, and extent of water infiltration in embankments dams or foundations. A section of the dam with a low seepage rate generally has a different temperature distribution to a section with a higher rate of seepage.

Two main approaches for detecting leaks with temperature measurements are used. The first, passive method uses natural temperature variations to detect and quantify any anomalies in the flow field, and the second, active method detects the presence and movement of water by assessing the thermal response in a heated fiber-optic cable; temperature variations caused by leaks are amplified.

32.4.6.5 Measuring water infiltration
The accumulation of water in the reservoir causes water infiltration through the dam and its foundations (even with the presence of a grout curtain). Depending on the type of dam and the geological nature of the foundation, expected flow rates can vary greatly. They may vary depending on the water level and temperature of the reservoir and can also be influenced by weather conditions or snowmelt. It is important to track the total flow through the dam. An unusual increase in seepage flow may indicate the presence of a potential hazard.

The seepage measurement system must be designed so that it can separate seepage issuing from the dam and seepage issuing from the foundations. Separating the collection of water into sections or panels of the dam is recommended in order to identify a particular zone in cases of abnormal behavior (Figure 32.23). Water flow arriving into a drainage gallery and issuing from drainage boreholes must also be measured. A decrease in flow may indicate a blocked drain.

---

[7] This system can also be used for detecting water infiltration in embankment dams.

(a)

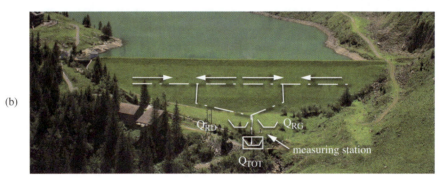

(b)

Figure 32.23 Theoretical diagram for measuring partial and total seepage flow in sections of (a) a concrete dam and (b) an embankment dam. Key: Q_{TOT} total flow; Q_{RG} flow from the left bank; Q_{RD} flow from the right bank (Courtesy H. Pougatsch and Elektrizitätswerk Nidwalden).

For concrete dams, seepage flow is generally concentrated along the least watertight zones in the concrete. Water, in particular, may find preferential seepage paths, especially along vertical joints and horizontal concrete lift joints, as well as at the concrete-rock interface. Seepage water is collected by the dam's network of internal galleries. As for embankment dams that do not have an upstream watertightness system, water gradually penetrates the upstream shell and is intercepted by drains and filters. It is then conveyed downstream for flow to be measured before being evacuated.

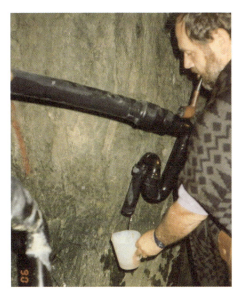

Figure 32.24 Timed volumetric measurement with a container (note the S-shaped pipe end to avoid blockage) (from SwissCoD, 2005a)

Figure 32.25 Weir with measuring needle (from SwissCoD, 2005a)

Flow measurement can be volumetric or be carried out by using a weir. Volumetric measurements are carried out manually with a tared container and a chronometer (Figure 32.24). With a calibrated weir, measuring water height upstream of a weir placed in the water evacuation channel enables flow to be calculated (Figure 32.25). Measurements can be carried out using an ultrasound sensor that may be automated and data transferred by computer link. Required accuracy is in the order of ± 5% of the measurement range, that is, presumed maximum flow + 100%.

Flow measurements are completed with water quality checks, including the measurement of turbidity and periodic chemical analyses. Turbidity measurements, especially for embankment dams, ensures that adequate knowledge is gathered concerning the movement of fine particles. As for the chemical analysis, this provides information about the origin of the water and also about the existence of a possible interaction between seepage water and materials in the dam or the grouting. Water sample collection occurs where measuring stations are located and at the outlets for open piezometers.

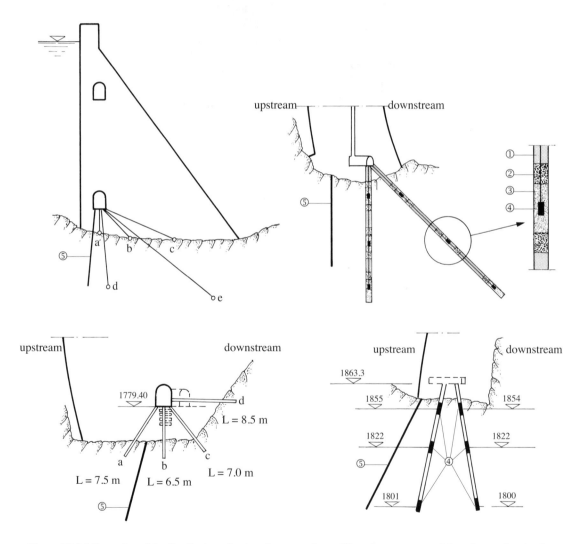

Figure 32.26 Examples of the distribution of points for measuring uplift at the concrete-rock interface and at depth: ① injection grout; ② clay balls; ③ quartz sand; ④ piezometer; ⑤ grout curtain (from SwissCoD, 2005a)

32.4.6.6 Measuring pressure

Concrete dam

Water infiltration under a concrete dam leads to the development of uplift, which counters the stabilizing effect of self weight. It is generally measured at the level of the concrete-rock interface at several locations along the surface of the foundation to track the reduction in uplift in an upstream to downstream direction. In some cases, measuring uplift at different depths is recommended (Figure 32.26).

Measuring uplift in the foundation of a concrete dam takes place most often from an inspection gallery, either through open or closed boreholes.

Open boreholes enable pressure to be measured when the level of pressure is lower than the measuring point. To find out the amount of load at a given point, the borehole must be sleeved impermeably up to the pressure tap (generally coated in sand). It is possible to place in one tube several piezometers that are attached to taps placed at different elevations and separated by watertight mortar. Measurement is carried out using a light or sound sensor. Accuracy is in the order of ± 0.05 m.

Pressure measurement in a closed borehole is carried out using a manometer. Boreholes for measuring uplift are usually placed in a series so as to establish the distribution of pore pressure in one section (Figure 32.27). Accuracy is in the order of ± 0.5 m or ± 1% of the entire height between the manometer and

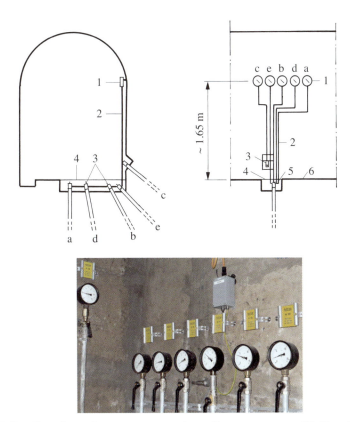

Figure 32.27 Coupling of tap tubes to manometers in a gallery to measure uplift. Key: 1 manometers; 2 flexible pipe or metal tube; 3 T-piece as closure and connection with a pressure tube; 4 metal cover; 5 gutter; 6 apron (from SwissCoD, 2005a).

the crest. It is directly related to the quality of the manometer. It is possible to place several pressure taps at different elevations in one borehole.

Embankment dam
Water infiltration through an embankment dam and its foundation is the source of pore pressure, which is hugely important for dam stability. Variation in pore pressure and its progression must therefore be monitored. These parameters are determined using pneumatic, hydraulic, or electric pressure cells. The cells are placed into the embankment during construction. To ensure adequate quality, a considerable number of cells must be installed. Special care must be paid to the cables, which are sensitive to differential settlement. Required accuracy is in the order of 1% of maximum measurable load, but with a minimum of ± 50 cm. Piezometers installed in the downstream shell enable the presence of groundwater to be observed (Figure 32.38).

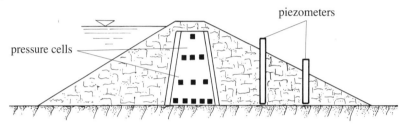

Figure 32.28 Zoned embankment dam: distribution of pressure cells in the core and piezometers in the downstream shell

Fluctuations in the saturation curve that become established in a homogeneous embankment dam may be tracked using piezometers (Figure 32.29).

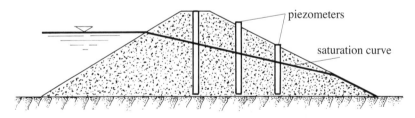

Figure 32.29 Homogeneous embankment dam: survey of the saturation curve using piezometers

32.4.6.7 Tracking foundation behavior

Deformation
For concrete dam foundations, measurements can be made using inverted pendulums and extensometers. The borehole micrometer, with or without an inclinometer, may be useful in some specific cases.

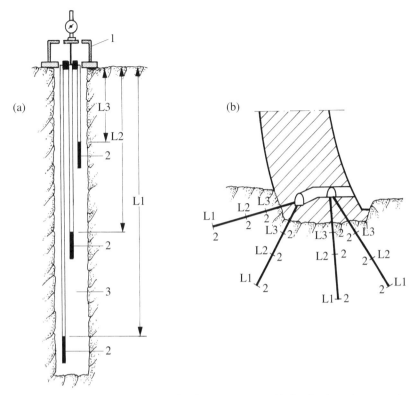

Figure 32.30 Cross-section and distribution of rod extensometers: (a) distribution of a measuring head fitted with 3 rods; (b) possible distribution of extensometers in the foundation of a concrete dam. Key: 1 measuring head; 2 rod anchoring point; 3 injection grout (from SwissCoD, 2005a).

Rod extensometers are installed from the base gallery of a concrete dam or from the downstream face (Figure 32.30). A rod is placed in a borehole in the rock; one end is sealed and the rest of the rod can slide. Measurements using a dial gauge enable the relative deformation between the sealing point and the measuring point to be established. Several rods of various lengths may be placed in the same borehole. Rod extensometers enable deformation in the foundation near the dam to be established. Required accuracy is at ± 0.2 mm. Measurements can easily be automated and sent by computer link.

The borehole micrometer (sliding micrometer) is a mobile measuring device that enables measurement in successive 1-meter sections of, for example, differential variations in length along a borehole (Figure 32.31). It is sometimes fitted with an inclinometer that enables displacement in three orthogonal directions to be determined. The borehole is fitted with a plastic, grooved tube with guides that is attached to its surroundings (rock, concrete, soil) with grout. It can be used in boreholes up to 150 m in length.

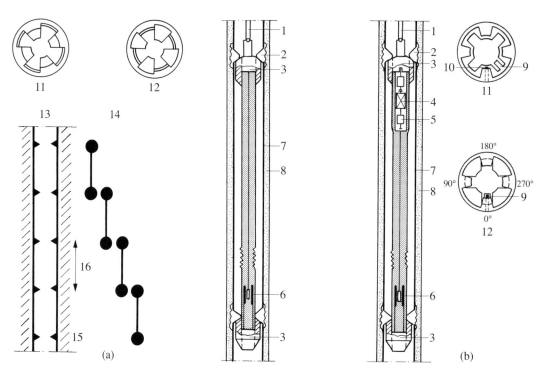

Figure 32.31 Example of the ISETH borehole micrometer: (a) measurement process and (b) Trivec with inclinometer (from SwissCoD, 2005a). Key: 1 rod; 2 conical measuring seats; 3 spherical head; 4 inclinometer; 5 positioning motor; 6 induction measuring sensor; 7 HPVC pipe; 8 grout; 9 grooves; 10 bolts; 11 maneuvering position; 12 measuring position (instrument rotation 45°); 13 casing with measuring mm; 14 position of successive sliding micrometer; 15 sensor position; 16 section to be measured.

Leveling, settlement gauges, and hydraulic settlement markers are some of the available methods for measuring settlement in loose material. These measurements are made in a gallery (providing one exists), in the transversal or longitudinal direction of an embankment dam. Abutment movement can be tracked by points that are installed in the immediate vicinity of the dam and connected to the network of geodesic measurements.

Pore pressure and piezometric level
A tube fitted with a manometer and pressure cells is used to measure pressure in the foundation.

Seepage and drainage flow
This can include water seeping through the foundation rock mass or water from springs that may or may not be collected. Measurement of flow can be volumetric or may be carried out by using a weir.

Table 32.32 summarizes the instruments used to track foundation behavior.

Table 32.32 Instruments and inspections for foundations

Types of measurements	Equipment
Deformation	Inclinometer
	Extensometer
	Borehole micrometer
	Geodesy
	Leveling
	Polygonal
Specific movements (joints)	Borehole micrometer
Uplift and pore pressure	Manometer
	Pressure cell
Groundwater level	Piezometer
Seepage and drainage flow rates	Weir
	Volumetric measurement
Turbidity	Venturi
Chemical analysis of seepage water	

32.4.6.8 Surveillance of areas outside of the dam

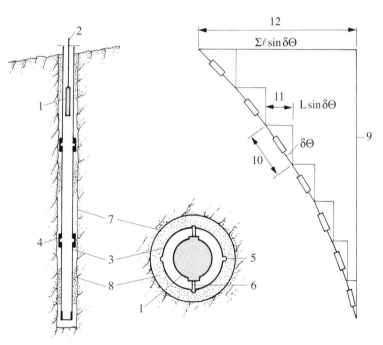

Figure 32.33 Approach for inclinometer measurements in a borehole (from SwissCoD, 2005a): 1 sensor; 2 cable; 3 grooved pipe; 4 sleeve; 5 groove; 6 sensor wheel; 7 borehole; 8 grout; 9 reference measurement; 10 measurement section; 11 horizontal displacement of measured section; 12 total horizontal displacement

This surveillance includes measurements (ground deformation, seepage) and visual observations. It includes upstream and downstream contact zones, the downstream toe of the dam, the sides of the reservoir, and the entire catchment area for the reservoir.

Downstream resurgence flow must also be recorded, as any variation in these rates may indicate an anomaly in groundwater flow networks. The measurement of flow rates may be volumetric and may possibly be carried out using a calibrated weir (for the measurement method, see § 32.4.6.5).

Lastly, recording fluctuations in groundwater levels (for example, downstream of an embankment dam) is sometimes recommended. Measurement of levels may be carried out using a sensor with an indicator, slid into an open borehole, or a pressure sensor with a recorder.

In the area right at the downstream toe of the dam, possible scouring should be recorded (topographical or bathymetric should there be water) periodically (every 5 to 10 years) or after an exceptional flood.

Should there be a marked tendency for sediment to build up in the reservoir, it is important to record its level, especially in the zone near the dam. Care should be taken that the water intake and water-release structures do not become blocked. Monitoring should be carried out, the frequency of which will depend on the amount of sediment build up (see § 32.4.6.1).

Lastly, in the area around the reservoir, unstable zones (which in the case of a slip could cause a wave and overtopping) should be monitored. Triangulation measurements (spatial variation of the displacement of points at the surface), distance measurements between several points, and inclinometer measurements (Figure 32.33) to track deep deformation can be employed.

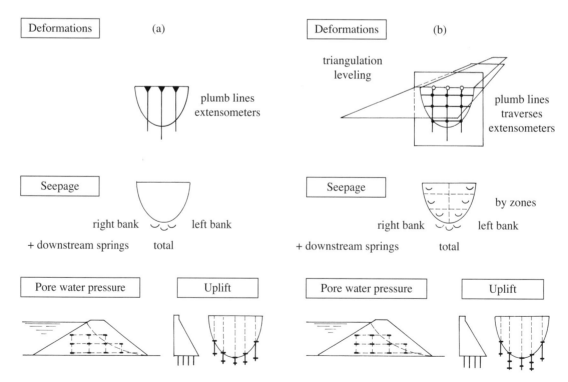

Figure 32.34 (a) Ordinary surveillance and (b) increased surveillance (from Biedermann, 1997)

32.4.6.9 Operating checks for measuring devices

Careful maintenance of measuring devices, as well as their base supports, reduces the risk of false measurements. Inspection and calibration provide information early on about the operating ability of these devices (SwissCoD, 2013).

When not in use, the devices must be stored in a safe and secure place. They must be checked prior to use. Some have integrated testing systems. Their condition must also be verified visually.

Measurement entries that have been sent by computer link must be checked monthly or annually using a back-up measurement taken with an independent manual method (see § 32.4.5).

An unexpected result does not automatically indicate abnormal behavior. Should this occur, the first step is to check whether there is a technical issue. Having spare devices available for use should a device become faulty is also recommended.

32.4.7 Frequency of measurements

The measuring program must provide enough information to assess dam behavior. It must be adapted to the type and size of the dam. The frequency will differ if measurements are taken during construction, first impoundment, or normal operation. For the latter, the program will also depend on the water level of the reservoir (a distinction may be made between a lowered or full reservoir) and the nature of dam behavior (normal or abnormal).

The most frequent measurements (weekly, biweekly, monthly) involve parameters that provide information about the overall behavior of the dam (for example, pendulums, concrete temperatures, drainage flow, pressure). Other complementary parameters (for example, joint deformation, rotation measurements) should be recorded once or twice a year. Full geodesic measurement programs, which involve specific measurements, are generally carried out every 5 years. For embankment dams, leveling and possibly polygonal measurements are made once or twice a year to track the development of deformation.

Periodic measurements must also be carried out after an extraordinary event, such as an earthquake or a flood (measurements of the most relevant parameters). Furthermore, in the case of an anomaly or specific behavior, the frequency of measurements will increase (Figure 32.34).

Automatic measurements occur on a daily basis.

32.5 Analysis and Interpretation of Measurements

The analysis and interpretation of surveillance data are part of a vital exercise enabling dam behavior to be assessed. The logging of surveillance data is generally undertaken by the dam operator, bearing in mind that a geodesic measurement program is usually entrusted to specialized experts. In accordance with accepted practice in Switzerland, the various interpretation and validation tasks are carried out by operators, experienced engineers, and senior experts (see Section 32.1). It is very important that analysis occurs soon after the measurements have been taken to ensure first of all the plausibility of raw measurements and later to verify the correct behavior of the dam or possibly to rapidly detect any noncompliant behavior (SFOE (SFWG), 2002b). Figure 32.35 provides an illustrated diagram of the paths of measurement data for the purpose of interpretation and validation.

An anomaly in one series of measurements can appear in the form of a sudden jump, a discontinuity, or a deviation. When particularities occur, the appropriate action should be taken. This could include repeating measurements and/or increasing their frequency, checking devices, installing new instruments, undertaking strengthening works, imposing limited operation, etc.

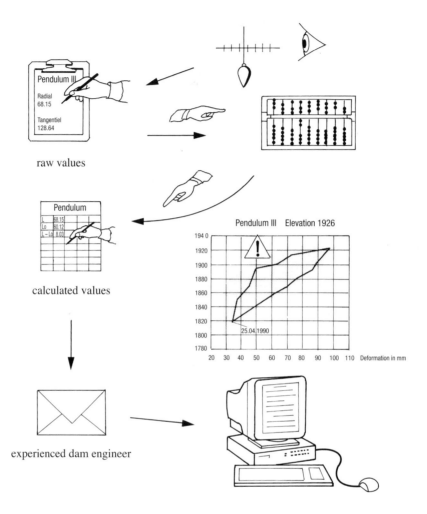

Figure 32.35 Path of measurement data for interpretation and validation (from SNCOLD, 1993)

32.5.1 Checking raw data

Once measurements have been carried out, it is very important to first check the plausibility of the raw data before transferring them to an expert for interpretation. It is possible to establish a set of plausibility criteria by using a mathematical model and physical laws or by using a historical database. This check, which is generally undertaken by operating staff, usually involves comparing the obtained values with

SURVEILLANCE AND MAINTENANCE 697

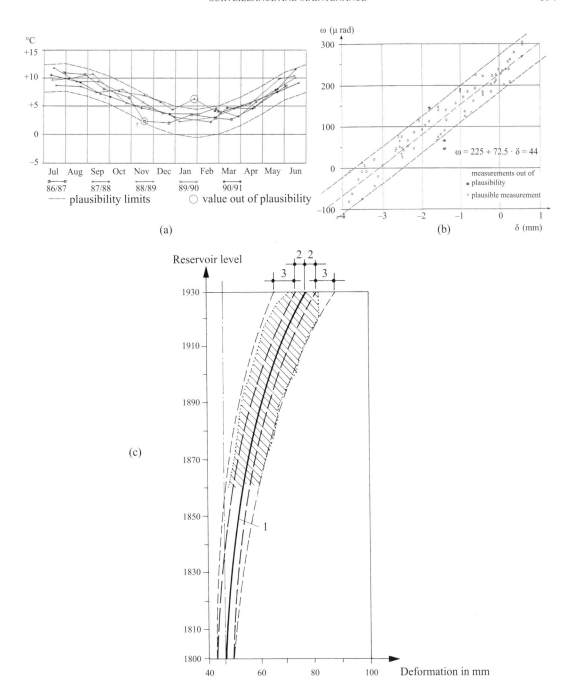

Figure 32.36 Ways of checking plausibility (from Lombardi, 1992): (a) historical plausibility checking (concrete temperature); (b) correlational plausibility checking (relation between readings from inclinometers ω and the radial deformation of a pendulum δ); (c) functional plausibility (Key: 1 theoretical model; 2 tolerance of the device measurements; 3 tolerance of physical value)

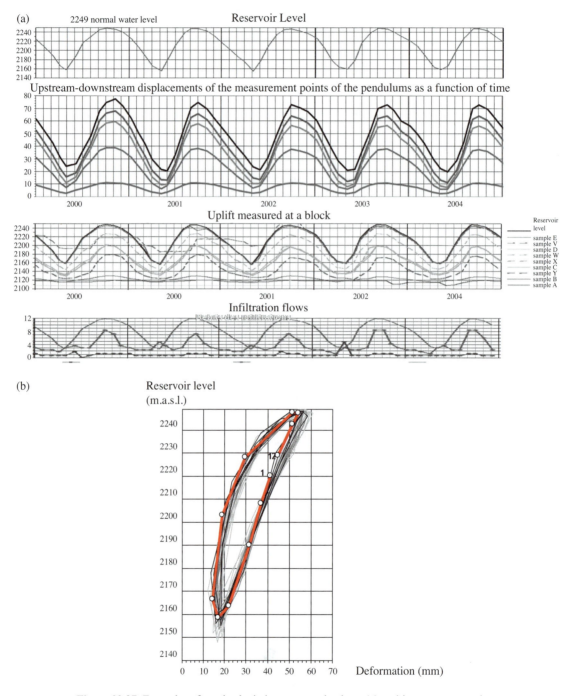

Figure 32.37 Examples of graphs depicting measured values: (a) multiyear representation of values from different parameters; (b) envelope curve: representation of the value of a parameter as a function of reservoir water level

the previous measurements, while taking into account the dam's state of load and the measurement period.

There are different types of plausibility checks: historical, correlational, and functional (Lombardi, 1992). Historical plausibility introduces a range of values whose limits are defined by the observed variation of a parameter, while taking into account the reservoir water level and/or the period (for example, concrete temperature measurements) (Figure 32.36 (a)). Correlational plausibility relates two parameters that have been measured differently (for example, pendulum and inclinometer measurements, radial and tangential deformation); a correlation and confidence interval are thus established. Functional plausibility involves setting tolerance limits (Figure 32.36 (c)).

Note that plausibility checks can be carried out at the dam site. Once the measurements are complete, the dam warden verifies if the observed value falls within a predetermined tolerance range based on the conditions at the time of the measurements. If there is any doubt, one or several measurements should be carried out again without delay.

The graph illustrating the measurement data from various parameters (deformation, pressure, flow) is also helpful for an initial evaluation but is not sufficient for an overall assessment of dam behavior (see § 32.5.2). These graphs can depict one value or a group of values as a function of time across the year in question or across several years (Figure 32.37 (a)) or even as a function of the reservoir water level. A graph showing parameters as a function of reservoir water level can also be established by drawing an envelope curve that includes results from several years (Figure 32.37 (b)). These graphs can demonstrate a possible non-cyclical evolution.

32.5.2 Processing of measurement data

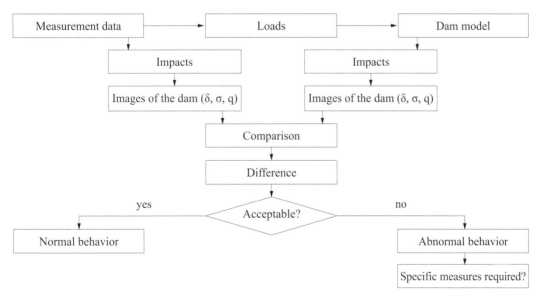

Figure 32.38 Flowchart for studying dam behavior (from Lombardi SA)

32.5.2.1 General principles[8]

As previously mentioned, an expert must be engaged to analyze the measurement data in order to confirm them and ratify them, providing overall dam behavior is in line with expectations.

The approach used to analyze data is based on the comparison at moment t between the observed values M (depicting measured reality) and theoretical or predicted values P (reflecting expected behavior), while taking into account environmental conditions (water pressure, air and concrete temperatures, etc.). This comparison will always lead to a difference D between the M and P values. The following general equation can be written:

$$M(t, env) = P(t, env) + D(t, env),$$

hence

$$D(t, env) = M(t, env) - P(t, env).$$

The observed value or behavior indicator M(t, env) corresponds to the representative size of one aspect of dam behavior and issues from one of the surveillance measurements. The expected value P(t, env) under the effect of load borne by the dam is given by a predictive model that forms the basis for reference. Then the D(t, env) differences between two series of values are evaluated. The issue lies in determining if the differences are acceptable or not. The lower the difference, the more the behavior is in line with expectations. Some differences are due to coincidental reasons (imprecise measurements) and others may reflect a dysfunction in the dam or possibly in an instrument. Decisions can then be made on the basis of this analysis. Figure 32.38 illustrates a flowchart for studying dam behavior.

It is important to develop a predictive model that is specific to the dam in order to verify its behavior. This may sometimes need to be adjusted or adapted during the lifetime of the dam. The model compares the elements of the predicted indicators and the crucial variables that are associated with them. The crucial variables that determine dam behavior are primarily reservoir water level, concrete temperature (related to air and water temperatures), and the age of the dam (time since its construction). Behavioral modeling can be done using a deterministic, statistical, hybrid, or mixed approach.

The deterministic model indicates whether the observed dam behavior corresponds to that predicted a priori. As for the statistical model (a posteriori), it answers the question of whether dam behavior is normal, based on its previous behavior.

32.5.2.2 Deterministic model

In the deterministic approach, the mathematical model is established a priori, while considering the characteristics specific to the dam and its foundations as well as physical laws (especially statistical laws and those relating to material strength). The model may be simple or more complex (calculation using finite elements). The model relates the causes (including water level, temperature, age) to the effect in question (for example, deformation of a concrete dam). Measurements are generally not required to create the model. The model must be able to quantify the effects of the various loads on the behavior of the dam. The analysis is not limited to the service conditions already produced but may be applied to all

---

[8] Further information can be found in the publication by Swiss Committee on Dams "Methods of analysis for the prediction and the verification of dam behaviour" (SwissCoD, 2003b).

future loading. For the calculation, the model takes into account the current loading conditions (SFOE (SFWG), 2002b).

In practice, this model is particularly appropriate for analyzing deformation δ(t) in concrete dams. It enables the contribution of static load $f_1(h)$ and thermal load $f_2(\theta)$ (instantaneous elastic effects) to be determined, as well as the deferred nonreversible effects $f_3(T)$ (creep, swelling) to be distinguished. The general behavior model used to predict measured deformation at one point in the dam at time t can be expressed using the equation:

$$\delta(t) = f_1(h) + f_2(\theta) + f_3(t).$$

The deterministic model is the only possible approach during first impoundment and the first years of operation. Later, it can be useful to regularly update it, in particular due to developments in material characteristics. As previously mentioned, it is very useful for deformation, but it cannot be used, for example, for seepage flow, which is practically impossible to predict a priori.

32.5.2.3 Statistical model

The statistical model is established a posteriori on the basis of observed behavior. It relates two series of values (causes + effects) by operating in a purely computational and mathematical way using a statistical treatment of available data (behavior and explanatory variables). As a result, it will only become available after a few years of operation and its reliability will depend on the quality of historical data available. This type of model is therefore not appropriate for tracking first impoundment and early years of operation. As it turns out, it cannot analyze conditions of exceptional load (including emptying, extreme temperatures) due to the absence of a statistical basis.

As with the deterministic model, deformation in the dam results from the hydrostatic effects due to variations in water level, thermal effects following variations in the thermal state of the dam body, and a nonreversible effect as a function of time due to various causes.

By using specific functions to reproduce behavior, the general model for predicting measured deformation at one point in the dam at time t can be expressed using the equation:

$$\delta(t) = f_1(h) + f_2(\theta) + f_3(t) + \varepsilon.$$

$f_1(h)$ is a function that represents the effect of hydrostatic load at point h, and it is acceptable to simulate deformation by a polynomial development at the level of the reservoir, the degree of which will depend on the type of dam and required accuracy, that is,

$$f_1(h) = a_1 + a_2 \cdot h + a_3 \cdot h^2 + a_4 \cdot h^3 + a_5 \cdot h^4.$$

$f_2(\theta)$ is a function that represents the effect related to the internal temperatures of the dam that are considered to have a linear influence and act instantaneously; often, seasonal functions are introduced, that is,

$$f_2(\theta) = b_1 \cdot \sin s + b_2 \cdot \cos s + b_3 \cdot \sin 2s + b_4 \cdot \cos 2s,$$

where $s = 2\pi (t - t_0)/365$ and t_0 = origin of time in question and t = time in days.

When concrete temperature measurements are available, the reversible thermal effect can be expressed as a function of internal temperatures

$$f_2(\theta) = \Sigma_n \cdot b_{k\theta k},$$

where b_k are the influence coefficients for the temperatures and θ_k the temperatures prevailing at the points of thermometric measurements.

$f_3(t)$ is a function that considers the nonreversible effects as a function of time (adaptation, foundation consolidation, creep, swelling, concrete degradation, etc.); it can, for example, be a combination of exponential functions

$$f_3(t) = c_1 \cdot e^{-t} + c_2 \cdot e^{t}$$

or decreasing and linear exponential

$$f_3(t) = c_1 \cdot e^{-t} + c_2 \cdot t.$$

There are other forms of writing to simulate these nonreversible phenomena.

As for ε, it represents effects due to other secondary causes and other experimental errors.

Finally, we obtain the following equation:

$$\delta(t) = a_1 + a_2 \cdot h + a_3 \cdot h^2 + a_4 \cdot h^3 + a_5 \cdot h^4$$
$$+ b_1 \cdot \sin s + b_2 \cdot \cos s + b_3 \cdot \sin 2s + b_4 \cdot \cos 2s$$
$$+ c_1 \cdot e^{-t} + c_2 \cdot e^{t}$$
$$+ \varepsilon.$$

The last equation defines the mathematical model to be adjusted to all of the observations by determining the coefficient values a_i, b_i, and c_i. The statistical method that may be employed involves searching for multiple correlations by using the least squares method.

On this subject, as part of its research programs, the SFOE (SFWG), Dam section, was involved in the development of the DamReg software. It can explain or predict a given variable that indicates the behavior of a dam (dependent variable) using a certain number of observed variables and functions of these variables (independent variables), which include water level, time, and observed temperatures. The statistical principle is that of multiple linear regression. The software can, within certain limits, be freely configured by the user. Thus, any combination of water level to the power of 1, 2, 3, 4, and 5, sine, or cosine of the season (position of the observation day in the cycle brought to 360 degrees), various deviation models, and observed temperatures can be used

In addition to the statistical treatment of concrete dam deformation that has just been outlined, this statistical modeling can be applied to other situations, including embankment displacement, seepage, piezometric levels, and joint movement (see SwissCoD, 2003). In some cases, the delayed impact between the cause (water level) and effect (e.g., piezometric level) must be taken into account.

32.5.2.4 Hybrid and mixed models

Hybrid and mixed models rely on an approach combining the characteristics of the two previously described methods.

One feature of the hybrid model is to take into account both the specific properties of the dam and its past behavior. It therefore relies on a deterministic mathematical model that has been improved by observations carried out during the first years of dam operation. Thereafter, it can still be improved upon on the basis of observed behavior. Predicting deformation can be expressed by the equation

$$\delta(t) = \beta_h f_1(h) + \beta \theta f_2(\theta) + \beta_t f_3(t).$$

The mixed model takes into account the deterministic prediction of the effect of water level and the statistical treatment of thermal effects. Predicting deformation (without considering nonreversible effects) can be expressed by the equation

$$\delta(t) = \beta_h f_1(h) + b_1 \cdot \sin s + b_2 \cdot \cos s + b_3 \cdot \sin 2s + b_4 \cdot \cos 2s.$$

32.5.2.5 Comparing deterministic and statistical models

Deterministic and statistical models both have advantages and disadvantages that must be weighed up when deciding on the approach to be adopted for interpreting surveillance measurements.

Table 32.39 presents in a qualitative manner the key points from both models. As demonstrated in the table, the two models are relatively complementary and it is therefore a good idea to combine them, especially for dams of a certain size.

It should be noted that the use of a deterministic model is particularly suited for the analysis of concrete dam deformation, while for other measurements, as well as for embankment dams, only a statistical analysis is possible. Within the scope of their application, deterministic models enable the detailed analysis of dam behavior, especially in the case of very slow, nonreversible phenomena or for exceptional service conditions (SFOE (SFWG), 2002).

Table 32.39 Qualitative comparison between the statistical and deterministic models for analyzing monitoring data (from SFOE (SFWG), 2002)

Model characteristics	Statistical model	Deterministic model
Model development	relatively simple	relatively complex
Use related to dam type	any type of dam	limited to concrete dams
Use related to measurement type	any type of measurement	limited to deformation measurements
Level of detail of the analysis	variable depending on available data	high
Use in cases of exceptional situations	limited	possible

32.5.2.6 Other types of analytical methods

Other analysis methods have been developed, whose main principle is still based on comparing a measured value with a calculated value, before verifying if the difference between the two values remains within an acceptable margin (Gicot H., 1976; Gicot O., 1985).

One of these methods brings all the measurements to the same conditions. To refine the gap, a displacement measured with a simple model is corrected, so as to shift the dam's effective load during measurement to a reference load, for which the water level and temperature state are fixed (Figure 32.40). This is also known as compensated displacement. Results from other methods of analysis are provided in Figure 32.41.

32.5.3 Comments regarding the interpretation of data from geodesic measurements

A series of geodesic measurements will enable the coordinates for various points in an outside network to be determined, as well as plane and/or elevation movements of monitoring points integrated into the general monitoring system. After having processed all the measurements, the surveyor provides a series of results that the senior dam engineer must interpret. Once the results are available, the question that is

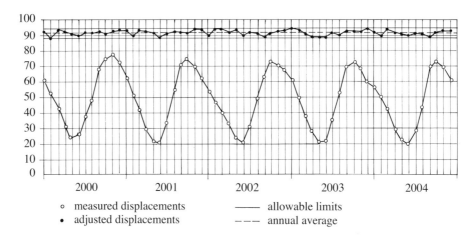

Figure 32.40 Comparison between measured and compensated displacement

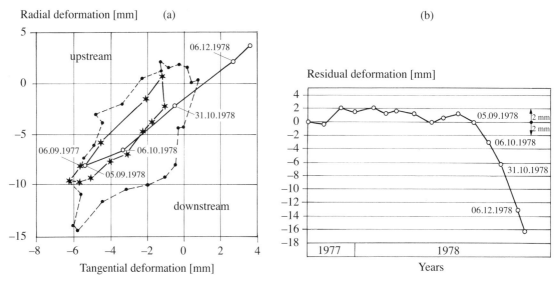

Figure 32.41 Analysis of exceptional deformation at the Zeuzier dam: (a) development in pendulum readings before and during settlement in late 1978, according to raw measurement data (graph prepared by O. Gicot); (b) development in residual deformation before and during settlement in late 1978 (graph prepared by Dr. G. Lombardi (SNCOLD, 1982))

generally asked in geodesy is: To what extent is the observed displacement real, or even, has it been influenced by errors. It should be noted that geodesic measurements may contain errors from various sources: environmental conditions, human error (including reading or sighting errors), systematic errors (in particular from faulty measuring devices), and also random (or accidental) errors.

A systematic error, whose causes may be known or unknown, is an error which, over several measurements of the same value of a given quantity carried out under the same conditions, either remains constant in absolute value and sign or varies according to a definite law with changing conditions. Random error varies in an unpredictable manner in absolute value and sign when a large number of measurements of the same value of a quantity are made under essentially identical conditions (same measuring device and same observer, same environmental conditions, etc.).

While measurements are being processed, it is possible to eliminate systematic errors of measurement through correction. With regard to random errors in a series of measurements, the limits within which this type of error are found at a given probability must be set. Random errors (or random variables) follow a normal distribution, which reinforces the importance of standard deviation σ, which is a good indication of the dispersion of a variable. The percentage of cases where observations are found in a given distribution interval can then be calculated. For example, the interval limited by standard deviations –σ and +σ around the average account for 68% of the observations.

The confidence interval or level of confidence P = 1 – α, in which the sought-after parameters are found with a given probability, can in theory be freely set. In practice, level P = 0.68 is used for the appreciation of a network, P = 0.95 or 0.99 for the analysis of deformation, and lastly, P = 0.997 for the search for a gross measurement error (Figure 32.42).

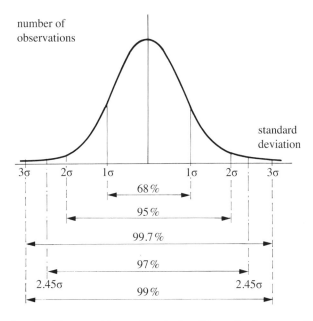

Figure 32.42 Distribution based on the normal law and level of confidence P = 1 – α according to standard deviation σ

Mean error (case with 1 dimension) and the error ellipse (case with 2 dimensions) are simple and practical indications for analyzing results from geodesic measurements. In the technical language of geodesy, mean error is associated with a standard deviation. Mean errors will identify the domain in which the determined displacement can be considered as real with a given probability. In the case with 1 dimension (for example, leveling), the probability that a parameter lies within the simple mean error is at 68% (Table 32.43); it is at 97.7% for the triple mean error.

For coordinates, the observations related to direction and distance are taken into account independently. The geodesic points from all the points with an identical probability form an ellipse. The confidence interval with 2 dimensions at 39.35% corresponds to the mean error ellipse (Table 32.41). If the mean error ellipse is multiplied by 2.15 or 3.04, the probability that the calculated points are found in the error ellipse increases to 90% and 99% respectively (Egger and Graf, 2002).

Table 32.43 summarizes the different levels of confidence depending on the type of measurement.

Table 32.43 Level of confidence for cases with 1, 2, and 3 dimensions

Dimension	Examples	Confidence limit field	Probability (level of confidence)
1D	Distance, altitude, vertical movements (changes in altitude)	Mean error Confidence interval	$P = 0.683$ P
2D	Coordinates, horizontal displacement	Error ellipse Confidence ellipse	$P = 0.394$ P
3D	Point in space, vector in the space	Error ellipsoid Confidence ellipsoid	$P = 0.199$ P

To be able to compare results on the basis of a homogeneous level of confidence (for example, $P = 0.95$), a factor of conversion must be introduced. To assess the difference between the measurements carried out at different times (determination of a difference in height or in displacement of a vector), a factor equal to $\sqrt{2}$ must be taken into account, due to the repetition of errors (Table 32.44).

Table 32.44 Factor of conversion: passage from the field of the mean error to the level of confidence $P = 1 - \alpha$

Level of confidence $1 - \alpha$	Dim = 1 interval		Dim = 2 ellipse		Dim = 3 ellipsoid	
	Initial position	Difference	Initial position	Difference	Initial position	Difference
0.199					1	1.41
0.394			1	1.41		
0.683	1	1.41				
0.950	1.95	2.77	2.45	3.46	2.70	3.82
0.990	2.58	3.85	3.04	4.30	3.37	4.76

After having verified whether the displacement is real, the next step involves ensuring that the kinematic of the displacement of the monitoring points – both those installed in the field and those installed on the dam – is coherent and plausible. In cases of doubt, it is a good idea to verify the state of the measuring points and/or investigate local conditions (position of the monitoring point, state of the dam, geology).

32.6 Reporting System

As part of normal operation, it is usual practice to put in place a series of reports that match the various tasks. These reports are drawn up by the operator and the engineer and contain data from measurements and their assessment (annual report relating to behavior), the condition of the dam (visual and annual checks), and operating tests. All reports must be filed in the dam file.

The engineer is responsible for writing the annual report on dam behavior. This report includes all measurement data in a graph form (annual and multiyear, parameter values as a function of reservoir water level, maximum/minimums of key parameters, etc.) and numeric tables. The senior expert is required to analyze dam behavior and give his or her point of view concerning the condition of the dam and its installations, as well as any recommendations. It is also important to include a summary of all maintenance work carried out and a list of all incidents (specific behavior, functioning errors in any ancillary structures, etc.) that marked the year, as well as to note the scale and frequency of any natural hazards that occurred (floods, earthquakes). The annual visit report and the operating test report can be included in the annual report (SFOE (SFWG), 2002b).

With regard to checking the condition of the dam, it is important that the information gathered is clearly described (written description: what, when, extent) and situated (indicated on a drawing). For retaining structures, it can be useful to draw up an illustration of specific points such as cracks, sweating, damp areas, damage, etc. With regard to the immediate surroundings of the dam, the recording of observations can include information pertaining to rockfalls, quality of seepage water, ground movement, the state of a glacier, etc. (SFOE (SFWG), 2002b).

For operating tests (gates, emergency unit), it is possible to prepare forms that outline the process, with gaps in which to write the observed values and incidents. Once completed, these forms serve as a test report (SFOE (SFWG), 2002b).

33. Emergency Planning and Public Safety

Malvaglia arch dam in Switzerland, height 92 m, year commissioned 1959 (Courtesy D. Grassi)

33.1 The Importance of the Emergency Plan

Along with structural safety, surveillance, and maintenance, the emergency plan is a cornerstone of the safety concept. It must enter into force as soon as the safety of a water-retaining structure is or could be in danger. It is due to permanent surveillance and visual monitoring that the existence of a hazard can be quickly identified. A broad outline of the emergency plan, as well as the description and consequences of the considered hazards has been covered in Section 5.4. This chapter primarily describes the measures that are implemented in Switzerland (SFOE, 2015c). However, the system in place and the provisions taken may obviously evolve over time as technology and experience gained from operation continues to develop.[1]

The emergency plan includes all the necessary preparations for dealing with a possible exceptional or extraordinary event. It sets out the organizational steps, that is, "who does what when," especially in a situation of crisis. It requires the development of strategies and the prior implementation of alarm systems, including the flood siren. If the situation is such that members of the public must be evacuated, the various means and structures for ensuring public safety will thus be implemented. Table 33.1 outlines the key points to consider for emergency planning.

Lastly, it should be noted that in the case of an accepted threat to the safety of a water-retaining structure, the dam operator's first priority is to take initial safety measures but also to inform the relevant authorities.

Table 33.1 Key action points for emergency planning

Goals	Situation and actions
When to sound the alarm?	After the uncontrolled release of a mass of water following the total or partial failure of a water-retaining structure, a landslide, or large volume of rock falling into the reservoir.
Where to sound the alarm?	Calculation of the flood wave (generally by assuming total failure) to establish a map of the flooded zone.
Who and how to evacuate?	Establishment of an evacuation map for inhabited areas in the flood zone.
How to sound the alarm?	Installation of an alarm system (including sirens, stations for setting off the alarm near the dam and outside of the flood zone).
How to communicate?	Ensure lines of communication between all parties (radio, telephone).
How to inform the public?	Distribute information leaflets with instructions for members of the public (letterbox drop). Information and instructions for the public on what to do during an event, broadcast by national and local radio/TV, and through the press. Press reports on the situation.

---

[1] Other publications cover measures to be taken in the case of an emergency: Biedermann, 1987, 1997; Darbre et al., 2009; Pougatsch et al., 1998; Wieland and Müller, 2009.

33.2 Strategy

Table 5.19 defines the various levels in an emergency strategy depending on the degree of control over the danger. Emergency regulations (§ 33.4.1) set criteria for the danger levels and indicate the elements that trigger a specific danger level. As a reminder, here are the 5 levels:

- Level 1   No or low danger
- Level 2   Limited danger
- Level 3   Significant danger
- Level 4   Serious danger
- Level 5   Extreme danger

In the case of a normal situation (danger level 1), the dam operator carries out the usual surveillance program (regular visual monitoring, measurements, and operating tests) in accordance with the surveillance and operating plans. The aim of this procedure is to quickly identify any abnormality in dam behavior so that further interpretation can occur as soon as the necessary actions have been carried out.

Depending on the nature of the specific observations, the level of surveillance may increase. The actions to be undertaken depend on the analysis of the situation and the resulting danger levels. These range from increasing surveillance to operating restrictions and may also include safeguard measures such as rehabilitation works, for example. Level 4 corresponds to the possible or even inevitable arrival of an uncontrolled mass of water, that is, the possibility of partial or total failure of the water-retaining structure, or the occurrence of a landslide or rockfall into the reservoir.

Emergency regulations set criteria for the danger levels and indicate the elements that trigger a specific danger level.

33.3 Possible Actions in the Event of Abnormal Dam Behavior[2]

Surveillance is increased as soon as an obvious abnormality in the water-retaining structure (non-compliant measurement results, observation of an operating fault in water-release structures, blockage of water intake points in water-release structures) or a worrying visual observation (risk of a rockfall) or outside phenomenon (flood, earthquake, sabotage) is detected.

This increased surveillance must ensure permanent analysis of the situation by the engineer responsible for the dam, as well as enable any measurements to be taken in a timely manner.

In the simplest situation, safeguard measures may be taken quickly, without it being necessary to move to a higher danger level. For example, measurements from the monitoring system can be repeated, additional measurements or a denser series of measurements can be taken, maintenance works can be carried out, or a temporary operating limit can be imposed.

Depending on the set danger level, technical measures may be necessary. A preventive drawdown (partial or total) may possibly be required if an extraordinary event means that entirely safe operation can no longer be guaranteed for the water-retaining structure and if there is a threat for the public and property. The water level may also be lowered to avoid overtopping in the case of an avalanche, landslide, rockfall or icefall into the reservoir. The water level must be maintained at a low level for as long as the risk remains. If (rapid) drawdown may lead to the risk of a rockfall or landslide in the submerged part of the basin, specific instructions must be planned (for example, limiting the speed of drawdown).

---

[2]   See also Section 5.4.3.

33.4 Preparation

33.4.1 Emergency regulations

In order to manage emergency situations, emergency regulations must be established by the dam operator. These must include the following elements:
- Emergency strategy
- Flood map (extent of the submerged zone)
- Danger analysis
- Emergency organization
- Emergency action plan

33.4.2 Extent of the submerged zone

One of the key tasks in the preparatory phase involves determining the extent of the submerged zone following the total and sudden failure of a dam. For some small dams whose spillway may become blocked by floating debris, the starting water level is accepted as being the same as the crest. This enables the near zone to be established. It then becomes possible to draw up an evacuation plan, which is distributed to local communities (Figure 33.2). This document contains various information: the boundaries of the submerged zone, escape routes, assembly points, and traffic rules. It must also explain the alarm sirens and give instructions on what members of the public should do.

Section 5.4.4 gives some hypotheses for the assessment of a flood wave. Calculations are based on topographical maps with contour lines that enable a series of cross profiles to be defined, if possible, in a regular shape and without any islands (1D model). In this case, a simple calculation method is applied that enables determination of the extent of the flood zone into characteristic points downstream of the dam. Wave flow, speed, and depth are results that must be calculated for the various cross profiles. Flood maps can then be established, which provide the boundaries for the flood zone, as well as the energy line.

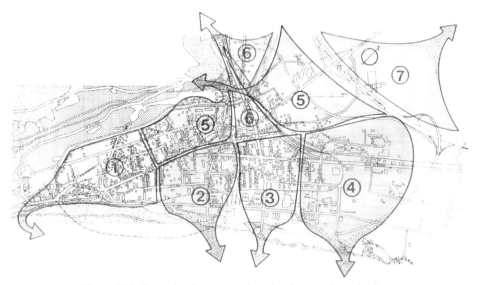

Figure 33.2 Example of an evacuation plan for a medium-sized town

Digital topographical models (e.g., the MNT25 base model from the Office fédéral de la topographie) are well suited to 2D calculations. In this case, the topography is defined by a large number of points (coordinates and elevations). These are connected in such a way that they form triangles and thus represent the ground surface. Excellent ground modeling is thus obtained. Wave flow, speed, and depth are given in 2 directions for each cell. The wave's extent over time and space can be tracked for any basin with contours.

The results obtained enable the creation of a flood map that shows the boundary of the flood zone and the arrival time of the face of the wave. The flood boundaries and maximum elevation of the energy line are also indicated on flood maps. The boundaries are defined by the energy line (water height h_w + kinetic energy $v^2/2g$). If the flow of the flood wave remains confined to the riverbed, only the water level is considered. In the case of a 2D model, the calculations enable the speed of flow of the wave to be determined below each cell in 2 directions, as well as the water depth. Water level is no longer horizontal in a cross profile and the energy line is no longer represented by a straight line but by an irregular curve (SFWG, 2002).

The flood map (see the example in Figure 5.22), as well as the emergency action plan must be transferred to the appropriate cantonal department and to the CENAL (OCENAL, 2007).

33.5 Alarm Systems and Sirens Available in Switzerland

33.5.1 Facilities in submerged zones

Various means must be implemented to ensure public safety, and, when required, to guarantee evacuation from danger zones due to the probable or actual uncontrolled release of a large quantity of water.

Audio means such as stationary sirens (general alarm) and flood sirens, or the use of mobile sirens, ensure that the alarm is sounded across a wide area (OAL, 2018). The main system, known as the flood alarm system, includes the flood siren and the general alarm siren. It is used for dams with a reservoir capacity greater than 2 mio m³, and also for dams with a capacity lower than 2 mio m³ if the potential danger in case of failure is considered high. A difference is drawn between a near zone that may be reached in fewer than 2 hours and is equipped with two types of sirens and a far zone equipped with only general alarm sirens (Figure 33.3). Simpler systems can be put in place for smaller dams.

The different sounds and tones of the sirens indicate to the public the different actions they must take. The general alarm signals to the public that they must listen to the radio. All the necessary information and actions to take will be broadcast. The flood siren, however, instructs people to leave the danger zone; information is also broadcast on the radio. The flood siren is preceded by the general alarm, so that all members of the public are informed and take the correct course of action.

A organization chart illustrating the flow of information and alarm warning to the public in the event of a possible uncontrolled release of water is provided in Figure 33.4.

Generally speaking, the alarm must be set off if the uncontrolled release of a large quantity of water cannot be avoided due to the (probable) partial or total failure of a water-retaining structure or a possible landslide or rockfall causing an impulse wave leading to overtopping of the dam. The dam operator is responsible for the sounding of the flood siren. This is manually activated from a flood alarm station or emergency station in the event of a proven partial or total failure of the water-retaining structure. Cantonal authorities are responsible for activating the general alarmsiren. They give the order to sound the general alarm via the dispatch center run by the cantonal police. The sirens may be remotely activated or activated manually on site.

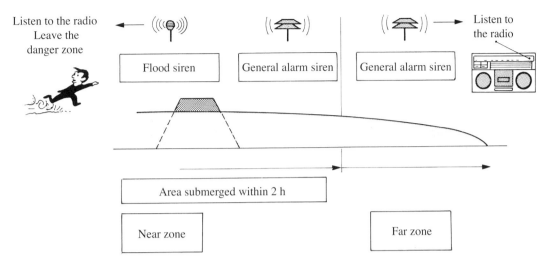

Figure 33.3 Diagram of a flood alarm system

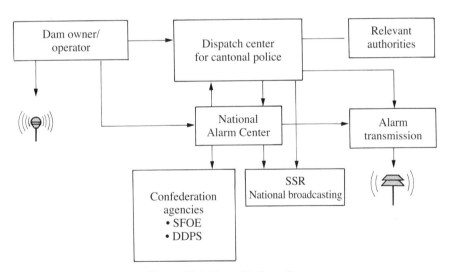

Figure 33.4 Flow of information

33.5.2 Analysis of hazards affecting key points in the flood alarm system

It is important to ensure, through the use of risk analysis, that the actions taken to control an emergency will not be disrupted or even prevented by external events. This analysis includes access (all locations must be accessible in the case of an emergency), system and water-release controls (power supply and gate operation guaranteed), and communication (guaranteed operation of internal and external communications, as well as means of alarm). The results will enable the dam operator to select the location for the sirens, the emergency and observation stations, and the moment when emergency staff must be sent to the dam.

33.6 Communication and Information

All alarm systems must allow those involved in emergency management to communicate among themselves and pass on information. As such, spoken lines of communication are vital.

Lastly, an alarm system would be useless if affected communities did not receive information and specific instructions on what to do. The authorities will therefore make use of national and private radio stations as an additional means of communication.

The dam operator must ensure communication with its own emergency team, especially between the observation stations, alarm stations, and the flood alarm station; the same goes for cantonal authorities for civil protection and the dispatch center at the cantonal police, as well as for supervisory authority for water-retaining structures.

33.7 Emergency Organization

To successfully ensure control of an event, the dam operator must have a highly structured organization with a clear description of the respective functions and tasks of each staff member. As a general rule, the dam operator must designate an emergency action manager, a flood alarm manager to lead interventions and activate the flood alarm (if the dam is equipped with such a system), an employee to be responsible for updating the regulations, and staff to be present at the activating and observation stations. Furthermore, involvement is required from experienced staff members, experts, and other specialists.

33.8 Emergency Action Plan

The documents required by the dam operator to control the situation in the event of an emergency are kept together in an emergency action plan. This contains the following elements, among others (SFOE, 2015c and 2015d):

- Organization chart of the dam operator's emergency team
- Description of tasks for each role in the emergency team
- Staff members involved and external organizations to be informed or alerted
- Contact lists for emergency staff, means of communication, and contact numbers
- Forms for informing the dispatch center at the cantonal police, including predefined messages for the alerts and lowering of each danger level
- Procedure forms for recording analyses of the situation, alarms, messages, and actions taken during an emergency.

IX. RESERVOIR SEDIMENTATION

This part contains a single chapter 34, which deals primarily with the issue of sedimentation in reservoirs and lakes. Sedimentation can lead to a considerable reduction in storage capacity and can also disrupt the functioning of water-release structures such as the bottom outlet and the water intake.

The sedimentation of reservoirs is specifically due to the arrival of sediment and suspended material into the reservoir by waterways. Flow determines how this material is deposited. Turbidity currents are often a major cause of sediment transport along deep, narrow reservoirs.

Sediment (coarse material that is transported along the bed of the waterway) and suspended materials come from the erosion of soil in the reservoir catchment, as well as from the disintegration and abrasion of coarse elements. Sediment accumulates at the head of the reservoir forming a delta.

The average rate of sedimentation for all artificial reservoirs worldwide is estimated at 1–2%. Switzerland's 0.2% rate of annual sedimentation in reservoirs is considerably lower than the global average. With regard to storage capacity, Swiss reservoirs are therefore highly sustainable. However, sediment transported regularly by turbidity currents can disrupt the functioning of water-release structures such as the bottom outlet and water intake after only a few decades of operation.

There are several preventive and retroactive measures that can be taken to mitigate sedimentation. These affect the catchment area, the reservoir, or the dam, and are outlined in chapter 34. This chapter also provides more detailed consideration on the control of turbidity currents and gives the example of studies carried out as part of the heightening project at the Grimsel reservoir.

34. Issues Created by Reservoir Sedimentation and Their Mitigation

Punt dal Gall arch dam in Switzerland, height 130 m, year commissioned 1968
(Courtesy © Swiss Air Force)

34.1 Overview

Sedimentation is a process that leads to the siltation of reservoirs with sediment and that in its final stages results in an expanse of sediment that replaces the free water surface. This phenomenon affects all still bodies of fresh water, as well as many sea bays. More specifically, sediment transported by waterways feeding the reservoirs accelerates the development of sedimentation.

Similarly, artificial lakes used to store water for irrigation, drinking water, the production of hydropower, and protection against floods are also affected by gradually increasing deposits of sediment. As sedimentation slowly reduces storage capacity in these reservoirs, the question of their sustainable development then arises (Schleiss et al., 2010, 2016).

Storage reservoirs have a specific morphology due to the fact that their lowest point is almost always located at the dam and, what's more, near the water-release structures. Water flowing into a reservoir may be regulated to a certain degree, which may in turn affect the water level. For most dams, water-release structures allow the near total drawdown of water level and a complete emptying of the reservoir. The morphological and hydrological specificities of these storage reservoirs compared to natural lakes means that more effective actions can be taken to counter sedimentation.

The sedimentation of reservoirs is specifically due to the arrival of sediment and suspended material into the reservoir by waterways. Flow determines how this material is deposited.

For shallow reservoirs, sediment may gather through a biological process. In addition, encroaching vegetation can form marshes.

Sediment and suspended material come from the erosion of the catchment area around the reservoir. Sediment is coarse material transported along the bed of waterways; it accumulates at the head of the reservoir forming a delta. Suspended material comes from surface erosion, as well as the disintegration and abrasion of coarse material. It is deposited beyond the delta and spreads throughout the whole reservoir depending on its weight. During a flood, fine sediment can also occasionally be transported through the reservoir right to the dam as a turbidity current.

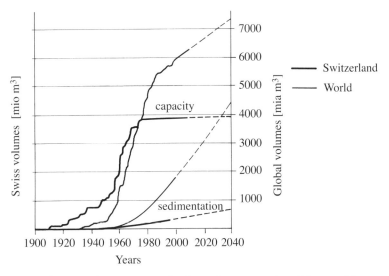

Figure 34.1 Increase in reservoir storage capacity due to the construction of dams and decrease in volume due to sedimentation in Switzerland and the world (Schleiss and Oehy, 2002)

The average rate of sedimentation in artificial reservoirs worldwide is estimated to be around 1–2%, which means that annual loss in storage capacity is between 1 and 2 % (Jacobsen, 1999; Oehy et al., 2000; Schleiss et al., 2016). If we assume that the annual global increase in the volume of stored water by new built reservoirs is closer to 1% than 2%, this clearly highlights the problem in sustainable development (Figure 34.1). If no effective measures against sedimentation in reservoirs are taken, most of the world's storage capacity will have disappeared by the end of the 21st century. The volume of sedimentation in each reservoir obviously differs and depends heavily on climatic conditions, as well as on the design of reservoirs and water-release structures. Switzerland's 0.2% rate of annual sedimentation in reservoirs is considerably lower than the global average. With regard to storage capacity, Swiss reservoirs can be deemed largely sustainable (Figure 34.1). However, the issue of sedimentation in Swiss or alpine reservoirs is also extremely important, as it has the potential to disrupt the functioning of water-release structures such as the bottom outlet and water intake after only a few decades of operation. In many regions around the world, the problem of reservoir sedimentation is increasing considerably due to the effects of climate change (Schleiss et al., 2016).

34.2 Surface Erosion in Alpine Catchment Areas

The inflow of sediment into a reservoir primarily depends on the predisposition of the catchment area to erosion. As part of a research study, 19 Swiss reservoirs were analyzed (Figure 34.2) and the annual proportion of sediment was determined from the bathymetry of the reservoir bottom or by measuring the quantity of sediment evacuated during flushing operations (Beyer Portner, 1999; Beyer Portner and Schleiss, 2000).

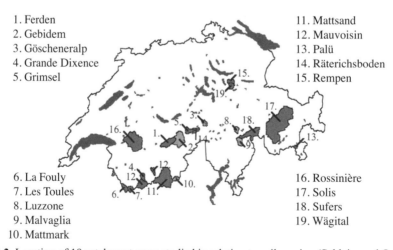

Figure 34.2 Location of 19 catchment areas studied in relation to soil erosion (Schleiss and Oehy, 2002)

Thanks to the data set resulting from the assessment of these 19 reservoirs, a statistical analysis was carried out that compared soil erosion to the characteristics of each catchment area. At the same time, an empirical model of erosion was developed by considering the following parameters:

- H_{som} average rainfall in summer (June–September) (mm)
- OV proportion of surfaces not covered with vegetation (%)

- EB   proportion of surfaces comprising erodible soil (%)
- ΔGL   annual modification in the average relative length of glaciers (%)

This erosion model, developed on the basis of a genetic algorithm that considers ground surfaces as much as possible, is defined by the following empirical equation of annual loss of volume V_A in m^3/km^2:

$$V_A = 0.2112 \cdot 1.10683^{OV} - 5.684 \cdot OV + 0{,}2112 \cdot (H_{som} \cdot \Delta GL + OV \cdot EB) + 11.$$

This equation based on the above-mentioned genetic algorithm, with a correlation coefficient of 0.925, describes the 19 observed catchment areas better than an equation based on traditional regression. Given that the annual modification in average relative length of glaciers ΔGL, as well as the proportion of surfaces comprising an erodible surface OV are included in the model, this can then take into account the effects of climate change.

34.3 Turbidity Current as the Principal Cause of Sediment Transportation in Reservoirs

Most solid materials transported into a reservoir are generally composed of suspended material at a proportion of 80 to 90% for small and medium reservoirs, and at practically 90 to 100% for large reservoirs. Coarse alluvium is a secondary concern. Large quantities of suspended material are mostly transported in waterways during a flood. Inflow heavily loaded with fine sediment coming from the catchment area has a higher unit weight than that of the calm water of the reservoir. The mass of turbid water that flows into the reservoir first pushes the lake's clear water away from the mouth of the waterway until a balance in the quantity of movement is established. Then, the water heavily loaded with suspended material plunges into the lighter water of the reservoir. This creates a water current in the reservoir, known as the *turbidity current*, which comprises a mix of water and fine, suspended sediment. This turbidity current can be compared physically to an avalanche of powder snow down a slope. The turbidity current moves relatively quickly along the sloped bottom of the reservoir toward the lowest part located near the dam (Figure 34.3).

Depending on the slopes of the thalweg, turbidity currents may reach high speeds. Due to this, sediment that has already been deposited may be once again picked up and transported toward the dam. The inflow of additional fine material by the erosion of the reservoir bottom increases the density of the turbidity current and increases its speed. However, it does slow down in flat sections, where suspended material is deposited, eventually leading to the die out of the turbidity current.

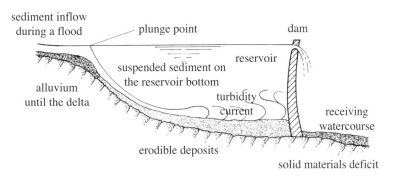

Figure 34.3 Diagram of a turbidity current in a reservoir (from De Cesare and Boillat, 1998)

These turbidity currents have been observed in the Luzzone reservoir, located in southern Switzerland, as part of a measuring campaign (De Cesare and Boillat, 1998; De Cesare, 1998). Using three measuring chains each fitted with two current meters, direction and speed were measured respectively at a height of 2 and 4 m from the bottom of the reservoir (Figures 34.4 and 34.5). These measurements were carried out in series over a relatively short time span during a flood. At points A and B (Figure 34.4), the currents were both heading toward the dam. At point C, the currents mainly came from the north and east of the catchment area (Figure 34.5). The maximum measured speed was 80 cm/s (at 2 m above the reservoir bottom) and 55 cm/s (at 4 m above the reservoir bottom).

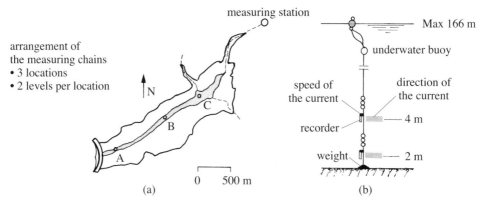

Figure 34.4 (a) Location of measuring chains at three points in the Luzzone reservoir; (b) arrangement of measuring devices fitted with two current meters (from De Cesare, 1998)

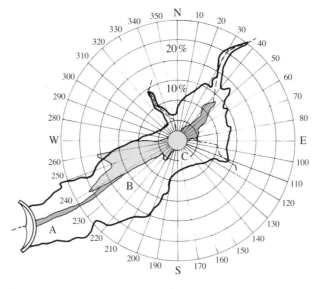

Figure 34.5 Direction and speed of turbidity currents measured in the Luzzone reservoir at point C (based on Figure 34.4), 4 m above the reservoir bottom: the velocity direction is indicated along with its intensity. The current clearly follows the thalweg (from De Cesare and Boillat, 1998).

A computational model, which was calibrated on the basis of laboratory tests and on-site measurements, can today simulate turbidity currents with great accuracy for a reservoir of any geometry (De Cesare et al., 2001). Assuming a hypothetical flood (peak flow 137 m^3/s) at the Luzzone reservoir, a calculation has shown that the turbidity current is capable of shifting an impressive amount of sediment. For example, some 9,000 m^3 of fine sediment were brought into the reservoir during a flood. Furthermore, 35,000 m^3 of solid material were transported along the thalweg toward the dam by the turbidity current, due to the effect of erosion. Lastly, material was deposited to a depth of 53 cm in a zone stretching 300 m upstream of the dam (Figure 34.6).

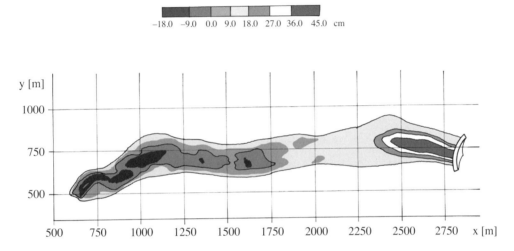

Figure 34.6 Calculated erosion and deposits caused by a turbidity current in the Luzzone reservoir after the total settlement of fine sediment. Huge amounts of sediment were transported from the upper part of the reservoir to the lower part (maximum erosion: 27 cm, maximum deposits 53 cm) (De Cesare and Boillat, 1998).

Turbidity currents form naturally during annual floods. In the case of dams in alpine environments, with respect to deep and long reservoirs, they are the principal cause of sediment transport along the reservoir thalweg to the dam, where the sediment is deposited. If the amount of solid material is low in comparison with the volume of the reservoir, the deposits are often concentrated near to the dam and, after some years of operation, they may have a major impact on water-release structures such as the bottom outlet and water intake. This phenomenon was observed in Switzerland some 40 to 60 years after dam reservoirs began operating.

34.4 Measures Against Sedimentation

Possible actions against sedimentation in reservoirs can in principle be divided into preventive and retroactive measures. The former mitigate sedimentation, while the latter serve to partially reduce it. Measures can be further separated into those occurring in the catchment area, the reservoir, and at the dam (Figure 34.7).

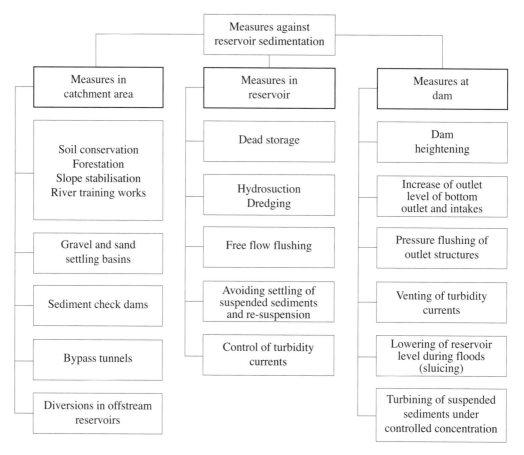

Figure 34.7 Overview of preventive and retroactive measures for mitigating sedimentation (adapted from Schleiss and Oehy, 2002)

34.4.1 Measures in the catchment area

Protecting against erosion in the catchment area is the most effective way of preventing sedimentation. In regions where climatic conditions allow it, vegetation is one way of protecting surfaces against erosion. Over the past century, reforestation in catchment basins has been one of the main tasks around reservoirs at risk from sedimentation, particularly in Asia. Unfortunately, the effect of reforestation on alleviating sedimentation only becomes apparent over the long term. Soil conservation measures in catchment basins are also necessary to guarantee cultivatable ground appropriate for agriculture, as well as to ensure protection against floods, debris flow, and slope failure. In catchment areas with bare slopes, such as high-altitude alpine catchments, only technical measures, such as slope stabilization, and protective actions for the beds and banks of waterways are able to limit erosion.

Sediment and gravel settling basins can be constructed in mountain streams and rivers in order to prevent unwanted deposits in downstream sections. These basins intercept practically all sediment transported during a flood, which must then be evacuated. Settling basins only locally affect a waterway's ability to transport sediment and are generally only capable of holding back a small quantity of suspended material. Furthermore, they are too small to have any great impact on the sedimentation in a reservoir. Larger traps are therefore necessary; these include sediment check dams constructed in the delta zone formed at the mouth of incoming waterways. These hold back practically all alluvium and also act as a sand trap. A high degree of sediment or material retention can only be obtained through large check dams. A degree of 90% retention is, for example, reached if the relationship between the volume of the reservoir and that of annual sediment inflow is in the order of 1:10. In order to prevent the check dam itself from filling up, the deposits must be continually evacuated. The material can be evacuated through the bottom outlet or through a bypass tunnel that circumvents the reservoir, releasing it into the river downstream of the main dam. For smaller reservoirs, sediment bypass tunnels have been successfully implemented, particularly in Switzerland (Vischer et al., 1997) and Japan (Sumi et al., 2004; Sumi, 2017). Controlling abrasion in these tunnels is a challenge (Boes, 2015; Boes et al., 2018, Hager et al., 2021). To limit the length of sediment bypass tunnels, their entrance can be installed in the reservoir. However, for the bypass tunnel to work, the reservoir must be lowered to the site of the tunnel entrance.

A reservoir's predisposition for sedimentation depends directly on the size of its catchment. Reservoirs fed by a small catchment and by water coming from neighboring catchment areas fill up with alluvium at a much slower rate. The condition is, however, that feed-in intakes only capture and divert water with a low suspended sediment load.

34.4.2 Measures in the reservoir

When sediment arrives in the reservoir, only passive or retroactive measures are worth considering to prevent or limit the negative effects of sedimentation. A method that has been used around the world to increase the storage capacity of a reservoir involves overdesigning its total capacity, allowing an additional volume that will contain sediment for a certain amount of time, usually set at 50 years. If this zone is not accessible, this is known as dead storage. In view of sustainability and high investment costs, such a concept can now no longer be justified (Schleiss et al., 2016).

Sedimentation can be slowed down or prevented in cases where the deposits are regularly evacuated. Dredging operations are carried out from the sides of the full or lowered reservoir or from a floating pontoon. Depending on the particle size of the sediment to be removed and the depth, suction dredging or usual mechanical dredging may be used.

Hydraulic dredging can be used to suck sediment through mobile pipes on the bottom of the reservoir. These are fitted with orifices through which the sediment penetrates once a valve at the end of the pipe is opened (SPSS – *Slotter Pipe Sediment Sluice*) (Jacobsen, 2000).

Flushing operations are an extremely effective way of hydraulically emptying the reservoir, (which, if possible, should be totally lowered). This process can cause environmental issues and silting up downstream of the dam. A high suspended material content is critical for river aqua system during short flushing operations (Boillat et al., 2000a; Boillat et al., 2000b, Schleiss et al., 2016, Hager et al., 2021).

Suspended material is the main source of sedimentation in a reservoir. If it can be prevented from being deposited, it can be evacuated continuously through water-release systems that tolerate water passing through the turbines with a limited concentration of solid material. Indeed, turbines are increasingly resistant to abrasion thanks to the use of new materials (Grein and Schachenmann, 1992, Felix et al., 2017).

To continually maintain the fine materials in suspension in the reservoir, these must be moved about by turbulence that is sufficiently strong, which, in the example of alpine reservoirs, may be generated by underwater jets powered by the energy from feed-in water transfer tunnels (Jenzer et al., 2011). A mechanically produced eddy is also possible by using a large movable mixer in the accumulation zone.

Lastly, the negative effects of sedimentation can be drastically limited by controlling turbidity currents in the reservoir, should these be the main cause of sedimentation, as is often the case in long, deep alpine reservoirs.

34.4.3 Measures at the dam

If a large volume of the storage capacity has already disappeared, it can be compensated for by heightening the dam, as long as this structural measure is possible. This option has been implemented in several dams in North Africa (Cornut, 1992).

If, in the first place, only the water-release structures are affected by sedimentation, and flushing is not possible, these structures must be heightened to ensure that operation can continue. The transformation of the bottom outlet and water intake at the Mauvoisin dam is an example of this (Hug et al., 2000; Schleiss et al., 1996).

During pressure flushing to free up the entrance to water-release structures, an erosion cone may form, whose slopes are equal to the angle of internal friction of the deposited sediment. For fine sediment, the slopes of the cone may reach 30° (Sinniger et al., 2000). Should the entrance to the water-release structure be totally covered by sediment, there is a danger that the deposits may consolidate and the cone-shaped erosion cannot occur once the gate is opened. This issue can be addressed by constructing a shaft with an injector to provide clear water at the beginning of the process (Krumdieck and Chamot, 1981). To regularly ensure that the entrance to the water intake is clear, its lower part must be fitted with an independent flushing system such as a flushing tunnel (Hug et al., 2000; Schleiss et al., 1996).

If the bottom outlet is sufficiently large, turbidity currents can theoretically be sucked up during a flood and directly released downstream. For alpine reservoirs, turbidity currents form during frequent floods that do not require release through the spillway. Under these conditions, the passage of density currents through the bottom outlet, known as *"venting,"* is rarely implemented due to the substantial loss of water that results. However, to improve the morphology of waterways downstream of dams and to reinvigorate solid transport, the regular release of artificial floods alongside the venting of turbidity currents, as well as the supply of sediment downstream is a promising measure for the reservoir from an environmental and sustainable point of view (Döring et al., 2018; Stähly et al., 2019).

Sedimentation can theoretically be impeded on a long-term basis provided that during a flood, with its high sediment load, the water level in the reservoir can be lowered and the flow regime remains close to natural conditions without the dam. Such an operation is necessary for reservoirs with relatively low storage capacity compared to annual inflow. In order to enable sustainable use of the reservoir, the Three Gorges dam system in China is based on this operating mode. Drawing down during a flood is not, of course, possible in annual storage alpine reservoirs for practical and operating reasons.

Lastly, the sediment that ends up in a reservoir can be put into motion by suction and incorporated with water passing through the turbines, while keeping an eye on its concentration (Heigerth et al., 2000). The maximum concentration depends firstly on the risk of damaging the headrace tunnel and secondly on wear and tear on the turbines due to abrasion (Felix et al., 2017).

34.5 Controlling Turbidity Currents

As mentioned in Section 34.3, turbidity currents are often a determining factor for the process of sediment being deposited in a reservoir. The following conditions can lead to the formation of a turbidity current (Oehy et al., 2000):

- Water inflow with a high concentration of suspended material
- Deep water at the mouth of the waterway
- Almost still water in the reservoir
- Reservoir bottom is steeply sloped
- Geometry of the reservoir is in a rectilinear, prism shape

Alpine reservoirs almost always meet these conditions; turbidity currents are therefore activated during small annual floods.

A parametric study of the Luzzone reservoir demonstrated that the speed and capacity for transport of a turbidity current are practically independent of flow conditions (Oehy et al., 2000). In other words, the phenomenon of an underwater avalanche is triggered by an inflow of water, in such a way that a movement develops independently of the flow conditions. That said, the characteristic diameter of the suspended particles in the flow plays a decisive role in the ability of the turbidity current to flow. Calculations taking into account, for example, a diameter of 0.01 mm, show that the turbidity current does not form at all or disappears rapidly independently of flow conditions. The fact that no particle diameter greater than 0.1 mm was found at the bottom of the reservoir near to the Luzzone dam confirms this observation (Sinniger et al., 2000). Characteristic turbidity currents can reach a travel speed of 0.5 to 0.8 m/s, which enables them to transport particles with a maximum diameter of 0.015 to 0.025 mm (Figure 34.8).

It is possible to greatly improve the sustainability of the reservoir if turbidity currents can be stopped or modified in such a way that they no longer transport sediment toward critical areas near the dam water-release structures. Due to the importance of turbidity currents for the sustainable operation of storage reservoirs in the Alps, technical measures, which should enable sedimentation through turbidity currents to be controlled, have been identified and assessed as part of a research project.

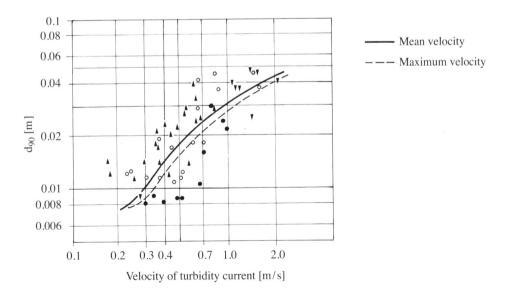

Figure 34.8 Maximum diameter of transportable particles as a function of development of the turbidity current (from Fan, 1986)

Technical measures seek to stop, dilute, and divert turbidity currents (Table 34.9). Obstacles and roughness and dissipation elements installed on the bottom of the reservoir, as well as the introduction of an auxiliary source of energy, are some of the possible measures for controlling turbidity currents.

Table 34.9 Morphological distribution of technical measures for controlling turbidity currents and their effects (from Schleiss and Oehy, 2002)

Measures Effects	Obstacles	Roughness and dissipation elements	Introduction of an auxiliary energy source
Stop	Dike or wall Floating or fixed geotextile curtain	Screen, ladder Groyne Sill, weir	Air curtain Water curtain
Dilute	Dissipation grids	Screen, ladder	Air curtain Water curtain Mechanical mixer Nozzle
Divert	Dike or directional wall Guiding geotextile Floating geotextile curtain	Guiding geotextile Floating geotextile curtain Screen, ladder	Local injections of water

If they are of a certain height, obstacles in the reservoir thalweg can stop turbidity currents and force the suspended material to be deposited. It is possible to install small dikes and walls with an opening to let water flow through when the reservoir level is very low (Figure 34.10 (a)). Turbidity currents can also be intercepted by a geotextile curtain, which is either stretched and set into the ground underneath the lowest section or hung by floating buoys (Figure 34.10 (b)). The curtains must also have an opening at ground level and be able to move and adapt to fluctuations in water level. For curtains to stop turbidity currents, they must not be completely watertight but meshed.

If the obstacles are highly permeable, for example, a screen, they will not directly stop turbidity currents but will create turbulence and dilute them so that they disappear.

With obstacles such as dikes and directional walls, the turbidity current can, for example, be diverted so that it peters out in a zone that will not create any problems (such as a bay). Its path can also be modified by a light geotextile attached to the ground that directs it toward a specific zone. In theory, geotextile curtains can also be used to create a local diversion.

Roughness elements are also able to stop the turbidity current in a given section along the thalweg. This involves creating small embankment groynes or installing screens and rakes in the thalweg. The latter also create turbulence by reducing the turbidity current. Lastly, these roughness elements can also be used to divert the current.

A third measure essentially involves stopping, reducing, or diverting the turbidity current by introducing an external source of energy. If an air curtain and water curtain provide enough resistance, the turbidity current will stop entirely (Figures 34.10 (c) and (d)). Otherwise, they will at least heavily dilute the turbidity current, causing it to slow down and die out. A mechanical mixer or nozzles can be used to create a turbidity current. By permanently suspending fine particles in the reservoir, the final goal is to evacuate them through the water intakes and turbines. The concentration of fine particles obviously needs to be limited to avoid abrasion of hydraulic machinery and prevent environmental issues in the river downstream. The

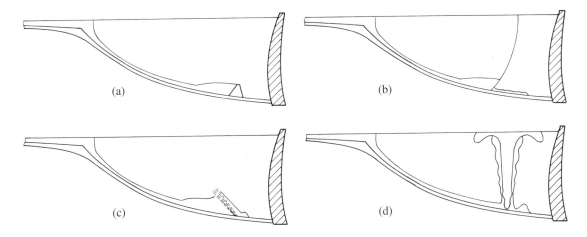

Figure 34.10 Technical measures to control a turbidity current in a reservoir: (a) dike used as an obstacle; (b) floating geotextile curtain; (c) water jet curtain aiming upward; (d) air bubble curtain (from Oehy and Schleiss, 2003)

concentration of sediment in turbidity currents can reach a few percent per volume. Dilution by a vortex created by a mixer and a greater amount of water limits the concentration, with the aim of preventing abrasion problems in hydraulic machinery. In addition, turbine runners are today covered with a new material capable of withstanding a concentration of sediment containing between 2 and 5 mg/l for more than one maintenance cycle (3 to 5 years) (Grein and Schachenmann, 1992). Mechanical mixers or nozzles are also able to set in motion fine particles already deposited at crucial spots, for example, near water-release structures, thus mixing them with water passing through the turbines.

Lastly, local injections of water using directed jets can be used to push the turbidity current into a specific direction or even to stop it. For local water injections, besides using pumps, water energy from feed-in tunnels can, for example, be introduced above the normal reservoir level (see § 34.4.2).

34.6 Example taken from the Grimsel Reservoir

As part of the Grimsel Plus project, the heightening by 23 m of both retaining structures at Grimsel, planned as the 3rd stage in the 1st phase of work, will increase the volume from 95 mio m³ to 170 mio m³ (for the geographical situation, see Figure 34.2). To ensure an adequate connection between the new concrete in the dam and the old, part of the existing dams must be demolished. There is estimated to be approximately 150,000 m³ of excavated material, including foundation excavations. As the reservoir has to be emptied for a short period of time, the question arises of whether the excavated material and material from the demolition of the dam could be used to create obstacles in the reservoir thalweg. Turbidity currents could thus be stopped in the future. As part of a research study, the reservoir was reproduced as a computational model and the turbidity current from the October 2000 flood (peak flow 40 m³/s, sediment concentration assumed at 15 g/l) was calculated. The calculations demonstrated that the turbidity current was directed toward the intakes after only 36 hours and deposits of 10 cm were shifted into the area during this single event (Figure 34.12, top).

In order to impact turbidity currents in the Grimsel reservoir in the future, the following obstacles were examined (Figure 34.11):

- Alternative 1: 15-meter-high dike, at an altitude of 1845 m.a.s.l.; length 150 m, embankment slope 20°, fill volume 100,000 m^3.
- Alternative 2: Two 10-meter-high dikes, at an altitude of 1860 m.a.s.l.; length 210 m, embankment slope 20°, fill volume 130,000 m^3.

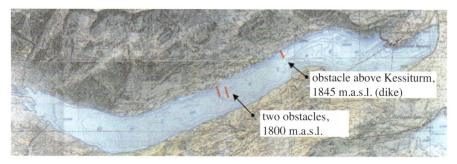

Figure 34.11 Grimsel reservoir with location of obstacles designed to slow turbidity currents (from Schleiss and Oehy, 2002)

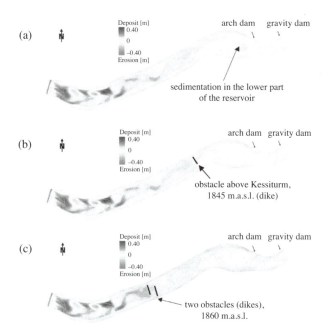

Figure 34.12 Simulated deposits and erosion due to a turbidity current in the Grimsel reservoir during the October 2000 flood (peak flow 40 m^3/s, sediment concentration assumed at 15 g/l); (a) without measures; (b) obstacle above Kessiturm at an altitude of 1845 m.a.s.l. (variant 1); (c) 2 obstacles at an altitude of 1860 m.a.s.l. (variant 2), (from Schleiss and Oehy, 2002)

The calculations demonstrated that the two alternatives stop the turbidity current and cause most of the material to settle upstream of the obstacles (Figure 34.12). Alternative 1 stopped the turbidity current almost entirely; in alternative 2, it continued to flow with reduced energy and disappeared near the "Kessiturm" narrowing, that is, more than 1 kilometer above the intakes. The volume behind the obstacles will be filled up in approximately 20 to 50 years, at which point the obstacles will no longer be effective.

34.7 Venting Turbidity Currents

As previously mentioned, turbidity currents are one of main processes that transport sediment into long and narrow reservoirs. They are able to transport suspended sediment from the plunge point near the delta right to the dam. Venting turbidity currents through the bottom outlets is an attractive option for reducing sedimentation in reservoirs. The venting of turbidity currents through bottom outlets with a limited capacity has been systematically studied as part of a research project (Chamoun et al., 2016, 2017, 2018). The effectiveness of venting in relation to the quantity of sediment evacuated through the reservoir water was assessed. The steeper the slope, the more effectively the turbidity currents are evacuated, given that the current is less rebounded at the dam. Turbidity currents should be vented through the dam at the very beginning of reservoir operation so that a cone upstream of the bottom outlet can be kept free of sediment and therefore maintain the steepest slope possible near the dam.

The venting degree, defined by the relationship between the flow exiting the bottom outlet and the flow of the turbidity current, has a considerable impact on the effectiveness of venting. In the experimental configuration, when the current reached the bottom outlet on a horizontal bed, a venting degree of about 100% resulted in the highest venting efficiency. For steeper slopes in the reservoir's thalweg near the dam (i.e., 2.4% and 5.0%), the optimum efficiency was obtained with a venting degree of around 135%. In cases where the bottom outlets have limited capacity compared to the flow of the turbidity currents, the effectiveness of venting can be estimated using Figure 34.13 (Chamoun et al., 2018).

Venting turbidity currents is more effective when it is synchronized with the arrival of the current at the dam. Installing a station to measure concentration approximately 300 m upstream of the bottom outlet is therefore recommended to observe the arrival of the current. In addition, the venting of sediment must continue while the turbidity current is present and for a minimum duration that will depend on the concentration of sediment in the evacuated current. Venting can be stopped once the muddy zone, which forms on the reservoir bottom upstream of the dam through suspended sediment, has been evacuated. At this point, the exiting flow is clear. This also enables the fine sediment deposited in the river downstream during operation to be flushed again.

In order to optimize effective venting of turbidity currents and minimize "dead" storage, the bottom outlet must be placed at the lowest possible level. The height and width of the entry to the release structure must be selected so that the limits of the intake cone in the reservoir match the dimensions of the body of the turbidity current. With the aim of ensuring that the bottom outlet dimensions remain reasonable, several release structures can be designed to ensure a sufficiently large intake cone.

34.8 Evacuating Suspended Fine Sediment through the Water Intake Using Jets

A new method of effectively and continuously evacuating fine sediment from the reservoir through the water intake has been developed (Jenzer et al., 2011, 2012, 2015). Emphasis was placed on evacuating fine sediment from around structures located near the dam where sediment is regularly deposited by density currents forming quite deep layers. For sediment to enter the water intake, it must be kept suspended in front of the intake entry. An upward flow of water and respectively an equivalent turbulence is required

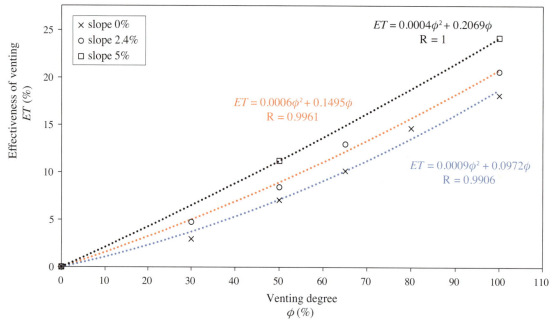

Figure 34.13 Effectiveness of venting sediment in turbidity currents as a function of the degree of venting i.e., evacuation through bottom outlets for the three different slopes tested: 0%, 2.45%, and 5% of the reservoir bed upstream of the bottom outlets

near the water intake to prevent the fine sediment brought by density currents from depositing. An upward mixer-like flow of water can successfully be generated by water jets arranged in an optimum layout. The jets can be fed by water collected by feed-in galleries, if available. Currently, in many cases, this water is released into the reservoir without any benefit being gained from its fall.

During an experimental systematic study, a circular arrangement with four jets proved to be very effective. With this optimum configuration, double the amount of sediment was evacuated compared to without jets, over the duration of the test. The flow field created was identical to that of an axial mixer, which in the literature is known for being conducive to the suspension of sediment. For practical applications, these results are highly promising, as considerable volumes of fine sediment will be able to be continually evacuated through the water intake. An initial analysis at the Mauvoisin dam showed that a circular arrangement of jets was cost-effective, even though only 7% of the sediment annually deposited into the dead storage area of the reservoir was evacuated (Jenzer et al., 2011). Even with this relatively modest evacuation, the zone near the water intake will be kept free of sediment. The maximum concentrations will not exceed those of density currents, i.e., approximately 5 g/l. These will instead decrease with the flow introduced by the water jets, so that abrasion in the turbines and environmental issues downstream can be controlled. For example, for 100 million m^3 of water turbined annually, approximately 100,000 t of fine sediment will be evacuated from the reservoir, as long as in the zone around the water intake a concentration of 1 g/l can be permanently suspended by a circular jet arrangement. A prototype installation with a circular arrangement of four jets, called SEDMIX, is currently being developed as part of a research and development project (De Cesare et al., 2018).

34.9 Effect of Cycles of Pumped Storage Operation on Sedimentation in Reservoirs

The question then arises of whether the cycles of pumped storage operation affect the process of sedimentation in upper and lower reservoirs. During a research project, study was made for the first time on whether fine particles transported into the reservoir close to the water intakes could be suspended by the turbulence caused by cycles of pumped storage operation (Müller et al., 2013, 2014, 2018). In this way, the depositing of fine sediment could be delayed, and the material could later be evacuated from the lower reservoir by the tailrace waterway system for the lower hydropower stage.

Prototype and laboratory measurements have shown that cycles of pumped storage operation significantly affect flow conditions in reservoirs. The resulting turbulence enables the fine sediment to be kept in suspension, which in turn means that considerable quantities are transported within the system. Overall, the sediment budget remains balanced as long as the two reservoirs have similar concentrations. However, the interaction between several water intakes, which feed the pumped storage power stations can create turbulence that prevents the deposit of fine sediment if the entry and exit points of these water intakes are placed in the best possible manner. Reservoirs in cascade favor the evacuation of fine sediment from the upstream reservoir toward the downstream reservoir (De Cesare et al., 2018; Guillen-Ludeña et al., 2018).

Figure 34.14 Sedimentation at the Sufers reservoir in the Grisons, Switzerland
(Courtesy Kraftwerke Hinterrhein AG)

34.10 Summary

Sedimentation is a crucial element in the sustainability of reservoirs. On a global scale, the average degree of sedimentation in reservoirs created by humans is practically in the same order as the increase in storage capacity provided by the construction of new dams. Over the course of the 21st century, the primary role of dam designers and operators will be to mitigate this process by implementing effective measures. One important measure is to prevent soil erosion by reforestation and to adopt soil conservation measures protecting agriculture wherever climatic conditions allow it. Similarly, the extent and mode of reservoir operation can together significantly affect the sedimentation of reservoirs with appropriate water-release and flushing structures. The selected options must take into account local conditions relating to the arrival of sediment. Computational modeling and forecasting of sedimentation can be of great assistance in choosing the right design for a reservoir.

In alpine regions, the rate of sedimentation is around ten times lower than the global average. Although the loss in volume is not yet major, this does not mean that sedimentation does not create problems. For alpine reservoirs, the conditions for the formation of turbidity currents are almost always met. These appear regularly during annual floods and carry large volumes of the sediment already deposited along the thalweg right up to the dam. Here, the layers of sediment increase relatively quickly, so much so that release structures such as water intakes and bottom outlets are affected after only 40 to 50 years of operation. Conditions for the sustainable operation of alpine reservoirs relate first of all to the control of turbidity currents through appropriate measures. There are several highly successful options, such as obstacles, roughness elements, and the introduction of an exterior source of energy, which may divert, reduce, and stop turbidity currents. To avoid sediment being deposited near the dam, venting turbidity currents through the bottom outlet, when possible, is both environmentally and economically beneficial. Artificial flooding combined with replenishment of sediment downstream of dams can work together with this venting of fine sediment and give momentum to a dynamic sediment transport in the river downstream (Döring et al., 2018; Stähly et al., 2019, 2020). Another promising option for the management of fine sediment is by using water jets near the dam to ensure its suspension before evacuating it at controlled concentrations through the water intake and the powerhouse.

Bibliography

ABCN, 2010. Ordonnance sur l'organisation des interventions en cas d'événement ABC et d'évènement naturel (Ordonnance sur les interventions ABCN). RS 520.17 du 20 octobre 2010 (Etat le 1er janvier 2013).

Amberg F., Bremen R., Droz P., Leroy R., Maier J. et Otto B., 2018. "Betonquellen bei Staumauern in der Schweiz". *wasser, energie, luft – eau, énergie, air*, 110. Jahrgang, Heft 4, 252-256.

Ammann E., 1987. "Die Sanierung der Stauanlage Gübsen". *wasser, energie, luft – eau, énergie, air*, 79. Jahrgang, Heft 5/6, 77-81.

Ammann E., Balissat M., Pougatsch H., Steiger K., Thomann P., Viret R. et De Vries F., 2003. "Recent Swiss experience with ageing and rehabilitation of concrete dams". Proceedings of 21th ICOLD Congress, Montréal, Q.82 - R.14, 219-245.

André S., 2004. *High velocity aerated flows on stepped chutes with macroroughness elements*. Communication n° 20 du Laboratoire de constructions hydrauliques, LCH-EPFL, Ed. A. Schleiss, Lausanne.

Babbitt D. H., 2003. "Improvement of Seismic Safety of Dams of California", Proceedings of 21th ICOLD Congress, Montréal, Vol. III, Q.83 - R.32, 537-551.

Bachmann H., 1994. "Erdbebenberechnung von Staumauern mit Stausee". *wasser, energie, luft – eau, énergie, air*, 86. Jahrgang, Heft 9, 284-293.

Balissat M., 1985. "Le traitement des fondations de grands barrages en relation avec les infiltrations". *wasser, energie, luft – eau, énergie, air*, 77e année, cahier 10, 307-314.

Barton N., Choubey. V., 1977. "The Shear Strength of Rock Joints in Theory and Practice". *Rock Mech. Suppl.*, 10(1-2): 1-54.

Bauer C., Hirschberg S. (Eds.), Bäuerle Y., Biollaz S., Calbry-Musyka A., Cox B., Heck T., Lehnert M., Meier A., Prasser H.-M., Schenler W., Treyer K., Vogel F., Wieckert H.C., Zhang X., Zimmermann M., Burg V., Bowman G., Erni M., Saar M., Tran M. Q., 2017. *Potentials, costs and environmental assessment of electricity generation technologies*. PSI, WSL, ETHZ, EPFL, Paul Scherrer Institut PSI, Villigen, Switzerland.

Baumer A., 2012. "Innalzamento diga Luzzone. Une sguardia su 15 anni d'esercizio". *wasser, energie, luft – eau, énergie, air*, 104e année, cahier 3, 204-208.

Beffa C., 2000. "Ein Parameterverfahren zur Bestimmung der flächigen Ausbreitung von Breschenabflüssen". *wasser, energie, luft – eau, énergie, air*, 93. Jahrgang, Heft 3/4, 2000.

Beyer Portner N., 1999. *Erosion de bassins versants alpins suisses par ruissellement de surface*. Communications n° 8 du Laboratoire de constructions hydrauliques, LCH-EPFL, Ed. A. Schleiss, Lausanne.

Beyer Portner N., Schleiss A., 2000. "Bodenerosion in alpinen Einzugsgebieten in der Schweiz". *Wasserwirtschaft* 90(2), 88-92.

BFE, 2014a. Bundesamt für Energie: *Diagramme zur Bestimmung der flächigen Ausbreitung von Breschenabflüssen (Verfahren "Beffa")*. BFE Hilfsmittel, Version 2.0, 26.06.2014.

BFE, 2014b. Bundesamt für Energie: *Vereinfachtes Verfahren zur Berechnung einer Flutwelle mit primär eindimensionaler Ausbreitung (Verfahren « CTGREF »)*. BFE Hilfsmittel, Version 2.0, 18.06.2014.

BFE, 2015. Office fédéral de l'énergie: *Stratégie d'urgence en cas de montée exceptionnelle du plan d'eau*. Etat 1.5.2015.

Biedermann R., 1981. "Bedeutung und Möglichkeiten der geodätischen Deformationsmessung in Rahmen der Talsperrenüberwachung". *wasser, energie, luft – eau, énergie, air*, 73. Jahrgang, Heft 9, 189-191.

Biedermann R., 1987a. "Anforderung an die Messeinrichtungen von Talsperren". *wasser, énergie, luft – eau, énergie, air*, 79. Jahrgang, Heft 1/2, 10-11.

Biedermann R., 1987b. "Talsperren; Planung für Notfälle". *wasser, energie, luft – eau, énergie, air*, 79. Jahrgang, Heft 5/6, 71-76.

Biedermann R., 1997. "Concept de sécurité pour les ouvrages d'accumulation: évolution du concept suisse depuis 1980". *wasser, energie, luft – eau, énergie, air*, 89e année, cahiers 3/4, 63-72.

Biedermann R., Delley P., Flury K., Hauenstein W., Lafitte R. et Lombardi G., 1988. "Safety of Swiss dams against floods; Design criteria and design flood". Proceedings of 16th ICOLD Congress, San Francisco, CA, Q.63 - R.22, 345-369.

BIEDERMANN R. et al., 1996. "Aménagements hydroélectriques à accumulation et protection contre les crues". *wasser, energie, luft – eau, énergie, air*, 88ᵉ année, cahier 10, 220-266.

BISCHOF R. et al., 2000. "205 dams in Switzerland for the welfare of the population". Q.77 - R.64. Proceedings of 20th ICOLD Congress, Beijing 2000, 997-1018.

BOES R. 1999. "Gewichtsstaumauer aus Walzbeton". *wasser, energie, luft – eau, énergie, air*, 91. Jahrgang, Heft 1/2, 11-15.

BOES R. M. (ed.), 2015. *Proc. 1st Intl. Workshop on Sediment Bypass Tunnels*. VAW Mitteilung 232. ETH Zurich, Zurich, Switzerland.

BOES R. M., MÜLLER-HAGMANN M., ALBAYRAK I., MÜLLER B., CASPESCHA L., FLEPP A., JACOBS F., AUEL C., 2018. "Sediment bypass tunnels: Swiss experience with bypass efficiency and abrasion-resistant invert materials". Proc. 26th ICOLD Congress, Vienna, Austria, Q.100 - R.40, 625-638.

BOILLAT J.-L., DE CESARE G., SCHLEISS A., OEHY C., 2000a. "Successful sediment flushing conditions in Alpine reservoirs". Proceedings International Workshop and Symposium on Reservoir Sedimentation Management, 26-27 October, Tokyo, Japan, 155-167.

BOILLAT J.-L., DUBOIS J., DE CESARE G., BOLLAERT E., 2000b. "Sediment management examples in Swiss Alpine reservoirs". Proceedings International Workshop and Symposium on Reservoir Sedimentation Management, 26-27 October, Tokyo, Japan.

BOILLAT J.-L. et SCHLEISS A., 2002. "Détermination de la crue extrême pour les retenues alpines par une approche PMF-PMP". *wasser, energie, luft – eau, énergie, air*, 94ᵉ année, cahiers 3/4, 107-116.

BOSSONEY C. L., 1994. "Monitoring and back analysis: the importance of the temperature load case". *Hydropower & Dams*, November 1994.

BREMEN F., COMO G., SALZGEBER H., 1999. "Wiederinstandsetzung der Staumauer Ferden". *wasser, energie, luft – eau, énergie, air*, 91. Jahrgang, Heft 11/12, 291-294.

BREMEN R., 1997. "The use of addiditives in cement grouts". *Hydropower & Dams*, Issue One, 71-76.

BRÜHWILER E., 1988. "Bruchmechanik von Staumauern unter quasi-statischen und erdbebendynamischen Belastung". Thèse EPFL, Département des matériaux.

BRÜHWILER E. et FREY P., 2002. *Alexandre Sarrasin, structure en béton armé, audace et invention*. Presses polytechniques et universitaires romandes, 2002.

BURY K. V. and KREUZER H., 1986. "The assessment of risk for a gravity dam". *Water Power & Dam Construction*, December 1986, 36-40.

BUSSARD T., WOHNLICH A., KOLIJI A. et LEROY R., 2015. "Surveillance des barrages-voûtes de l'Hongrin en Suisse, problématiques de venues d'eau et de stabilité des appuis". Colloque CFBR – Fondations des Barrages 8 et 9 avril 2015, Chambéry, France.

CFBR, 2013. "Recommandations pour le dimensionnement des évacuateurs de crues de barrages". Comité Français des Barrages et Réservoirs (CFBR).

CFGB, 1994. Barrages & réservoirs. "Les crues de projet des barrages: Méthode du Gradex". Bulletin du Comité Français des Grands Barrages n° 2, novembre 1994.

CFGB, 1997. *Petits barrages, recommandations pour la conception, la réalisation et le suivi*. Comité Français des Grands Barrages, Degoutte G. (éd.), Cemagref Editions.

CHAMOUN S., DE CESARE G. et SCHLEISS A. J., 2016. "Managing reservoir sedimentation by venting turbidity currents: A review". *International Journal of Sediment Research*, Vol. 31(3), 195-204.

CHAMOUN S., DE CESARE G. et SCHLEISS A. J., 2017. "Management of turbidity current venting in reservoirs under different bed slopes". *Journal of Environmental Management*, Vol. 204, 519-530.

CHAMOUN S., DE CESARE G. et SCHLEISS A. J., 2018. "Le transit des courants de turbidité, une technique pour réduire l'alluvionnement des réservoirs de barrages". *wasser, energie, luft – eau, énergie, air*, 110ᵉ année, cahier 1, 7-12.

CHAPPUIS Ph., 1987. "Modélisation non linéaire du comportement du béton sous des sollicitations dynamiques". IBK ETH-Zurich, Diss. N° 155.

CHAPUIS J., REBORA B., ZIMMERMANN Th., 1985. "Numerical approach of cracks propagation analysis in gravity dam during earthquake", Proceedings of 15th ICOLD Congress, Lausanne, Vol 2, Q.57 - R.26, 451-474.

CHEN H. Q., 2009. "Consideration on seismic safety of dams in China after the Wenchuan Earthquake". *Eng. Sci.* 44(2), 19-24.
CHOPRA, A. K., 2020. *Earthquake Engineering for Concrete Dams. Analysis, Design and Evaluation*. Wiley-Blackwell.
COMFORT G., 2000. "Ice loads against dams faces". *Journal of Hydraulic Engineering*, December 2000, 880-882.
COOKE B. J., 1997. "Developments in high concrete face rockfill dams". *Hydropower & Dams*, Issue Four, 69-73.
COOKE B. J., 2000. "The plinth of the CFRD". *Hydropower & Dams*, Issue Six, 61-64.
CORNUT R., 1992. "Surélévation du barrage de Lalla Takerkoust (Maroc). Conception, réalisation, auscultation". *wasser, energie, luft – eau, énergie, air*, 84e année, cahiers 7/8, 156-160.
COTTIN L., 1992. "L'informatique dans la surveillance des barrages en France". *wasser, energie, luft – eau, énergie, air*. 84e année, cahiers 1/2. 8-11.
CRESPO F., 1967. "Calcul des barrages". Cours de travaux hydrauliques, Technicum du soir, Lausanne.
CRUZ P. T., FREITAS Jr M. S., 2007. "Cracks and Flows in Concrete Face Rock Fill Dams (CFRD)". 5th International Conference on Dam Engineering, February 2007, 1-11.
CTGREF, 1978. Centre technique du génie rural des eaux et des forêts (CTGREF)[1] : "Appréciation globale des difficultés et des risques entraînés par la construction des barrages", note n° 5, juin 1978.
DARBRE, G. R., 1993a. "Strong-Motion Instrumentation of Concrete Dams". Proceedings of the 6th International Conf. Of Soils Dynamics and Earthquake Eng. Bath.
DARBRE G. R., 1993b. "Tremblement de terre : modèles de calculs". *wasser, energie, luft – eau, énergie, air*, 85e année, cahiers 1/2, 7-16.
DARBRE G. R., 1998a. *Dam Risk Analysis*. Swiss Federal Office for Water Management.
DARBRE G. R., 1998b. "Probabilistic assessment of current requirements on uncontrolled overtopping of dams during floods". Closing Report. Seminar on "Risk and safety of technical system".
DARBRE G. R., 2000. "Probabilistic safety assessment of dams", Proceedings of 20th ICOLD Congress, Beijing 2000, Vol. I. Q.76 – R.13, 185-196.
DARBRE G. R., 2004. "Réseaux d'accélérographes pour barrages". *IAS* 11/94, 182-185 et *IAS* 13/94, 234-240.
DARBRE G. R., DE SMET CAM, KRAEMER C., 2000. Natural frequencies measured from ambient vibration response of the arch dam of Mauvoisin. *Earthquake Engineering and Structural Dynamics* 29 :577–586.
DARBRE, G. R. et POUGATSCH H., 1993. "L'équipement de barrages dans le cadre du réseau national d'accélérographe". *wasser, energie, luft – eau, énergie, air*, 85e année, cahiers 11/12, 368-373.
DARBRE G. R. et PROULX J., 2002. Continuous ambient-vibration monitoring of the arch dam of Mauvoisin. *Earthquake Engineering and Structural Dynamics* 31 :475–480.
DARBRE G. R., SCHWYTER M., CONSTANTIN C., SCHLEGEL T. 2009. "Emergency Planning for Large Dams : Swiss Experience in Water Alarm". Proceedings of 23th ICOLD Congress, Brasilia, Mai 2009, Q91.
DARBRE G. R., SCHWAGER M.V. et PANDURI R., 2018. "Seismic safety evaluation of all large dams in Switzerland : lessons learned". Proceedings of 26th ICOLD Congress, July 1-7, 2018, Vienna, Austria, Q.101 – R.12, 198-216.
DE CESARE G., 1998. "Alluvionnement des retenues par courants de turbidité". Communication n° 7 du Laboratoire de constructions hydrauliques, LCH-EPFL, Ed. A. Schleiss, Lausanne.
DE CESARE G. et BOILLAT J.-L., 1998. "Trübeströme im Stausee Luzzone. Vergleich zwischen numerischer Modellierung und Naturmessungen. Symposium Planung und Realisierung im Wasserbau, Garmisch-Partenkirchen". *Mitteilungen der Versuchsanstalt für Wasserbau und Wasserwirtschaft* Nr. 8, TU München, 213-224.
DE CESARE G., SCHLEISS, A. et HERMANN, F., 2001. "Impact of turbidity currents on reservoir sedimentation". *Journal of Hydraulic Engineering* 127(1), 6-16.
DE CESARE G., MANSO P., CHAMOUN S., GUILLEN-LUDENA. S., AMINI A. et SCHLEISS A. J., 2018. "Innovative methods to release fine sediments from seasonal reservoirs". Proceedings of the 26th International Congress on Large Dams, ICOLD, Vienna, Austria, July 1-7, 2018, Q.100 – R.39, 612-624, CRC Press/Balkema.
DEINUM Ph. J., 1987. "Versuchsinstallation des « Sperry-Tilt Sensing » Systems zum Erfassen der Durchbiegung der Bogenmauer Emosson". *wasser, energie, luft – eau, énergie, air*, 79. Jahrgang, Heft 1/2, 17-19.

---

[1] Since 2011 : Institut national de recherche en sciences et technologies pour l'environnement et l'agriculture (IRSTEA).

DITCHEY E. J. et CAMPBELL D. B., 2000. "Roller compacted concrete and stepped spillways". *Hydraulics of Stepped Spillways*. Minor & Hager (eds), Balkema.

DOLCETTA M., MARAZIO A. et BAVESTRELLO, F., 1991. "The peripheral joint at the arch dams. Design, behavior and constructive aspects". *Felsbau* 9, Nr. 2, 79-89.

DÖRING M., TONOLLA D., ROBINSON CH. T., SCHLEISS A., STÄHLY S., GUFLER CH., GEILHAUSEN M. et DI CUGNO N., 2018. "Künstliches Hochwasser an der Saane – Eine Massnahme zum nachhaltigen Auenmanagement". *wasser, energie, luft – eau, énergie, air*, 110. Jahrgang, Heft 2, 119-127.

DROZ P., 1987. *Modèle numérique du comportement non linéaire d'ouvrages massifs en béton*. Thèse de doctorat, EPF Lausanne, IENER.

DUNGAR R. et ZAKERZADEH N., 1992. "Critical temperature laoding in arch dams. A review of Stucky-Derron formulation". *Dam Engineering*, Vol. III, Issue 2, 161-165.

DUNSTAN M. R. H., 1989. "Recent Developments in Roller Compacted Concrete Dam Construction". *Water Power & Dam Construction Handbook 1989*, 39-47.

DUNSTAN M. R. H., 1994. "The state-of-the-art of RCC dams". *Hydropower & Dams*, March 1994, 44-54.

DUNSTAN M. R. H., 1999. "Recent development in RCC dams". *Hydropower & Dams*, Issue One, 40-45.

DURAND J.-M., DEGOUTTE G., ROYET P. et JENSEN M., 1998. "La technique du béton compacté sur rouleau (B.C.R.). Possibilités d'application pour les barrages en Afrique". *Sud Science & Technologie*, N°4, janvier 1998.

DUSSART J., DESCHARD B. et PENEL F., 1992. "Petit Saut: an RCC dam in a wet tropical climate". *Water Power & Dam Construction*, February 1992, 30-32.

EGGER K., 1982. "Geodätische Deformationsmessungen. Eine zeitgemässe Vorstellung". *wasser, energie, luft – eau, énergie, air*, 74. Jahrgang, Heft 1/2, 1-4.

EGGER K. et GRAF A., 2002. "Die mittleren Fehler in der Geodäsie als Grundlage für die Interpretation von Verschiebungen". *wasser, energie, luft – eau, énergie, air*, 94. Jahrgang, Heft 11/12, 331-336.

EINSTEIN H. H. et DESCOEUDRES F., 1972. "Inventaire des essais in situ de mécaniques des roches (principes et critiques)". *Bulletin technique de la Suisse romande*, 98e année, n° 22, 28 octobre 1972, 345-356.

EWERT F. K., 1992. "The individual groutability of rocks". *Water Power & Dams Construction*, January 1992, 23-30.

FABER M. H. et STEWART M. G., 2001. "Risk Analysis for Civil Engineering Facilities: Overview and Discussion". Discussion Paper for the Joint Committee on Structural Safety, August 2001.

FAN J., 1986. "Turbidity currents in reservoirs". *Water International*, 107-116.

FELIX D., ALBAYRAK I. et BOES R. M., 2017. "Weiterleitung von Feinsedimenten via Triebwasser als Massnahme gegen die Stauraumverlandung". *wasser, energie, luft – eau, énergie, air*, 109e Jahrgang, Heft 2, 85-90.

FERC, 1997. *Guidelines*. Chapter X, Other Dams. October 1997.

FERC, 2002. *Guidelines*. Chapter III Gravity Dams. Revised October 2002.

FERN I., JONNERET A., KOLLY J.-C., 2019. "Travaux de confortement de l'appui rive gauche du barrage de Lessoc". *wasser, energie, luft – eau, énergie, air*, 111. Jahrgang, Heft 3, 173-180.

FOEV (SAEFL), 1998. *EIE des aménagements hydroélectriques. Mesures de protection de l'environnement* (information concernant l'étude d'impact sur l'environnement n° 8 de l'OFEFP (OFEV), Berne 1998. In French.

FOEV, 2009. Manuel EIE. *Directive de la Confédération sur l'étude de l'impact sur l'environnement*, 2009. In French.

FRCOLD, 1997. Bulletin du Comité Français des Grands Barrages. *Erosion interne; typologie, détection et réparation*, mai 1997.

FRISCHKNECHT R., ITTEN R. et FLURY K., 2012. "Treibhausgas-Emissionen der Schweizer Strommixe (Greenhouse gas emissions of the Swiss electricity mixes)". Studie im Auftrag des Bundesamtes für Umwelt (BAFU). ESU-Services Ltd., Uster, Switzerland.

FUCHS H., 2013. "Solitary Impulse Wave Run-up and Overland Flow". *Mitteilungen 221, Versuchsanstalt für Wasserbau, Hydrologie und Glaziologie (VAW)*, R. M. Boes, Hrsg., ETH Zürich.

GAGNON L., 2002. "IRN Statement on emissions from hydro reservoirs, a case of misleading science". *Hydropower & Dams*, Issue 4, 115-120.

GARCÍA HERNÁNDEZ J., SCHLEISS A. J. et BOILLAT J.-L., 2011. "Decision Support System for the hydropower plants management: the MINERVE project". *Dams and Reservoirs under Changing Challenges* – Schleiss & Boes (Eds), Taylor & Francis Group, London, 459-468. ISBN 978-0-415-68267-1.

GARCÍA HERNÁNDEZ J., CLAUDE A., PAREDES ARQUIOLA J., ROQUIER B. et BOILLAT J.-L., 2014. "Integrated flood forecasting and management system in a complex catchment area in the Alps – Implementation of the MINERVE project in the Canton of Valais". *Swiss Competences in River Engineering and Restoration*, Schleiss, Speerli & Pfammatter Eds, 87-97. Taylor & Francis Group, London, ISBN 978-1-138-02676-6, doi:10.1201/b17134-12.

GHRIB F., LÉGER P., TINAWI R., CUPIEN R. et VEILLEUX. 1997. "Seismic safety evaluation of gravity dams". *Hydropower & Dams*, Issue Two, 126-138.

GICOT H., 1970. "Le barrage de Gebidem de l'aménagement hydroélectrique de Bitsch". *Technique suisse des barrages*, Association Suisse pour l'Aménagement des Eaux, Publication n° 42, 45-66.

GICOT H., 1976. "Une méthode d'analyse des déformations des barrages". Proceedings of 12th ICOLD Congress, Mexico 1976, Vol. IV, report C1, 787-790.

GICOT O., 1985. "Barrages-voûtes de grande portée. Schiffenen. Barrages suisse. Surveillance et entretien". *SNCOLD*, Lausanne, 139-147.

GIOVAGNOLI M., SCHRADER E. et ERCOLI F., 1992. "Concepcion dam: a pratical solution to RCC problems". *Water Power & Dam Construction*, February 1992, 48-51.

GREIN H. et SCHACHENMANN A., 1992. "Solving problems of abrasion in hydroelectric machinery". *Water Power & Dam Construction*, 8, 19-24.

GRÜTTER F. et SCHNITTER N. J., 1982. "Analytical Risk Assessment for Dams". Proceedings of 14th ICOLD Congress, Rio de Janeiro 1982, Q.52 – R.32, 611-625.

GUERRIN, A., 1960-1961. « La Méthode calcul des conduites enterrées du Syndicat national des fabricants de travaux centrifugée en béton armé ». *Béton armé*, N° 29 et 20.

GUILLEN-LUDEÑA S., MANSO P. et SCHLEISS A. J., 2018. "Multidecadal sediment balance modelling of a cascade of alpine reservoirs and perspectives based on climate warming". *Water*, Vol. 19, n.1759.

GUNN R. M., 2001. "Non-linear design and safety analysis of arch dams using damage mechanics". *Hydropower & Dams*, Issues 2 and 3, 67-74 et 72-80.

GUNN R. M. et BOSSONEY C., 1996. "Creep analysis of mass concrete dams". *Hydropower & Dams*, Issue 6, 73-79.

HAGER, W.H.; SCHLEISS, A.J.; BOES, R.M.; PFISTER, M., 2021. *Hydraulic Engineering of Dams*. CRC Press Tylor & Francis Group, London, UK, ISBN 978-0-47197289-1.

HAMMERSCHLAG J.-G. et MERZ Ch., 2000. "Réaction alcali-granulat". *TFB Bulletin du ciment*, mai et septembre 2000.

HEIGERTH G., NIEDERMÜHLBICHLER H. et KNOBLAUCH H., 2000. "Sedimenttransport zwischen Hochdruckspeichern". Proceedings Wasserbau Symposium « Betrieb und Überwachung wasserbaulicher Anlagen », 19.-21. Oktober 2000, Graz, Österreich, *Mitteilung des Instituts für Wasserbau und Wasserwirtschaft*, Nr. 34, 347-356.

HEIN B. R., 1986. "Neuere Entwicklungen bei Felsschüttdämmen mit Betonflächendichtung". *Wasserwirtschaft* 76 (1986) 7/8, 320-327.

HELLER V., 2008. "Landslide generated impulse waves: Prediction of near field characteristics". *Mitteilungen 204, Versuchsanstalt für Wasserbau, Hydrologie und Glaziologie (VAW)*, H.-E. Minor, Hrsg. ETH Zürich.

HELLER V., HAGER W. H. et MINOR H.-E., 2009. "Landslide generated impulse waves in reservoirs: Basics and computation". *Mitteilungen 211, Versuchsanstalt für Wasserbau, Hydrologie und Glaziologie (VAW)*, R. Boes, Hrsg., ETH Zürich.

HERTIG J., 2005. *Etudes d'impacts sur l'environnement*. Traité de génie civil, vol. 23, Presses polytechniques et universitaires romandes.

HERTIG J.-A., FALLOT J.-M. et BRENA A., 2007. Etablissement des cartes de précipitations extrêmes pour la Suisse, Méthode d'utilisation des cartes de PMP pour l'obtention de la PMF, Projet Cruex, Directives crues de l'OFEN.

HERZOG M., 1993. "Die Bau und Konsolidationssetzungen von Staudämmen". *wasser, energie, luft – eau, énergie, air*, 85. Jahrgang, Heft 1/2, 35-37.

HOHBERG J.-M., WEBER B. et BACHMANN H., 1992. "Erdbebeneinwirkung bei Stauseen". *Schweizer Ingenieur und Architekt*, Nr. 19, 361-365.

HOLDERBAUM R. E. et ROARABAUGH D. P., 2001. "Trends and innovation in RCC dam design and construction". *Hydropower & Dams*, Issue 3, 63-69.

HOLLINGWORTH F., HOPPER D. J. et GERINGER J. J., 1989. "Roller compacted concrete arched dams". *Water Power & Dam Construction*, November 1989, 29-34.

HUBER A., 1980. "Schwallwellen in Seen als Folge von Felsstürzen". *Mitteilung der Versuchsanstalt für Wasserbau, Hydrologie und Glaziologie an der ETHZ*, Nr. 47.

HUG Ch., BOILLAT J.-L. et SCHLEISS A., 2000. "Hydraulische Modellversuche für die neue Wasserfassung der Stauanlage Mauvoisin". Proceedings Wasserbau Symposium «Betrieb und Überwachung wasserbaulicher Anlagen», 19.-21. Oktober 2000, Graz, Osterreich, *Mitteilung des Instituts für Wasserbau und Wasserwirtschaft*, Nr. 34, 367-376.

Hydropower & Dams, 2021. *2021 World Atlas & Industry Guide*, Aqua Media Int. Ltd.

HØEG K., 1992. "An evaluation of asphaltic concrete cores for embankment dams". *Water Power & Dams Construction*, July 1992, 32-34.

ICOLD, 1992. *Bituminous Core for Fill Dams. State of the art.* Bulletin 84.

ICOLD, 1995. *Dams Faillures. Statical analysis.* Bulletin 99.

ICOLD, 1997. "Position Paper on Dams and the Environment", may 1997.

ICOLD, 2000. *The Gravity Dam. A dam for the Future. Review and recommendations.* Bulletin 117.

ICOLD, 2003. *Roller-compacted concrete dams. State of the art and case histories.* Bulletin 126.

ICOLD, 2011. *Reservoirs and seismicity – State of knowledge.* Bulletin 137.

ICOLD, 2012. *Dams and the Environment from a Global Perspective.* Supplementary Paper 2012.

ICOLD, 2016. *Inspection of Dams. Following earthquake guidelines.* Bulletin 166.

ICOLD, 2017. *Internal Erosion of Existing Dams.* Bulletin 164.

ICOLD, 2018a. *Cutoffs for Dams.* Bulletin 150.

ICOLD, 2018b. *Roller-Compacted Concrete Dams.* Bulletin 177.

ICOLD, 2018c. *Dam Surveillance Guide.* Bulletin 158.

ICOLD, 2018d. *Asphalt Concrete Cores for Embankment Dams.* Bulletin 179.

ICOLD, 2020. *Roller-Compacted Concrete Dams.* Bulletin 177 (supersedes ICOLD bulletins 77, 126)

JACOBSEN T., 1999. "Sustainable reservoir development: the challenge of reservoir sedimentation. Hydropower into the next century" – III, 1999, Gmunden, Austria, Conference Proceedings, 719-728.

JACOBSON T., 2000. "Khimi pressurized tunnel sandtrap sediment removal during operation by use of Slotted Pipe Sediment Sluicer." Proceedings of Hydro 2000 – Making hydro more competitiv. 2-4 October, Berne, Switzerland, 153-159.

JANSEN R. B., 1988. *Advanced Dam Engineering for Design, Construction and Reabilitation.* Van Nostrand Reinhold, England.

JENSSEN L., 1997. Incorporating risk analysis in dam emergency planning. *Workshop «Risk-based dam safety evaluation»*, Trondhein, June 1997.

JENZER ALTHAUS J., DE CESARE G. et SCHLEISS A. J., 2011. "Entlandung von Stauseen über Triebwasserfassungen durch Aufwirbeln der Feinsedimente mit Wasserstrahlen". *wasser, energie, luft – eau, énergie, air*, 103. Jahrgang, Heft 2, 105-112.

JENZER ALTHAUS J., DE CESARE G. et SCHLEISS A. J., 2012. "Innovative and low cost continuous evacuation of sediment through the intake structure using multiple jets at the reservoir bottom", Proceedings of the 24th ICOLD Congress, 2-8 June 2012, Kyoto, Japan, Q.92 – R.17, 242-248.

JENZER ALTHAUS J. M. I., DE CESARE G. et SCHLEISS A. J., 2015. "Sediment Evacuation from Reservoirs through Intakes by Jet-Induced Flow", *Journal of Hydraulic Engineering*, 141(2): 04014078-1–04014078-9.

JIA J., LINO M., JIN F. et ZHENG C., 2016. "The cemented material dam: a new, environmentally friendly type of dam". *Engineering* 2(4), 490-497.

JIN F., ZHOU H., LI F. et AN X., 2017. "Application cases of rock-filled concrete dams". Proceedings of 2nd ICOLD Workshop on Cemented Material Dams (CMDs), Prague, July 3, 2017, 100-108.

JIN F., ZHOU H. et HUANG D., 2018a. "Research on rock-filled concrete dams: a review". *Dam Engineering*, 29, 101-112.

JIN F., ZHOU H. et HUANG D., 2018b. "Recent advances in rock-filled concrete dams and self-protected underwater concrete". Proceedings of 5th International Symposium on Dam Safety, Istanbul, Turkey, October 27-31, Hasan Tosun *et al.* (Eds), Vol. 1, 77-84

JOHANSEN P. M., VICK C. K. et RIKARTSEN C., 1997. "Risk analysis of three Norwegian rockfil dams". *Hydropower*, Balkema, Rotterdam, 431-451.

JOOS B. et KOLLY J.-C., 1995. "D'un simple modèle de détermination des déformations d'un barrage-poids sous l'influence de la température". Colloque «Research and Development in the Field of Dams», Crans-Montana.

JORDAN F., GARCÍA HERNÁNDEZ J., DUBOIS J. et BOILLAT J.-L., 2008. "Minerve – Modélisation des intempéries de nature extrême du Rhône valaisan et de leurs effets". *Communication n° 38 du Laboratoire de constructions hydrauliques*, LCH-EPFL, Ed. A. Schleiss, Lausanne.

JULLIARD H., 1961. "Le développement de la construction des barrages-poids en Suisse. *Cours d'eau et énergie*, Nr. 6/7, 174-177.

KREUZER H., 1983. "The design of El Cajon: a retrospective view". *Water Power & Dams Construction*, October 1983, 16-23.

KREUZER H., 2000. "The use of risk analysis to support dam safety decision and management". Proceedings of the 20th ICOLD Congress, Beijing 2000, General Report Q.76, 769-896.

KRUMDIECK A. et CHAMOT Ph., 1981. "Spülung von Sedimenten in kleinen und mittleren Staubecken". *Mitteilung der Versuchsanstalt für Wasserbau, Hydrologie und Glaziologie der ETH Zürich*, Nr. 53, 257-270.

LAFITTE R., 1985. "Utilisation des tirants de précontraintes dans les barrages". *wasser, energie, luft – eau, énergie, air*. 77e année, cahiers 1/2, 9-13.

LAFITTE R., 1989. "Le béton de barrage: perspective de recherche en relation avec la théorie de la mécanique de la rupture". *wasser, energie, luft – eau, énergie, air*, 81e année, cahiers 7/8, 161-168.

LAFFITE R., 1993. "Probabilistic risk analysis of large dams: its value and limits". *Water Power & Dam Construction*. March 1993, 13-16.

LAFITTE R., 1996. "Classes of risk for dams". *Hydropower & Dams*, Issue 6, 59-65.

LAFITTE R. et DE CESARE G., 2005. "Quantified criteria for electricity generation systems". Proceedings of Hydro 2005 (on CD-Rom). Policy into Practice, 17-20 October 2005, Villach, Austria, n° 13.01.

LAMBE T. W., WHITMAN R. V., 1969. "Soils Mechanics". Massachusetts Institute of Technology. John Wyley and Sons, New York, 1969.

LAZARO Ph. et GOLLIARD D., 1999. "Confortement de l'appui rive gauche du barrage de Montsalvens". *wasser, energie, luft – eau, énergie, air*. 91e année, cahiers 11/12, 295-302.

LCH, 1996. "Recherche dans le domaine des barrages. Crues extrêmes". Séminaire à l'EPFL 15-16.10.1996, Communication 5.

LE DELLIOU P., 1998. "Sécurité des barrages. Barrages et évaluation des risques". *Barrages, Bulletin du STEEGB* n° 2, 4e trimestre 1998, 1-3.

LECLERC M., LÉGER P. et TINAWI R., 2003. CADAM & RS DAM. "Analyses de stabilité statiques et dynamiques assistées par ordinateur pour barrages-poids", Juin 2003.

LÉGER P., TINAWI R., BHATTACHARJEE S. S. et LECLERC M., 1997. "Failure Mechanisms of Gravity Dams subjected to hydrostatic overload: influence of weak lift joints". Proceedings of the 19th ICOLD Congress, Florence, Vol. IV, Q.75 – R.2, 11-37.

LINK H., 1967. "Zur Beurteilung und Bestimmmung der Gleitsicherheit von Gewicht- und Pfeilermauern". *Die Wasserwirtschaft*. 1/1967, 35-46.

LINO M., AGRESTI P. and DELORME F., 2017. "The faced symmetrical hardfill dam (FSHD) – Feedback from 25 years of experience". *Proceeding of 2nd ICOLD Workshop on Cemented Material Dams (CMDs)*. Prague, July 3, 2017, pp. 16-26.

LMC EPFL, 1996. "Etude expérimentale sur les fluages des bétons âgés. Barrage de Luzzone. Fluage isotherme à changement constant". Lausanne, mars 1996.

LOFTI V., KOSHRANG G., MALLA S. et WIELAND M., 1995. "Seismic analysis and earthquake-resistant design of arch-gravity dam". 2nd Int. Conference on Seismology and Earthquake Engineering (SEE-2), Tehran, May 15-17, 1995.

LOMBARDI, G., 1988. "Querkraftbedingte Schäden in Bogensperren". *wasser, energie, luft – eau, énergie, air*, 80. Jahrgang, Heft 5/6, 119-125.

LOMBARDI G., 1992. "L'informatique dans l'auscultation des barrages". *wasser, energie, luft – eau, énergie, air*, 84e année, cahiers 1/2, 2-8.

LOMBARDI G., 1996. "Selecting the grouting intensity". *Hydropower & Dams*, Issue 4, 62-66.

LOMBARDI G., 2001. "Sécurité des barrages – Auscultation. Interprétation des mesures". Commentaires généraux, 25 pages.

LOMBARDI G., 2003. "L'injection des masses rocheuses". Conférence donnée à Rabat le 6 juin 1993.

LOMBARDI G., 2007. "3-D analysis of gravity dams". *Hydropower & Dams*, Issue 1, 98-102.

LOMBARDI G. et DEERE D., 1993. "Grouting design and control using the GIN principle". *Water Power & Dam Construction*, June 1993, 15-22.

LONDE P. et LINO M., 1992. "The faced symmetrical hardfill dams: a new concept for RCC". *Water Power and Dams Construction*, February 1992, 19-24.

LUGEON M., 1933/1979. *Barrages et géologie. Méthodes de recherches. Terrassement et imperméabilisation*. Réédité en 1979.

MAKDISI F. I. et SEED H. B., 1978. "Simplified procedure for estimating dam and embankment earthquake-induced deformations", *Journal of the Geotechnical Engineering Division*, ASCE, Vol. 104, No. GT7, 849-867.

MALLA S. et WIELAND M., 1995. "Effect of friction in vertical contraction joints of arch dams". Symposium on Research and Development in the Field of Dams, Crans-Montana, Switzerland, Sept. 7-9, 1995.

MARULANDA A., AMAYA F. et RAMIREZ C. A., 1991. "Colombian experience with concrete face rockfill dams". *Water Power & Dam Construction*, January 1991, 25-32.

MASON P. J., 1997. "The Evolving Dam". *Hydropower & Dams*, Issue 5, 69-73.

MASON P. J. et DUNLOP C. C., 2003. "Roller compacted concrete dams: a graphical summary". *Hydropower & Dams*. Issue 5, 124-125.

MATERÓN B., 1992. "Evolution in slab construction for the highest CFRDs". *Water Power and Dam Construction*. April 1992, 10-15.

MATERÓN B., 2007. "State of the art of compacted concrete face rockfill dams (CFRD'S)". 5th International Conference on Dam Engineering, February 2007, 21-27.

MATSUZAWA H. et al., 1985. "Dynamic soils and water pressures of submerged soils". *Journal of the Geotechnical Engineering*, vol. 111, n° 10, October 1985, 1161-1776.

MINOR H.-E., 1988. "Design of spillways for large dams". *Indian Journal of Power & River Valley Development*. March 1988, 89-116.

MOUVET L. et DARBRE G. R., 2000. "Probabilistic treatment of uncertainties: malfunctionning of discharge works and sliding of concrete dam under earthquake". Proceedings of the 20th ICOLD Congress, Beijing 2000, Vol. I, Q.76 – R.14, 197-212.

MÜLLER D. et HUBER A., 1992. "Auswirkung von Schwallwellen auf Stauanlagen". *wasser, energie, luft – eau, énergie, air*, 84. Jahrgang, Heft 5/6, 96-100.

MÜLLER M., DE CESARE G. et SCHLEISS A. J., 2013. "Einfluss von Pumpspeichersequenzen auf die Strömungsverhältnisse und das Absetzverhalten von Feinsedimenten in Stauseen", *wasser, energie, luft – eau, énergie, air*, 105. Jahrgang, Heft 3, 181-190.

MÜLLER M., DE CESARE G. et SCHLEISS A. J., 2014. "Continuous Long-Term Observation of Suspended Sediment Transport between Two Pumped-Storage Reservoirs", *Journal of Hydraulic Engineering*, 140 (5).

MÜLLER M., DE CESARE G. et SCHLEISS A. J., 2018. "Flow field in a reservoir subject to pumped-storage operation – in situ measurement and numerical modeling", *Journal of Applied Water Engineering and Research*, Vol. 6, N° 2, 109-124.

MÜLLER R. W., 2001. "Abschätzung von Dammbruchflutwellen für kleine Stauanlagen". *wasser, energie, luft – eau, énergie, air*, 93. Jahrgang, Heft 3/4, 71-72.

NEOCO, 2007. *National Emergency Operations Centre Ordinance*, in RS 520.18 of April 17, 2018, in French, German, Italian.

NEWMARK N. M., 1965. "Effects of earthquakes on dams and embankments", *Geotechnique* 15, N° 2, 139-160.

OAL, 2010. *Ordinance on Alert and Alarm*. RS 520.12 of August 18, 2010. In French, German, Italian. (Ordinance on the alarm and the security radio network adapted on 1st January 2019).

OBERNHUBER P., 2014. "Internationale Übersicht über die Anforderungen an die Gleit- und Kippsicherheitsnachweiese von Gewichtsmauern". SFOE, avril 2014.

OEHY Ch., DE CESARE G. et SCHLEISS A., 2000. "Einfluss von Trübeströmen auf die Verlandung von Staubecken". Symposium Betrieb und Überwachung wasserbaulicher Anlagen, 19.-21. Oktober 2000, Graz, Österreich, *Mitteilung des Instituts für Wasserbau und Wasserwirtschaft*, Nr. 34, 413-422.

OEHY Ch. et SCHLEISS A., 2003. "Beherrschung von Trübeströmen in Stauseen mit Hindernissen, Gitter, Wasserstrahl- und Luftblasenschleier". *wasser, energie, luft – eau, énergie, air*, 95. Jahrgang, Heft 5/6, 143-152.

OTTO B., BALISSAT M. et RICCIARDI A., 2016. "The Muttsee dam, Switzerland: Highest dam location in the Alps". Proceedings – HYDRO 2016, Achievements, Opportunities and Challenges, 10-12 October 2016, Montreux, Switzerland.

PAILLEX E., 1985. *Le béton des barrages*. Publication SIKA SA.

PANCHAUD F., 1962. "Le barrage de Tourtemagne". *Schweizerische Bauzeitung*, vol. 80, 401-402.

PINTO N. L. de S., 2001. Questions to ponder on designing very high CFRDs. *Hydropower & Dams*, Issue Five.

PINTO N. L de S., 2007. "A challenge to very high CFRD dams: very high concrete face compressive stresses". 5th International Conference on Dam Engineering, February 2007, 441-446.

PINTO N. L de S. et MARQUES FILHO P. L., 1998. "Estimating the maximum face deflection in CFRDs". *Hydropower & Dams*, Issue 6, 28-31.

PIRCHER W. et SCHWAB H., 1988. Design, construction and behaviour of the asphaltic concrete core wall of the Finstertal dam. Proceedings of the 16th ICOLD Congress, San Fransisco, 1998, Q. 61, R. 49, 901-924.

PIRCHER W., 1993. "36 000 dams and still more needed". *Water Power & Dam Construction*, May 1993, 15-18.

POST G. et LONDE P., 1953. *Les barrages en terre compactées. Pratiques américaines*. Gauthier – Villars, éditeur.

POUGATSCH H., 1975. "Aménagement de la Haute-Sarine. Palier de Lessoc". Société générale pour l'industrie, Genève, Rapport interne (non publié).

POUGATSCH H., 1983. "Manœuvre et essai de fonctionnement des organes mobiles". *wasser, energie, luft – eau, énergie, air*, 75e année, cahiers 11/12, 273-275.

POUGATSCH H., 1990. "Barrage de Zeuzier. Rétrospective d'un événement particulier". *wasser, energie, luft – eau, énergie, air*, 82e année, cahier 9, 195-208.

POUGATSCH H., 1993. "Le réseau sismique national d'accélérographe". *wasser, energie, luft – eau, énergie, air*. 85e année, cahiers 5/6, 111-113.

POUGATSCH H., 2002. "Surveillance des ouvrages d'accumulation. Conception générale du dispositif d'auscultation". *wasser, energie, luft – eau énergie, air*, 94e année, cahiers 9/10, 267-271.

POUGATSCH H. et MÜLLER R. W., 2000. "Aspects constructifs relatifs aux basins d'accumulation pour l'enneigement artificiel et aux digues contre avalanche". *wasser, energie, luft – eau, énergie, air*, 92e année, cahiers 11/12, 349-354.

POUGATSCH H. et SONDEREGGER T., 2003. "Stauanlagenüberwachung. Ausgestaltung der Messanlage". *wasser, energie, luft – eau énergie, air*, 95 Jahrgang, Heft 5/6. 153-157.

POUGATSCH H., MÜLLER R. W. et KOBELT A., 1998. "Water Alarm Concept in Switzerland". Proc. of International Symposium on New Trends and Guidelines on Dam Safety. Barcelona, Spain, June 1998, Berga *et al.* (eds), Balkema, Rotterdam, 235-242.

POUGATSCH H., MÜLLER R. W., SONDEREGGER T. ET KOBELT, A., 2011. "Improvement of Swiss dams on the basis of experience". Proc. Int. Symposium on Dams and Reservoirs under Changing Challenges, 77th Annual Meeting of ICOLD, Lucerne, Switzerland June 1, 2011, 145-152.

PRUSKA M.-L., 1963. "Une nouvelle méthode de calcul des pressions d'un remblai sur une conduite". *Béton armé*, 6e année, N° 48, mai 1963, 10-18.

RABOUD P.-B., DUBOIS J., BOILLAT J.-L., COSTA S. et PITTELOUD P.-Y., 2001. "Projet Minerve – Modélisation de la contribution des bassins d'accumulation lors des crues en Valais". *wasser, energie, luft – eau, énergie, air*, 93e année, cahiers 11/12, 313-317.

RAPHAEL J.-M., 1970. "The optimum gravity dam". Proceedings of the Conference on rapid Construction of Concrete Dams, March 1-5, Pacific Grove, CA, USA. New York: ASCE, 221-244.

RECHSTEINER G., 1994. "Assainissement du barrage de Cleuson". *IAS* n° 23, 26 octobre 1994, 422-428.

Rescher O.-J., 1993. "Arch dams with an upstream base joint". *Water Power & Dam Construction*, March 1993, 17-25.

Sabarly F., 1968. "Les injections et les drainages de fondation de barrages en roches peu perméables". *Géotechique*, 18, 229-249.

Sägesser R. et Mayer-Rosa D., 1978. "Erdbebengefährdung in der Schweiz". *Schweizerische Bauzeitung*, 96. Jahrgang, Heft 7, 16. Februar 1978, 107-123.

Salmon G. M. et Hartford D. N. D., 1995. "Risk analysis for dam safety". *International Water Power & Dam Construction*, Part I, March 1995, 42-47 and Part II, April 1995, 38-39.

Sarkaria G. S. et Andriolo F. R., 1995. "Special factors in design of high RCC gravity dams". *International Water Power & Dam Construction*, April and August 1995.

Sarrasin A., 1939. "Notes sur les barrages à arches multiples". *Schweizerische Bauzeitung*, 13 mai 1939, 231-235.

Scherrer M., 2007. "Hydroélectricité et climat global". Projet SIE, EPFL, Février 2007 (non publié).

Schewe L., 1990. "Design and construction of concrete facings for embankment dams". *Hydro Power and Dam Construction*, March 1990, 34-40.

Schleiss A., 1999. "Constructions hydrauliques. Facteur clé de la prospérité économique et du développement durable au XXIe siècle", *IAS* n° 11, 9 juin 1999, 198-205.

Schleiss. A. J., 2000. "The importance of hydraulic schemes for the sustainable development in the 21st century". *Hydropower & Dams*, volume 7, Issue 1, 19-24.

Schleiss, A., 2012. "Talsperrenerhöhungen in der Schweiz: energiewirtschaftliche Bedeutung und Randbedingungen". *wasser, energie, luft – eau, énergie, air*, 104. Jahrgang, Heft 3, 209-215.

Schleiss A., De Cesare G. et Jenzer Althaus J., 2010. "Verlandung der Stauseen gefährdet die nachhaltige Nutzung der Wasserkraft". *wasser, energie, luft – eau, énergie, air*, 102. Jahrgang, Heft 1, 31-40.

Schleiss A., Feuz B., Aemmer M. et Zünd B., 1996. "Verlandungsprobleme im Stausee Mauvoisin. Ausmass, Auswirkungen und mögliche Massnahmen". Int. Symposium «Verlandung von Stauseen und Stauhaltungen». *Mitteilung der Versuchsanstalt für Wasserbau, Hydrologie und Glaziologie der ETH Zürich*, Nr. 142, 37-58.

Schleiss A. et Oehy Ch., 2002. "Verlandung von Stauseen und Nachhaltigkeit". *wasser, energie, luft – eau, énergie, air*, 94. Jahrgang, Heft 7/8, 227-234.

Schleiss A. J., 2002. "Scour evaluation in space and time – the challenge of dam designers". Proc. Int. Workshop Rock scour due to falling high-velocity jets, EPFL Lausanne, Switzerland; Schleiss & Bollaert (eds), Swets & Zeitlinger, Lisse, 3-22.

Schleiss A. J., 2016a. "Dams and reservoirs as security belt around the world to ensure water, food and energy". *World Atlas & Industry Guide*, 12-13, Aqua Media Intl., London, UK.

Schleiss A. J., 2016b. "Talsperren und Speicher als lebenswichtige Infrastrukturanlagen für den weltweiten Wohlstand (Dams and reservoirs as vital infrastructure for the global welfare)". *Wasserwirtschaft* 106(6), 12-15 (in German).

Schleiss A. J., 2016c. "Dams are a question of heart – better dams for a better world". *The Dams Newsletter* CIGB-ICOLD, N° 15, May 2016.

Schleiss A. J., 2017a. ICOLD welcomes his 100th member country supporting his mission "Better Dams for a Better World". *World Atlas & Industry Guide*, 10-11, Aqua Media Intl., London, UK.

Schleiss A. J., 2017b. "Better water infrastructures for a better world – the important role of water associations". *Hydrolink*, 86-87.

Schleiss A. J., 2018. "Sustainable and safe development of dams and reservoirs as vital water infrastructures in this century - The important role of ICOLD". Proc. International Dam Safety Conference, 23 and 24 January 2018 Thiruvananthapuram, Kerala, 3-16.

Schleiss A. J., Franca M. J., Juez C. et De Cesare G., 2016. "Reservoir sedimentation". *Journal of Hydraulic Research*, 54:6, 595-614, DOI:10.1080/00221686.2016.1225320.

Schmidt R., 2007. "Feasibility study for completion of the Rogun scheme, Tadjikistan". *Hydropower & Dams*, Issue 3, 102-107

Schneebeli G., 1971. *Les parois moulées dans le sol. Techniques de réalisation, méthodes de calculs*. Eyrolles.

SCHNEIDER T. R., 1987. "Mise en évidence des glissements et éboulements potentiels". *wasser, energie, luft – eau, énergie, air*, 79e année, cahier 9, 203-207.

SCHNITTER G., 1956. "Theorie zur Berechnung von Staumauern und Staudämmen". *Wasser- und Energiewirtschaft*. Nr. 7/8/9, 183-186.

SCHNITTER N. J., 1985. Le développement de la technique des barrages en Suisse. Comité national suisse des grands barrages, *Barrages suisses – Surveillance et entretien*, publié à l'occasion du 15e Congrès international des grands barrages, Lausanne 1985, 11-23.

SCHNITTER N. J., 1988. "The evolution of embankment dams", *Water Power & Dams Handbook*, 27-35.

SCHNITTER N. J., 1993. "Die Sprengung des Steindammes Peruča in Kroatien am 28. Januar 1993". *wasser, energie, luft – eau, énergie, air*, 85 Jahrgang, Heft 5/6, 102-104.

SCHNITTER N. J., 1994. *A History of Dams – the Useful Pyramids*, Balkema, Rotterdam.

SCHNITZLER E., 1985. "Les pionniers suisses ou le goût du risque calculé". EPFL, *Polyrama*, 1985, 6-12.

SCHOBER W., 2003. Embankment Dams. Research and development, construction and operation. *Large Dams in Austria*/Volume 34. Austrian National Committee on Large Dams, Innsbruck, September 2003.

SCHWAGER M. V., PANDURI R., DARBRE G. R., 2016. "Concept et aspects particuliers de la révision de la directive suisse sur la sécurité des ouvrages d'accumulation". Colloque CFBR "Sécurité des barrages et enjeux", Chambéry, 23 et 24 novembre 2016, 187-197.

SED/SSS, 2018. "Le nouveau réseau accélérométrique suisse". http://www.seismo.ethz.ch

SEED H. B. et IDRISS I. M., 1982. "Ground Motions and Soil Liquefaction during Earthquakes", *Engineering Monograph on Earthquake Criteria*, Structural Design and Strong Motion Records, EERC.

SERAFIM L. J., 1989. "Four decades of concrete arch design". *Water Power & Dam Construction Handbook*, 27-32.

SERAFIM L. J. et CLOUGH R. W. (editors), 1990. *Spanish arch dam engineers*, Arch Dams, 546-554, Rotterdam.

SFOE (SFWG), 2002a. Sécurité des ouvrages d'accumulation. *Documentation de base relative aux critères d'assujettissement. Sicherheit der Stauanlagen Basisdokument zu den Unterstellungskriterien*, Version 1.0, 01.06.2002.

SFOE (SFWG), 2002b. Sécurité des ouvrages d'accumulation. *Directives de l'OFEG (SFWG). Sicherheit der Stauanlagen. Richtlinien des BWG*, Version 1.1., 01.11.2002.

SFOE (SFWG), 2002c. Sécurité des ouvrages d'accumulation. *Documentation de base relative à la sécurité structurale*. Sicherheit der Stauanlagen. *Basisdokument zur konstruktiven Sicherheit*, Version 1.0., 01.08.2002.

SFOE (SFWG), 2003. Sécurité des ouvrages d'accumulation. *Documentation de base pour la vérification des ouvrages d'accumulation aux séismes*. Sicherheit der Stauanlagen. *Basisdokument zu dem Nachweis der Erdbebensicherheit*, Version 1.2., mars 2003.

SFOE, 2014a. *Directive sur la sécurité des ouvrages d'accumulation. Partie B : Risque potentiel particulier comme critère d'assujettissement. Richtlinie über die Sicherheit der Stauanlagen. Teil B: Besonderes Gefährdungspotenzial als Unterstellungskriterium.*

SFOE, 2014b. Calcul de l'onde de submersion selon Beffa (document d'aide SFOE, www.bfe.admin.ch). In French, German.

SFOE, 2014c. Calcul de l'onde de submersion selon CTGREF (document d'aide SFOE, www.bfe.admin.ch).

SFOE, 2014d. Internationale Übersicht über die Anforderungen an die Gleit- und Kippsicherheitsnachweise von Gewichtsmauern (in German), étude du Dr. Pius Obernhuber sur mandat de l'OFEN, avril 2014.

SFOE, 2015a. *Directive sur la sécurité des ouvrages d'accumulation. Partie A : Généralités. Richtlinie über die Sicherheit der Stauanlagen Teil A: Allgemeines.* In French, German.

SFOE, 2015b. *Directive sur la sécurité des ouvrages d'accumulation. Partie D : Mise en service et exploitation* (Mise en service – Maintenance – Surveillance). *Richtlinie über die Sicherheit der Stauanlagen. Teil D; Inbetriebnahme und Betrieb - Inbetriebnahme / Unterhalt / Überwachung.* In French, German.

SFOE, 2015c. *Directive sur la sécurité des ouvrages d'accumulation. Partie E : Plan en cas d'urgence. Richtlinie über die Sicherheit der Stauanlagen. Teil E: Notfallkonzept.* In French, German.

SFOE, 2015d. *Exemple de règlement en cas d'urgence Ouvrage d'accumulation sans dispositif d'alarme-eau. Richtlinie über die Sicherheit der Stauanlagen. Hilfsmittel Beispiel Notfallreglement ohne Wasser Alarm.* In French, German.

SFOE, 2016. *Directive sur la sécurité des ouvrages d'accumulation. Partie C3 : Sécurité aux séismes. Richtlinie über die Sicherheit der Stauanlagen Teil C3: Erdbebensicherheit. Direttiva sulla sicurezza degli impianti di accumulazione. Parte C3: Sicurezza sismica.* In French, German, Italian.

SFOE, 2017. *Directive sur la sécurité des ouvrages d'accumulation. Partie C1 : Dimensionnement et construction. Richtlinie über die Sicherheit der Stauanlagen. Teil C1: Planung und Bau, Direttiva sulla sicurezza degli impianti di accumulazione, Parte C1: Dimensionamento e costruzione.* In French, German, Italian.

SFOE, 2018. *Directive sur la sécurité des ouvrages d'accumulation. Partie C2 : Sécurité en cas de crue et abaissement de la retenue. Richtlinie über die Sicherheit der Stauanlagen. Teil C2: Hochwassersicherheit und Stauseeabsenkung , Direttiva sulla sicurezza degli impianti di accumulazione. Parte C2: Sicurezza contro le piene e abbassamento della ritenuta.* In French, German, Italian.

SFWG, 2002. Federal Office for Water and Geology, Sécurité des ouvrages d'accumulation, Documentation de base relative à la sécurité structurale, rapports de l'OFEG, série Eaux, Version 1.0, Bienne, 2002. In French, German.

SFWG, 2003. Federal Office for Water and Geology, Sécurité des ouvrages d'accumulation, Documentation de base pour la vérification des ouvrages d'accumulation aux séismes, rapports de SFWG, série Eaux, Version 1.2, Bienne, 2003. In French, German.

SHARMA S. K., 1979. "Response and stability of earth dams during strong earthquakes", miscellaneous papers, GL-79-13, US Army Engineer WES, CE, Vivksburg, Miss, 1979.

SHERARD J. I. and DUNNIGA L. P., 1989. "Critical filters for impervious soils". *J. Geotech. Eng. ASCE*, Vol. 115 N° 7, 927-947.

SIA, 2013a. Swiss Society of Engineers and Architects. Norme 267, Géotechnique, Geotechnik.

SIA, 2013b. Swiss Society of Engineers and Architects. Norme 267/1, Géotechnique – Spécifications complémentaires, Geotechnik – Ergänzende.

SIA, 2014. Swiss Society of Engineers and Architects. Norme 261, Actions sur les structures porteuses, Einwirkungen auf Tragwerke, Azioni sulle strutture portanti.

SINNIGER R., 1987. Observation des versants d'une retenue. *wasser, energie, luft – eau, énergie, air*, 79. Jahrgang, Heft 9. 209-210.

SINNIGER R., 1985. "L'histoire des barrages". EPFL, *Polyrama*, 1985, 2-5.

SINNIGER R. et HAGER W. H., 1984. "Retentionsvorgänge in Speicherseen". *Schweizer Ingenieur und Architekt*, 104(26), 535-539.

SINNIGER R., DE CESARE G. et BOILLAT J.-L., 2000. "Eigenschaften junger Sedimente in Speicherseen". *wasser, energie, luft – eau, énergie, air*; 92. Jahrgang, Heft 1/2, 9-12.

SKERMER N. A., 1993. "Homer M. Hadley and the conception of the RCC dam. An historical note". *Geotechnical News*.

SKRIKERUD P., 1983. "Modelle und Berechnungensverfahren für das Rissverhalten von unarmierten Betonbau unter Erdbeenbeanspruchung". Dissertation ETH Zürich, *IBK-Bericht* n° 139, Birkhäuser, Basel.

SMIT P., 1996. "Datenerfassung und Bestimmung der Abminderung der Bodenbewegung bei Erdbeben in der Schweiz". Diss. ETH Nr. 11396, Juris Druck + Verlag AG.

SNCOLD, 1964. Swiss National Committee on Large Dams. *Behavior of Large Swiss Dams*.

SNCOLD, 1982. Swiss National Committee on Large Dams. "Abnormal Behaviour of Zeuzier Arch-Dam (Switzerland)", special issue to XIVᵉ ICOLD-Congress in Rio de Janeiro. *wasser, energie, luft – eau, énergie, air*, cahier 3, 65-112.

SNCOLD, 1985. Swiss National Committee on Large Dams, "Swiss Dams. Monitoring and Maintenance". Edition for the 15ᵉ International Congress on Large Dams in Lausanne. *wasser, energie, luft – eau, énergie, air.*

SNCOLD, 1988. Swiss National Committee on Large Dams, "Measuring Installations for Dam Monitoring. Concepts. Reliability. Redundancy". *wasser, energie, luft – eau, énergie, air*, 80. Jahrgang 1988, Heft 1/2, 9-20.

SNCOLD, 1992. Swiss National Committee on Large Dams. "Auftrieb bei Betonsperren". in German, Working Group "Uplift", April 1992.

SNCOLD, 1993. Swiss National Committee on Large Dams. Working Group for Dam Monitoring: *L'informatique dans la surveillance des barrages, saisie et traitement des mesures*, in French.

SNCOLD, 1994. Swiss National Committee on Large Dams. Working Group for monitoring of dams: "The use of electronic-based systems in dam surveillance. The capture and processing of readings". *wasser, energie, luft – eau, énergie, air*, 86. Jahrgang, Heft 9, 260-278.

SNCOLD, 1997a. Swiss National Committee on Large Dams. Working Group for Dam Observations: "Surveillance de l'état des barrages et check-lists pour les contrôles visuels", In French, German, Italian.

SNCOLD, 1997b. Swiss National Committee on Large Dams. Working Group for Dam Observations: "Mesures de déformations géodésiques et photogrammétriques pour la surveillance des barrages", 82 pages.

Sobrinho J. A. et al., 2007. Performance and concrete face repairs at Campos Novos. *Hydropower & Dams*, Issue 2.

Stähly S., Franca M. J., Robinson C. T., Schleiss A. J. 2019. "Sediment replenishment combined with an artificial flood improves river habitats downstream of a dam", *Scientific Reports*, 9(1):51-79. doi: 10.1038/s41598-019-41575-6.

Stähly, S., Franca, M. J., Robinson, C. T., and Schleiss, A. J. 2020. "Erosion, transport and deposition of a sediment replenishment under flood conditions. *Earth Surf. Process. Landforms*, 45: 3354–3367. https://doi.org/10.1002/esp.4970.

Stark T. D. et Mesri G., 1992. "Undrained Shear Strenght of Liquefied Sands for Stability Analysis", *Journal of Geotechnical Engineering Division*, ASCE, 118 (11), 1727-1747.

Stojnic, I., Pfister, M., Matos, J., and Schleiss A. J., 2021. "Influence of smooth and stepped chute approach flow on the performance of a stilling basin". *Journal of Hydraulic Engineering*, 147(2), 04020097.

Stucky A., 1937. "Le barrage de Beni-Bahdel". *Bulletin technique de la Suisse romande*, 5 juin 1937, N°12, 141-153.

Stucky A., 1970. "Barrages en Afrique du Nord". *Technique suisse des barrages*, Association suisse pour l'aménagement des eaux, Publication N° 42, Baden, 111-114.

Stucky A. et Derron M.-H., 1957. *Problèmes thermiques posés par la construction des barrages-réservoirs*. EPUL, Publication n° 38.

Stucky A., Panchaud F. et Schnitzler E., 1951. *Contribution à l'étude de barrages-voûtes. Effet de l'élasticité des appuis*, EPUL, Publication n°13.

Stucky J.-P., 1956. *Technologie et contrôle des barrages en béton*. EPUL, Publication n° 39 (tiré à part du BTSR. n° 19 du 15 septembre 1956).

Stucky J.-P., 1997. *Barrages en béton*, 3e partie, EPFL-Repro, Lausanne.

Sumi, T., Okano, M., Takata, Y., 2004. "Reservoir sedimentation management with bypass tunnels in Japan". Proc. 9th Intl. Symp. River sedimentation Yichang, China, 1036-1043.

Sumi, T. (ed.), 2017. Proc. 2nd Intl. Workshop on Sediment Bypass Tunnels. Kyoto University, Kyoto, Japan. http://ecohyd.dpri.kyoto-u.ac.jp/en/index/SBTworkshop.html

SwissCoD, 2000a. Swiss Commitee on Dams, Working Group for Dam Observations: "Concrete of Swiss dams: experiences and synthesis". *wasser, energie, luft – eau, énergie, air*, 92(7/8), 205-233.

SwissCoD, 2000b. Swiss Commitee on Dams, Working Group for Dam Observations: *Sécurité structurale des barrages : Plan d'utilisation et plan de sécurité*, in French, December 2000.

SwissCoD, 2001a. Swiss Commitee on Dams, Working Group for Dam Observations: *Monographie de barrage – Recommandations pour la rédaction, (contenu et structure)*, in French, German, Italian, November 2001.

SwissCoD, 2001b. Swiss Committee on Dams, *Le béton des barrages suisses*, in French, unpublished, August 2001.

SwissCoD, 2003a. Swiss Commitee on Dams, *Analyses de risques dans le domaine de la sécurité des ouvrages d'accumulation suisses*. Recommandations pour des projets de recherches, in French.

SwissCoD, 2003b. Swiss Committee on Dams, "Methods of analysis for the prediction and the verification of dam behavior". *wasser, energie, luft – eau, énergie, air*, 95(3/4), 74-98.

SwissCoD, 2005. Swiss Committee on Dams, Working Group for Dam Observations: *Dossier de l'ouvrage d'accumulation. Recommandations*, in French, German, Italian.

SwissCoD, 2006. Swiss Committee on Dams, Working Group for Dam Observations: "Dam Monitoring instrumentation - concepts, reliability and redundancy". *wasser, energie, luft – eau, énergie, air*, 98(2), 143-180.

SwissCoD, 2013. Comité suisse des barrages, Groupe de travail pour l'observation des barrages. *Géodésie pour la surveillance des ouvrages d'accumulation*.

SwissCoD, 2015. Swiss Committee on Dams, Working Group for Dam Observations: "Role and duties of Dam Warden - Level 1 surveillance of water retaining facilities". *wasser, energie, luft – eau, énergie, air*, 107(2), 99-110.

SwissCoD, 2017a. Swiss Committee on Dams, *Floating debris at reservoir dam spillways*. Report of the Swiss Committee on Dams on the state of floating debris issues at dam spillways. Working group on floating debris at dam spillways, November 2017.

SwissCoD, 2017b. Swiss Committee on Dams, *Concrete swelling of dams in Switzerland*. Report of the Swiss Committee on Dams on the state of concrete swelling in Swiss Dams. AAR Working Group. May 2017.

Terrier, S., Pfister, M., Schleiss, A.J., 2022. "Performance and design of a stepped spillway aerator". *Water*, 14(2), 153.

Terzaghi K. et Peck R. B., 1965. *Mécanique des sols appliquée aux travaux publics et aux bâtiments*. Dunod, Paris, 565 pages.

Tognola F. et Balissat M., 2011. "The new Muttsee Dam (Switzerland)". Proc. Int. Symposium on Dams and Reservoirs under Changing Challenges, 77th Annual Meeting of ICOLD, Lucerne, Switzerland, June 1, 2011, 867-874.

Tschernutter P. et Nackler K., 1991. "Construction of Feistritzbach dam with central asphaltic concrete membrane and the influence of the poor quality rock of fill behaviour". Proceedings of the 17th ICOLD Congress, Vienne 1991, Q.67 – R.27, 443-464.

USACE, 1982. Ice Engineering, US Army Corps of Engineers, EM-1110-2-1612, 15 october 1982.

USACE, 1983. Design of gravity dams on rock foundation: sliding stability assessment by limit equilibrium and selection of shear strength parameters. US Army Corps of Engineers. Technical Report GL-83-13. October 1983.

USACE, 1983. Design of gravity dams on rock foundation: sliding stability. Technical Report GL-83-13. October 1983.

USACE, 1994a. Considerations for Earth and Rock-Fill Dams. US Army Corps of Engineers, *General Design and Construction*. EM 1110-2-2300. 31 July 1994.

USACE, 1994b. *Standard Practice for concrete for civil works structure*. US Army Corps of Engineers. EM 1110-2-2000. 1 February 1994.

USACE, 1994c. US Army Corps of Engineers. Engineering and Design. *Rock Foundations*. EM 1110-1-2908, 30 November 1994.

USACE, 1994d. US Army Corps of Engineers. *Grouting Technology*. EM 1110-2-3506. January 1984.

USACE, 1995. *Engineering and Design. Gravity dam Design*. US Army Corps of Engineers. EM 1110-2-2200. 30 June 1995.

USACE, 2000. *Engineering and Design. Roller-Compacted Concrete*. US Army Corps of Engineers. EM 1110-2-2006. 15 January 2000.

USACE, 2005. Stability analysis of concrete structures, US Army Corps of Engineers, EM 1110-2-2100, Dec 05.

USBR, 1981. U.S. Department of the Interior, Bureau of Reclamation, *Freeboard criteria and guidelines for computing freeboard allowance for storage dams*, Denver, Colorado, (revised 1992).

USBR, 1987. U.S. Department of the Interior, Bureau of Reclamation. *Design of Small Dams*, Denver, Colorado, third Edition.

USBR, 1990. U.S. Department of the Interior, Bureau of Reclamation. *Earth Manual*, Part 2. Materials Engineering Branch, Research and Laboratory Service Division, Denver, Colorado.

Valloton, O., 2012. "Surélévation du barrage de Vieux-Emosson". *wasser, energie, luft – eau, énergie, air*, 104e année, cahier 3, 209-215.

VAW-ETHZ, 2009. Eidgenössische Technische Hochschule Zürich, Versuchsanstalt für Wasserbau, Hydrologie und Glaziologie: *Landslide generated impulse waves in reservoirs – Basics and computation*. Bericht VAW 4257, 27.02.2009.

Vick S., 1997. "New Directions". *International Water Power & Dam Construction*, May 1997, 40-42.

Viret R., Gigot O., Lazaro P. et Amberg F., 2003. "Confortement du barrage de la Maigrauge, Fribourg". Proceedings of the 21th ICOLD Congress, Montréal, Q.82 – R.7, 95-119.

VISCHER D. L., HAGER W. H., CASANOVA C., JOOS B., LIER P. et MARTINI O., 1997. "Bypass tunnels to prevent reservoir sedimentation", Q.74a Performance of reservoirs sedimentation, Proc. of 19th Congress on Large Dams, Florence, Italy, 605-624.

VON MATT U., 1997. "Überprüfung und Erhaltung von Verankerung". *wasser, energie, luft – eau, énergie, air*, 89. Jahrgang, Heft 11/12, 291-293.

Water Power & Dam Constructions, 1985. "Four major dam failures re-examined". November 1985, 33-46.

WCD, 2000. World Commission on Dams, *Dams and Development: A New Framework for Decision-Making*, November 2000.

WEBER B., 2002. Prédiction du comportement des barrages par comparaison mesuré-calculé – Aspects statistiques. SFOE (SFWG), document interne non publié.

WEIDMANN M., 2003. *Les tremblements de terre en Suisse*. Verlag Desertina. Coire, Suisse.

WIELAND M. et MÜLLER R. W., 2009. "Dam safety, emergency action plans and water alarm systems". *International Water Power & Dam Construction*. January 2009, 34-38.

WIELAND M., 2002. "Erdbeben und Talsperren". *wasser, energie, luft – eau, énergie, air*, 94. Jahrgang, Heft 9/10, 277-285.

WIELAND M., 2003. "Aspects sismiques relatifs aux barrages". Proceedings of the 21th ICOLD Congress, Montréal, General Report Q.83, 1243-1362.

WIELAND M., 2006. "Earthquake safety of existing dams". First European Conference on Earthquake Engineering and Seismology. Geneva, Switzerland, 3-8 september 2006, 16 pages.

WIELAND M., 2014. "Seismic Hazard and Seismic Design and safety Aspects of Large Dam Projects". Second European Conference on Earthquake Engineering and Seismology, Istanbul Aug. 25-29, 2014.

WIELAND M. et CHEN H., 2009. "Lessons Learnt from the Wenchuan Earthquake". *Int. Journal Water Power and Dam Construction*, September, 36-40.

WRFA, 2010. *Water Retaining Federal Act* (in French, German, Italian). *RS 721.101 of 1st october 2010 (State in 1st January 2013)*.

WRFO, 1998. *Water retaining Federal Ordinance* (in French, German, Italian). December 7, 1998.

WRFO, 2012. *Water retaining Federal Ordinance* (in French, German, Italian). RS 721.101.1 of October 17, 2012 (state in 1st April 2018).

XU Z., 2008. "Overview of CFRD Construction in China". *International Journal on Hydropower & Dams*, Vol. 15, Issue 4, 68-72.

ZANGAR C. N., 1952. *Hydrodynamic pressures on dams due to horizontal earthquake effects*, Engineering Monograph N° 11, Bureau of Reclamation, May 1952.

ZEIMETZ F., 2017. "Development of a methodology for extreme flood estimations in alpine catchments for the verification of dam safety". In: SCHLEISS A. J. (Ed.), *Communication N° 68*. Laboratoire de Contstructions Hydrauliques (LCH), Ecole Polytechnique Fédérale de Lausanne (EPFL), Switzerland, ISSN 1661-1179.

ZEIMETZ F., GARCÍA HERNÁNDEZ J., JORDAN F., FALLOT J.-M. et SCHLEISS A., 2017. "Abschätzung von Extremhochwassern bei Talsperren nach der Methode CRUEX++". *wasser, energie, luft – eau, énergie, air*, 109. Jahrgang, Heft 4, 261-270.

ZWAHLEN R., 2003. "Identification and mitigation of environmental impacts of dams projects". In: Ambasht, R.S. (ed): *Modern Trends in Applied Aquatic Ecology*. Kluwer Academic Publishers, New York, 281-370.

Notations

Symbol	Unit	Description	Reference (Chapter/Section/§)
A_{TT}	–	Maximum movement amplitude in case of earthquake	10.4.1
A	m^2, m^2/m	Surface, unit sliding surface	13.4.1, fig. 13.7, 17.3.1
C	–	Volumetric coefficient	10.3.1
C	kg/m^3	Cement content of full concrete	18.2.2
C, C_i	CHF, £, US $	Magnitude of consequences (risk analysis)	9.3
C_e	–	Westergaard coefficient	11.5.2
C_u	–	Uniformity coefficient	22.4.1
D	mm	Maximum grain diameter of the aggregate	17.4.1.2, 18.2.1, 18.2.2
D	kg/m^3	Cement content	13.8.2.1
Dyn	–	Dynamic loading effects	11.1; fig. 11.2, 13.2
E	–	Mean square deviation	18.6.2
E	dm^3/m^3	Total moisture content (concrete, grout)	17.4.1.2, 18.2.3, 18.2.4
E	kN, kN/m	Hydrostatic pressure	13.3
E_{us}	kN, kN/m	Upstream horizontal water pressure	11.1; fig. 11.2, 13.2
E_{us-h}	kN, kN/m	Upstream horizontal pressure	11.1; fig. 11.2, 13.2
E_{us-v}	kN, kN/m	Upstream vertical pressure	11.1; fig. 11.2, 13.2
E_{ds}	kN, kN/m	Downstream hydrostatic pressure	11.1; fig. 11.2
E_{ds-h}	kN, kN/m	Downstream horizontal water pressure	11.1; fig. 11.2
E_{ds-v}	kN, kN/m	Downstream vertical water pressure	11.1; fig. 11.2, 13.2
E_B	kN/mm^2 MPa	Elastic modulus of concrete	15.4.2
E_d	kN/mm^2 MPa	Dynamic elastic modulus of concrete	13.7.2.3
E_e	kN, kN/m	Inertia force of water	11.5.2, 13.6.2
E_h	kN, kN/m	Horizontal substitution seismic force	26.7.5
E_v	kN, kN/m	Vertical substitution seismic force	26.7.5
E_R	kN/mm^2 MPa	Deformation modulus of rock	15.4.3.4
E_r	kN/mm^2 MPa	Deformation modulus of embankment	25.4.1
E_s	kN/mm^2 MPa	Deformation modulus of concrete	13.7.2.3
E_w	kN, kN/m	Weight of the water acting on the dam structure	13.4.4, 24.5, fig. 24.14
E_z	kN, kN/m	Water pressure at level z	13.5.1
F	km	Fetch	27.1.2
F_e	kN, kN/m	Inertia force of the dam	11.5.2, 13.6.2
FS	–	Factor of safety	13.4.1, 13.4.4
F_{sed}	kN, kN/m	Sediment pressure	11.1, fig. 11.2; 11.4, fig. 11.5, 13.2
F_T	kN, kN/m	Earth pressure	11.1; fig. 11.2, 13.2
G	kN, kN/m	Ice load	fig. 11.2
H_{alt}	m.s.m.	Altitude	11.4.7
H	m	Height above the thalweg (bed of the watercourse) Height of subjection of a dam (according to Swiss legislation)	2.1.7, fig. 11.4
H	m	Static height of the dam	15.3.4, fig. 15.17
H_f	m	Dam height above foundation	10.1.2, 15.3.5
$H_{curtain}$	m	Depth of grout curtain	11.4.3
I	–	Intensity of an earthquake at a given place	10.4.1
I_{MM}	–	Intensity of an earthquake according to MM scale (modified Mercalli)	10.4.1

Symbol	Unit	Description	Reference
I_{MSK}	–	Intensity of an earthquake according to MSK scale (Medvedev, Sponheuer, Karnik)	10.4.1
I_0	–	Intensity of an earthquake at the epicenter	10.4.1
JRC	–	Joint Roughness Coefficient	13.4.2
JCS	MPa	Joint wall Compressive Strength	13.4.2
K	m/s	Permeability coefficient	22.2
K_h	m/s	Horizontal permeability coefficient	21.3
K_r	m/s	Permeability of the embankment	25.3.4; fig. 25.12
K_v	m/s	Vertical permeability coefficient	21.3
L_{arc}	m	Developed length of an arch	15.4.3.7
L_c	m	Developed length of the crest	10.1.2, 11.1; fig. 11.2, 15.3.4, fig. 15.17
L_{ca}	m	Chord length of an arch (distance between abutments)	15.3.2, fig. 15.6, 15.3.4, fig. 15.17
L_{wave}	m	Wave length	27.1.2
M	kNm/m	Moment at the crown, respectively at the springing	15.4.3.6
M	–	Magnitude	10.4.1
M_E	kNm/m	Moment under the effect of water pressure	15.4.3.5, fig. 15.38
M_P	kNm/m	Moment under the effect of self-weight	15.4.3.5, fig. 15.37
M_R	kNm/m	Moment, shear force at the toe of the canteliver	15.4.3.4
N	kN/m	Normal load at the crown, respectively at the springing	15.4.3.6
N_R	kN/m	Normal load at the toe of the canteliver	15.4.3.6
P	kN/m	Self weight	11.1; fig. 11.2, 13.2, 15.4.3.5, fig. 15.37
P	%	Percentage of particles with a diameter d	18.2.1
P, P_i		Probability of an event occuring	9.3
P_m	kN/m	Total weight of an embankment	24.5, fig. 24.14
P_s	kN/m	Weight of the structure and ancillary equipment	13.4.4
P_z	kN/m	Self weight at level z	13.5.1
Q	kN/m	Shear force	
Q_E	kN/m	Shear force under the effect of water pressure	15.4.3.5, fig. 15.38
QH_i			
QH_{tot}	kN/m	Horizontal substitution for seismic forces	13.7.2.5
Q_R	kN/m	Moment, shear force at the toe of the canteliver	15.4.3.4
Q_{RB}	l/s	Total seepage flow from the right bank	34.4.6.5, fig. 32.23
Q_{LB}	l/s	Total seepage flow from the left bank	34.4.6.5, fig. 32.23
Q_{TOT}	l/s	Total seepage flow of infiltrations	34.4.6.5, fig. 32.23
QV_{tot}			
QV_i	kN/m	Vertical substitution for seismic forces	13.7.2.3
R_{TT}	km	Distance between the epicenter and a given place	10.4.1
R	MPa	Shear strength Sliding resistant shear force	13.4.1, fig. 13.7, 17.3.1
R	MPa	Resistance to sliding	13.4.1, fig. 13.7
R	CHF, £, US $	Risk	9.3
R_a	m	Outer radius of an arch	15.4.2, fig. 15.22
R_{car}	N/mm^2	Characteristic strength of a concrete	18.6.2
R_{ex}	N/mm^2	Required strength of a concrete	18.6.2
R_i	m	Inner radius of an arch	15.4.2, fig. 15.22

Symbol	Unit	Description	Reference (Chapter/Section./§)
R_{mean}	m	Mean radius of an arch	15.2.2, fig. 15.4, 15.3.2, fig. 15.6, 15.4.2
R_m	N/mm²	Average compressive cube strength	18.6.2
R_{wave}	m	Wave runup height	13.10.5, 27.1.2
S	kN kN/m	Uplift	13.2, 13.3, 13.4, 25.1.1.1
S	m²	Average surface of dam	15.3.5
S_R	–	Overturning safety factor	13.3
S_G	–	Sliding safety factor	13.4
T	s	Period	10.4.1
T	°C	Thermal effects	11.1 ; fig. 11.2, 13.2
T_a T_{air}	°C	Air temperature	fig. 11.2
$T_{upstream}$	°C	Concrete temperature on the upstream face	15.5.2
$T_{downstream}$	°C	Concrete temperature on the downstream face	15.5.2
T_b	°C	Dam body temperature	fig. 11.2
T_e T_{eau}	°C	Water temperature	fig. 11.2
T_{min}	kN	Minimal post-tensioning force	13.11.3.1
T_s	s	Stationary duration	11.5.2
T_{TT}	s	Period of motion during an earthquake	10.4.1
U	kN kN/m	Uplift, Pore pressure	13.9, 32.4.2
V	m³	Concrete volume	15.3.5
V	cm³	Volume of the sample	25.1.1.1
W	kJ/m³	Heat from cement hydration	13.8.2.1

Symbol	Unit	Description	Reference (Chapter/Section./§)
a	cm/s²	Earthquake acceleration	13.6.2
a_c	cm/s²	Critical acceleration of a slip surface	26.7.6
a_D	cm/s²	Maximum acceleration at the crest	26.7.4
a_G	cm/s²	Maximum acceleration at the center of gravity	26.7.4
ah	cm/s2	Ground peak horizontal acceleration	11.5.2, 26.7.4
a_v	cm/s²	Peak vertical acceleration	11.5.2, 26.7.4
b	m	Base width	fig. 3.4, 11.4.3, fig. 11.14, 13.2, 15.3.4, fig. 15.17
b	m²/m	Base unit area (A)	13.7.2.3
c		Slenderness coefficient	15.3.5
c	MPa	Cohesion	17.3.1, 20.4
c	MPa	Concrete / rock cohesion on the contact surface and possibly on the foundation excavation	13.7.2.3
c_B	kJ/°C·kg	Heat specific to the concrete	13.8.2.1
c'	MPa	Effective cohesion on the considered compressed sliding surface	13.4.1
d	mm	Particle diameter	10.3.1, 18.2.1
d	m	Thickness	13.6.3, 15.4.3.7, 15.4.3.4, 15.5.2, 15.6.1
d_a	m	Thickness ot arch abutment	
d_b	m	Base thickness	15.3.4, fig. 15.17
d_c	m	Crest thickness	15.3.4, fig. 15.17, 15.4.3.4
d_{max}	mm	Maximum diameter of particles	22.4.1
e	–	Void ratio	25.1.1.1
e	m	Thickness of an upstream facing	24.2.21

e_p	m	Distance between the upstream facing and the upstream wall of a gallery	15.7.2
e_R	m	Eccentricity of self-weight relative to the center of gravity of the canteliver	13.6.3, 15.4.3.5, fig. 15.37, 15.6.1
$fc(90)$	N/mm^2	Concrete compressive strength at 90 days	18.6
$fc(x)$	N/mm^2	Concrete compressive strength at x days	18.6
f_D	m	Freeboard	13.10.5, fig. 13.48, 27.1.3
f_s	Hz	First natural frequency (base frequency)	13.7.2.5
g	m/s^2	Gravity acceleration	26.7.4
h	m	Water pressure height	11.1; fig. 11.2, 13.2
h	m	Dam height and hydrostatic pressure for the case of a triangular dam	13.2, 13.3
h_{us}	m	Upstream hydrostatic head	11.4.1
h_{ds}	m	Downstream hydrostatic head	11.4.1
h_e	m	Application height of the water inertia force	11.5.2
h_i	m	Height of the headwater	fig. 5.4
h_l	m	Height of a concrete lift	fig. 13.38
h_{sed}	m	depth of sediment layer	11.4, fig. 11.5
h_T	m	height of a downstream embankment	fig. 11.2, fig. 11.3
h_{TT}	km	Depth of an earthquake	10.4.2
h_{wave}	[m]	Wave height	27.1.2
$h_w\ h_e$	m	Water height	fig. 13.29, 13.7.2.5
h'	m	Height of the headwater	11.4.1
i	degré [°]	Angle of dam/rock dilatancy on a sliding surface	13.7.2.3
k	–	Uplift coefficient	13.3
k_a	m/s	Lowering speed	25.3.4; fig. 25.12
l	m	Block width	fig. 13.38
m	t/m	Mass of the sliding block	26.7.5.2
m_i	–	Slope of the upstream or downstream face of a gravity dam	13.3
m_1	–	Slope of the upstream face	3.2.3
m_2	–	Slope of the downstream face	3.2.3
m_s	kg/m	Dam mass	13.7.2.5
m_w	kg/m	Water mass	13.7.2.5
n_i	–	Slope of an embankment dam	24.1
n	%	Porosity of the soil in situ	26.7.4.2
n	%	Void percentage	25.1.1.1
n_{max}	%	Porosity at maximum compactness	26.7.4.2
n_{min}	%	Porosity at minimum compactness	26.7.4.2
p_E	kN/m	Water pressure	15.4.3.3
p_A	kN/m	Proportion of p_E taken up by the arch	15.4.3.3
p_C	kN/m	Proportion of p_E taken up by the cantilever	15.4.3.3
p_m	kN/m	Uniform water pressure at the arch axis	15.4.2
p_s	kg/cm^2	Snow pressure	11.4.7
q	l/s	Seepage	32.4.2
q_f	kN/m^3	Pressure exerted by an avalanche, debris flow	11.5.3, 11.5.4
u	kN kN/m	Pore pressure	25.1.1.1
v_f	m/s	Avalanche velocity, debris flow	11.5.3, 11.5.4
v_s	m/s	Speed of shear wave	26.7.5
v_v	km/h	Wind speed	27.1.2

NOTATIONS

v	m/s	Speed	5.4.5
v	mm	Volume of a grain	10.3.1
w	kg, g	Wet weight of the sample	25.1.1.1
w_e	%	Final moisture content	25.1.1.1
w_{opt}	%	Optimum moisture content	25.1.1.1
z	m	Depth	26.7.4.5

Symbol	Unit	Description	Reference (Chapter/Section./§)
α	cm/s²	Acceleration on the ground (due to an earthquake)	10.4.1, 11.5.2
α_h	cm/s²	Horizontal acceleration	13.6.2
α_{TT}	–	Attenuation coefficient (case of earthquake)	10.4.2
β_a	°	Horizontal angle at abutment between the average line of the foundations and the tangent to the average line of an arch	15.3.2, fig. 15.6
β_b	degré [°]	Inclination of a foundation from the horizontal	13.4.3, fig. 13.15
β_i	degré [°]	Vertical angle of the downstream face of the dam with respect to the horizontal	11.3.2
β	°C⁻¹	Thermal dilatation coefficient of the concrete	15.5.3, 18.3.2.2
γ	g/m³ kg/m³	Unit weight	20.4
γ_b	t/m³	Concrete unit weight	Box Chapter 13
γ_d	g/cm³ kg/m³	Apparent dry unit weight	25.1.1.1
γ_d	g/cm³ kg/m³	Unit weight of dry soil	26.7.4.2
$\gamma_{d\ max}$	g/cm³ kg/m³	Unit weight of dry soil at maximum compactness	26.7.4.2
$\gamma_{d\ min}$	g/cm³ kg/m³	Unit weight of dry soil at minimum compactness	26.7.4.2
γ_e, γ_w	t/m³	Water unit weight	Box Chapter 13, Chapter 11
γ_i	g/cm³ kg/m³	Unit weight of submerged sediment	11.4.2
γ_T	g/cm³ kg/m³	Unit weight	11.3.2
ΔH	m	Height of dam heigthening	13.11.3.1
Δ_{TT}	mm	Displacement (due to an earthquake)	10.4.1
ΔT	°C	Uniform change of concrete temperature	15.5.3, 15.5.4
Δ	N/mm² MPa	Shear stress	13.41
τ_a	N/mm² MPa	Shear strength	13.4.1
$\Delta\tau$	°C	Linear temperature difference between the faces	15.5.3, 15.5.4
δ	cm, mm	Deformation	10.4.2, 15.4.2, 32.4.2
δ_A	cm, mm	Radial deformation of the arch	15.4.3.4, fig. 15.29
δ_C	cm, mm	Radial deformation of the cantilever	15.4.3.4, fig. 15.29
δ_c	cm	Deformation due to consolidation	25.3.2
δ_R	cm, mm	Transversal deformation of the rock at the toe of the cantilever	15.4.3.4
δ_r	cm	Settlement of an embankment	25.3.2
$\delta(t)$	cm, mm	Deformation at time t	32.5.2.2

ζ	–	Damping of the material	13.7.2.3
θ_b	°C	Concrete temperature	32.4.2
λ	–	Slenderness of a dam	3.2.4, 10.1.2
λ	–	Uplift coefficient	11.4.3
ν, ν_B	–	Poisson's ratio	13.7.2.3
ρ	–	Density	13.7.2.3, 26.7.5
ρ_B	kg/m³	Unit density of concrete	13.3
ρ_E	kg/m³	Unit density of water	13.3
ρ_w	kg/m³	Water density	13.7.2.5
ΣH	kN	Sum of horizontal static forces	13.7.2.7, 17.3.1
$\Sigma S, \Sigma U$	kN	Sum of vertical uplift forces	17.3.1
ΣV	kN	Sum of vertical static forces	13.7.2.7, 17.3.1
ΣV	kN	Resultant of all forces perpendicular to the foundation	13.4.1
σ	N/mm² MPa	Normal stress	17.3.1
σ_v	N/mm² MPa	Normal vertical stress	26.7.4
σ_ϕ	N/mm² MPa	Uniform stress in the arch	15.4.2
$\sigma_{h,us}$	N/mm² MPa	Upstream horizontal stresses	15.4.3.6
$\sigma_{h,ds}$	N/mm² MPa	Downstream horizontal stresses	15.4.3.6
σ_m	N/mm² MPa	Average effective normal stress at contact surface	13.7.2.3
σ'_n	N/mm² MPa	Effective normal stress	13.4.2
$\sigma_{v,us}$	N/mm² MPa	Upstream stresses	15.4.3.5, fig. 15.37
$\sigma_{v,ds}$	N/mm² MPa	Downstream stresses	15.4.3.5, fig. 15.37
$\sigma_{z,us}$	N/mm² MPa	Upstream vertical stress at level z	13.5.1
$\sigma_{z,ds}$	N/mm² MPa	Downstream vertical stress at level z	13.5.1
σ_I, σ_{II}	N/mm² MPa	Principal stresses	13.5.2
τ	N/mm² MPa	Shear stress	13.4.1, 26.7.4
τ_a	N/mm² MPa	Available shear strength determined by Coulomb's equation	13.4.1
τ_m	N/mm² MPa	Average shear stress at contact surface	13.7.2.3
ϕ, ϕ'	Degré [°]	Angle of internal friction	11.3.2, 13.4.1, 17.3.1, 20.4
ϕ_r	Degré [°]	Residual friction angle or on smooth joint	13.4.2
ϕ_w	degré [°]	Friction angle between embankment and concrete	11.3.2
ϕ'	degré [°]	effective angle of internal friction of the considered slip surface	13.4.1

φ	degré [°]	Angle of concrete/rock friction on a sliding surface	13.7.2.3
ϕ$_i$	degré [°]	Angle of concrete/rock friction on a sliding surface	fig. 11.4
2α	degré [°]	Opening angle of an arch	15.2.2, fig. 15.4

List of abbreviations of cited organizations

ICOLD	International Committee on Large Dams
NEOC	National Emergency Operations Centre
WCD	World Commission on Dams
SNCOLD	Swiss National Committee on Large Dams
SwissCoD	Swiss Committee on Dams
DDPS	Federal Department of Defense, Civil Protection and Sport
FOCP	Federal Office for Civil Protection
EIA	Environmental Impact Assessment
EPFL	Swiss Federal Institute of Technology, Lausanne
ETHZ	Swiss Federal Institute of Technology, Zurich
EMPA LFEM	Swiss Federal Laboratories for Material Science and Technology
ICID	International Commission on Irrigation and Drainage
ICOLD	International Commission on Large Dams
IHA	International Hydropower Association
LCH	Laboratory of Hydraulic Constructions at EPFL
SFWG	Swiss Federal Office for Water and Geology
SFOE	Swiss Federal Office for Energy
FOEV	Federal Office for the Environment
WRFO	Water Retaining Federal Act
SAEFL	Swiss Agency for the Environment, Forests and Landscape
SIA	Swiss Society of Engineers and Architects
SED	Swiss Seismological Service (SED)
USACE	US Corps of Engineers

Abbreviations of Swiss Cantons

ZH	Zurich	GL	Glarus	AR	Appenzell Outer-Rhodes	VD	Vaud
BE	Bern	ZG	Zug	AI	Appenzell Inner-Rhodes	VS	Valais
LU	Lucerne	FR	Fribourg	SG	St. Gallen	NE	Neuchâtel
UR	Uri	SO	Solothurn	GR	Grisons	GE	Geneva
SZ	Schwyz	BS	Basel	AG	Aargau	JU	Jura
OW	Obwalden	BL	Basel Disctrict	TG	Thurgau		
NW	Nidwalden	SH	Schaffhausen	TI	Ticino		

Index

A

Abrasion 165, 477, 727
Abutment 56, 149, 150, 346
   Abutment consolidation 650
   Abutment strengthening 650
Accelerogram 172
Accelerometer 669
Additive 563, 630
Admixture 428, 445
Ageing 126, 181, 313, 562
Aggregate 164, 428, 444
   distribution 470
Air
   entrainers 468
   temperature 443, 667, 673
Alarm 91, 102, 714
   system 97, 715
Alignment 670, 678
Alkali-Aggregate Reaction (AAR) 165, 504
Amplitude of movement 169, 173
Analysis
   Analysis of measurement results 655
   Dynamic analysis 180, 269, 271
   Risk analysis 125, 126, 130, 141, 715
Anchoring forces 199, 202
Anchor sealing 310
Angle of friction 20, 204, 255, 447, 540
Appearance of cracks 51, 107, 398, 501, 509, 523
Appurtenant structure 87, 147, 180
Artificial cooling 64, 292
Artificial flood release 34, 74
Asphalt
   concrete 62, 181, 545
   core 519, 545
   facing 29, 520, 551
Assessment criteria 9, 100
Atterberg limits 167
Authorization 87, 236
Avalanche 7, 93, 109, 228

B

Backward erosion 601
Base joint 402
Batching plant 481
Bed load 33, 183
Behavioral anomalies 92
Bending moment 289, 393, 397, 424
Bentonite 520, 563, 648

Berms 614
Biaxial failure diagram 489
Biaxial strength 488
Blastfurnace cement 468
Blast furnace slag 293
Block rocking analysis 272
Blocks 51
Block stability 294, 297
Borehole 156, 160, 312, 630, 641
   micrometer 691
Borrow
   material 167
   pits 75, 430
Borrow areas 531
Borrowing zones 63
Borrow pits 75
Bottom outlet 32, 34, 86, 130, 220, 524
Brazilian test 474
Breach 100, 116, 600
Buttress
   head 53, 320
   web 331
Bypass tunnels 154, 726

C

Cantilever 56, 278, 327, 349
Canyon 148, 341
Cavitation 477, 478
Cement
   content 298, 438
   gel 473, 648
Cemented Sand Gravel Rock Dam 61
Cementitious materials 444
Central adjustment 376
Chemical attack 477
Chimney drains 63
Circular arches 26, 345
Clay core 62, 180, 519, 537
Climate change 45, 722
Climatic conditions 147, 286, 545
Closure temperature 389, 406
Cofferdams 428, 454
Cohesion 20, 157, 166, 252
Collapse 113, 114, 599
Compaction 20, 35, 63, 167, 201, 450, 569
Compensation flow 73
Compressive strength 260, 276, 293, 441, 447
Concentration of stresses 54, 328
Concession 233

Concrete
 cooling 291, 317
 facing 551, 554
 -rock contact 64
 shrinkage 229, 406, 501
 shrinking 51
 temperature 199, 213, 443
 vibration 479
Consolidation 20, 63, 157, 161, 523, 573, 637
Construction material 147
Contact erosion 601
Contraction joints 51, 438, 443
Cooling pipes 213, 214, 292, 405
Crack
 opening 501, 632
 origin 499
 propagation 492, 500
Cracking corrosion 313
Creep 229, 502, 702
Crushed aggregate 164, 485
Curvature 56, 345, 348
Curved beams 56
Cut-off
 curtain 604
 wall 551, 604

D

Dam
 categories 37, 222
 heightening 29, 41, 301, 407
 shell 519, 537
Damping 224, 276
Danger 99, 138, 185, 712
 analysis 713
Debris flow 109, 229
Decommissioning 88
Deformability 159, 161, 365
 measurements 159
Deformation
 compatibility 362
 modulus 162, 276, 289, 559, 560, 573
Deformeter 669
Delayed deformation 499
Deposited sediment 33
Deposits 88, 645, 673, 721, 728
Design flood 183, 192, 218, 300
Deterministic
 approach 128, 700
 concepts 125
Diaphragm wall 519, 545, 604, 615, 648
Differential settlement 65, 108, 161, 524, 563
Digital imagery 670, 683

Dikes 61, 99, 531, 730
Dilatometer 158, 669
 test 157
Direct
 measurements 89
 pendulum 669
 shear test 157, 572
Diversion tunnel 421, 524
Documentation
 Dam file 659
 Dam monograph 660
 Surveillance guidelines 659
Double curvature 49, 57, 342
Downstream face 52, 201, 243, 250, 262
Drainage
 blanket 63, 521, 533, 604
 boreholes 257, 268, 335
 curtain 212, 458
 effect 33, 212
 gallery 212, 521
 holes 208, 637, 645
 system 52, 107, 212, 519, 532, 627
Drains 20, 107, 200, 212, 303, 318
Drawdown 92, 100, 133, 200, 578
 of the reservoir 29, 200, 399, 538
Drilling investigations 158
Drinking water 39, 69, 87, 721
Dry (or cold) joint 452
Durability 167, 181, 305, 448, 631
Dynamic
 behavior 21, 357, 594
 method 271

E

Earth pressure 199, 201, 256, 277
Earthquake 32, 64, 92, 95, 121, 168
Elastic filler 557
Elliptical arches 26
Elliptic spiral arches 57
Emergency
 action plan 713, 716
 organization 713, 716
 plan 99
 planning 83, 711
 regulations 103, 712
 strategy 99, 712, 713
Energy production 7, 344
Environmental impact assessment (EIA) 8, 69, 235
Epicenter 168, 172, 177
Epigenetic valley 154
Epoxy 33, 477

Erosion 101, 107, 114, 133, 216, 461, 477
Error ellipse 705
Evacuation plan 100, 713
Evaporation 473, 502
Event tree 131
Exceptional drawdown 499
Exploratory
  boreholes 157
  galleries 156, 301
Extensometer 669
Extraordinary event 132, 659, 695

F

Face 56, 510, 538, 553, 603
Factor of safety 249, 256, 268, 284, 294, 441, 484, 583
Failure 37, 74, 92, 98, 246
  circles 581
  rate 113, 668
Fault tree 131
Fauna 34, 70, 72, 237
Feasibility 9, 153, 480, 519, 540
Fetch 609
Fiber-optic
  cables 443
  sensor 36, 669, 671
Filter 20, 168, 521, 533, 537
  criteria 538, 541, 577
Finite Element Method (FEM) 21, 271
First impoundment 73, 85, 87, 113, 116, 163, 181, 524, 598, 605
Fish
  farming 7
  ladders 71
  population 72
Floating debris 87, 109, 187
Flood
  control 7, 147, 181
  forecasting 41
  map 100, 713
  retention 41, 194
  routing 195
  safety 29, 183, 192
  wave calculations 102
Flora 34, 70, 72, 237
Fluidifiers 468
Flushing 34, 86, 88, 92, 154, 237, 722, 727, 728
Fly ash 293, 444, 455
Force of inertia 267
Formwork 319, 323, 341, 415, 428
Foundation surface 293, 623
Freeboard 107, 187, 518, 597, 609, 612

Fresh concrete 290, 293, 450, 469, 472
Fresh (or hot) joint 451
Friction angles 253
Frost 298, 452, 471
Frost damage 512
Frost resistance 471, 476
FSHD (Faced Symmetrical Hardfill Dam) 61

G

Gate operation 86, 131, 660, 715
Geodesic networks 680, 681
Geodetic measurements 33, 506, 678, 680
Geological surveys 147, 163
Geomembrane 460, 520, 551, 562
Geophysical
  investigation 157
  methods 157
Geotechnical tests 157, 158
Geotextiles 562, 730
GIN curve 633, 636
Gorge 110, 148
Gradex method 189, 191
Grading curves 469, 585
Granulometric analyses 164
Gravel pits 61, 164, 467
Gravitational acceleration 172, 204
Greenhouse gases 43, 71
Ground movement 33, 220, 271
Groundwater table 156, 522, 674
Grout
  curtain 87, 109, 130, 152, 208, 268, 637, 639
  cut-off 627, 636
  holes 208

H

Hardening 291, 293, 388, 443
Hardfill 459, 460
Hardness 165, 480
Hazard identification 128
Hazardous event conditions 126, 134
Heightening 42, 301, 407, 462
Heritage protection 74
Horizontal struts 54
Hydration 51, 468, 501, 504, 684
Hydration heat 51, 298, 319, 336, 444
Hydraulic fill 531
Hydraulic jacking test 159
Hydrodynamic pressure 218, 220, 269, 392
Hydroelectric schemes 24, 25, 32, 40, 560
Hydrogen embrittlement 313
Hydrograph 182, 191
Hydropeaking 69, 74

Hydropower plant 7, 74, 85, 183
Hypocenter 168,169

I

Ice load 199, 217, 245, 277, 667
Ice pressure 200
Impulse wave 92, 109, 228, 612
Incidents 95
Inclinometer 36, 669
Individual risk 139
Induced earthquake 73, 96, 180
Inflow 5
Intensity 169
Intensity maps 222, 585
Internal
  angle of friction 157, 441, 447
  erosion 20, 107, 116, 428, 517, 532, 596, 598
  friction 202, 252, 295
Inundation 7
Inverted pendulum 669, 676
Irrigation 7, 13, 19, 39, 69, 721

J

Jet grouting 646
Joint
  closure 351, 371, 375, 389, 405
  grouting 57, 351, 406, 456
Jointmeter 669, 671
JRC 255

L

Landslides 73, 92, 109, 611, 684
Laser beam 683, 685
Laser scanning 36, 670, 683
Lefranc test 157, 160
Leisure activities 7, 22, 41
Levelling 36, 506
Level of danger 39, 100
Lévy's rule 207, 249, 253, 261
Life cycle analysis 43
Lifetime 8, 34, 200, 668
Likelihood of occurrence 126, 131, 136, 260
Liquefaction 107, 540, 582, 588, 603
Liquidity limit 540
Load-bearing 56, 264, 334, 357
  test 157
Load cell 669
Load scenarios 200
Logarithmic spiral arches 57
Low heat cement 293

Lugeon
  test 157, 161, 628
  value 628
Lugeon, Maurice 160

M

Magnitude
  earthquake 115, 117, 169, 170, 172, 176, 177
  of consequences 126
  of damage 127
Maintenance 42, 562, 659
  gate 86, 665
Major accidents 125, 140, 238
Manometer 669, 689, 693
Masonry 14, 21, 49, 201
Maximum Credible Earthquake (MCE) 181, 220
Measured uplift 213
Measurement program 659
Measuring program 659, 695
Membrane theory 357
Microbar
  test 165
  testing 509
Microcracks 500
Micrometer 669, 671, 691, 693
Mixing water 445, 468
Mobile water release structures 186
Modal dynamic method 271
Moisture swelling 504
Monitoring system measurements 659
Monitoring systems 36, 87
Monolithic 57, 299, 388, 437, 452
Monte Carlo simulation 128, 135
Mortar layer 438, 476
Multiple arch dams 55, 65, 415
  Arch 416
  Arch grow 417
  Bracing 418
  Buttress 415, 417, 418
  Center angle 416
Multipurpose reservoirs 41

N

Narrow valley 148, 294, 352, 524
Natural event 95, 126, 199
Natural frequencies 265, 280, 385, 386
Number of passes 450, 531
Numerical analysis
  2D analysis 270, 294, 422
  3D analysis 294
Numerical modeling 21

O

Operating Basis Earthquake (OBE) 220
Operating tests 99, 659, 663
Operational checks 89
Optical alignment measurements 677
Optimum moisture content 517, 567
Overflow height 187
Overflowing 116
Overpressure 267, 422, 598
Overtopping 86, 131, 138, 181, 203, 300, 461, 609
Overturning 52, 246, 249, 284, 310, 330

P

Parabolic
  arches 26
  spiral arches 57
Peak flow 188, 189
Peak Ground Acceleration 173, 220
Pendulum 36, 669, 676
Penetration test 157, 588
Percentage of impurity 165
Peripherical rings 527
Permafrost 33, 154
Permeability test 157, 480, 639, 641
Petrographic test 467, 509
Photogrammetry 147, 683, 670
Piezometers 161, 605, 688, 690
Piping 107, 133, 532, 577, 601, 605
Placement tests 450
Plasticity
  index 540
  limit 540
Plasticizers 468
Plinth 66, 460, 551, 556
Plunge pool 187, 431
PMF (Probable Maximum Flood) 185, 192, 194, 613
PMP (Probable Maximum Precipitation) 185, 192, 193
Poisson's ratio 276, 375, 447, 476
Polygonal measurements 675
Pore pressure 108, 199, 216, 516, 689, 692
Post-tensioning
  anchors 304
  cables 308
Pozzolanic cement 293, 468
Preferential flow path 601
Pressure
  balance 669, 673
  cell 669, 693
Prestressed rock anchors 29, 309, 650
Preventive actions 121
Probability density functions 128

Proctor
  curve 567
  test 167, 567, 569
Pseudo dynamic method 271
Pseudo-static method 282, 584
Public
  health 74
  liability insurance 141
Pulvino 400
Pumped-storage projects 28, 40
Pumping test 157, 161
PVC membrane 434

Q

Quantitative risk assessment 128
Quarries 153, 164, 467, 539

R

Rapid drawdown 109, 200, 538, 582
Reconstruction 42
Refrigeration 292, 402, 456, 468
Rehabilitation 29, 42, 88, 92, 399
Reinforced concrete slab 460, 519
Reinforcement (*see* rehabilitation)
Reservoir
  sedimentation 34, 301, 722
  volume 34, 42, 301
Residual flow 73
Resin mortar 477
Response spectrum 222, 227, 271, 275
  analysis 271
  method 281
Return period 185, 190, 222, 594, 612
Riprap 571, 581, 610, 616, 617
Risk
  -based approach 125
  estimation 129, 136
  evaluation 129, 137
  management 83, 127, 129
Risk acceptance 126, 138
  criteria 139
Rock bolts 162, 652
Rockfall 33, 109, 531, 664
Rock foundation 63, 65, 161, 264, 428, 551
Rock Quality Designation (RQD) 158, 628
Rod extensometers 572, 691
Rounded aggregate 164, 444, 469

S

Sabotage 92, 97, 110, 582
Safety flood 183, 185, 218

Sand traps 154
Satellite 36, 681
Saturation line 532, 568
Schistosity 156, 165
Scouring 109, 114, 187, 694
Sealing system 327, 336, 438, 577
Seasonal storage 499
Sediment
  bypass tunnels 727
  pressure 199, 203, 243
  replenishment 34
Sedimentary deposits 85
Sedimentation level 673
Seepage
  flow 156, 207, 598, 627, 685
  gradient 64, 537, 563
  paths 527, 599, 627
  pressure 199, 216
Segregation 428
Seismic
  hazard 173, 178, 222, 590
  reflection 157
  refraction 157
  shock 33
Seismicity 147, 168, 174
Seismogram 169, 172
Seismometer 669
Service gate 86, 665
Settlement 63, 64, 107, 116, 157, 374, 523, 559, 576
  mark 679
Settling basins 154, 726
Shear
  box 180
  keys 271, 294, 297, 298, 407
  strength 157, 253, 260, 296, 386, 441, 447
  test 157, 255, 572
  vane test 157
Shotcrete 30, 162, 623, 651
Shrinkage joints 52
Simple compression test 157
Simulation 135, 189, 192, 195
Single curvature 56, 341, 454
Sirens 38, 102, 713
Site selection 155
Ski jump 187
Slenderness 59, 149
  coefficient 355
  ratio 149
Sliding surface 108, 205, 252
Slip surface 253, 581
Slope
  protection 537
  stability 20, 517

Sloped layers 452
Slump 438, 472
  test 472
Societal risk 139
Soil erosion 7, 33, 154, 722, 736
Solar radiation 293, 388
Sources of uncertainty 135
Spectral acceleration 279, 282, 592
Spillway 64, 86, 130, 524
Stable grout 630, 648
Statistical analysis 131, 270, 352, 704
Steel
  fibers 477
  girders 162
Stiffness 56
Stratification joints 107, 156
Strengthening works 35, 122, 154, 650
Stress concentrations 403, 430, 442, 461
Strong Motion stations 174
Structural safety 83, 200, 221, 693
Submersion 75, 524, 576
Suffusion 107, 541, 602
Surface curing 452
Surveillance and maintenance 83
Suspended
  load 33
  material 721, 723, 727, 729
  solids 154
Sustainable development goals 44
Swelling 108, 165, 229
Symmetrical profile 430, 459

T

Temperature
  effects 51, 54
  rise 290
Tensile strength 276, 394, 449
Test pits 156
Thermal
  effects 199, 423, 456, 701
  expansion coefficient 476
  swelling 504
Thermometers 36, 392, 684
Threats 85, 92, 97, 663
Top event 131
Torsion moment 393
Tourism 7, 8
Transition zones 537, 546, 570, 581
Trenches 156, 527
Trial-load method 358
Triangular uplift diagram 207
Triaxial
  compression test 157

test 159
Turbidity current 34, 721

U

Uncertainty 129, 134
Unstable grout 630
Uplift
 coefficient 206, 246, 294
 distribution 207, 210, 320, 420
Upstream facing 62, 181, 432, 520, 551

V

Valley
 flanks 50, 56, 153, 335, 419
 shape 59, 147, 524, 560
 shape factor 560
 width 59
Vegetation 70, 108, 601, 605, 618
Venting 728
Vibrating rollers 428, 444, 451, 532, 571
Virtual work 362, 365, 369
Visual checks 89, 663
Visual inspections 659, 663
Vogt 288, 374

W

Water
 alarm 100
 -energy-food nexus 45
 intake 29, 524, 712, 722
 jet 34, 552, 731
 Regulation Act 36
Waterstop 52, 292, 299, 336, 557
Watertight cut-off 115, 650
Watertightness 61, 63, 92, 161, 298, 325, 406, 476, 546
Wave
 calculations 101
 run-up 610
Weathering 153, 168, 298, 471
Web 54, 181, 319, 331
Wedge 253, 310, 356
Weirs 22, 86, 182, 227
Westergaard coefficient 220
Wide valley 151, 352, 430
Wire alignment 677
Workability 164, 428, 448, 546
World Register of Dams 18

Index of Dams

A

Abanco 353
Aguamilpa 560
Albarellos 353
Albigna 93, 318, 485
Al Massira 320, 336
Almendra 353, 638
Alpe Gera 427
Alto Cavado 353
Alvide 353
Alvito 353
Anchicaya 560
Ancipa 330
Anderson Ranch 17
Arrowrock 17
Asse-Dasse 353
Assouan 96
Avene 353

B

Baells 353
Baijia 460
Bannalp 38, 529
Barberine 34
Barra Grande 552, 560
Baslerweier 24
Bastyan 560
Belesar 353
Beni-Bahdel 416, 417, 418, 421
Bennet embankment dam 17
Bermellar 353
Bernina Pass 25
Beznar 353
Bubal 353
Buchholz 25
Buffalo Bill 17, 21

C

Cabril 683
Cabrum 353
Cachi 353
Caldeirão 353
Cambambe 353
Campos Novos 552, 559, 560
Candim 353
Caneiro 353
Capanda 353
Carcoar 353

Cavagnoli 67, 353
Cethana 560
Chambon 17
Changuinola 1 455, 456
Chatelôt 58, 150, 197, 353
Cheeseman 17
Chicoasen 17
Ciervos 353
Cleuson 25, 30, 31, 321
Collegats 353
Contra 26, 27, 58, 195, 239, 355, 485
Curnera 26, 81, 353, 354, 485

D

Daniel-Johnson 398, 399, 416
Dáo 353
Deriner 353
Dez 465
Diablo 17
Dixence 25, 35, 53, 330
Dix River 17
Drossen 353

E

Edertalsperre 97
Eiras 353
El Atazar 353
El Cajón 355, 640
El Makhazzine 535
El Vellon 353
Emosson 26, 58, 149, 355, 402, 403, 485

F

Faux-la-Montagne 415
Feistritzbach 546
Ferden 84, 94, 485
Finstertal 546
Florence Lake 417
Foz do Areia 560
Funcho 353

G

Gebidem 26, 34, 58, 348, 402, 485, 651
Gelmer 657
Gigerwald 353
Girabolhos 353
Glashütte 94

Golillas 560
Gomal Zam 455
Göscheneralp 26, 28, 152, 518, 573, 575
Grande Dixence 17, 18, 25, 26, 179, 243, 259, 298, 299, 300, 404, 485
Gran Suarna 353
Granval 416
Gries 209, 241
Grimsel 39, 731
Grimsel Ouest 353
Grimsel reservoir 732
Guara 353
Gübsen 24
Gübsensee 30, 60

H

Hongrin 26, 27, 402, 413, 645
Hoover 17, 355
Hsinfengkiang 96
Illsee 30, 32

I

Isola 485

J

Janneh 454
Janovas 353
Jawa 14
Jinping I 17
Jiroft 353
Joux-Verte 23

K

Kakavakia 353
Kalaritikos 353
Kardeh 353
Kariba 96
Karun III 19
Kebar 14
Kerville 428
Khao-Laem 560
Kholong Tha 431
Klöntal 24, 25, 38
Kölnbrein 355, 398, 399
Kopperton 17
Kops 355
Köroglu Kotanli I 455, 456
Kotanli II 455, 456
Koyna 96
Kurit 14, 17

L

La Carcovada 353
Lago Bianco Sud 30
La Remolina 353
Las Portas 353
Le Gage 355, 398, 399
Les Marécottes 415, 416, 417
Les Olivettes 436
Lessoc 324, 652
Les Toules 58, 148, 355, 402
Limberg 353
Limmern 26, 28, 353, 640
Limmernboden 195, 643
Lindoso Alto 353
Linth-Limmern 153, 354
List 30, 31
Longtan 427, 455
Lower Pieman 560
Lower San Fernando 116, 117
Lucendro 25, 315, 320, 321, 330
Luzzone 26, 28, 34, 58, 239, 353, 355, 407, 409, 496, 724, 725

M

Mackintosh 560
Maigrauge 29, 30, 308, 309
Mala'a dam 13
Malpasset 114
Malvaglia 353, 709
Manuel Lorenzo Pardo 353
Manzaneda 353
Maoping 548
Marathia 457, 458
Marécottes 55
Marib dam 13, 14
Marmorera (Castiletto) 28, 515, 518
Mattmark 26, 28, 32, 179, 195, 518, 525, 579
Mauvoisin 17, 26, 27, 28, 33, 34, 58, 179, 239, 353, 355, 407, 408, 485, 496, 638
McKays Point 349
Meffrouch 416
Mica 17
Miel I 427, 455
Mikri Santa 353
Mohale 552, 560
Möhnetalsperre 97
Moiry 26, 58, 347, 353, 485, 497
Montanejos 353
Montsalvens 11, 21, 25, 33, 34, 36, 58, 353, 653
Mratinje 355
Mud Moutain 17
Murchison 560
Muslen 25, 30, 31, 304
Muttsee 28, 244

N

Nalps 26, 353, 354, 485
Naret I 353
New Croton 17
New Spicer Meadow 549
Nurek 17, 96, 513, 518

O

Oberaar 25, 485
Oberems 55, 416
Oberhasli 35
Odéáxere 353
Oroville 17
Osiglietta 400
Ova Spin 661
Oymapinar 640

P

Palagnedra 32, 34, 95, 187
Panix 85, 123
Pathfinder 21
Paute 355
Peruča 97, 98
Petit-Saut 439
Pfaffensprung 33, 58, 653
Pigniu 28
Pine Flat 273
Platanovryssi 353
Portugues 455, 456
Puerto Seguro 353
Punt dal Gall 26, 179, 353, 376, 377, 378, 637, 719

Q

Qwyhee 17

R

Räterichsboden 25, 94, 111
Rempen 34, 35
Rialp 59, 425
Robiei 119
Roggiasca 355
Rogun 17, 518
Rossens 25, 58, 214, 355, 621
Rossinière 154, 155

S

Sadd-el Kafara 13
Salanfe 153, 640
Salmon Creek 21
Salto Funil 353
Salt Springs 17
Salvajina 560
Sambuco 47
Sanetsch 153, 640
San Gabriel 17
San Leonardo 17
Santa Maria 148, 353, 355, 485
Santótis 353
San Vincente 462
Satna Euralia 353
Schiffenen 231, 355, 402
Schlegeis 355
Schmalwasser 543
Schräh 17, 30, 32, 35
Seeuferegg 39
Sefid Ruud 333
Segredo 560
Serra 30, 56
Sfikia 353
Shahid Rajaee 353
Shih Kang 96, 115, 457
Shoukoubao 460
Shuibuya 565
Solis 3, 28
Songlin 460
Spitallamm 36, 39
St-Barthélémy C 56, 151
Steno 353
Stevenson 355
Stillwater 17
Storvatn 546
Sufers 735
Susqueda 353
Sussundenga 353
Swift 17

T

Tabellout 454, 455, 456
Taham 542
Tang E Soleyman 353
Tarbela 427
Ten Mile Creek 415
Teton 116, 598
Three Gorges 74, 548, 728
Tianshengquiao 552
Tolla 355
Torogh 353
Tourtemagne 60, 61
Trinity 17

U

Upper Gotvand 607

V

Vajont 17, 73, 110, 355
Valdecanás 353
Valle di Lei 339, 350, 353, 355
Valsamiotis 458
Varosa 353
Vasasca 485
Verbois 25, 72
Vidraru 353
Vieux-Emosson 28, 41, 407, 410, 411, 625
Villa Franca 353

W

Waldhalde 24
Willow Creek 427
Wolwedans 429

X

Xiaowan 17
Xibing 454
Xingo 560

Z

Zayandehrood, Zayandeh-Rud 145, 353
Zervreilla 26, 58, 485
Zeuzier 26, 32, 33, 58, 90, 91, 148, 355, 399, 704
Zeyzoun 117
Ziller 355
Zimapan 148
Zipingpu 180
Z'Mutt 353

Anton J. Schleiss graduated in Civil Engineering from the Swiss Federal Institute of Technology (ETH) in Zurich, Switzerland, in 1978. After joining the Laboratory of Hydraulic, Hydrology and Glaciology at ETH as a research associate and senior assistant, he obtained a Doctorate of Technical Sciences on the topic of pressure tunnel design in 1986. After that he worked for 11 years for Electrowatt Engineering Ltd. (now Pöyry-AFRY) in Zurich and was involved in the design of many hydropower projects around the world as an expert on hydraulic engineering and underground waterways. Until 1996 he was Head of the Hydraulic Structures Section in the Hydropower Department at Electrowatt. In 1997, he was nominated full professor and became Director of the Laboratory of Hydraulic Constructions (LCH) in the Civil Engineering Department of the Swiss Federal Institute of Technology Lausanne (EPFL). The LCH activities comprise education, research and services in the field of both fundamental and applied hydraulics and design of hydraulic structures and schemes. The research focuses on the interaction between water, sediment-rock, air and hydraulic structures as well as associated environmental issues and involves both numerical and physical modeling of water infrastructures. From 1999 to 2009 he was Director of the Master of Advanced Studies (MAS) in Water Resources Management and Hydraulic Engineering held in Lausanne in collaboration with ETH Zurich and the universities of Innsbruck (Austria), Munich (Germany), Grenoble (France) and Liège (Belgium). In May 2018, he became Professor Emeritus at EPFL. Until today more than 50 PhD and Postdoc research projects have been carried out under his guidance. From 2006 to 2012 he was the Head of the Civil Engineering program of EPFL and chairman of the Swiss Committee on Dams (SwissCoD). In 2006, he obtained the ASCE Karl Emil Hilgard Hydraulic Price as well as the J. C. Stevens Award. He was listed in 2011 among the 20 international personalities that "have made the biggest difference to the sector Water Power & Dam Construction over the last 10 years". Between 2014 and 2017 he was Council member of International Association for Hydro-Environment Engineering and Research (IAHR) and he was chair of the Europe Regional Division of IAHR until 2016. For his outstanding contributions to advance the art and science of hydraulic structures engineering he obtained in 2015 the ASCE-EWRI Hydraulic Structures Medal. The French Hydrotechnical Society (SHF) awarded him with the Grand Prix SHF 2018 and IAHR as honorary member in 2021. After having served as vice-president between 2012 and 2015 he was president of the International Commission on Large Dams (ICOLD) from 2015 to 2018. With more than 40 years of experience he is regularly involved as a consultant and expert in large water infrastructures projects including hydropower and dams all over the world. On behalf of ICOLD, he coordinated from 2019 to 2022 the Hydropower Europe Forum funded by the EU Horizon 2020 research and innovation program.

Henri Pougatsch graduated in civil engineering in 1966 from the Swiss Federal Institute of Technology in Lausanne (EPFL). He began his professional career as a project engineer in the consulting office of Professors Alfred and Jean-Pierre Stucky, internationally well recognized dam specialists. This activity allowed him to quickly become familiar with the design of concrete dams and their monitoring. In 1968, he entered an engineering company in Geneva as project manager for studies and implementation of major projects in Switzerland and abroad in the field of dams and hydraulic structures. In 1980, he joined the dam's supervision section of the Swiss Federal Office for Water and Geology. In 1997, he was named Swiss commissioner responsible for dam safety, an officer position he held until the end of 2004. As part of his responsibilities, he initiates and supervises the preparation of the early versions of the Swiss guidelines for dam safety. Under his lead, several experts in the specific field of dams, including professors at the Swiss Federal Institute of Technology, members of the Swiss Committee on Dams and officers of federal and cantonal administrations, as well as the Swiss dam's supervision section, have contributed to the edition of these guidelines. The guidelines give a detailed interpretation of legal bases and consider the latest developments concerning the design, construction and monitoring of dams. He served also as a member of the Franco-Swiss Permanent Supervision Commission of the Emosson Dam. As a permanent member of the technical commission of the Swiss Dams Committee (SwissCoD), he chaired the working group for dam supervision and actively participated in the other working groups. In 2019 he became honorary member of SwissCoD. From 1999 to 2009 he was invited, as a lecturer, to teach dam safety concepts at the Master of Advanced Studies (MAS) in Water Resources Management and Hydraulic Engineering coordinated by the Laboratory of Hydraulic Constructions at EPFL.